Leif B.G. Andersen and Vladimir V. Piterbarg

Interest Rate Modeling

Volume I: Foundations and Vanilla Models

Atlantic Financial Press
London New York

Leif B.G. Andersen
Bank of America Merrill Lynch
One Bryant Park
New York, NY 10036
USA
leif.andersen@baml.com

Vladimir V. Piterbarg
Barclays
5 The North Colonnade
London, E14 4BB
UK
vladimir.piterbarg@barclays.com

Mathematics Subject Classification (2010): 60H10, 60H35, 62P05, 65C05, 65C20, 91G20, 91G30, 91G60, 91G80

JEL Classification: G12, G13, E43

LSI Subjects: BUS036540 (Business & Economics : Investments & Securities - Options), BUS091000 (Business & Economics : Business Mathematics), MAT003000 (Mathematics : Applied)

Library of Congress Control Number: 2010905508

Suggested reference:

Interest Rate Modeling, Volume I: Foundations and Vanilla Models, by Leif B.G. Andersen and Vladimir V. Piterbarg, 1st edition, Atlantic Financial Press, 2010

The present volume is the original first hardcover edition, third printing.

Errata is available at www.andersen-piterbarg-book.com

Submit typos and corrections at www.andersen-piterbarg-book.com

ISBN-10: 0-9844221-0-2
ISBN-13: 978-0-9844221-0-4

© 2010 Leif B.G. Andersen, Vladimir V. Piterbarg.
All rights reserved. This work may not be translated or copied in whole or in part without the written permission of the authors or Atlantic Financial Press (publisher@andersen-piterbarg-book.com), except for brief excerpts in connection with reviews or scholarly analysis. Use in connection with any form of information storage and retrieval, electronic adaptation, computer software, or by similar or dissimilar methodology now known or thereafter developed is forbidden.

Printed by Lightning Source Ltd.

ref v1.hc.003

To our families

— *L.B.G.A, V.V.P*

Preface

For quantitative researchers working in an investment bank, the process of writing a fixed income model usually has two stages. First, a theoretical framework for yield curve dynamics is specified, using the language of mathematics (especially stochastic calculus) to ensure that the underlying model is well-specified and internally consistent. Second, in order to use the model in practice, the equations arising from the first step need to be turned into a working implementation on a computer. While specification of the theoretical model may be seen as the difficult part, in quantitative finance applications the second step is technically and intellectually often more challenging than the first. In the implementation phase, not only does one need to translate abstract ideas into computer code, one also needs to ensure that the resulting numbers being produced are meaningful to a trading desk, are stable and robust, are in line with market observations, and are produced in a timely manner. Many of these requirements are, as it turns out, extremely challenging, and not only demand a strong knowledge of actual market practices (which tend to deviate in significant ways from "textbook" theory), but also require application of a large arsenal of techniques from applied mathematics, chiefly approximation methods and numerical techniques.

While there are many good introductory books on fixed income derivatives on the market, when we hire people who have read them we find that they still require significant training before they become productive members of our quantitative research teams. For one, while existing literature covers some aspects of the first step above, advanced approaches to specifying yield curve dynamics are typically not covered in sufficient detail. More importantly, there is simply too little said in the literature about the process of getting the theory to work in the real world of trading and risk management. An important goal of our book series is to close these gaps in the literature.

As we write this in early 2010, financial markets are still reeling from a severe crisis that has, at least in part, been blamed on over-the-counter

(OTC) options markets, the venue where complex derivative securities are transacted. Stricter regulation of some types of OTC derivatives currently seems all but inevitable, and many common OTC securities may in the future either be outlawed or traded only on public exchanges. In the wake of the crisis, opinion of financial engineers and bankers has hit an all-time low, with many in the public convinced that they are peddlers of toxic waste or "weapons of financial destruction". All things considered, the present may therefore seem like an inauspicious moment to launch a series of monographs on the pricing and risk management of interest rate derivatives. We disagree, for several reasons. First, in defense of OTC derivatives we note that although they certainly can be used inappropriately to create excessive leverage and risk, many complex (or "exotic") derivatives serve as innovative and cost-effective vehicles for bank clients to reduce their financial risk. Second, irrespective of what will ultimately transpire on the regulatory front, it has become obvious that going forward both regulators and market participants need a better grasp of the management and characterization of complex financial risk. This is perhaps particularly true for the quantitative research professionals (the "quants", in common parlance) who recently have been taken to task by the press for the failure of their models and their inability to predict the credit crisis. While this simplistic characterization is actually quite unfair, there is no doubt that many derivatives models that worked well enough before the credit crisis are no longer adequate. Indeed, even the simple task of pricing a basic interest rate swap — possibly the simplest of all interest rate derivatives — has recently required major methodology revisions[1]. If nothing else, a severe crisis serves to expose weaknesses in the foundation on which models are built, allowing one to reinforce it for future storms. In this light, we feel that the time is just about right for a comprehensive, practical, and up-to-date exposition of interest rate modeling and risk management[2].

The three volumes of *Interest Rate Modeling* are aimed primarily at practitioners working in the area of interest rate derivatives, but much of the material is quite general and, we believe, will also hold significant appeal to researchers working in other asset classes. Students and academics interested in financial engineering and applied work will find the material particularly useful for its description of real-life model usage and for its expansive discussion of model calibration, approximation theory, and numerical methods. In preparing the books we have drawn on nearly 30 years of combined industry experience, and much of the material has never been exposed in book form before.

Quantitative finance attracts students and practitioners from many different academic fields, and with varying levels of preparation in mathematics

---

[1] We cover this in Chapter 6.
[2] We ought to note that interest rate derivatives (unlike *credit* derivatives) so far have not been directly implicated in the financial crisis.

and computation. (Case in point: L.B.G.A was originally a robotics engineer and V.V.P a probabilist.) To cater to a broad audience, we have kept the exposition fairly informal; graduate students in applied fields such as engineering and physics should feel at home with the level (or lack) of rigor used in the book. We have relied on a proposition-proof format throughout, largely because this facilitates easier cross-referencing in a long text, but acknowledge that the format is occasionally more formal than the results themselves. For instance, we tend to skip over technical regularity conditions in our proofs and also frequently list approximate results in propositions without explicitly specifying the sense in which they approximate true values. Although the exposition is largely self-contained, some previous knowledge of basic option pricing principles (e.g., at the level of Hull [2006]) may be useful.

Interest Rate Modeling divides into three separate volumes. *Volume I* provides the theoretical and computational foundations for the series, emphasizing the construction of efficient grid- and simulation-based methods for contingent claims pricing. Numerical methods serve an extremely important role in the text, so we develop this topic to an advanced level suitable for professional-quality model implementations. Placing this material early in the text allows us to incorporate it into our discussion of individual models in subsequent chapters. The second part of Volume I is dedicated to local-stochastic volatility modeling and to the construction of vanilla models for individual swap and Libor rates. Although the focus is eventually turned toward fixed income securities, much of the material in this volume applies to a broad capital market setting and will be of interest to anybody working in the general area of asset pricing.

Volume II is dedicated to in-depth study of term structure models of interest rates. While providing a thorough analysis of classical short rate models, the primary focus of the volume is on multi-factor stochastic volatility dynamics, in the setups of both the separable HJM and Libor market models. Implementation techniques are covered in detail, as are strategies for model parameterization and calibration to market data.

The first half of *Volume III* contains a detailed study of several classes of fixed income securities, ranging from simple vanilla options to highly exotic cancelable and path-dependent trades. The analysis is done in product-specific fashion, covering, among other subjects, risk characterization, calibration strategies, and valuation methods. In its second half, Volume III studies the general topic of derivative portfolio risk management, with a particular emphasis on the challenging problem of computing smooth price sensitivities to market input perturbations.

Although much of the material in *Interest Rate Modeling* is focused on the technical and theoretical issues surrounding model implementation on a computer, it is impractical for us to delve into the exercise of writing actual computer routines. Fortunately, there are several specialized books on how to write good quant code, see, e.g., Hyer [2010] and Joshi [2004]. Both of these books work with C++ which is still the most common computer language used in professional quant libraries. For those that choose to work with C++, we wholeheartedly endorse books by Scott Meyers (see, e.g., Meyers [2005]) and Andrei Alexandrescu (see, e.g., Sutter and Alexandrescu [2004]) as guides to sound and maintainable code.

During the six year process of writing this book series, we have received encouragement and constructive criticism from many people. We particularly wish to thank Peter Carr, Peter Forsyth, Alexandre Antonov, Peter Jäckel, Dominique Bang, Martin Dahlgren, Neil Oliver, Patrick Roome, Regis van Steenkiste, Natasha Bushueva, Chenxu Li, Paul Cloke and many members of the research teams at Barclays Capital and Bank of America Merrill Lynch. Natalia Kryzhanovskaya meticulously proofread our first draft, and contributed greatly to the harmonization of notation across what turned out to be a very long manuscript. All remaining errors are, of course, entirely our own. Speaking of errors: with nearly 20,000 equations, it is probable that a few typos remain, despite our best efforts to weed them out. A list of errata is maintained on www.andersen-piterbarg-book.com where supplemental material and news will also be posted on a running basis. We greatly appreciate reporting of typos or factual errors to our web address, and will list the names of all those who contribute to error spotting in future editions of *Interest Rate Modeling*. The following readers have contributed so far: Hamid Arian, Xin Li, Anonymous From Australia, Kou Toukairin, Dominique Bang, Matthias Lutz, Zhiyong Duan, Pierre Gauthier, Ziggy Jonsson, Sebastian Poloczek, Wojciech Slusarski, Mike Staunton, Wenbin Kong, Andrew S. Campbell, Rune Ljungmann, Patrick Morin, Claudio Moni, Julien Warmberg, Wendong Qu, Numa Lescot, Fredrik Åkesson, Liang Wu, Ronnie Barnes, Carsten Vocke, Oleg Butkovskiy, Bing Song, Alan Brace, Vivien Bégot, Qi Shen, Amit Kumar Soni, Mark Joshi, Julien Pantz, Igor Bluestein, Liu Chaoxie, Adrien Treccani, and Soren Kold Hansen.

Lastly, we owe a great debt of gratitude to our families for their support and patience, even when our initial plans for a brief book on tips and tricks for working quants ballooned into something more ambitious that consumed many evenings and weekends over the last six years.

London, New York, *Leif B.G Andersen*
June 2004 — August 2010 *Vladimir V. Piterbarg*

Table of Contents for All Volumes

VOLUME I Foundations and Vanilla Models

Part I Foundations

1 **Introduction to Arbitrage Pricing Theory** 3
   1.1 The Setup 3
   1.2 Trading Gains and Arbitrage........................... 7
   1.3 Equivalent Martingale Measures and Arbitrage 8
   1.4 Derivative Security Pricing and Complete Markets 11
   1.5 Girsanov's Theorem 12
   1.6 Stochastic Differential Equations 14
   1.7 Explicit Trading Strategies and PDEs 17
   1.8 Kolmogorov's Equations and the Feynman-Kac Theorem . 18
   1.9 Black-Scholes and Extensions 21
      1.9.1 Basics... 22
      1.9.2 Alternative Derivation 25
      1.9.3 Extensions..................................... 27
          1.9.3.1 Deterministic Parameters and Dividends 27
          1.9.3.2 Stochastic Interest Rates 28
   1.10 Options with Early Exercise Rights 30
      1.10.1 The Markovian Case 32

XII   Contents

| | | 1.10.2 | Some General Bounds | 34 |
| | | 1.10.3 | Early Exercise Premia | 36 |

2 Finite Difference Methods 43
    2.1 1-Dimensional PDEs: Problem Formulation 43
    2.2 Finite Difference Discretization 45
        2.2.1 Discretization in x-Direction. Dirichlet Boundary Conditions 45
        2.2.2 Other Boundary Conditions 48
        2.2.3 Time-Discretization 49
        2.2.4 Finite Difference Scheme 50
    2.3 Stability 52
        2.3.1 Matrix Methods 52
        2.3.2 Von Neumann Analysis 53
    2.4 Non-Equidistant Discretization 56
    2.5 Smoothing and Continuity Correction 58
        2.5.1 Crank-Nicolson Oscillation Remedies 58
        2.5.2 Continuity Correction 59
        2.5.3 Grid Shifting 59
    2.6 Convection-Dominated PDEs 61
        2.6.1 Upwinding 62
        2.6.2 Other Techniques 63
    2.7 Option Examples 63
        2.7.1 Continuous Barrier Options 64
        2.7.2 Discrete Barrier Options 66
        2.7.3 Coupon-Paying Securities and Dividends 67
        2.7.4 Securities with Early Exercise 68
        2.7.5 Path-Dependent Options 69
        2.7.6 Multiple Exercise Rights 71
    2.8 Special Issues 72
        2.8.1 Mesh Refinements for Multiple Events 72
        2.8.2 Analytics at the Last Time Step 76
        2.8.3 Analytics at the First Time Step 78
    2.9 Multi-Dimensional PDEs: Problem Formulation 79
    2.10 Two-Dimensional PDE with No Mixed Derivatives 80
        2.10.1 Theta Method 81
        2.10.2 The Alternating Direction Implicit (ADI) Method 82
        2.10.3 Boundary Conditions and Other Issues 85
    2.11 Two-Dimensional PDE with Mixed Derivatives 85
        2.11.1 Orthogonalization of the PDE 86
        2.11.2 Predictor-Corrector Scheme 89
    2.12 PDEs of Arbitrary Order 91

3 Monte Carlo Methods 95
3.1 Fundamentals 95
3.1.1 Generation of Random Samples 97
3.1.1.1 Inverse Transform Method 98
3.1.1.2 Acceptance-Rejection Method 99
3.1.1.3 Composition 101
3.1.2 Correlated Gaussian Samples 103
3.1.2.1 Cholesky Decomposition 103
3.1.2.2 Eigenvalue Decomposition 104
3.1.3 Principal Components Analysis (PCA) 105
3.2 Generation of Sample Paths 106
3.2.1 Example: Asian Basket Options in Black-Scholes Economy 106
3.2.2 Discretization Schemes, Convergence, and Stability 108
3.2.3 The Euler Scheme 110
3.2.3.1 Linear-Drift SDEs 112
3.2.3.2 Log-Euler Scheme 112
3.2.4 The Implicit Euler Scheme 113
3.2.4.1 Implicit Diffusion Term 114
3.2.5 Predictor-Corrector Schemes 115
3.2.6 Ito-Taylor Expansions and Higher-Order Schemes . 116
3.2.6.1 Ordinary Taylor Expansion of ODEs ... 117
3.2.6.2 Ito-Taylor Expansions 118
3.2.6.3 Milstein Second-Order Discretization Scheme 119
3.2.7 Other Second-Order Schemes 121
3.2.8 Bias vs. Monte Carlo Error 122
3.2.9 Sampling of Continuous Process Extremes 124
3.2.10 PCA and Bridge Construction of Brownian Motion Paths 128
3.2.10.1 Brownian Bridge and Quasi-Random Sequences 128
3.2.10.2 PC Construction 130
3.3 Sensitivity Computations 131
3.3.1 Finite Difference Estimates 132
3.3.1.1 Black-Scholes Delta 132
3.3.1.2 General Case 133
3.3.2 Pathwise Estimate 135
3.3.2.1 Black-Scholes Delta 135
3.3.2.2 General Case 136
3.3.2.3 Sensitivity Path Generation 138
3.3.3 Likelihood Ratio Method 139
3.3.3.1 Black-Scholes Delta 139
3.3.3.2 General Case 140
3.3.3.3 Euler Schemes 140

		3.3.3.4	Some Remarks	142
3.4	Variance Reduction Techniques			142
	3.4.1	Variance Reduction and Efficiency		143
	3.4.2	Antithetic Variates		144
		3.4.2.1	The Gaussian Case	144
		3.4.2.2	General Case	145
	3.4.3	Control Variates		145
		3.4.3.1	Basic Idea	145
		3.4.3.2	Non-Linear Controls	147
	3.4.4	Importance Sampling		149
		3.4.4.1	Basic Idea	149
		3.4.4.2	Density Formulation	149
		3.4.4.3	Importance Sampling and SDEs	151
		3.4.4.4	More on SDE Path Simulation	152
		3.4.4.5	Rare Event Simulation and Linearization	154
3.5	Some Notes on Bermudan Security Pricing			158
	3.5.1	Basic Idea		158
	3.5.2	Parametric Lower Bound Methods		159
	3.5.3	Parametric Lower Bound: An Example		160
	3.5.4	Regression-Based Lower Bound		161
	3.5.5	Upper Bound Methods		163
	3.5.6	Confidence Intervals		164
	3.5.7	Other Methods		164
3.A	Appendix: Constants for Φ^{-1} Algorithm			165

4 Fundamentals of Interest Rate Modeling 167
4.1	Fixed Income Notations		167
	4.1.1	Bonds and Forward Rates	167
	4.1.2	Futures Rates	169
	4.1.3	Annuity Factors and Par Rates	170
4.2	Fixed Income Probability Measures		171
	4.2.1	Risk Neutral Measure	172
	4.2.2	T-Forward Measure	174
	4.2.3	Spot Measure	175
	4.2.4	Terminal and Hybrid Measures	176
	4.2.5	Swap Measures	178
4.3	Multi-Currency Markets		178
	4.3.1	Notations and FX Forwards	178
	4.3.2	Risk Neutral Measures	179
	4.3.3	Other Measures	180
4.4	The HJM Analysis		181
	4.4.1	Bond Price Dynamics	181
	4.4.2	Forward Rate Dynamics	182
	4.4.3	Short Rate Process	183
4.5	Examples of HJM Models		184

| | | | Contents | XV |

	4.5.1	The Gaussian Model	184
	4.5.2	Gaussian HJM Models with Markovian Short Rate	187
	4.5.3	Log-Normal HJM Models	189

5 Fixed Income Instruments ... 191
5.1 Fixed Income Markets and Participants ... 191
5.2 Certificates of Deposit and Libor Rates ... 194
5.3 Forward Rate Agreements (FRA) ... 195
5.4 Eurodollar Futures ... 196
5.5 Fixed-for-Floating Swaps ... 197
5.6 Libor-in-Arrears Swaps ... 200
5.7 Averaging Swaps ... 201
5.8 Caps and Floors ... 201
5.9 Digital Caps and Floors ... 203
5.10 European Swaptions ... 203
5.10.1 Cash-Settled Swaptions ... 205
5.11 CMS Swaps, Caps and Floors ... 206
5.12 Bermudan Swaptions ... 207
5.13 Exotic Swaps and Structured Notes ... 208
5.13.1 Libor-Based Exotic Swaps ... 209
5.13.2 CMS-Based Exotic Swaps ... 210
5.13.3 Multi-Rate Exotic Swaps ... 210
5.13.4 Range Accruals ... 211
5.13.5 Path-Dependent Swaps ... 212
5.14 Callable Libor Exotics ... 213
5.14.1 Definitions ... 213
5.14.2 Pricing Callable Libor Exotics ... 215
5.14.3 Types of Callable Libor Exotics ... 216
5.14.4 Callable Snowballs ... 216
5.14.5 CLEs Accreting at Coupon Rate ... 216
5.14.6 Multi-Tranches ... 217
5.15 TARNs and Other Trade-Level Features ... 217
5.15.1 Knock-out Swaps ... 218
5.15.2 TARNs ... 218
5.15.3 Global Cap ... 219
5.15.4 Global Floor ... 219
5.15.5 Pricing and Trade Representation Challenges ... 220
5.16 Volatility Derivatives ... 220
5.16.1 Volatility Swaps ... 220
5.16.2 Volatility Swaps with a Shout ... 221
5.16.3 Min-Max Volatility Swaps ... 222
5.16.4 Forward Starting Options and Other Forward Volatility Contracts ... 222
5.A Appendix: Day Counting Rules and Other Trivia ... 224
5.A.1 Libor Rate Definitions ... 224
5.A.2 Swap Payments ... 225

Part II Vanilla Models

6 Yield Curve Construction and Risk Management 229
   6.1 Notations and Problem Definition 230
      6.1.1 Discount Curves................................ 230
      6.1.2 Matrix Formulation 232
      6.1.3 Construction Principles and Yield Curves........ 232
   6.2 Yield Curve Fitting with N-Knot Splines 234
      6.2.1 C^0 Yield Curves: Bootstrapping 234
          6.2.1.1 Piecewise Linear Yields 235
          6.2.1.2 Piecewise Flat Forward Rates 236
      6.2.2 C^1 Yield Curves: Hermite Splines................ 238
      6.2.3 C^2 Yield Curves: Twice Differentiable Cubic Splines 240
      6.2.4 C^2 Yield Curves: Twice Differentiable Tension Splines 243
   6.3 Non-Parametric Optimal Yield Curve Fitting 245
      6.3.1 Norm Specification and Optimization 245
      6.3.2 Choice of λ 248
      6.3.3 Example 249
   6.4 Managing Yield Curve Risk............................ 250
      6.4.1 Par-Point Approach............................ 251
      6.4.2 Forward Rate Approach 252
      6.4.3 From Risks to Hedging: The Jacobian Approach .. 254
      6.4.4 Cumulative Shifts and other Common Tricks 256
   6.5 Various Topics in Discount Curve Construction.......... 258
      6.5.1 Curve Overlays and Turn-of-Year Effects 258
      6.5.2 Cross-Currency Curve Construction............. 259
          6.5.2.1 Basic Problem 259
          6.5.2.2 Separation of Discount and Forward Rate Curves 260
          6.5.2.3 Cross-Currency Basis Swaps 262
          6.5.2.4 Modified Curve Construction Algorithm 263
      6.5.3 Tenor Basis and Multi-Index Curve Group Construction.................................. 265
   6.A Appendix: Spline Theory 270
      6.A.1 Hermite Spline Theory 270
      6.A.2 C^2 Cubic Splines 273
      6.A.3 C^2 Exponential Tension Splines 274

7 Vanilla Models with Local Volatility 277
   7.1 General Framework................................... 278
      7.1.1 Model Dynamics 278
      7.1.2 Volatility Smile and Implied Density 278
      7.1.3 Choice of φ 279

7.2		CEV Model	280
	7.2.1	Basic Properties	280
	7.2.2	Call Option Pricing	282
	7.2.3	Regularization	284
	7.2.4	Displaced Diffusion Models	285
7.3		Quadratic Volatility Model	287
	7.3.1	Case 1: Two Real Roots to the Left of $S(0)$	287
	7.3.2	Case 2: One Real Root to the Left of $S(0)$	291
	7.3.3	Extensions and Other Root Configurations	291
7.4		Finite Difference Solutions for General φ	292
	7.4.1	Multiple λ and T	293
	7.4.2	Forward Equation for Call Options	293
7.5		Asymptotic Expansions for General φ	295
	7.5.1	Expansion around Displaced Log-Normal Process	296
	7.5.2	Expansion around Gaussian Process	298
7.6		Extensions to Time-Dependent φ	299
	7.6.1	Separable Case	300
	7.6.2	Skew Averaging	301
		7.6.2.1 Examples	304
		7.6.2.2 A Caveat About the Process Domain	306
	7.6.3	Skew and Convexity Averaging by Small-Noise Expansion	307
	7.6.4	Numerical Example	311

8 Vanilla Models with Stochastic Volatility I 315

8.1	Model Definition	315
8.2	Model Parameters	317
8.3	Basic Properties	319
8.4	Fourier Integration	324
	8.4.1 General Theory	324
	8.4.2 Applications to SV Model	327
	8.4.3 Numerical Implementation	330
	8.4.4 Refinements of Numerical Implementation	332
	8.4.5 Fourier Integration for Arbitrary European Payoffs	336
8.5	Integration in Variance Domain	339
8.6	CEV-Type Stochastic Volatility Models and SABR	343
8.7	Numerical Examples: Volatility Smile Statics	345
8.8	Numerical Examples: Volatility Smile Dynamics	348
8.9	Hedging in Stochastic Volatility Models	353
	8.9.1 Hedge Construction, Delta and Vega	353
	8.9.2 Minimum Variance Delta Hedging	355
	8.9.3 Minimum Variance Hedging: an Example	357
8.A	Appendix: General Volatility Processes	359

XVIII Contents

9 Vanilla Models with Stochastic Volatility II 363
- 9.1 Fourier Integration with Time-Dependent Parameters 363
- 9.2 Asymptotic Expansion with Time-Dependent Volatility .. 366
- 9.3 Averaging Methods 371
  - 9.3.1 Volatility Averaging 371
  - 9.3.2 Skew Averaging 373
  - 9.3.3 Volatility of Variance Averaging 374
  - 9.3.4 Calibration by Parameter Averaging 377
- 9.4 PDE Method 381
  - 9.4.1 PDE Formulation 382
  - 9.4.2 Range for Stochastic Variance 382
  - 9.4.3 Discretizing Stochastic Variance 383
  - 9.4.4 Boundary Conditions for Stochastic Variance..... 385
  - 9.4.5 Range for Underlying 386
  - 9.4.6 Discretizing the Underlying 387
- 9.5 Monte Carlo Method................................. 387
  - 9.5.1 Exact Simulation of Variance Process 388
  - 9.5.2 Biased Taylor-Type Schemes for Variance Process 389
    - 9.5.2.1 Euler Schemes 389
    - 9.5.2.2 Higher-Order Schemes 390
  - 9.5.3 Moment Matching Schemes for Variance Process.. 390
    - 9.5.3.1 Log-normal Approximation 390
    - 9.5.3.2 Truncated Gaussian 391
    - 9.5.3.3 Quadratic-Exponential 392
    - 9.5.3.4 Summary of QE Algorithm 394
  - 9.5.4 Broadie-Kaya Scheme for the Underlying 395
  - 9.5.5 Other Schemes for the Underlying 396
    - 9.5.5.1 Taylor-Type Schemes 396
    - 9.5.5.2 Simplified Broadie-Kaya............. 396
    - 9.5.5.3 Martingale Correction 398
- 9.A Appendix: Proof of Proposition 9.3.4.................. 398
- 9.B Appendix: Coefficients for Asymptotic Expansion........ 402

VOLUME II Term Structure Models

Part III Term Structure Models

10 One-Factor Short Rate Models I 407
  10.1 The One-Factor Gaussian Short Rate Model 407
    10.1.1 The Ho-Lee Model 408
      10.1.1.1 Notations and First Steps 408
      10.1.1.2 Fitting the Term Structure of Discount Bonds 409
      10.1.1.3 Analysis and Comparison with HJM Approach 411
    10.1.2 The Mean-Reverting GSR Model 413
      10.1.2.1 The Vasicek Model 413
      10.1.2.2 The General One-Factor GSR Model ... 415
      10.1.2.3 Time-Stationarity and Caplet Hump ... 418
    10.1.3 European Option Pricing 420
      10.1.3.1 The Jamshidian Decomposition 420
      10.1.3.2 Gaussian Swap Rate Approximation ... 422
    10.1.4 Swaption Calibration 423
    10.1.5 Finite Difference Methods 424
      10.1.5.1 PDE and Spatial Boundary Conditions. 425
      10.1.5.2 Determining Spatial Boundary Conditions from PDE 426
      10.1.5.3 Upwinding 427
    10.1.6 Monte Carlo Simulation 427
      10.1.6.1 Exact Discretization 427
      10.1.6.2 Approximate Discretization 429
      10.1.6.3 Using other Measures for Simulation ... 430
  10.2 The Affine One-Factor Model 431
    10.2.1 Basic Definitions 431
      10.2.1.1 SDE 431
      10.2.1.2 Regularity Issues 432
      10.2.1.3 Volatility Skew 432
      10.2.1.4 Time-Dependent Parameters 433
    10.2.2 Discount Bond Pricing and Extended Transform .. 433
      10.2.2.1 Constant Parameters 434
      10.2.2.2 Piecewise Constant Parameters 436
    10.2.3 Discount Bond Calibration..................... 437
      10.2.3.1 Change of Variables 437
      10.2.3.2 Algorithm for $\omega(t)$ 438
    10.2.4 European Option Pricing 439

	10.2.5	Swaption Calibration	441
		10.2.5.1 Basic Problem	441
		10.2.5.2 Calibration Algorithm	442
	10.2.6	Quadratic One-Factor Model	443
	10.2.7	Numerical Methods for the Affine Short Rate Model	444

11 One-Factor Short Rate Models II ... 445
11.1 Log-Normal Short Rate Models ... 445
11.1.1 The Black-Derman-Toy Model ... 445
11.1.2 Black-Karasinski Model ... 447
11.1.3 Issues in Log-Normal Models ... 447
11.1.4 Sandmann-Sondermann Transformation ... 448
11.2 Other Short Rate Models ... 451
11.2.1 Power-Type Models and Empirical Model Estimation ... 451
11.2.2 The Black Shadow Rate Model ... 452
11.2.3 Spanned and Unspanned Stochastic Volatility: the Fong and Vasicek Model ... 454
11.3 Numerical Methods for General One-Factor Short Rate Models ... 455
11.3.1 Finite Difference Methods ... 456
11.3.2 Calibration to Initial Yield Curve ... 457
    11.3.2.1 Forward Induction ... 458
    11.3.2.2 Forward-from-Backward Induction ... 460
    11.3.2.3 Yield Curve and Volatility Calibration ... 461
    11.3.2.4 The Dybvig Parameterization ... 463
    11.3.2.5 Link to HJM Models ... 464
    11.3.2.6 The Hagan and Woodward Parameterization ... 465
11.3.3 Monte Carlo Simulation ... 469
    11.3.3.1 SDE Discretization ... 469
    11.3.3.2 Practical Issues with Monte Carlo Methods ... 470
11.A Appendix: Markov-Functional Models ... 472
11.A.1 State Process and Numeraire Mapping ... 472
11.A.2 Libor MF Parameterization ... 473
11.A.3 Swap MF Parameterization ... 475
11.A.4 Non-Parametric Calibration ... 476
11.A.5 Numerical Implementation ... 477
11.A.6 Comments and Comparisons ... 478

12 Multi-Factor Short Rate Models ... 479
12.1 The Gaussian Model ... 480
12.1.1 Development from Separability Condition ... 480
    12.1.1.1 Mean-Reverting State Variables ... 481

		12.1.1.2	Further Changes of Variables	485
	12.1.2	Classical Development		487
		12.1.2.1	Diagonalization of Mean Reversion Matrix	488
	12.1.3	Correlation Structure		490
	12.1.4	The Two-Factor Gaussian Model		491
		12.1.4.1	Some Basics	491
		12.1.4.2	Variance and Correlation Structure	492
		12.1.4.3	Volatility Hump	494
		12.1.4.4	Another Formulation of the Two-Factor Model	494
	12.1.5	Multi-Factor Statistical Gaussian Model		497
	12.1.6	Swaption Pricing		501
		12.1.6.1	Jamshidian Decomposition	502
		12.1.6.2	Gaussian Swap Rate Approximation	506
	12.1.7	Calibration via Benchmark Rates		507
	12.1.8	Monte Carlo Simulation		510
	12.1.9	Finite Difference Methods		511
12.2	The Affine Model			512
	12.2.1	Introduction		512
	12.2.2	Basic Model		512
	12.2.3	Regularity Issues		514
	12.2.4	Discount Bond Prices		515
	12.2.5	Some Concrete Models		517
		12.2.5.1	Fong-Vasicek Model	517
		12.2.5.2	Longstaff-Schwartz Model	517
		12.2.5.3	Multi-Factor CIR Models	519
	12.2.6	Brief Notes on Option Pricing		520
12.3	The Quadratic Gaussian Model			520
	12.3.1	Quadratic Gaussian Models are Affine		521
	12.3.2	The Basics		522
	12.3.3	Parameterization		524
		12.3.3.1	Smile Generation	524
		12.3.3.2	Quadratic Term	525
		12.3.3.3	Linear Term	527
	12.3.4	Swaption Pricing		528
		12.3.4.1	State Vector Distribution Under the Annuity Measure	528
		12.3.4.2	Exact Pricing of European Swaptions	529
		12.3.4.3	Approximations for European Swaptions	530
	12.3.5	Calibration		533
	12.3.6	Spanned Stochastic Volatility		534
	12.3.7	Numerical Methods		534
12.A	Appendix: Quadratic Forms of Gaussian Vectors			535

13 The Quasi-Gaussian Model ... 539
13.1 One-Factor Quasi-Gaussian Model ... 539
- 13.1.1 Definition ... 539
- 13.1.2 Local Volatility ... 541
- 13.1.3 Swap Rate Dynamics ... 542
- 13.1.4 Approximate Local Volatility Dynamics for Swap Rate ... 543
  - 13.1.4.1 Simple Approximation ... 544
  - 13.1.4.2 Advanced Approximation ... 544
- 13.1.5 Linear Local Volatility ... 547
- 13.1.6 Linear Local Volatility for a Swaption Strip ... 549
- 13.1.7 Volatility Calibration ... 550
- 13.1.8 Mean Reversion Calibration ... 552
  - 13.1.8.1 Effects of Mean Reversion ... 552
  - 13.1.8.2 Calibrating Mean Reversion to Volatility Ratios ... 554
  - 13.1.8.3 Calibrating Mean Reversion to Inter-Temporal Correlations ... 557
  - 13.1.8.4 Final Comments on Mean Reversion Calibration ... 559
- 13.1.9 Numerical Methods ... 560
  - 13.1.9.1 Direct Integration ... 560
  - 13.1.9.2 Finite Difference Methods ... 562
  - 13.1.9.3 Monte Carlo Simulation ... 565
  - 13.1.9.4 Single-State Approximations ... 565

13.2 One-Factor Quasi-Gaussian Model with Stochastic Volatility 569
- 13.2.1 Definition ... 569
- 13.2.2 Swap Rate Dynamics ... 571
- 13.2.3 Volatility Calibration ... 572
- 13.2.4 Mean Reversion Calibration ... 573
- 13.2.5 Non-Zero Correlation ... 574
- 13.2.6 PDE and Monte Carlo Methods ... 574

13.3 Multi-Factor Quasi-Gaussian Model ... 575
- 13.3.1 General Multi-Factor Model ... 575
- 13.3.2 Local and Stochastic Volatility Parameterization ... 576
- 13.3.3 Swap Rate Dynamics and Approximations ... 579
- 13.3.4 Volatility Calibration ... 584
- 13.3.5 Mean Reversions, Correlations, and Numerical Schemes ... 584

13.A Appendix: Density Approximation ... 585
- 13.A.1 Simplified Forward Measure Dynamics ... 585
- 13.A.2 Effective Volatility ... 587
- 13.A.3 The Forward Equation for Call Options ... 587
- 13.A.4 Asymptotic Expansion ... 588
- 13.A.5 Proof of Theorem 13.1.14 ... 590

14 The Libor Market Model I 591
14.1 Introduction and Setup 592
14.1.1 Motivation and Historical Notes 592
14.1.2 Tenor Structure 593
14.2 LM Dynamics and Measures 593
14.2.1 Setting 593
14.2.2 Probability Measures............................ 594
14.2.3 Link to HJM Analysis............................ 597
14.2.4 Separable Deterministic Volatility Function 598
14.2.5 Stochastic Volatility 600
14.2.6 Time-Dependence in Model Parameters 603
14.3 Correlation 603
14.3.1 Empirical Principal Components Analysis 604
14.3.1.1 Example: USD Forward Rates 605
14.3.2 Correlation Estimation and Smoothing 606
14.3.2.1 Example: Fit to USD Data 609
14.3.3 Negative Eigenvalues........................... 610
14.3.4 Correlation PCA 611
14.3.4.1 Example: USD Data 613
14.3.4.2 Poor Man's Correlation PCA 614
14.4 Pricing of European Options 615
14.4.1 Caplets... 615
14.4.2 Swaptions 616
14.4.3 Spread Options................................... 619
14.4.3.1 Term Correlation 620
14.4.3.2 Spread Option Pricing 621
14.5 Calibration 622
14.5.1 Basic Principles 622
14.5.2 Parameterization of $\|\lambda_k(t)\|$ 622
14.5.3 Interpolation on the Whole Grid................. 624
14.5.4 Construction of $\lambda_k(t)$ from $\|\lambda_k(t)\|$ 625
14.5.4.1 Covariance PCA 626
14.5.4.2 Correlation PCA 626
14.5.4.3 Discussion and Recommendation 627
14.5.5 Choice of Calibration Instruments 627
14.5.6 Calibration Objective Function.................. 630
14.5.7 Sample Calibration Algorithm................... 633
14.5.8 Speed-Up Through Sub-Problem Splitting 633
14.5.9 Correlation Calibration to Spread Options 635
14.5.10 Volatility Skew Calibration 637
14.6 Monte Carlo Simulation 637
14.6.1 Euler-Type Schemes 638
14.6.1.1 Analysis of Computational Effort 639
14.6.1.2 Long Time Steps 641
14.6.1.3 Notes on the Choice of Numeraire 642

	14.6.2	Other Simulation Schemes	643
		14.6.2.1 Special-Purpose Schemes with Drift Predictor-Corrector	643
		14.6.2.2 Euler Scheme with Predictor-Corrector	644
		14.6.2.3 Lagging Predictor-Corrector Scheme	644
		14.6.2.4 Further Refinements of Drift Estimation	646
		14.6.2.5 Brownian-Bridge Schemes and Other Ideas	647
		14.6.2.6 High-Order Schemes	650
	14.6.3	Martingale Discretization	650
		14.6.3.1 Deflated Bond Price Discretization	651
		14.6.3.2 Comments and Alternatives	652
	14.6.4	Variance Reduction	653
		14.6.4.1 Antithetic Sampling	653
		14.6.4.2 Control Variates	654
		14.6.4.3 Importance Sampling	655

15 The Libor Market Model II 657
15.1 Interpolation 657
    15.1.1 Back Stub, Simple Interpolation 658
    15.1.2 Back Stub, Arbitrage-Free Interpolation 659
    15.1.3 Back Stub, Gaussian Model 661
    15.1.4 Front Stub, Zero Volatility 662
    15.1.5 Front Stub, Exogenous Volatility 663
    15.1.6 Front Stub, Simple Interpolation 666
    15.1.7 Front Stub, Gaussian Model 666
15.2 Advanced Swaption Pricing via Markovian Projection 667
    15.2.1 Advanced Formula for Swap Rate Volatility 670
    15.2.2 Advanced Formula for Swap Rate Skew 672
    15.2.3 Skew and Smile Calibration in LM Models 674
15.3 Near-Markov LM Models 676
15.4 Swap Market Models 676
15.5 Evolving Separate Discount and Forward Rate Curves 678
    15.5.1 Basic Ideas 679
    15.5.2 HJM Extension 680
    15.5.3 Applications to LM Models 683
    15.5.4 Deterministic Spread 687
15.6 SV Models with Non-Zero Correlation 687
15.7 Multi-Stochastic Volatility Extensions 690
    15.7.1 Introduction 690
    15.7.2 Setup 690
    15.7.3 Pricing Caplets and Swaptions 691
    15.7.4 Spread Options 692
    15.7.5 Another Use of Multi-Dimensional Stochastic Volatility 693

VOLUME III Products and Risk Management

Part IV Products

16 Single-Rate Vanilla Derivatives 697
   16.1 European Swaptions 697
      16.1.1 Smile Dynamics 698
      16.1.2 Adjustable Backbone......................... 699
      16.1.3 Stochastic Volatility Swaption Grid 702
      16.1.4 Calibrating Stochastic Volatility Model to Swaptions 703
      16.1.5 Some Other Interpolation Rules 705
   16.2 Caps and Floors....................................... 706
      16.2.1 Basic Problem 706
      16.2.2 Setup and Norms 707
      16.2.3 Calibration Procedure........................ 708
   16.3 Terminal Swap Rate Models 709
      16.3.1 TSR Basics 709
      16.3.2 Linear TSR Model........................... 711
      16.3.3 Exponential TSR Model...................... 714
      16.3.4 Swap-Yield TSR Model 715
   16.4 Libor-in-Arrears...................................... 716
   16.5 Libor-with-Delay 719
      16.5.1 Swap-Yield TSR Model 720
      16.5.2 Other Terminal Swap Rate Models 721
      16.5.3 Approximations Inspired by Term Structure Models 721
      16.5.4 Applications to Averaging Swaps 722
   16.6 CMS and CMS-Linked Cash Flows 723
      16.6.1 The Replication Method for CMS.............. 724
      16.6.2 Annuity Mapping Function as a Conditional Expected Value 726
      16.6.3 Swap-Yield TSR Model 727
      16.6.4 Linear and Other TSR Models 728
      16.6.5 The Quasi-Gaussian Model 730
      16.6.6 The Libor Market Model 731
      16.6.7 Correcting Non-Arbitrage-Free Methods 733
      16.6.8 Impact of Annuity Mapping Function and Mean Reversion................................... 735
      16.6.9 CDF and PDF of CMS Rate in Forward Measure. 737
      16.6.10 SV Model for CMS Rate 739
      16.6.11 Dynamics of CMS Rate in Forward Measure 741

         16.6.12 Cash-Settled Swaptions 744
    16.7 Quanto CMS 745
         16.7.1 Overview 746
         16.7.2 Modeling the Joint Distribution of Swap Rate
                 and Forward Exchange Rate 748
         16.7.3 Normalizing Constant and Final Formula 749
    16.8 Eurodollar Futures 750
         16.8.1 Fundamental Results on Futures................ 750
         16.8.2 Motivations and Plan 752
         16.8.3 Preliminaries................................. 753
         16.8.4 Expansion Around the Futures Value 754
         16.8.5 Forward Rate Variances 757
         16.8.6 Forward Rate Correlations..................... 758
         16.8.7 The Formula 759
    16.9 Convexity and Moment Explosions 760

17 **Multi-Rate Vanilla Derivatives**......................... 765
    17.1 Introduction to Multi-Rate Vanilla Derivatives 765
    17.2 Marginal Distributions and Reference Measure 767
    17.3 Dependence Structure via Copulas..................... 768
         17.3.1 Introduction to Gaussian Copula Method 768
         17.3.2 General Copulas............................... 770
         17.3.3 Archimedean Copulas 772
         17.3.4 Making Copulas from Other Copulas............ 773
    17.4 Copula Methods for CMS Spread Options 776
         17.4.1 Normal Model for the Spread 776
         17.4.2 Gaussian Copula for Spread Options 777
         17.4.3 Spread Volatility Smile Modeling with the Power
                 Gaussian Copula 780
         17.4.4 Copula Implied From Spread Options 781
    17.5 Rates Observed at Different Times..................... 784
    17.6 Numerical Methods for Copulas 785
         17.6.1 Numerical Integration Methods................. 786
         17.6.2 Dimensionality Reduction for CMS Spread Options 789
         17.6.3 Dimensionality Reduction for Other Multi-Rate
                 Derivatives 791
         17.6.4 Dimensionality Reduction by Conditioning....... 793
         17.6.5 Dimensionality Reduction by Measure Change ... 797
         17.6.6 Monte Carlo Methods 799
    17.7 Limitations of the Copula Method 801
    17.8 Stochastic Volatility Modeling for Multi-Rate Options.... 803
         17.8.1 Measure Change by Drift Adjustment 803
         17.8.2 Measure Change by CMS Caplet Calibration 804
         17.8.3 Impact of Correlations on the Spread Smile 805
         17.8.4 Connection to Term Structure Models........... 806

17.9 CMS Spread Options in Term Structure Models 808
    17.9.1 Libor Market Model 808
    17.9.2 Quadratic Gaussian Model 810
17.A Appendix: Implied Correlation in Displaced Log-Normal
    Models 811
    17.A.1 Preliminaries 811
    17.A.2 Implied Log-Normal Correlation 812
    17.A.3 A Few Numerical Results 813

18 Callable Libor Exotics 817
18.1 Model Calibration for Callable Libor Exotics 817
    18.1.1 Risk Factors for CLEs 818
    18.1.2 Model Choice and Calibration 821
18.2 Valuation Theory 822
    18.2.1 Preliminaries 822
    18.2.2 Recursion for Callable Libor Exotics 823
    18.2.3 Marginal Exercise Value Decomposition 824
18.3 Monte Carlo Valuation 825
    18.3.1 Regression-Based Valuation of CLEs, Basic Scheme 825
    18.3.2 Regression for Underlying 827
    18.3.3 Valuing CLE as a Cancelable Note 829
    18.3.4 Using Regressed Variables for Decision Only . 830
    18.3.5 Regression Valuation with Boundary Optimization 832
    18.3.6 Lower Bound via Regression Scheme 833
    18.3.7 Iterative Improvement of Lower Bound 835
    18.3.8 Upper Bound 838
        18.3.8.1 Basic Ideas 838
        18.3.8.2 Nested Simulation (NS) Algorithm 839
        18.3.8.3 Bias and Computational Cost of NS
             Algorithm 842
        18.3.8.4 Confidence Intervals and Practical Usage 844
        18.3.8.5 Non-Analytic Exercise Values 845
        18.3.8.6 Improvements to NS Algorithm 847
        18.3.8.7 Other Upper Bound Algorithms 849
    18.3.9 Regression Variable Choice 850
        18.3.9.1 State Variables Approach 850
        18.3.9.2 Explanatory Variables 851
        18.3.9.3 Explanatory Variables with Convexity . 854
    18.3.10 Regression Implementation 856
        18.3.10.1 Automated Explanatory Variable
             Selection 856
        18.3.10.2 Suboptimal Point Exclusion 858
        18.3.10.3 Two Step Regression 859
        18.3.10.4 Robust Implementation of Regression
             Algorithm 860

18.4 Valuation with Low–Dimensional Models 864
    18.4.1 Single-Rate Callable Libor Exotics 864
    18.4.2 Calibration Targets for the Local Projection Method 864
    18.4.3 Review of Suitable Local Models 866
    18.4.4 Defining a Suitable Analog for Core Swap Rates .. 867
    18.4.5 PDE Methods for Path-Dependent CLEs 870
        18.4.5.1 CLEs Accreting at Coupon Rate 870
        18.4.5.2 Snowballs 872

19 Bermudan Swaptions 875
19.1 Definitions.. 875
19.2 Local Projection Method 876
19.3 Smile Calibration..................................... 878
19.4 Amortizing, Accreting, Other Non-Standard Swaptions ... 880
    19.4.1 Relationship Between Non-Standard and Standard Swap Rates 882
    19.4.2 Same-Tenor Approach......................... 883
    19.4.3 Representative Swaption Approach 884
    19.4.4 Basket Approach 887
    19.4.5 Super-Replication for Non-Standard Bermudan Swaptions 890
    19.4.6 Zero-Coupon Bermudan Swaptions 894
    19.4.7 American Swaptions 895
        19.4.7.1 American Swaptions vs. High-Frequency Bermudan Swaptions 896
        19.4.7.2 The Proxy Libor Rate Method 897
        19.4.7.3 The Libor-as-Extra-State Method 898
    19.4.8 Mid-Coupon Exercise 899
19.5 Flexi-Swaps... 900
    19.5.1 Purely Global Bounds......................... 901
    19.5.2 Purely Local Bounds 901
    19.5.3 Marginal Exercise Value Decomposition 903
    19.5.4 Narrow Band Limit 904
19.6 Monte Carlo Valuation 905
    19.6.1 Regression Methods........................... 905
    19.6.2 Parametric Boundary Methods 906
        19.6.2.1 Sample Exercise Strategies for Bermudan Swaptions 906
        19.6.2.2 Some Numerical Tests 909
        19.6.2.3 Additional Comments................ 911
19.7 Other Topics.. 912
    19.7.1 Robust Bermudan Swaption Hedging with European Swaptions 912
    19.7.2 Carry and Exercise 914

		19.7.3 Fast Pricing via Exercise Premia Representation	916

19.A Appendix: Forward Volatility and Correlation 919
19.B Appendix: A Primer on Moment Matching............. 920
    19.B.1 Basics... 920
    19.B.2 Example 1: Asian Option in BSM Model 922
    19.B.3 Example 2: Basket Option in BSM Model 924

20 TARNs, Volatility Swaps, and Other Derivatives 925
20.1 TARNs 925
    20.1.1 Definitions and Examples...................... 925
    20.1.2 Valuation and Risk with Globally Calibrated Models 927
    20.1.3 Local Projection Method 928
    20.1.4 Volatility Smile Effects 929
    20.1.5 PDE for TARNs............................... 931
20.2 Volatility Swaps 933
    20.2.1 Local Projection Method 934
    20.2.2 Shout Options 935
    20.2.3 Min-Max Volatility Swaps 938
    20.2.4 Impact of Volatility Dynamics on Volatility Swaps 940
20.3 Forward Swaption Straddles 945

21 Out-of-Model Adjustments 951
21.1 Adjusting the Model 952
    21.1.1 Calibration to Coupons 952
    21.1.2 Adjusters..................................... 954
    21.1.3 Path Re-Weighting 956
    21.1.4 Proxy Model Method 961
    21.1.5 Asset-Based Adjustments...................... 963
    21.1.6 Mapping Function Adjustments 965
21.2 Adjusting the Market 965
21.3 Adjusting the Trade 966
    21.3.1 Fee Adjustments 967
    21.3.2 Fee Adjustment Impact on Exotic Derivatives 968
    21.3.3 Strike Adjustment 969

Part V Risk Management

22 Introduction to Risk Management 975
22.1 Risk Management and Sensitivity Computations 976
    22.1.1 Basic Information Flow........................ 976
    22.1.2 Risk: Theory and Practice 978
    22.1.3 Example: the Black-Scholes Model 980

XXX   Contents

        22.1.4 Example: Black-Scholes Model with Time-Dependent Parameters 983
        22.1.5 Actual Risk Computations 985
        22.1.6 What about Θ_{prm} and Θ_{num}? 986
        22.1.7 A Note on Trading P&L and the Computation of Implied Volatility 987
    22.2 P&L Analysis 990
        22.2.1 P&L Predict 991
        22.2.2 P&L Explain 993
            22.2.2.1 Waterfall Explain 993
            22.2.2.2 Bump-and-Reset Explain 994
    22.3 Value-at-Risk 995
    22.A Appendix: Alternative Proof of Lemma 22.1.1 998

23 Payoff Smoothing and Related Methods 1001
    23.1 Issues with Discretization Schemes 1001
        23.1.1 Problems with Grid Dimensioning 1002
        23.1.2 Grid Shifts Relative to Payout 1002
        23.1.3 Additional Comments 1005
    23.2 Basic Techniques 1006
        23.2.1 Adaptive Integration 1006
        23.2.2 Adding Singularities to the Grid 1007
        23.2.3 Singularity Removal 1009
        23.2.4 Partial Analytical Integration 1010
    23.3 Payoff Smoothing For Numerical Integration and PDEs .. 1012
        23.3.1 Introduction to Payoff Smoothing 1012
        23.3.2 Payoff Smoothing in One Dimension 1014
            23.3.2.1 Box Smoothing 1015
            23.3.2.2 Other Smoothing Methods 1018
        23.3.3 Payoff Smoothing in Multiple Dimensions 1019
    23.4 Payoff Smoothing for Monte Carlo 1022
        23.4.1 Tube Monte Carlo for Digital Options 1022
        23.4.2 Tube Monte Carlo for Barrier Options 1024
        23.4.3 Tube Monte Carlo for Callable Libor Exotics 1029
        23.4.4 Tube Monte Carlo for TARNs 1029
    23.A Appendix: Delta Continuity of Singularity-Enlarged Grid Method 1030
    23.B Appendix: Conditional Independence for Tube Monte Carlo 1032

24 Pathwise Differentiation 1035
    24.1 Pathwise Differentiation: Foundations 1035
        24.1.1 Callable Libor Exotics 1035
            24.1.1.1 CLE Greeks 1036
            24.1.1.2 Keeping the Exercise Time Constant ... 1038
            24.1.1.3 Noise in CLE Greeks 1040

	24.1.2	Barrier Options	1041
24.2	Pathwise Differentiation for PDE Based Models		1044
	24.2.1	Model and Setup	1044
	24.2.2	Bucketed Deltas	1045
	24.2.3	Survival Density	1048
24.3	Pathwise Differentiation for Monte Carlo Based Models		1051
	24.3.1	Pathwise Derivatives of Forward Libor Rates	1051
	24.3.2	Pathwise Deltas of European Options	1054
		24.3.2.1 Pathwise Deltas of the Numeraire	1054
		24.3.2.2 Pathwise Deltas of the Payoff	1055
	24.3.3	Adjoint Method For Greeks Calculation	1056
	24.3.4	Pathwise Delta Approximation for Callable Libor Exotics	1058
24.4	Notes on Likelihood Ratio and Hybrid Methods		1060

25 Importance Sampling and Control Variates ... 1063

- 25.1 Importance Sampling In Short Rate Models ... 1063
- 25.2 Payoff Smoothing by Importance Sampling ... 1065
  - 25.2.1 Binary Options ... 1065
  - 25.2.2 TARNs ... 1068
  - 25.2.3 Removing the First Digital ... 1068
  - 25.2.4 Smoothing All Digitals by One-Step Survival Conditioning ... 1069
  - 25.2.5 Simulating Under the Survival Measure Using Conditional Gaussian Draws ... 1072
  - 25.2.6 Generalized Trigger Products in Multi-Factor LM Models ... 1074
- 25.3 Model-Based Control Variates ... 1077
  - 25.3.1 Low-Dimensional Markov Approximation for LM models ... 1078
  - 25.3.2 Two-Dimensional Extension ... 1081
  - 25.3.3 Approximating Volatility Structure ... 1082
  - 25.3.4 Markov Approximation as a Control Variate ... 1084
- 25.4 Instrument-Based Control Variates ... 1086
- 25.5 Dynamic Control Variates ... 1090
- 25.6 Control Variates and Risk Stability ... 1093

26 Vegas in Libor Market Models ... 1095

- 26.1 Basic Problem of Vega Computations ... 1095
- 26.2 Review of Calibration ... 1097
- 26.3 Vega Calculation Methods ... 1098
  - 26.3.1 Direct Vega Calculations ... 1098
    - 26.3.1.1 Definition and Analysis ... 1098
    - 26.3.1.2 Numerical Example ... 1101
  - 26.3.2 What is a Good Vega? ... 1102

		26.3.3	Indirect Vega Calculations	1105

- 26.3.3 Indirect Vega Calculations 1105
  - 26.3.3.1 Definition and Analysis 1105
  - 26.3.3.2 Numerical Example and Performance Analysis............................ 1107
- 26.3.4 Hybrid Vega Calculations...................... 1111
  - 26.3.4.1 Definition and Analysis 1111
  - 26.3.4.2 Numerical Example.................. 1113
- 26.4 Skew and Smile Vegas... 1114
- 26.5 Vegas and Correlations 1115
  - 26.5.1 Term Correlation Effects 1115
  - 26.5.2 What Correlations should be Kept Constant? 1116
  - 26.5.3 Vegas with Fixed Term Correlations 1118
  - 26.5.4 Numerical Example 1119
- 26.6 Deltas with Backbone 1120
- 26.7 Vega Projections 1122
- 26.8 Some Notes on Computing Model Vegas.................. 1124

Appendix

A Markovian Projection 1129
- A.1 Marginal Distributions of Ito Processes 1129
- A.2 Approximations for Conditional Expected Values 1134
  - A.2.1 Gaussian Approximation 1134
  - A.2.2 Least-Squares Projection 1136
- A.3 Applications to Local Stochastic Volatility Models 1137
  - A.3.1 Markovian Projection onto an SV Model 1137
  - A.3.2 Fitting the Market with an LSV Model.......... 1139
  - A.3.3 On Calculating Proxy Local Volatility 1143
- A.4 Basket Options in Local Volatility Models 1145
- A.5 Basket Options in Stochastic Volatility Models 1149
- A.A Appendix: $\mathrm{E}(\sqrt{z_n(t) z_m(t)})$ and $\mathrm{E}(\sqrt{z_n(t)})$ 1152
  - A.A.1 Proof of Proposition A.A.1 1153
    - A.A.1.1 Step 1. Reduction to Covariance........ 1153
    - A.A.1.2 Step 2. Linear Approximation.......... 1154
    - A.A.1.3 Step 3. Coefficients 1154
    - A.A.1.4 Step 4. Order of Approximation 1155
  - A.A.2 Proof of Lemma A.A.2 1155

References i

Index xxi

Commonly Used Notations

Probability Notations

- $(\Omega, \mathcal{F}, \mathrm{P})$: probability space.
- $\mathcal{F}_t$, $\mathcal{B}_t$: filtrations of σ-algebras.
- P: probability measure.
- Q: risk-neutral measure.
- Q^B: spot Libor measure.
- Q^T, Q^n: forward measure for time T or T_n (given tenor structure).
- $Q^{n,m}$: swap measure for swap rate $S_{n,m}$ (given tenor structure).
- Q^N: measure for numeraire N.
- E, E^P, E^Q, E^T, E^n, $\mathrm{E}^{n,m}$, E^A, ...: expectations under various measures.
- E_t, E_t^P, E_t^Q, E_t^T, E_t^n, $\mathrm{E}_t^{n,m}$, E_t^A, ...: expectations conditional on $\mathcal{F}_t$ under various measures.
- $Z(t)$, $W(t)$, $W^T(t)$, $W^n(t)$, $W^{n,m}(t)$, $W^A(t)$, ...: Brownian motions under various probability measures.
- $\mathrm{Var}(X)$: variance of X.
- $\mathrm{Stdev}(X)$: standard deviation of X.
- $\mathrm{Cov}(X, Y)$: covariance of X, Y.
- $\mathrm{Corr}(X, Y)$: correlation of X, Y.
- $\mathcal{N}(\mu, \Sigma)$: Gaussian distribution with mean μ and variance-covariance matrix Σ.
- $\mathcal{LN}(\mu, \sigma^2)$: log-normal distribution with mean μ and variance σ^2.
- $\mathcal{U}(a, b)$: uniform distribution on an interval $[a, b]$.
- $\Phi(z)$: standard Gaussian CDF, $\phi(z)$: standard Gaussian PDF.
- $\Gamma(a, x)$: the (upper) incomplete Gamma function, $\Gamma(a, x) = \int_x^\infty u^{a-1} e^{-u}\, du$.
- $\Gamma(a)$: the Gamma function, $\Gamma(a) = \Gamma(a, 0)$.
- $\mathcal{E}(X(t))$: Doléans exponential martingale for the process $X(t)$.
- $\langle X(t) \rangle$, $\langle X(t), Y(t) \rangle$: quadratic variation and covariation.

Finance Notations

- $T_0 < T_1 < \ldots < T_N$: tenor structure.
- τ_n: year fraction between T_n and T_{n+1}.
- $\beta(t)$: continuously compounded money market account.
- $B(t)$: discretely compounded money market account.
- $P(t,T)$: zero-coupon (or discount) bond price at time t for maturity T.
- $P(t,T,S)$: forward bond price at time t, for delivery of S-maturity discount bond at time T, $T \leq S$.
- $y(t,T,S)$: continuously compounded forward yield at time t for the period $[T,S]$.
- $f(t,T)$: instantaneous forward rate at t for maturity T.
- $r(t)$: short rate at time t, $r(t) = f(t,t)$.
- $L(t,T,S)$: forward Libor rate at time t for the period $[T,S]$.
- $L_n(t)$: forward Libor rate at t for the period $[T_n, T_{n+1}]$, given a tenor structure, $L_n(t) = L(t, T_n, T_{n+1})$.
- $S_{n,m}(t)$: forward swap rate at time t, starting at T_n and with the final payment date at T_{n+m} (given a tenor structure).
- $A_{n,m}(t)$: annuity at time t, with the first payment date T_{n+1} and the final payment date T_{n+m} (given a tenor structure).
- $U_n(t)$: the n-th exercise ("underlying") value of a Bermudan swaption or a callable Libor exotic.
- $H_n(t)$: the n-th hold value of a Bermudan swaption or a callable Libor exotic.
- $\sigma_B(t, S; T, K)$: an implied Black volatility smile, parameterized by the time t spot S, strike K and expiry T.
- $c_B(t, S; T, K)$, $c_B(t, S; T, K, \sigma)$: price of a call option in the Black model with time t spot S, strike K, expiry T and Black volatility σ.
- $c_N(t, S; T, K)$, $c_N(t, S; T, K, \sigma)$: price of a call option in the Gaussian (Normal, or Bachelier) model with time t spot S, strike K, expiry T and Normal volatility σ.

Miscellaneous Notations

- $\mathrm{Re}\,(z)$, $\mathrm{Im}\,(z)$: real and imaginary part of a complex number z.
- $O(\cdot)$, $o(\cdot)$: "Big O" and "Little o" order symbols.
- $1_{\{A\}}$: indicator of A.
- L^1 and L^2: spaces of integrable and square-integrable random variables, vectors, or functions.
- C^n: space of functions with the n-th continuous derivative, i.e $C = C^0$ are continuous functions, C^1 are differentiable functions with continuous derivative, C^2 are twice-differentiable functions with continuous second-order derivative, etc.

- $\mathcal{L}$, $\mathcal{J}$: differential operators, e.g. $a\,\partial/\partial x + b\,\partial^2/\partial x^2$ or $\partial/\partial t + a\,\partial/\partial x + b\,\partial^2/\partial x^2$.
- $(\mathcal{F}f)(\omega), (\mathcal{F}^{-1}\varphi)(x)$: direct and inverse Fourier transforms.
- $\triangleq$: "is defined as", e.g. $f(x) \triangleq x^2$.
- x^+, x^-: maximum and minimum of x and 0, i.e. $x^+ = \max(x,0)$, $x^- = \min(x,0)$.
- $\lfloor x \rfloor$: integer part of real number x.
- $A^\top$: transpose of matrix A.
- $\det(A)$: the determinant of a square matrix A.
- $\operatorname{tr}(A)$: the trace of a square matrix A.
- $\operatorname{diag}(a)$: a square matrix with the vector a on the diagonal and zeros elsewhere.

Part I

Foundations

1
Introduction to Arbitrage Pricing Theory

For reference, this chapter reviews selected results from stochastic calculus and from the modern theory of asset pricing. The material in this chapter is well covered in existing literature, so we keep the chapter brief and the mathematical treatment informal. For a more rigorous treatment we refer to Duffie [2001] or Musiela and Rutkowski [1997]. Most of the necessary mathematical foundation for the theory is available in Karatzas and Shreve [1991], Øksendal [1992], and Protter [2005].

The treatment in this chapter focuses on asset pricing in general; we shall specialize it to interest rate securities in Chapter 4. Chapter 5 introduces fixed income markets in detail.

1.1 The Setup

Unless otherwise noted, in this book we shall always consider an economy with continuous and frictionless trading taking place inside a finite horizon $[0, T]$. We assume the existence of traded dividend-free assets with prices characterized by a p-dimensional vector-valued stochastic process $X(t) = (X_1(t), \ldots X_p(t))^\top$. Uncertainty and information arrival is modeled by a probability space $(\Omega, \mathcal{F}, \mathrm{P})$, with Ω being a sample space with outcome elements ω; $\mathcal{F}$ being a σ-algebra on Ω; and P being a probability measure on the measure space $(\Omega, \mathcal{F})$. Information is revealed over time according to a filtration $\{\mathcal{F}_t, \ t \in [0,T]\}$, a family of sub-$\sigma$-algebras of $\mathcal{F}$ satisfying $\mathcal{F}_s \subseteq \mathcal{F}_t$ whenever $s \leq t$. We can loosely think of $\mathcal{F}_t$ as the information available at time t. We assume that the process $X(t)$ is adapted to $\{\mathcal{F}_t\}$, i.e. that $X(t)$ is fully observable at time t. For technical reasons, we require that the filtration satisfies the "usual conditions"[1]. Let $\mathrm{E}^\mathrm{P}(\cdot)$ be the expectation

---

[1] To satisfy the "usual conditions", $\mathcal{F}_t$ must be right-continuous for all t, and $\mathcal{F}_0$ must contain all the null-sets of $\mathcal{F}$, i.e. all subsets of sets of zero P-probability.

operator for the measure P; when conditioning on information at time t, we will use the notation $E_t^P(\cdot) = E^P(\cdot|\mathcal{F}_t)$.

In all of the models in this book, we specialize the abstract setup above to the situation where information is generated by a d-dimensional vector-valued *Brownian motion* (or *Wiener process*) $W(t) = (W_1(t), \ldots, W_d(t))^\top$, where W_i is independent of W_j for $i \neq j$. Brownian motions are treated in detail in Karatzas and Shreve [1991]; here, we just recall that a scalar Brownian motion W_i is a continuous stochastic process starting at 0 (i.e. $W_i(0) = 0$), having independent Gaussian increments: $W_i(t) - W_i(s) \sim \mathcal{N}(0, t-s)$, $t \geq s$. The filtration we consider is normally always the one *generated* by W, $\mathcal{F}_t = \sigma\{W(u), 0 \leq u \leq t\}$, possibly augmented to satisfy the usual conditions. We will generally assume that the price vector $X(t)$ is described by a vector-valued *Ito process*:

$$X(t) = X(0) + \int_0^t \mu(s,\omega)\,ds + \int_0^t \sigma(s,\omega)\,dW(s), \quad (1.1)$$

or, in differential notation,

$$dX(t) = \mu(t,\omega)\,dt + \sigma(t,\omega)\,dW(t), \quad (1.2)$$

where $\mu : \mathbb{R} \times \Omega \to \mathbb{R}^p$ and $\sigma : \mathbb{R} \times \Omega \to \mathbb{R}^{p \times d}$ are processes of dimension p and $p \times d$, respectively. We assume that both μ and σ are adapted to $\{\mathcal{F}_t\}$ and are in L^1 and L^2 respectively, in the sense that for all $t \in [0,T]$,

$$\int_0^t |\mu(s,\omega)|\,ds < \infty, \quad (1.3)$$

$$\int_0^t |\sigma(s,\omega)|^2\,ds < \infty, \quad (1.4)$$

almost surely[2]. In (1.4), we have defined

$$|\sigma(t,\omega)|^2 = \operatorname{tr}\left(\sigma(t,\omega)\sigma(t,\omega)^\top\right). \quad (1.5)$$

We notice that the sample paths of X generated by (1.1) are almost surely continuous, with no jumps in asset prices.

A technical treatment of Ito processes and the Ito integral with respect to Brownian motion can be found in Karatzas and Shreve [1991]. For our needs, it suffices to think of the Ito integral as

$$\int_0^t \sigma(s,\omega)\,dW(s) = \lim_{n \to \infty} \sum_{i=1}^n \sigma\left((i-1)\delta, \omega\right)\left[W(i\delta) - W((i-1)\delta)\right], \quad (1.6)$$

---

[2] An event holds "almost surely" — often abbreviated by "a.s." — if the probability of the event is one.

where $\delta \triangleq t/n$. We note that the integrand σ is here always evaluated at the *left* of each interval $[(i-1)\delta, i\delta]$. Other choices are possible[3], but, as we shall see, the "non-anticipative" structure of the Ito integral gives rise to a number of useful results and makes it particularly useful as a model of trading gains (see Section 1.2).

We list a few relevant definitions and results below.

Definition 1.1.1 (Martingale). *Let $Y(t)$ be an adapted vector-valued process with $\mathrm{E}^{\mathrm{P}}(|Y(t)|) < \infty$ for all $t \in [0,T]$. We say that $Y(t)$ is a martingale under measure P if for all $s, t \in [0,T]$ with $t \leq s$,*

$$\mathrm{E}_t^{\mathrm{P}}(Y(s)) = Y(t), \quad a.s.$$

If we replace the equality sign in this equation with $\leq$ or $\geq$, $Y(t)$ is said to be a *supermartingale* or a *submartingale*, respectively.

Definition 1.1.2 (Space H^2). *Let $|\sigma(t,\omega)|^2$ be as defined in (1.5). We say that σ is in H^2, if for all $t \in [0,T]$ we have*

$$\mathrm{E}^{\mathrm{P}}\left(\int_0^t |\sigma(s,\omega)|^2 \, ds\right) < \infty.$$

The importance of Definition 1.1.2 becomes clear from the following result:

Theorem 1.1.3 (Properties of Ito Integral). *Define $I(t) = \int_0^t \sigma(s,\omega) \, dW(s)$ and assume that σ is in H^2. Then*

1. *$I(t)$ is $\mathcal{F}_t$-measurable.*
2. *$I(t)$ is a continuous martingale. In particular, $\mathrm{E}^{\mathrm{P}}(I(t)) = 0$ for all $t \in [0,T]$.*
3. *$\mathrm{E}^{\mathrm{P}}(|I(t)|^2) = \mathrm{E}^{\mathrm{P}}(\int_0^t |\sigma(s,\omega)|^2 \, ds) < \infty$.*
4. *$\mathrm{E}^{\mathrm{P}}(I(t)I(s)^\top) = \mathrm{E}^{\mathrm{P}}(\int_0^{\min(t,s)} \sigma(u,\omega)\sigma(u,\omega)^\top \, du)$.*

A proof of Theorem 1.1.3 can be found in, e.g., Karatzas and Shreve [1991]. The equality in the third item of Theorem 1.1.3 is known as the *Ito isometry*. Due to the inequality in the third item, we say that the martingale defined in the process is a *square-integrable martingale*.

While it is common in applied work to simply assume that Ito integrals are martingales, without technical regularity conditions on $\sigma(t,\omega)$ (such as the H^2 restriction in Theorem 1.1.3), we should note that Ito integrals involving general processes in L^2 can, in fact, only be guaranteed to be *local martingales*. A process X is said to be a local martingale if there exists a

---

[3] The *Stratonovich* stochastic integral evaluates σ at the mid-point of each interval.

sequence of stopping times[4] $\{\tau_n\}_{n=1}^{\infty}$, with $\tau_n \to \infty$ as $n \to \infty$, such that $X(\min(t, \tau_n))$, $t \geq 0$, is a martingale for all n. In other words, all "driftless" Ito processes of the type

$$dY(t) = \sigma(t, \omega) \, dW(t) \tag{1.7}$$

are local martingales, but not necessarily martingales. Interestingly, a converse result holds as well; all local martingales adapted to the filtration generated by the Brownian motion W can be represented as Ito processes of the form (1.7):

Theorem 1.1.4 (Martingale Representation Theorem). *If Y is a local martingale adapted to the filtration generated by a Brownian motion W, then there exists a process σ such that (1.7) holds. If Y is a square-integrable martingale, then σ is in H^2.*

The proof of Theorem 1.1.4 can be found in Karatzas and Shreve [1991].
In the manipulation of functionals of Ito processes, the key result is a famous result by K. Ito:

Theorem 1.1.5 (Ito's Lemma). *Let $f(t, x)$, $x = (x_1, \ldots, x_p)^\top$, denote a continuous function, $f : [0, T] \times \mathbb{R}^p \to \mathbb{R}$, with continuous partial derivatives $\partial f / \partial t = f_t$, $\partial f / \partial x_i = f_{x_i}$, $\partial^2 f / \partial x_i \partial x_j = f_{x_i x_j}$. Let $X(t)$ be given by the Ito process (1.2) and define a scalar process $Y(t) = f(t, X(t))$. Then $Y(t)$ is an Ito process with stochastic differential*

$$dY(t) = f_t(t, X(t)) \, dt + f_x(t, X(t)) \, \mu(t, \omega) \, dt + f_x(t, X(t)) \, \sigma(t, \omega) \, dW(t)$$
$$+ \frac{1}{2} \sum_{i=1}^{p} \sum_{j=1}^{p} f_{x_i x_j}(t, X(t)) \left(\sigma(t, \omega) \sigma(t, \omega)^\top \right)_{i,j} dt,$$

where $f_x = (f_{x_1}, \ldots, f_{x_p})$.

For easy reference, the result below lists Ito's lemma for the special case where $p = d = 1$.

Corollary 1.1.6. *For the case $p = d = 1$, Ito's lemma becomes*

$$dY(t) = \left(f_t(t, X(t)) + f_x(t, X(t)) \, \mu(t, \omega) + \frac{1}{2} f_{xx}(t, X(t)) \, \sigma(t, \omega)^2 \right) dt$$
$$+ f_x(t, X(t)) \, \sigma(t, \omega) \, dW(t).$$

Ito's lemma can be motivated heuristically from a Taylor expansion. For instance, for the scalar case in Corollary 1.1.6, we write informally

---

[4] Recall that a stopping time τ is simply a random time adapted to the given filtration, in the sense that the event $\{\tau \leq t\}$ belongs to $\mathcal{F}_t$.

$$f(t+dt, X(t+dt)) = f(t, X(t)) + f_t\, dt + f_x\, dX(t) + \frac{1}{2} f_{xx}\, (dX(t))^2 + \cdots. \tag{1.8}$$

Here, we have

$$(dX(t))^2 = \mu(t,\omega)^2\, (dt)^2 + \sigma(t,\omega)^2\, (dW(t))^2 + 2\mu(t,\omega)\sigma(t,\omega)\, dt\, dW(t).$$

As shown earlier, $(dW(t))^2 = dt$ in quadratic mean, whereas all other terms in the expression for $(dX(t))^2$ are of order $O(dt^{3/2})$ or higher and can be neglected for small dt. In the limit, we therefore have $(dX(t))^2 = \sigma(t,\omega)^2\, dt$ which can be inserted into (1.8). The result in Corollary 1.1.6 then emerges.

Remark 1.1.7. The quantity $(dX(t))^2$ discussed above is the differential of the *quadratic variation* of $X(t)$, often denoted by $\langle X(t), X(t) \rangle$. That is,

$$d\langle X(t), X(t) \rangle = (dX(t))^2 \quad \Rightarrow \quad \langle X(t), X(t) \rangle = \int_0^t (dX(u))^2.$$

For two different (scalar) Ito processes $X(t)$ and $Y(t)$, we may equivalently define the *quadratic covariation* process $\langle X(t), Y(t) \rangle$ by

$$d\langle X(t), Y(t) \rangle = dX(t)\, dY(t).$$

Sometimes we also write $d\langle X(t), Y(t) \rangle = \langle dX(t), dY(t) \rangle$. If $X(t)$ is a p-dimensional process and $Y(t)$ is a q-dimensional process, the quadratic covariation $\langle X(t), Y(t)^\top \rangle$ is a $(p \times q)$-dimensional matrix process whose (i,j)-th element is $\langle X_i(t), Y_j(t) \rangle$, $i = 1, \ldots, p$, $j = 1, \ldots, q$.

The so-called *Tanaka extension* (see Karatzas and Shreve [1991]) extends Ito's lemma to continuous but non-differentiable functions. At points where the function has a kink, the Tanaka extension (loosely speaking) justifies using the Heaviside (step-) function for the first-order derivative and the Dirac delta function for the second-order derivative. An application of the Tanaka extension can be found in Section 1.9.2 and in Chapter 7, along with further discussion and references.

1.2 Trading Gains and Arbitrage

Working in the setting of Section 1.1 with assets driven by Ito processes, we now consider an investor engaging in a trading strategy involving the p assets $X_1, \ldots, X_p$. Let the trading strategy be characterized by a predictable[5] adapted process $\phi(t, \omega) = (\phi_1(t, \omega), \ldots, \phi_p(t, \omega))^\top$, with $\phi_i(t, \omega)$ denoting

---

[5] A *predictable* process is one where we, loosely speaking, can "foretell" the value of the process at time t, given all information available up to, but not including, time t. All adapted continuous processes are thus predictable. For a technical definition of predictable processes, see Karatzas and Shreve [1991].

the holdings at time t in the i-th asset X_i. The value $\pi(t)$ of the trading strategy at time t is thus (dropping the dependence on ω in the notation)

$$\pi(t) = \phi(t)^\top X(t). \tag{1.9}$$

The gain from trading over a small time interval $[t, t+\delta]$ is (approximately) $\phi(t)^\top [X(t+\delta) - X(t)]$, suggesting (compare to (1.6)) that the Ito integral

$$\int_0^t \phi(s)^\top dX(s) = \int_0^t \phi(s)^\top \mu(s)\, ds + \int_0^t \phi(s)^\top \sigma(s)\, dW(s)$$

is a proper model for trading gains over $[0, t]$. An investment strategy is said to be *self-financing* if, for any $t \in [0, T]$,

$$\pi(t) - \pi(0) = \int_0^t \phi(s)^\top dX(s). \tag{1.10}$$

This relationship simply expresses that changes in portfolio value are solely caused by trading gains or losses, with no funds being added or withdrawn.

Self-financing trading strategies allow investors to turn a certain initial investment $\pi(0)$ into stochastic future wealth $\pi(t)$. Under natural assumptions on possible trading strategies (e.g., that there is finite supply of all assets) we would expect that there should be limitations to the profits that self-financing strategies can create. Most notably, it should be impossible to create "something for nothing", that is, to turn a zero initial investment into future wealth that is certain to be non-negative and may be positive with non-zero probability. To express this formally, we introduce the concept of an *arbitrage opportunity*:

Definition 1.2.1 (Arbitrage). *An arbitrage opportunity is a self-financing strategy ϕ for which $\pi(0) = 0$ and, for some $t \in [0, T]$,*

$$\pi(t) \geq 0 \text{ a.s., and } P(\pi(t) > 0) > 0, \tag{1.11}$$

with π given in (1.9).

In economic equilibrium, arbitrage strategies cannot exist and precluding (1.11) constitutes a fundamental consistency requirement on the asset processes.

1.3 Equivalent Martingale Measures and Arbitrage

We turn to the question of characterizing the conditions under which the trading economy is free of arbitrage opportunities. A concise way to state these conditions involves *equivalent martingale measures*, a concept we shall work our way up to in a number of steps. First, we recall that two probability measures P and $\widehat{\mathrm{P}}$ on the same measure space $(\Omega, \mathcal{F})$ are said to be *equivalent* if $\mathrm{P}(A) = 0 \Leftrightarrow \widehat{\mathrm{P}}(A) = 0$, $\forall A \in \mathcal{F}$; that is, the two measures have the same null-sets. An important result from measure theory states that equivalent measures are uniquely associated through a quantity known as a *Radon-Nikodym derivative*:

Theorem 1.3.1 (Radon-Nikodym Theorem). *Let* P *and* $\widehat{\mathrm{P}}$ *be equivalent probability measures on the common measure space* $(\Omega, \mathcal{F})$. *There exists a unique (a.s.) non-negative random variable* R *with* $\mathrm{E}^{\mathrm{P}}(R) = 1$, *such that*

$$\widehat{\mathrm{P}}(A) = \mathrm{E}^{\mathrm{P}}(R 1_{\{A\}}), \quad \text{for all } A \in \mathcal{F}.$$

For a proof of Theorem 1.3.1, see e.g. Billingsley [1995]. The random variable R in the theorem is known as a *Radon-Nikodym derivative* and is denoted $d\widehat{\mathrm{P}}/d\mathrm{P}$. In the theorem we have used an *indicator* $1_{\{A\}}$; this quantity is 1 if the event A comes true, 0 if not.

For later use, we associate any probability measure $\widehat{\mathrm{P}}$ with a *density process*

$$\varsigma(t) = \mathrm{E}_t^{\mathrm{P}}\left(\frac{d\widehat{\mathrm{P}}}{d\mathrm{P}}\right), \quad \forall t \in [0, T]. \tag{1.12}$$

Clearly, $\varsigma(t)$ is a P-martingale with $\varsigma(0) = 1$ and $\varsigma(t) = \mathrm{E}_t^{\mathrm{P}}(\varsigma(T))$. A simple conditioning exercise demonstrates that for any $\mathcal{F}_T$-measurable random variable $Y(T)$, with $R = d\widehat{\mathrm{P}}/d\mathrm{P}$,

$$\begin{aligned}
\mathrm{E}^{\widehat{\mathrm{P}}}(Y(T)|\mathcal{F}_t) &= \frac{1}{\mathrm{E}^{\mathrm{P}}(R|\mathcal{F}_t)} \mathrm{E}^{\mathrm{P}}(RY(T)|\mathcal{F}_t) \\
&= \varsigma(t)^{-1} \mathrm{E}^{\mathrm{P}}\left(\mathrm{E}^{\mathrm{P}}(R|\mathcal{F}_T) Y(T) | \mathcal{F}_t\right) \\
&= \mathrm{E}^{\mathrm{P}}\left(Y(T) \frac{\varsigma(T)}{\varsigma(t)} \bigg| \mathcal{F}_t\right).
\end{aligned} \tag{1.13}$$

We shall use this result on numerous occasions in this book.

We now introduce the important concept of a *deflator*, a strictly positive Ito process used to normalize the asset prices. Let the deflator be denoted $D(t)$, and define the normalized asset process $X^D(t) = (X_1(t)/D(t), \ldots, X_p(t)/D(t))^\top$. We say that a measure Q^D is an *equivalent martingale measure induced by* D if $X^D(t)$ is a martingale with respect to Q^D. If Q^D is a martingale measure, we say that a self-financing trading strategy is *permissible* if

$$\int_0^t \phi(s)^\top dX^D(s)$$

is a martingale. For the Ito setup discussed earlier, a permissible strategy[6] is obtained by, say, requiring that $\phi(t)^\top \sigma^D(t)$ is in H^2 per Theorem 1.1.3,

---

[6] The technical restriction on trading positions imposed by only considering permissible trading strategies rules out certain pathological strategies, such as the

where σ^D is the diffusion coefficient of X^D (compare to Corollary 1.5.2). An application of Ito's lemma combined with (1.9)–(1.10) implies that $\pi(t)/D(t)$ is a Q^D-martingale when the trading strategy is permissible.

For permissible trading strategies, the importance of equivalent martingale measures follows from the following theorem:

Theorem 1.3.2 (Sufficient Condition for No-Arbitrage). *Restrict attention to permissible trading strategies. If there is a deflator D such that the deflated asset price process allows for an equivalent martingale measure, then there is no arbitrage.*

For a proof we refer to Musiela and Rutkowski [1997]. We note that Theorem 1.3.2 only provides sufficient conditions for the absence of arbitrage, and known (and rather technical) counterexamples demonstrate that the existence of an equivalent martingale measure does not follow from the absence of arbitrage in a setting with permissible trading strategies. A body of results known as the *fundamental theorem of arbitrage* establishes the conditions under which the existence of an equivalent martingale measure is also a necessary condition for the absence of arbitrage. The results are rather technical, but generally state that absence of arbitrage and the existence of an equivalent martingale measure are "nearly" equivalent concepts. The exact notion of "nearly" equivalent is discussed in Duffie [2001] as well as in the authoritative reference[7] Delbaen and Schachermayer [1994]. For our purposes in this book, we ignore many of these technicalities and often simply treat the absence of arbitrage and the existence of a martingale measure as equivalent concepts.

Finally, if the deflator is one of the p assets, we call the deflator a *numeraire*. Let us, say, assume that X_1 is strictly positive and can be used as a numeraire. Also assume that a deflator D has been identified such that Theorem 1.3.2 holds. As $X_1(t)/D(t)$ is a Q^D-martingale, we can use the Radon-Nikodym theorem to define a new measure Q^{X_1} by the density $\varsigma(t) = (X_1(t)/D(t))/(X_1(0)/D(0))$. For an $\mathcal{F}_T$-measurable variable $Y(T)$, we then have, from (1.13),

$$X_1(t)E_t^{Q^{X_1}}\left(\frac{Y(T)}{X_1(T)}\right) = D(t)E_t^{Q^D}\left(\frac{Y(T)}{D(T)}\right). \quad (1.14)$$

In particular, if $Y(t)/D(t)$ is a Q^D-martingale, $Y(t)/X_1(t)$ must also be a Q^{X_1}-martingale. In practice, it normally suffices to only consider deflators from the set of available numeraires.

---

doubling strategy considered in Harrison and Kreps [1979]. A realistic resource-constrained economy will always bound the size of the positions one can take in an asset, sufficing to ensure that predictable trading strategies are permissible.

[7] In a nutshell, Delbaen and Schachermayer [1994] show that absence of arbitrage implies only the existence of a *local* martingale measure.

Remark 1.3.3. Some sources define $1/D(t)$ (rather than $D(t)$) as the deflator. The convention used in this book is more natural for our applications.

1.4 Derivative Security Pricing and Complete Markets

A T-maturity *derivative security* (also known as a *contingent claim*) pays out at time T an $\mathcal{F}_T$-measurable random variable $V(T)$, and makes no payments before T. We assume that $V(T)$ has finite variance, and say that the derivative security is *attainable* (or sometimes *redundant*) if there exists a permissible trading strategy ϕ such that $V(T) = \phi(T)^\top X(T) = \pi(T)$ a.s. The trading strategy is said to *replicate* the derivative security. Importantly, the absence of arbitrage dictates that the time 0 price of an attainable derivative security $V(0)$ must be equal to the cost of setting up the self-financing strategy, i.e. $V(0) = \pi(0)$. More generally, $V(t) = \pi(t)$, $t \in [0,T]$. This observation is the foundation of *arbitrage pricing* and allows us to price derivative securities as expectations under an equivalent martingale measure. Specifically, consider a deflator D and assume the existence of an equivalent martingale measure Q^D induced by D; the existence of Q^D guarantees that there are no arbitrages in the market, by Theorem 1.3.2. Now, from the martingale property of $\pi(t)/D(t)$ in the measure Q^D and the relation $V(t) = \pi(t)$ it immediately follows that

$$\frac{V(t)}{D(t)} = \mathrm{E}_t^{Q^D}\left(\frac{V(T)}{D(T)}\right)$$

or

$$V(t) = D(t)\mathrm{E}_t^{Q^D}\left(\frac{V(T)}{D(T)}\right). \tag{1.15}$$

If all finite-variance $\mathcal{F}_T$-measurable random variables can be replicated, the market is said to be *complete*. In a complete market, all derivatives are "spanned" and hence have unique prices. Interestingly, a similar uniqueness result holds for equivalent martingale measures:

Theorem 1.4.1. *In the absence of arbitrage, a market is complete if and only if there exists a deflator inducing a unique martingale measure.*

From (1.14) it follows that the martingale measures induced by all numeraires must then be unique as well.

In practical applications, we shall often manipulate the choice of numeraire asset to simplify computations. The following result is useful for this:

Theorem 1.4.2 (Change of Numeraire). *Consider two numeraires $N(t)$ and $M(t)$, inducing equivalent martingale measures Q^N and Q^M, respectively. If the market is complete, then the density of the Radon-Nikodym derivative relating the two measures is uniquely given by*

$$\varsigma(t) = \mathrm{E}_t^{Q^N}\left(\frac{dQ^M}{dQ^N}\right) = \frac{M(t)/M(0)}{N(t)/N(0)}.$$

Proof. As the market is complete, all derivatives prices are unique. Consider an integrable $\mathcal{F}_T$-measurable payout $V(T) = Y(T)M(T)$, with time t price $V(t)$. From Theorem 1.4.1 and (1.15) we must have

$$V(t) = N(t)\mathrm{E}_t^{Q^N}\left(\frac{M(T)Y(T)}{N(T)}\right) = M(t)\mathrm{E}_t^{Q^M}\left(\frac{M(T)Y(T)}{M(T)}\right)$$

or

$$\mathrm{E}_t^{Q^M}\left(Y(T)\right) = \mathrm{E}_t^{Q^N}\left(Y(T)\frac{M(T)/N(T)}{M(t)/N(t)}\right).$$

Comparison with (1.13), and the fact that the density must be scaled to equal 1 at time 0, reveals that the Radon-Nikodym derivative for the measure shift is characterized by the density in the theorem. $\square$

1.5 Girsanov's Theorem

The last two sections have demonstrated a close link between the concept of arbitrage and the existence and uniqueness of equivalent martingale measures. In this section, we consider i) the conditions on the asset prices that allow for an equivalent martingale measure; and ii) the effect on asset dynamics from a change of probability measure. We consider two measures P and P(θ) related by a density $\varsigma^\theta(t) = \mathrm{E}_t^P(d\mathrm{P}(\theta)/d\mathrm{P})$, where $\varsigma^\theta(t)$ is an *exponential martingale* given by the Ito process

$$d\varsigma^\theta(t)/\varsigma^\theta(t) = -\theta(t)^\top dW(t),$$

where $W(t)$ is a d-dimensional P-Brownian motion. The d-dimensional process θ is known as the *market price of risk*. By an application of Ito's lemma, we can write

$$\begin{aligned}\varsigma^\theta(t) &= \exp\left(-\int_0^t \theta(s)^\top dW(s) - \frac{1}{2}\int_0^t \theta(s)^\top \theta(s)\, ds\right) \\ &\triangleq \mathcal{E}\left(-\int_0^t \theta(s)^\top dW(s)\right)\end{aligned} \quad (1.16)$$

where $\mathcal{E}(\cdot)$ is the *Doléans exponential*. An often-quoted sufficient condition on $\theta(t)$ for (1.16) to define a proper martingale (and not just a local martingale) is the *Novikov condition*

$$\mathrm{E}^P\left[\exp\left(\frac{1}{2}\int_0^t \theta(s)^\top \theta(s)\, ds\right)\right] < \infty. \quad (1.17)$$

The Novikov condition can often be difficult to verify in practical applications.

Armed with the notation above, we are now ready to state the main result of this section.

1.5 Girsanov's Theorem

Theorem 1.5.1 (Girsanov's Theorem). *Suppose that $\varsigma^\theta(t)$ defined in (1.16) is a martingale under measure P. Then for all $t \in [0, T]$*

$$W^\theta(t) = W(t) + \int_0^t \theta(s)\, ds$$

is a Brownian motion under the measure $\mathrm{P}(\theta)$.

To discuss a strategy to prove Girsanov's theorem, assume for simplicity that the dimension of the Brownian motion is $d = 1$. One way to construct a proof for Theorem 1.5.1 is to demonstrate that the joint moment-generating function (mgf)[8] (under $\mathrm{P}(\theta)$) of the increments

$$W^\theta(t_1), W^\theta(t_2) - W^\theta(t_1), \ldots, W^\theta(t_n) - W^\theta(t_{n-1}), \quad 0 < t_1 < \ldots < t_n,$$

is the same as that of n independent Gaussian random variables with expectations 0 and variances $t_1, t_2 - t_1, \ldots$. That is, for any positive integer value of n and any set of values $\alpha_i \in \mathbb{R}$, $i = 1, 2, \ldots, n$, we need to show that,

$$\mathrm{E}^{\mathrm{P}^\theta}\left[\exp\left(\sum_{i=1}^n \alpha_i \left(W^\theta(t_i) - W^\theta(t_{i-1})\right)\right)\right] = \prod_{i=1}^n \exp\left(\alpha_i^2 (t_i - t_{i-1})/2\right),$$

where we have defined $t_0 = 0$. While carrying out such a proof is not difficult, we here merely justify the final result by examining the case $n = 1$ only. Specifically, we consider

$$\mathrm{E}^{\mathrm{P}(\theta)}\left[\exp\left(\alpha W^\theta(t)\right)\right],$$

where $\alpha \in \mathbb{R}$ and $t > 0$. Shifting probability measure, we get

$$\mathrm{E}^{\mathrm{P}(\theta)}\left[\exp\left(\alpha W^\theta(t)\right)\right] = \mathrm{E}^{\mathrm{P}(\theta)}\left[\exp\left(\alpha W(t) + \alpha \int_0^t \theta(s)\, ds\right)\right]$$

$$= \mathrm{E}^{\mathrm{P}}\left[\exp\left(\alpha W(t) + \alpha \int_0^t \theta(s)\, ds\right)\mathcal{E}\left(-\int_0^t \theta(s)\, dW(s)\right)\right]$$

$$= e^{\alpha^2 t/2}\mathrm{E}^{\mathrm{P}}\left[\exp\left(\int_0^t (\alpha - \theta(s))\, dW(s) - \frac{1}{2}\int_0^t (\alpha - \theta(s))^2\, ds\right)\right]$$

$$= e^{\alpha^2 t/2}\mathrm{E}^{\mathrm{P}}\left[\mathcal{E}\left(\int_0^t (\alpha - \theta(s))\, dW(s)\right)\right]$$

$$= e^{\alpha^2 t/2},$$

---

[8] Recall that the moment-generating function of a random variable Y in some measure P is defined as the expectation $\mathrm{E}^{\mathrm{P}}(\exp(\alpha Y))$, $\alpha \in \mathbb{R}$. Unlike the characteristic function, the moment-generating function is not always well-defined for all values of the argument α.

as desired. In the last step, we used the fact that the Doléans exponential is a martingale with initial value 1.

Girsanov's theorem implies that we can shift probability measure to transform an Ito process with a given drift to an Ito process with nearly arbitrary drift. Specifically, we notice that our asset price process (under P)

$$dX(t) = \mu(t)\,dt + \sigma(t)\,dW(t)$$

can be written

$$dX(t) = (\mu(t) - \sigma(t)\theta(t))\,dt + \sigma(t)dW^\theta(t),$$

where $W^\theta(t)$ is a Brownian measure under the measure $P(\theta)$. This process will be driftless provided that θ satisfies the "spanning condition" $\mu(t) = \sigma(t)\theta(t)$ for all $t \in [0, T]$. This gives us a convenient way to check for the existence of equivalent martingale measures:

Corollary 1.5.2. *For a given numeraire D, assume that the deflated asset process satisfies*

$$dX^D(t) = \mu^D(t)\,dt + \sigma^D(t)\,dW(t),$$

where $\sigma^D(t)$ is sufficiently regular to make $\int_0^t \sigma^D(s)\,dW(s)$ a martingale. Assume also that there exists a θ such that the density ς^θ is a martingale and (a.s.)

$$\sigma^D(t)\theta(t) = \mu^D(t), \quad t \in [0, T], \tag{1.18}$$

then D induces an equivalent martingale measure and there is no arbitrage.

Equation (1.18) is a system of linear equations and we can use rank results from linear algebra to determine the circumstances under which (1.18) will have solutions (no arbitrage) and when these are unique (complete market). For instance, a necessary condition for the market to be complete is that rank$(\sigma) = d$. Further results along these lines can be found in Musiela and Rutkowski [1997] and Duffie [2001].

We conclude this section by noting that while a change of probability measure affects the drift μ of an Ito process, it does not change the diffusion coefficient σ. This is sometimes known as the *diffusion invariance principle*.

1.6 Stochastic Differential Equations

So far we have defined the asset process vector to be an Ito process with general measurable coefficients $\mu(t, \omega)$ and $\sigma(t, \omega)$. In virtually all applications, however, we restrict our attention to the case where these coefficients are deterministic functions of time and the state of the asset process[9]. In other words, we consider a *stochastic differential equation* (SDE) of the form

$$dX(t) = \mu(t, X(t))\, dt + \sigma(t, X(t))\, dW(t), \quad X(0) = X_0, \tag{1.19}$$

with $\mu : [0, T] \times \mathbb{R}^p \to \mathbb{R}^p$; $\sigma : [0, T] \times \mathbb{R}^p \to \mathbb{R}^{p \times d}$; and X_0 an initial condition. A *strong solution*[10] to (1.19) is an Ito process

$$X(t) = X_0 + \int_0^t \mu(s, X(s))\, ds + \int_0^t \sigma(s, X(s))\, dW(s).$$

A number of restrictions on μ and σ are needed to ensure that the solution to (1.19) exists and is unique. A standard result is listed below.

Theorem 1.6.1. *In (1.19) assume that there exists a constant K such that for all $t \in [0, T]$ and all $x, y \in \mathbb{R}^p$,*

$$|\mu(t, x) - \mu(t, y)| + |\sigma(t, x) - \sigma(t, y)| \le K|x - y|, \quad \text{(Lipschitz condition)},$$
$$|\mu(t, x)|^2 + |\sigma(t, x)|^2 \le K^2 \left(1 + |x|^2\right), \quad \text{(growth condition)}.$$

Then there exists a unique solution to (1.19).

We notice that the dynamics of (1.19) do not depend on the past evolution of $X(t)$ beyond the state of X at time t. This lack of path-dependence suggests that X is a *Markov* process. We formalize this as follows.

Definition 1.6.2 (Markov Process). *The $\mathbb{R}^p$-valued stochastic process $X(t)$ is called a Markov process if for all $s, t \in [0, T]$ with $t \le s$,*

$$P(X(s) \in B | \mathcal{F}_t) = P(X(s) \in B | X(t)) \tag{1.20}$$

for all sets B in the p-dimensional σ-algebra of Borel sets $\mathfrak{B}^p$. If (1.20) holds with t replaced by a stopping time, the process is a strong Markov process.

Expressed verbally, the Markov property implies that the past and future become statistically independent when we condition on the present.

Theorem 1.6.3 (Markov Property of SDEs). *Let the coefficient of the SDE for $X(t)$ satisfy the conditions in Theorem 1.6.1. Then $X(t)$ is a strong Markov process.*

---

[9] In this section, the process X is generic and need not represent financial assets.

[10] In a strong solution, the Brownian motion is given and the solution is adapted to the filtration generated by it. If we are free to pick our own Brownian motion on some different probability space, we say that (1.19) holds in a *weak sense*. For financial applications where we normally only need the law of the underlying process, weak solutions are typically sufficient. The distinction between weak and strong solutions is of little importance for our purposes and we shall ignore it going forward.

Let us consider the explicit solutions of a few simple SDEs. First, consider a *linear* SDE

$$dX(t) = (AX(t) + B(t))\,dt + C(t)\,dW(t),$$

where A is a constant $p \times p$ matrix, and B and C are deterministic matrices of dimension $p \times 1$ and $p \times d$, respectively. The solution to this equation can, by Ito's lemma, be verified to be

$$X(t) = e^{At}X(0) + \int_0^t e^{A(t-s)}\left(B(s)\,ds + C(s)\,dW(s)\right).$$

The term

$$\int_0^t e^{A(t-s)}C(s)\,dW(s)$$

is distributed as a p-dimensional Gaussian random variable with mean 0 and, from Theorem 1.1.3, variance-covariance matrix

$$\Sigma \triangleq \mathrm{Var}\left(\int_0^t e^{A(t-s)}C(s)\,dW(s)\right) = \int_0^t e^{A(t-s)}C(s)C(s)^\top e^{A^\top(t-s)}\,ds.$$

Extensions to time-varying A are straightforward, and basically involve replacing the exponential matrix e^{At} with the solution of a homogeneous ODE with time-dependent coefficients. Details can be found in, e.g., Arnold [1974] and key results are listed in Chapter 12.

Now let us specialize to the scalar case with $p = 1$. An SDE of great importance is the *geometric Brownian motion with drift* (GBMD):

$$dX(t)/X(t) = \mu(t)\,dt + \sigma(t)\,dW(t),$$

where $\mu(t)$ and $\sigma(t)$ are *deterministic* (with $\sigma(t)$ having dimension $1 \times d$). An application of Ito's lemma to $\ln(X(t))$ reveals that

$$X(t) = X(0)\exp\left(\int_0^t \left(\mu(s) - \frac{1}{2}\sigma(s)\sigma(s)^\top\right)ds + \int_0^t \sigma(s)\,dW(s)\right)$$

$$= X(0)\exp\left(\int_0^t \mu(s)\,ds\right) \mathcal{E}\left(\int_0^t \sigma(s)\,dW(s)\right). \tag{1.21}$$

Being an exponential of a Gaussian random variable, $X(t)$ follows a *lognormal* distribution, with moments (see Karatzas and Shreve [1991])

$$\mathrm{E}^\mathrm{P}(X(t)) = X(0)\exp\left(\int_0^t \mu(s)\,ds\right), \tag{1.22}$$

$$\mathrm{E}^\mathrm{P}(X(t)^2) = \mathrm{E}^\mathrm{P}(X(t))^2 \exp\left(\int_0^t \sigma(s)\sigma(s)^\top\,ds\right). \tag{1.23}$$

1.7 Explicit Trading Strategies and PDEs

After the mathematical interlude of Section 1.6, we now return to financial markets and a more careful analysis of the trading strategies that replicate derivative securities. We have already established that in a complete market such strategies must exist for any given derivative, but it still remains to determine these strategies explicitly. Consider a Markovian setup where the asset vector X satisfies an SDE of the form (1.19). Let there be given a derivative security V paying out at time T an amount $V(T) = g(X(T))$, for some smooth payout function $g : \mathbb{R}^p \to \mathbb{R}$. The Markovian form of the asset dynamics suggests that the time t derivative price is a function of t and $X(t)$ only, $V(t) = V(t, X(t))$ for some deterministic function $V(t, x)$, $x \in \mathbb{R}^p$. Conjecturing that this function is smooth enough to allow for an application of Ito's lemma for all $t \in [0, T)$, Theorem 1.1.5 implies (suppressing dependence on $X(t)$ for brevity)

$$dV(t) = V_t(t)\, dt + \sum_{i=1}^{p} V_{x_i}(t)\mu_i(t)\, dt$$

$$+ \frac{1}{2} \sum_{i=1}^{p} \sum_{j=1}^{p} V_{x_i x_j}(t) \Sigma_{i,j}(t)\, dt + \sum_{i=1}^{p} V_{x_i}(t)\sigma_i(t)\, dW(t), \quad (1.24)$$

where σ_i is the i-th row of the $p \times d$ matrix σ and $\Sigma_{i,j}$ is the (i,j)-th element in $\sigma\sigma^\top$. We recall that subscripts like V_{X_i} denote partial differentiation, see Theorem 1.1.5.

If $V(t)$ can be replicated by a self-financing trading strategy ϕ in the p assets, we must also have, from (1.10),

$$dV(t) = \phi(t)^\top dX(t) = \sum_{i=1}^{p} \phi_i(t)\mu_i(t)\, dt + \sum_{i=1}^{p} \phi_i(t)\sigma_i(t)\, dW(t). \quad (1.25)$$

Comparing terms in (1.24) and (1.25) we see that both equations will hold, provided that for all $t \in [0, T]$

$$\phi_i(t) = \frac{\partial V(t, X(t))}{\partial x_i}, \quad i = 1, \ldots, p, \quad (1.26)$$

and

$$\frac{\partial V(t, x)}{\partial t} + \frac{1}{2} \sum_{i=1}^{p} \sum_{j=1}^{p} \frac{\partial^2 V(t, x)}{\partial x_i \partial x_j} \Sigma_{i,j}(t, x) = 0. \quad (1.27)$$

To the extent that the system above allows for a solution (it may not if the market is not complete), from (1.26) we see that the trading strategy that replicates the derivative V holds $\partial V(t, X(t))/\partial x_i$ units of asset X_i at time

t. The quantity $\partial V/\partial x_i$ is often known as the *delta* with respect to X_i[11]. Note that, from (1.9) and (1.26) we have that

$$V(t, X(t)) = \sum_{i=1}^{p} \frac{\partial V(t, X(t))}{\partial x_i} X_i(t). \qquad (1.28)$$

Besides identifying an explicit replication strategy, the arguments above have also produced (1.27), a partial differential equation (PDE) for the value function $V(t, x)$. The PDE is a second-order parabolic equation in p spatial variables, with known terminal condition $V(T, x) = g(x)$ (a so-called *Cauchy problem*). Solving this PDE provides an alternative way to price the derivative, as compared to the purely probabilistic expectations-based methods outlined earlier (see (1.15)). We shall investigate the link between expectations and PDEs in more detail in Section 1.8.

Inspection of the valuation PDE (1.27) reveals that the drifts μ_i of the asset price SDE (1.19) are notably absent, making the price of the derivative security independent of drifts. This is typical of derivatives in complete markets and follows from the fact that derivatives can be priced preference-free, by arbitrage arguments. In contrast, for the elements of the fundamental asset price vector, risk-averse investors would demand that assets with high volatilities $|\sigma_i|$ be rewarded with higher drifts (more precisely, higher rates of return) as compensation for the additional uncertainty.

1.8 Kolmogorov's Equations and the Feynman-Kac Theorem

In earlier sections, we have seen that derivatives prices can be expressed as expectations under certain probability measures or as solutions to PDEs. This hints at a deeper connection between expectations and PDEs, a connection we shall explore in this section. As part of this exploration, we list results for transition densities that will be useful later in model calibration.

As in Section 1.6, we consider a Markov vector SDE of the type (see (1.19))

$$dX(t) = \mu(t, X(t))\, dt + \sigma(t, X(t))\, dW(t), \quad X(0) = X_0, \qquad (1.29)$$

where the coefficients are assumed smooth enough to allow for a unique solution (see Theorem 1.6.1). Now define a functional

$$u(t, x) = \mathrm{E}^{\mathrm{P}}\left(g(X(T)) \,|\, X(t) = x\right),$$

---

[11] Note that taking a position in V and following a trading strategy with $\phi_i = -\partial V/\partial x_i$, $i = 1, \ldots, p$ will effectively remove any exposure to V (as we simultaneously take a long position in V and, through a trading strategy, a short position in V). This strategy is known as a *delta hedge*.

1.8 Kolmogorov's Equations and the Feynman-Kac Theorem

for a function $g: \mathbb{R}^p \to \mathbb{R}$. Under regularity conditions on g, it is easy to see that the process $u(t, X(t))$, being a conditional expectation, must be a martingale. Proceeding informally, an application of Ito's lemma gives, for $t \in [0, T]$ (suppressing dependence on $X(t)$),

$$du(t) = u_t(t)\, dt + \sum_{i=1}^{p} u_{x_i}(t)\mu_i(t)\, dt + \frac{1}{2}\sum_{i=1}^{p}\sum_{j=1}^{p} u_{x_i x_j}(t) \Sigma_{i,j}(t) dt + O(dW(t)),$$

where as before $\Sigma_{i,j}$ is the (i,j)-th element of $\sigma\sigma^\top$. From earlier results, we know that for $u(t, X(t))$ to be a martingale, the term multiplying dt in the equation above must be zero. Defining the operator

$$\mathcal{A} = \sum_{i=1}^{p} \mu_i(t, x) \frac{\partial}{\partial x_i} + \frac{1}{2}\sum_{i=1}^{p}\sum_{j=1}^{p} \Sigma_{i,j}(t, x) \frac{\partial^2}{\partial x_i \partial x_j},$$

we deduce that $u(t, x)$ satisfies the PDE

$$\frac{\partial u(t, x)}{\partial t} + \mathcal{A}u(t, x) = 0, \tag{1.30}$$

with terminal condition $u(T, x) = g(x)$. The equation above is known as the *Kolmogorov backward equation* for the SDE (1.29). The operator $\mathcal{A}$ is known as the *generator* or *infinitesimal operator* of the SDE, and can be identified as

$$\mathcal{A}u(t, x) = \lim_{h \downarrow 0} \frac{\mathrm{E}^{\mathrm{P}}\left(u\left(t, X(t+h)\right) | X(t) = x\right) - u(t, x)}{h}.$$

In arriving at (1.30) we made several implicit assumptions, most notably that the function $u(t, x)$ exists and is twice differentiable. Sufficient conditions for the validity of (1.30) can be found in Karatzas and Shreve [1991], for instance. A relevant result is listed below.

Theorem 1.8.1. *Let the process $X(t)$ be given by the SDE (1.29), where the coefficients μ and σ are continuous in x and satisfy the Lipschitz and growth conditions of Theorem 1.6.1. Consider a continuous function $g(x)$ that is either non-negative or satisfies a polynomial growth condition, meaning that for some positive constants K and q*

$$g(x) \leq K(1 + |x|^q), \quad x \in \mathbb{R}^p.$$

If $u(t, x)$ solves (1.30) with boundary condition $u(T, x) = g(x)$, and $u(t, x)$ satisfies a polynomial growth condition in x, then

$$u(t, x) = \mathrm{E}^{\mathrm{P}}\left(g\left(X(T)\right) | X(t) = x\right), \quad t \in [0, T]. \tag{1.31}$$

Conditions required to ensure existence of a solution to (1.30) are more involved, and we just refer to Karatzas and Shreve [1991] and the references therein.

A family of functions g of particular importance to many of our applications is
$$g(x) = e^{ik^\top x}, \quad k \in \mathbb{R}^p,$$
where $i = \sqrt{-1}$ is the imaginary unit. In this case $u(t,x)$ becomes the *characteristic function* of $X(T)$, conditional on $X(t) = x$. We refer to any standard statistics textbook (e.g. Ochi [1990]) for the many useful properties of characteristic functions.

For the Markov process $X(t)$ in (1.29), let us now introduce a *transition density*, given heuristically by
$$p(t,x;s,y)\,dy \triangleq \mathrm{P}\left(X(s) \in [y, y+dy] | X(t) = x\right), \quad 0 \le t \le s \le T.$$

We can loosely think of the transition density as a special case of the functional $u(t,x)$ above, with boundary condition $u(s,x) = \delta(x-y)$, where $\delta(\cdot)$ is the Dirac delta function. Sometimes $p(\cdot,\cdot;\cdot,\cdot)$ is called a *Green's function* or a *fundamental solution* to (1.30). Under certain regularity conditions discussed in Karatzas and Shreve [1991], the transition density solves the Kolmogorov backward equation
$$\frac{\partial p(t,x)}{\partial t} + \mathcal{A}p(t,x) = 0, \quad (s,y) \text{ fixed},$$
subject to the boundary condition $p(s,x;s,y) = \delta(x-y)$. Further, the general expectation $u(t,x) = \mathrm{E}^{\mathrm{P}}(g(X(T))|X(t)=x)$ in Theorem 1.8.1 can be written
$$u(t,x) = \int_{\mathbb{R}^p} g(y) p(t,x;T,y)\,dy, \quad t \in [0,T]. \tag{1.32}$$

In many applications, it is useful to have a result that produces transition densities at future times $s \ge t$ from a known state at time t, rather than vice-versa. For this, we first define an operator $\mathcal{A}^*$ by
$$\mathcal{A}^* f(s,y) = -\sum_{i=1}^p \frac{\partial\left[\mu_i(s,y) f(s,y)\right]}{\partial y_i} + \frac{1}{2} \sum_{i=1}^p \sum_{j=1}^p \frac{\partial^2\left[\Sigma_{i,j}(s,y) f(s,y)\right]}{\partial y_i \partial y_j}.$$

In the transition density $p(t,x;s,y)$ now consider (t,x) fixed and let $\mathcal{A}^*$ operate on the resulting function of s and y. Under additional regularity conditions, we then have the *forward Kolmogorov equation*
$$-\frac{\partial p(s,y)}{\partial s} + \mathcal{A}^* p(s,y) = 0, \quad (t,x) \text{ fixed}, \tag{1.33}$$
subject to the boundary condition $p(t,x;t,y) = \delta(x-y)$.

The forward Kolmogorov equation is sometimes known as the *Fokker-Planck* equation. We stress that the backward equation is more general than the forward equation, in the sense that the former holds for general terminal conditions $g(x)$, whereas the latter only holds for δ-type initial conditions.

We round off this section by a useful extension to the Kolmogorov backward equation. Specifically, consider extending the PDE (1.30) to

$$\frac{\partial u(t,x)}{\partial t} + \mathcal{A}u(t,x) + h(t,x) = r(t,x)u(t,x), \tag{1.34}$$

where $h, r : [0,T] \times \mathbb{R}^p \to \mathbb{R}$. Given the boundary condition $u(T,x) = g(x)$, the *Feynman-Kac solution* to (1.34), should it exist, is given by

$$u(t,x) = \mathrm{E}^{\mathrm{P}}\left(\psi(t,T)g\left(X(T)\right) + \int_t^T \psi(t,s)h\left(s,X(s)\right)\,ds \,\bigg|\, X(t) = x\right), \tag{1.35}$$

where

$$\psi(t,T) = \exp\left(-\int_t^T r\left(s,X(s)\right)\,ds\right), \quad t \in [0,T].$$

The result is easily understood from an application of Ito's lemma, similar to the one used above to motivate the backward Kolmogorov equation. Sufficient regularity conditions for the Feynman-Kac result to hold are identical to those of Theorem 1.8.1, supplemented with the requirement that r be nonnegative and continuous in x; and the requirement that h be continuous in x and either be nonnegative or satisfy a polynomial growth requirement in x. See Duffie [2001] for further details about the often delicate regularity issues surrounding the Feynman-Kac result.

For later use, let us finally note that when $g(x) = \delta(x-y)$ and $h(t,x) = 0$, $u(t,x)$ in (1.35) will equal

$$G(t,x;T,y) \triangleq \mathrm{E}^{\mathrm{P}}\left(e^{-\int_t^T r(s,X(s))\,ds}\delta\left(X(T) - y\right)|X(t) = x\right).$$

The function G is known as a *state-price density* or as an *Arrow-Debreu security price* function. In particular, notice that for an arbitrary $g(x)$, we then have

$$\mathrm{E}^{\mathrm{P}}\left(e^{-\int_t^T r(s,X(s))\,ds}g\left(X(T)\right)|X(t) = x\right) = \int_{\mathbb{R}^p} G(t,x;T,y)\,g(y)\,dy. \tag{1.36}$$

Comparison with (1.32) shows that the state-price density is, essentially, equivalent to a Green's function with built-in discounting.

1.9 Black-Scholes and Extensions

In reviews of asset pricing theory, a discussion of the seminal *Black-Scholes-Merton* model (sometimes just known as the *Black-Scholes* model) of Black and Scholes [1973] and Merton [1973] is nearly mandatory. As the Black-Scholes-Merton (BSM) model constitutes a well-behaved setting in which to tie elements of previous sections together, our text is no exception. To provide a smoother transition to material that follows, we do, however, extend the usual analysis to include a simple case of stochastic interest rates.

1.9.1 Basics

In the basic BSM economy, two assets are traded: a money market account β and a stock S. In previous notations, $X(t) = (\beta(t), S(t))^\top$ and $p = 2$. The money market account value is 1 at time 0 and accrues risk-free interest at a continuously compounded, non-negative rate of r, initially assumed constant. The dynamics for β are thus given by an ordinary differential equation (ODE)

$$d\beta(t)/\beta(t) = r\,dt, \quad \beta(0) = 1,$$

implying that simply $\beta(t) = \beta(0)e^{rt}$.

The stock dynamics are assumed to satisfy GBMD under measure P:

$$dS(t)/S(t) = \mu\,dt + \sigma\,dW(t), \tag{1.37}$$

where W is a Brownian motion of dimension $d = 1$, and μ and σ are constants.

Taking first a probabilistic approach, we notice that β is positive and can be used as a numeraire. Let $S^\beta(t) = S(t)/\beta(t)$ be the stock price deflated by β. By Ito's lemma,

$$dS^\beta(t)/S^\beta(t) = (\mu - r)\,dt + \sigma\,dW(t).$$

Applying Girsanov's theorem (see Theorem 1.5.1) and Corollary 1.5.2, we see that if $\sigma \neq 0$, β will induce a unique equivalent martingale measure, with the measure shift characterized by the density process[12]

$$d\varsigma(t)/\varsigma(t) = -\theta\,dW(t), \quad \theta = \frac{\mu - r}{\sigma}.$$

Clearly, $\varsigma(t)$ defines an exponential martingale. The probability measure induced by the money market account β is called the *risk-neutral martingale measure* and is traditionally denoted Q. Under Q, $W^\beta(t) = W(t) + \theta t$ is a Brownian motion, and

$$dS^\beta(t)/S^\beta(t) = \sigma\,dW^\beta(t),$$
$$dS(t)/S(t) = r\,dt + \sigma\,dW^\beta(t), \tag{1.38}$$

or, from (1.21),

$$S(T) = S(t)e^{(r-\frac{1}{2}\sigma^2)(T-t)+\sigma\left(W^\beta(T)-W^\beta(t)\right)}, \quad t \in [0,T]. \tag{1.39}$$

We note that under Q, the drift μ of the stock process is replaced by the risk-free interest rate r. That is, under Q agents in the economy will

---
[12] The reader may recognize the market price of risk θ as the *Sharpe ratio* of the stock S, a measure of how well the risk of stock (represented by σ) is compensated by excess return (represented by $\mu - r$).

appear to be indifferent ("neutral") to the risk of the stock, content with an average growth rate of the stock equal to that of the money market account.

Before proceeding with the BSM analysis, we wish to emphasize that the drift restriction imposed on the stock in the risk-neutral measure Q is a general result. In a larger setting with a *p*-dimensional vector asset process X, if the Q-dynamics of the components of X are all of the form

$$dX_i(t) = rX_i\, dt + O(dW(t)), \quad i = 1, \ldots, p,$$

there is no arbitrage. This result holds unchanged if the interest rate is random (see Section 1.9.3).

Returning to the BSM setting, we note that the risk-neutral measure is unique, whereby the market is complete and all derivative securities on S (and β) are attainable. Let us consider a few such securities. First, we consider a security paying at time T $1 for certain. Such a security is a *discount bond* and we shall denote its time t price by $P(t, T)$, $t \in [0, T]$. If the interest rate is positive, we would expect $P(t, T) \leq 1$ as a reflection of the time value of money, with equality only holding for $t = T$. Application of the basic derivative pricing equation (1.15) immediately gives

$$P(t,T) = \beta(t) E_t^Q \left(\frac{1}{\beta(T)} \right) = E_t^Q \left(e^{-r(T-t)} \right) = e^{-r(T-t)}.$$

This result is trivial, as it is easily seen that the amount $e^{-r(T-t)}$ invested in the money market account at time t will grow to exactly $1 at time T.

Second, consider a derivative V paying $V(T) = S(T) - K$ at time T, with K being an arbitrary constant. Proceeding as above, at time $t \leq T$ the arbitrage-free price must be

$$V(t) = E_t^Q \left(e^{-r(T-t)} (S(T) - K) \right)$$
$$= e^{-r(T-t)} \left(E_t^Q (S(T)) - K \right) = S(t) - KP(t,T), \quad (1.40)$$

where the last equality follows from property (1.22) of GBMD. We notice that $V(t) = 0$ if $K = S(t)/P(t,T)$. This value of K is known as the time t *forward price of* $S(T)$[13].

Third, consider the derivative that was the main focus of the original BSM analysis, a *European call option* paying[14] $c(T) = (S(T) - K)^+$, with K being a positive *strike price*. Following (1.40), we can write

$$c(t) = P(t,T) E_t^Q \left((S(T) - K)^+ \right). \quad (1.41)$$

From the representation (1.39), basic probability theory allows us to write this expectation as

---

[13] We shall touch on the closely related concept of a *futures price* in Section 4.1.2.
[14] We use the notations $x^+ = \max(x, 0)$, $x^- = \min(x, 0)$ throughout this book.

$$c(t) = P(t,T) \int_{-\infty}^{\infty} \left(S(t) e^{(r-\frac{1}{2}\sigma^2)(T-t)+z\sigma\sqrt{T-t}} - K \right)^+ \phi(z)\, dz, \quad (1.42)$$

where $\phi(z) = (2\pi)^{-1/2} \exp(-z^2/2)$ is the standard Gaussian density. A straightforward evaluation of the integral leads to the famous *Black-Scholes-Merton call pricing formula*:

Theorem 1.9.1. *In the BSM economy, the arbitrage-free time t price of a K-strike call option maturing at time T is*

$$c(t) = S(t)\Phi(d_+) - KP(t,T)\Phi(d_-), \quad (1.43)$$

$$d_\pm \triangleq \frac{\ln(S(t)/K) + (r \pm \sigma^2/2)(T-t)}{\sigma\sqrt{T-t}}, \quad t < T,$$

where $\Phi(\cdot)$ is the Gaussian cumulative distribution function.

A formula for a *European put option* $p(t)$ paying $(K - S(T))^+$ can be obtained from (1.43) by *put-call parity*:

$$c(t) - p(t) = V(t),$$

where $V(t)$ is the forward contract defined above.

Remark 1.9.2. At time t, call and put options with strikes equal to $S(t)$ are said to be *at-the-money* (ATM). If $S(t) > K$, the call option is *in-the-money* (ITM) and the put option is *out-of-the-money* (OTM). If $S(t) < K$, the call is OTM and the put is ITM. The ATM, ITM, and OTM monikers are sometimes used to refer to the ordering of the *forward value* $\mathrm{E}_t(S(T)) = S(t)e^{r(T-t)}$ (for a T-maturity option) rather than the spot $S(t)$, relative to the strike K.

In deriving (1.43), the choice of β as numeraire was arbitrary. If we instead use S (which is also strictly positive) as numeraire, we can write

$$c(t) = S(t)\mathrm{E}_t^{\mathrm{Q}^S}\left(\frac{(S(T)-K)^+}{S(T)} \right) = S(t)\mathrm{E}_t^{\mathrm{Q}^S}\left((1 - K/S(T))^+ \right), \quad (1.44)$$

where Q^S is the martingale measure induced by S. To identify the measure shift involved in moving from P to Q^S, consider that $\beta^S(t) = \beta(t)/S(t)$ must be a martingale in Q^S. By Ito's lemma, in measure P we have

$$d\beta^S(t)/\beta^S(t) = (r - \mu + \sigma^2)\, dt - \sigma\, dW(t),$$

such that $dW^S(t) = ((r-\mu)/\sigma + \sigma)\, dt - dW(t)$ is a Brownian motion under Q^S. Application of Ito's lemma on $1/S(t)$ yields, after a few rearrangements,

$$dS(t)^{-1}/S(t)^{-1} = -r\, dt + \sigma\, dW^S(t),$$

which is a GBMD as before. Evaluation of the expectation (1.44) can be verified to recover the BSM formula (1.43).

Our derivation of the BSM formula was so far entirely probabilistic. Writing $c(t) = c(t, \beta, S)$, the arguments in Section 1.7 allow us to write c as a solution to the PDE (see (1.27))

$$\frac{\partial c}{\partial t} + \frac{1}{2}\sigma^2 S^2 \frac{\partial^2 c}{\partial S^2} = 0, \qquad (1.45)$$

subject to the boundary condition $c(T, \beta, S) = (S-K)^+$. From (1.28) we also have that the replication positions in β and S are $\frac{\partial c}{\partial \beta}$ and $\frac{\partial c}{\partial S}$, respectively. That is,

$$c(t, \beta, S) = \frac{\partial c}{\partial \beta}\beta + \frac{\partial c}{\partial S}S. \qquad (1.46)$$

As β is deterministic, we can actually eliminate c-dependence on this variable by a change of variables $\tilde{c}(t, S) = c(t, \beta, S)$. By the chain rule

$$\frac{\partial \tilde{c}}{\partial t} = \frac{\partial c}{\partial t} + \frac{\partial c}{\partial \beta}\frac{\partial \beta}{\partial t} = \frac{\partial c}{\partial t} + \frac{\partial c}{\partial \beta}r\beta = \frac{\partial c}{\partial t} + rc - \frac{\partial c}{\partial S}rS$$

where the last equation follows from (1.46). Inserting this into (1.45) yields the original *Black-Scholes PDE*

$$\frac{\partial \tilde{c}}{\partial t} + rS\frac{\partial \tilde{c}}{\partial S} + \frac{1}{2}\sigma^2 S^2 \frac{\partial^2 \tilde{c}}{\partial S^2} = r\tilde{c}, \qquad (1.47)$$

with $\tilde{c}(T, S) = (S - K)^+$. We can solve this equation by classical methods (see Lipton [2001] for several techniques), or we can use the Feynman-Kac result to write it as an expectation. We leave it as an exercise to the reader to verify that Feynman-Kac leads to the same expectation as derived earlier by probabilistic means (see (1.41)).

A final note: the derivation of the Black-Scholes PDE above was somewhat non-standard due to the initial assumption of option price being a function of the deterministic numeraire β. A more conventional (but entirely equivalent) argument sets up a portfolio of the call option and a position in the stock, and demonstrates that the stock position can be set such that the total portfolio growth is deterministic (risk-free) on $[t, t+dt]$. Equating the portfolio growth with the risk-free rate yields the Black-Scholes PDE (1.47). See Hull [2006] for details of this approach.

1.9.2 Alternative Derivation

We have already demonstrated several different ways of proving the BSM call pricing formula, but as shown in Andreasen et al. [1998] there are many more. One particularly enlightening proof is based on the concept of *local time* and shall briefly be discussed in this section. The proof, which borrows from the results in Carr and Jarrow [1990], will also allow us to demonstrate the Tanaka extension of Ito's lemma, mentioned earlier in Section 1.1.

As above, we assume that the stock price process is as in (1.38), and define the forward stock price $F(t) \triangleq S(t)/P(t,T)$. Clearly,

$$dF(t)/F(t) = \sigma \, dW^\beta(t), \quad t \leq T, \tag{1.48}$$

where W^β is a Brownian motion in the risk-neutral measure. Define the random variable $I(t) = (F(t) - K)^+$. The first derivative of I with respect to F is an indicator function $1_{\{F(t)>K\}}$ and the second derivative can be interpreted as the Dirac delta function, $\delta(F(t) - K)$. As I is clearly not twice differentiable, Ito's lemma formally does not apply, but the Tanaka extension nevertheless gives us permission to write

$$dI(t) = 1_{\{F(t)>K\}} \, dF(t) + \frac{1}{2}\sigma^2 F(t)^2 \delta\left(F(t) - K\right) dt$$

$$= 1_{\{F(t)>K\}} \, \sigma F(t) \, dW^\beta(t) + \frac{1}{2}\sigma^2 K^2 \delta\left(F(t) - K\right) dt.$$

In integrated form,

$$I(T) = I(t) + \int_t^T 1_{\{F(u)>K\}} \, \sigma F(u) \, dW^\beta(u) + \frac{1}{2}\sigma^2 K^2 \int_t^T \delta\left(F(u) - K\right) du.$$

The second integral in this expression is a random variable known as the *local time of F spent at the level K*, on the interval $[t, T]$. Taking expectations, it follows that

$$\mathrm{E}_t^Q\left(I(T)\right) = I(t) + \frac{1}{2}\sigma^2 K^2 \int_t^T \mathrm{E}_t^Q\left(\delta\left(F(u) - K\right)\right) du.$$

Here, if $p(t, y; u, x)$ is the density of $F(u)$ given $F(t) = y$, $u \geq t$, then obviously

$$\mathrm{E}_t^Q\left(\delta\left(F(u) - K\right)\right) = p(t, F(t); u, K).$$

By the definition of $F(T)$ we have $F(T) = S(T)$, such that $I(T) = (S(T) - K)^+$. From (1.41), we may therefore write the time t European call option price as

$$c(t) = P(t,T)\mathrm{E}_t^Q\left(I(T)\right)$$

$$= (S(t) - KP(t,T))^+ + \frac{P(t,T)}{2}\sigma^2 K^2 \int_t^T p(t, F(t); u, K) \, du. \tag{1.49}$$

The formula (1.49) decomposes the call option into a sum of two terms, the *intrinsic value* and the *time value*, respectively. The time value can be made more explicit by observing from the representation (1.39) that[15]

---

[15] This also follows directly from the fact that $F(u)$ is a log-normal random variable with moments given by (1.22) and (1.23).

$$p(t, F(t); u, K) = \frac{1}{K\sigma\sqrt{u-t}\sqrt{2\pi}} \exp\left(-\frac{1}{2}d(u)^2\right),$$

$$d(u) \triangleq \frac{\ln(F(t)/K) - \frac{1}{2}\sigma^2(u-t)}{\sigma\sqrt{u-t}}.$$

In other words, we have arrived at the following result.

Proposition 1.9.3. *The European call option price $c(t)$ on the process (1.38) can be written as*

$$c(t) = (S(t) - KP(t,T))^+ + \frac{P(t,T)\sigma K}{2}\int_t^T \frac{\phi(d(u))}{\sqrt{u-t}}\,du, \tag{1.50}$$

where $\phi(x)$ is the Gaussian density.

Explicit evaluation of the integral in (1.50) can be verified to produce the BSM formula in Theorem 1.9.1. We leave this as an exercise to the reader.

1.9.3 Extensions

1.9.3.1 Deterministic Parameters and Dividends

In our basic BSM setup, consider now first a simple extension to a deterministic interest rate $r(t)$ and a deterministic volatility $\sigma(t)$. Carrying out the analysis as before, we see that discount bond prices now become

$$P(t,T) = e^{-\int_t^T r(s)\,ds}. \tag{1.51}$$

The BSM call pricing formula (1.43) holds unchanged provided $P(t,T)$ is changed according to (1.51), and we redefine

$$d_\pm \triangleq \frac{\ln(S(t)/K) + \int_t^T (r(s) \pm \sigma(s)^2/2)\,ds}{\sqrt{\int_t^T \sigma(s)^2\,ds}}.$$

Let us further assume that the stock pays dividends at a deterministic rate of $q(t)$. Our framework so far, however, has assumed that assets pay no cash over $[0,T]$. To salvage the situation, consider a fictitious asset S^* obtained by reinvesting all dividends into the stock S itself. It is easily seen that

$$S^*(t) = S(t)e^{\int_0^t q(s)\,ds},$$

and clearly $S^*(t)$ satisfies the requirements of generating no cash flows on $[0,T]$. Stating the call option payout as

$$c(T) = (S(T) - K)^+ = \left(S^*(T)e^{-\int_0^T q(s)\,ds} - K\right)^+$$

and performing the pricing analysis of Section 1.9.1 on $S^*(t)$, rather than $S(t)$, results in a dividend-extended BSM call option formula:

$$c(t) = S(t)e^{-\int_t^T q(s)\,ds}\Phi(d_+) - KP(t,T)\Phi(d_-),$$

$$d_\pm \triangleq \frac{\ln(S(t)/K) + \int_t^T (r(s) - q(s) \pm \sigma(s)^2/2)\,ds}{\sqrt{\int_t^T \sigma(s)^2\,ds}}.$$

When the stock pays a dividend rate of $q(t)$, note that the risk-neutral process for $S(t)$ is

$$dS(t)/S(t) = (r(t) - q(t))\,dt + \sigma(t)\,dW^\beta(t),$$

which extends (1.38). Note that for the special case where $r(t) = q(t)$, $S(t)$ becomes a martingale and the call option price formula simplifies to

$$c(t) = P(t,T)\left(S(t)\Phi(d_+) - K\Phi(d_-)\right), \tag{1.52}$$

where now

$$d_\pm \triangleq \frac{\ln(S(t)/K) \pm \frac{1}{2}\int_t^T \sigma(s)^2\,ds}{\sqrt{\int_t^T \sigma(s)^2\,ds}}.$$

Remark 1.9.4. The martingale call formula (1.52) typically emerges when pricing options on futures and forward prices (see (1.48)) and is often called the *Black formula*, in honor of the work in Black [1976].

1.9.3.2 Stochastic Interest Rates

We now get even more ambitious and wish to consider call option pricing in the case where the interest rate r is stochastic. The money market account β becomes

$$\beta(t) = e^{\int_0^t r(s)\,ds},$$

and is now assumed an $\mathcal{F}_t$-measurable random variable. Proceeding as in Section 1.9.1, we find that under the risk-neutral measure $\mathbb{Q}$, the call option price expression is (assuming that the stock pays no dividends)

$$c(t) = \beta(t)\mathrm{E}_t^\mathbb{Q}\left(\frac{1}{\beta(T)}(S(T) - K)^+\right) = \mathrm{E}_t^\mathbb{Q}\left(e^{-\int_t^T r(s)\,ds}(S(T) - K)^+\right). \tag{1.53}$$

In (1.53), we emphasize that the numeraire no longer can be pulled out from the expectation. Still, to simplify call option computations, it would be convenient to somehow remove the term $\exp(-\int_t^T r(s)\,ds)$ from the expectation in (1.53). By substituting 1 for $(S(T) - K)^+$ in the expression above, we first notice that

1.9 Black-Scholes and Extensions

$$P(t,T) = \mathrm{E}_t^Q \left(e^{-\int_t^T r(s)\,ds} \right).$$

This inspires us to perform a new measure shift, where we use the discount bond $P(t,T)$, rather than $\beta(t)$, as our numeraire. Let the martingale measure induced by $P(t,T)$ be denoted Q^T, often termed the *T-forward measure*. By the standard result (1.15) we have

$$c(t) = P(t,T)\mathrm{E}_t^{Q^T}\left(P(T,T)^{-1}(S(T)-K)^+\right)$$
$$= P(t,T)\mathrm{E}_t^{Q^T}\left((S(T)-K)^+\right),$$

where we have used that $P(T,T) = 1$. From Theorem 1.4.2, Q^T and Q are related by the density

$$\varsigma(t) = \mathrm{E}_t^Q\left(\frac{dQ^T}{dQ}\right) = \frac{P(t,T)/P(0,T)}{\beta(t)}. \tag{1.54}$$

To proceed, we need to add more structure to the model by making assumptions about the stochastic process for $P(t,T)$. We shall spend considerable effort in subsequent chapters on this issue, but for this initial application we simply assume that $P(t,T)$ has Q dynamics

$$dP(t,T)/P(t,T) = r(t)\,dt - \sigma_P(t,T)\,dW_P(t), \tag{1.55}$$

where $\sigma_P(t,T)$ is deterministic and $W_P(t)$ is a Brownian motion correlated to the stock Brownian motion. Notice that the drift of $P(t,T)$ under Q is not freely specifiable and must be equal to the risk-free rate; see the discussion following (1.38). For clarity, let the stock Brownian motion be renamed $W_S(t)$, and assume that the correlation between $W_P(t)$ and $W_S(t)$ is a constant ρ. In the setting of vector-valued Brownian motion with independent components used in earlier sections, we can introduce correlation by writing $W(t) = (W_1(t), W_2(t))^\top$ and setting, say,

$$W_P(t) = W_1(t),$$
$$W_S(t) = \rho W_1(t) + \sqrt{1-\rho^2}\,W_2(t).$$

The filtration $\{\mathcal{F}_t\}$ of our extended BSM setting is the one generated by the 2-dimensional $W(t)$.

Under Q^T, the deflated process $S^P(t) = S(t)/P(t,T)$ is a martingale. An application of Ito's lemma combined with the Diffusion Invariance Principle shows that the Q^T process for $S^P(t)$ is

$$dS^P(t)/S^P(t) = \sigma_P(t,T)\,dW_1(t) + \sigma(t)\left(\rho\,dW_1(t) + \sqrt{1-\rho^2}\,dW_2(t)\right), \tag{1.56}$$

where $\sigma(t)$ as before is the deterministic volatility of the stock S. We recognize $S^P(t)$ as a drift-free geometric Brownian motion with instantaneous variance of

$$\sigma_P(t,T)^2 + \sigma(t)^2 + 2\rho\sigma(t)\sigma_P(t,T).$$

Exploiting the convenient fact that $S^P(T) = S(T)$ and $c(T) = (S^P(T)-K)^+$ (as $P(T,T) = 1$), we get

$$\begin{aligned} c(t) &= P(t,T)\mathrm{E}_t^{Q^T}\left((S^P(T) - K)^+\right) \\ &= P(t,T)\int_{-\infty}^{\infty}\left(S^P(t)e^{-\frac{1}{2}v(t,T)+z\sqrt{v(t,T)}} - K\right)^+ \phi(z)\,dz, \end{aligned} \quad (1.57)$$

where we have defined the "term", or total, variance

$$v(t,T) \triangleq \int_t^T \left(\sigma_P(s,T)^2 + \sigma(s)^2 + 2\rho\sigma(s)\sigma_P(s,T)\right)ds. \quad (1.58)$$

Completing the integration (compare with (1.42)) and using $S^P(t) = S(t)/P(t,T)$, we arrive at a modified BSM-type call option formula:

Proposition 1.9.5. *Consider a BSM economy with stochastic interest rates evolving according to (1.55). Define term variance $v(t,T)$ as in (1.58). Then, the T-maturity European call option price is*

$$c(t) = S(t)\Phi(d_+) - KP(t,T)\Phi(d_-),$$
$$d_\pm = \frac{\ln\left(S(t)/(KP(t,T))\right) \pm \frac{1}{2}v(t,T)}{\sqrt{v(t,T)}}.$$

Proposition 1.9.5 was originally derived in Merton [1973], using PDE methods. Extensions to dividend-paying stocks are straightforward and follow the arguments shown in Section 1.9.3.1.

1.10 Options with Early Exercise Rights

In our previous definition of a contingent claim, we assumed that the claim involved a single $\mathcal{F}_T$-measurable payout at time T. In reality, a number of derivative contracts may have intermediate cash payments from, say, scheduled coupons or through "rebates" for barrier-style options. Mostly, such complications are straightforwardly incorporated; see for instance Section 2.7.3. Of particular interest from a theoretical perspective are the claims that allow the holder to accelerate payments through *early exercise*. Derivative securities with early exercise are characterized by an adapted payout process $U(t)$, payable to the option holder at a stopping time (or *exercise policy*) $\tau \leq T$, chosen by the holder. If early exercise can take place at any time in some interval, we say that the derivative security is an *American option*; if exercise can only take place on a discrete set of dates, we say that it is a *Bermudan option*.

1.10 Options with Early Exercise Rights

Let the allowed (and deterministic) set of exercise dates larger than or equal to t be denoted $\mathcal{D}(t)$, and suppose that we are given at time 0 a particular exercise policy τ taking values in $\mathcal{D}(0)$, as well as a pricing numeraire N inducing a unique martingale measure Q^N. Let $V^\tau(0)$ be the time 0 value of a derivative security that pays $U(\tau)$. Under some technical conditions on $U(t)$, we can write for the value of the derivative security

$$V^\tau(0) = \mathrm{E}^{Q^N}\left(\frac{U(\tau)}{N(\tau)}\right), \tag{1.59}$$

where we have assumed, with no loss of generality, that $N(0) = 1$. Let $\mathcal{T}(t)$ be the time t set of (future) stopping times taking value in $\mathcal{D}(t)$. In the absence of arbitrage, the time 0 value of a security with early exercise into U must then be given by the *optimal stopping problem*

$$V(0) = \sup_{\tau \in \mathcal{T}(0)} V^\tau(0) = \sup_{\tau \in \mathcal{T}(0)} \mathrm{E}^{Q^N}\left(\frac{U(\tau)}{N(\tau)}\right), \tag{1.60}$$

reflecting the fact that a rational investor would choose an exercise policy to optimize the value of his claim.

We can extend (1.60) to future times t by

$$V(t) = N(t) \sup_{\tau \in \mathcal{T}(t)} \mathrm{E}_t^{Q^N}\left(\frac{U(\tau)}{N(\tau)}\right), \tag{1.61}$$

where $\sup_{\tau \in \mathcal{T}(t)} \mathrm{E}_t^{Q^N}(U(\tau)/N(\tau))$ is known as the *Snell envelope* of U/N under Q^N. The process $V(t)$ must here be interpreted as the value of the option with early exercise, *conditional* on exercise not having taken place before time t. To make this explicit, let $\tau^* \in \mathcal{T}(0)$ be the optimal exercise policy, as seen from time 0. We can then write, for $0 < t \leq T$,

$$V(0) = \mathrm{E}^{Q^N}\left(1_{\{\tau^* \geq t\}} V(t)/N(t)\right) + \mathrm{E}^{Q^N}\left(1_{\{\tau^* < t\}} U(\tau^*)/N(\tau^*)\right), \tag{1.62}$$

where we break the time 0 value into two components: one from the time t value of the option, should it not have been exercised before time t; and one from the right to exercise on $[0, t]$. As we can always elect — possibly suboptimally — to never exercise on $[0, t]$, from (1.62) we see that

$$V(0) \geq \mathrm{E}^{Q^N}\left(V(t)/N(t)\right),$$

which establishes that $V(t)/N(t)$ is a *supermartingale* under Q^N. This result also follows directly from known properties of the Snell envelope; see, e.g., Musiela and Rutkowski [1997].

For later use, focus now on the Bermudan case and assume that $\mathcal{D}(0) = \{T_1, T_2, \ldots, T_B\}$, where $T_1 > 0$ and $T_B = T$. For $t < T_{i+1}$, define $H_i(t)$ as the time t value of the Bermudan option when exercise is restricted to the dates $\mathcal{D}(T_{i+1}) = \{T_{i+1}, T_{i+2}, \ldots, T_B\}$. That is

$$H_i(t) = N(t)\mathrm{E}_t^{Q^N}\left(V(T_{i+1})/N(T_{i+1})\right), \quad i = 1, \ldots, B-1.$$

At time T_i, $H_i(T_i)$ can be interpreted as the *hold value* of the Bermudan option, that is, the value of the Bermudan option if not exercised at time T_i. If an optimal exercise policy is followed, clearly we must have at time T_i

$$V(T_i) = \max\left(U(T_i), H_i(T_i)\right), \quad i = 1, \ldots, B, \tag{1.63}$$

such that

$$H_i(t) = N(t)\mathrm{E}_t^{Q^N}\left(\max\left(U(T_{i+1}), H_{i+1}(T_{i+1})\right)/N(T_{i+1})\right), \quad i = 1, \ldots, B-1. \tag{1.64}$$

Starting with the terminal condition $H_B(T) = 0$, (1.64) defines a useful iteration backwards in time for the value $V(0) = H_0(0)$. We shall use this later for the purposes of designing valuation algorithms in Chapter 18, and for computing price sensitivities (deltas) in Chapter 24.

We note that the idea behind (1.63) is often known as *dynamic programming* or the *Bellman principle*. Loosely speaking, we here work "from the back" to price the Bermudan option. As we shall see later (in Chapter 2), this idea is particularly well-suited for numerical methods that proceed backwards in time, such as finite difference methods.

1.10.1 The Markovian Case

We now specialize to the Markovian case where $U(t) = g(t, x(t))$, where $g: [0, T] \times \mathbb{R}^n \to \mathbb{R}$ is continuous and

$$dx(t) = \mu\left(t, x(t)\right) dt + \sigma\left(t, x(t)\right) dW(t) \tag{1.65}$$

is an n-dimensional Markovian process, where μ and σ satisfy the regularity conditions of Theorem 1.6.1. The n-dimensional process[16] $x(t)$ here defines the state of the exercise value $U(t)$, so we say that $x(t)$ is a *state variable process*. For concreteness let our numeraire $N(t)$ be the money market account

$$N(t) = \beta(t) = e^{\int_0^t r(u, x(u))\, du},$$

where the short interest rate $r: [0, T] \times \mathbb{R}^n \to \mathbb{R}$ is here assumed a function of time and the state variable vector x. In (1.65), $W(t)$ is understood to be a d-dimensional Brownian motion in the risk-neutral measure Q.

Writing $V(t) = V(t, x(t))$, we have from (1.61)

$$V(t, x) = \sup_{\tau \in \mathcal{T}(t)} \mathrm{E}^Q\left(\left. e^{-\int_t^\tau r(u, x(u))\, du} g(\tau, x(\tau)) \right| x(t) = x\right). \tag{1.66}$$

---

[16] Note that $x(t)$ is an abstract construct, and does not necessarily coincide with any asset price process.

1.10 Options with Early Exercise Rights

For dates $t \in \mathcal{D}(0)$, clearly $V(t,x) \geq g(t,x)$, with equality holding only when time t exercise is optimal. This leads us to define the concept of an *exercise region* as

$$\mathcal{X} = \{(t,x) \in \mathcal{D}(0) \times \mathbb{R}^n : V(t,x) = g(t,x)\}.$$

Similarly, we define the complement of $\mathcal{X}$,

$$\mathcal{C} = \{(t,x) \in [0,T] \times \mathbb{R}^n : (t,x) \notin \mathcal{X}\},$$

to be the *continuation region*, i.e. the region where we wait (either because exercise is not optimal or because it is not allowed, $t \notin \mathcal{D}(0)$) rather than exercise the option.

For Markovian systems, rather than solving the optimization problem (1.66) directly, it is often particularly convenient to invoke the Bellman principle. Extending the ideas presented earlier, let us, somewhat loosely, state the Bellman principle as follows: for any $t \in \mathcal{D}(0)$,

$$V(t,x) = \lim_{\Delta \downarrow 0} \max\left(g(t,x), \mathrm{E}^Q_t\left(e^{-\int_t^{t+\Delta} r(u,x(u))\,du} V(t+\Delta, x(t+\Delta))\right)\right). \tag{1.67}$$

Again, this simply says that the option value at time t is the maximum of the exercise value and the hold value, that is, the present value of continuing to hold on to the option for a small period of time. As we have seen above, for a Bermudan option, (1.67) also holds for finite Δ (namely up to the next exercise date).

The Bellman principle provides us with a link between present (time t) and future (time $t + \Delta$) option values that we can often exploit in a numerical scheme. For this, however, we need further characterization of $V(t,x)$ in the continuation region. By earlier arguments, we realize that $V(t,x)/\beta(t)$ must be a Q-martingale on the continuation region. Assuming sufficient smoothness for an application of Ito's lemma, this leads to a PDE formulation, to hold for $(t,x) \in \mathcal{C}$,

$$\mathcal{J}V(t,x) = 0, \tag{1.68}$$

where

$$\mathcal{J} = \frac{\partial}{\partial t} + \mu(t,x)\frac{\partial}{\partial x} + \frac{1}{2}\sum_{i=1}^n \sum_{j=1}^n \left(\sigma(t,x)\sigma(t,x)^\top\right)_{i,j} \frac{\partial^2}{\partial x_i \partial x_j} - r(t,x).$$

Assume first that our option is of the Bermudan type, and let T_i and T_{i+1} be subsequent exercise dates in the exercise schedule. For any function f of time, define $f(t\pm)$ to be the limits $\lim_{\varepsilon \downarrow 0} f(t \pm \varepsilon)$, and assume that $V(T_{i+1}-, x)$ is known for all x. As all values of $t \in (T_i, T_{i+1})$ by definition must be in the continuation region, we can use (1.68) to solve for $V(T_i+, x)$. Applying the Bellman principle (1.67) at time T_i then leads to the condition

$$V(T_i-, x) = \max\left(g(T_i, x), V(T_i+, x)\right).$$

In PDE parlance, this is a so-called *jump condition* which is straightforward to incorporate into a numerical solution; see Section 2.7.4 for details.

For American-style options, (1.68) continues to apply on $\mathcal{C}$. The Bellman principle here leads to the characterization that

$$\mathcal{J}V(t,x) < 0,$$

for $(t,x) \in \mathcal{X}$, i.e. we exercise when the rate of return from holding the option strictly fails to match $r(t,x)$. The American option pricing problem is often conveniently summarized in a *variational inequality*, to hold on $\mathcal{X} \cup \mathcal{C}$,

$$V(t,x) \geq g(t,x), \quad \mathcal{J}V(t,x) \leq 0, \quad (V(t,x) - g(t,x))\,\mathcal{J}V(t,x) = 0, \tag{1.69}$$

and subject to the boundary condition $V(T,x) = g(T,x)$. The first of these three conditions expresses that the option is always worth at least its exercise value; the second expresses the supermartingale property of $V(t,x)$; and the third implies (after a little thought) that $\mathcal{J}V(t,x) = 0$ on $\mathcal{C}$ and $\mathcal{J}V(t,x) < 0$ on $\mathcal{X}$. The system (1.69) is discussed more carefully in Duffie [2001], where additional discussion of regularity issues may also be found.

1.10.2 Some General Bounds

In many cases of practical interest, solving PDEs and/or variational inequalities is not computationally feasible. In such situations, we may be interested in at least bounding the value of an option with early exercise rights. Providing a lower bound is straightforward: postulate an exercise policy τ and compute the price $V^\tau(0)$ by direct methods. From (1.60), clearly this provides a lower bound

$$V^\tau(0) \leq V(0). \tag{1.70}$$

The closer the postulated exercise policy τ is to the optimal exercise policy τ^*, the tighter this bound will be. We shall later study a number of numerical techniques to generate good exercise strategies for fixed income options with early exercise rights, see Chapter 18.

To produce an upper bound, we can rely on duality results established in Rogers [2001], Haugh and Kogan [2004] and Andersen and Broadie [2004]. Let $\mathcal{K}$ denote the space of adapted martingales M for which $\sup_{\tau \in [0,T]} E^{Q^N}|M(\tau)| < \infty$. For a martingale $M \in \mathcal{K}$, we then write

$$V(0) = \sup_{\tau \in \mathcal{T}(0)} E^{Q^N}\left(\frac{U(\tau)}{N(\tau)}\right)$$

$$= \sup_{\tau \in \mathcal{T}(0)} E^{Q^N}\left(\frac{U(\tau)}{N(\tau)} + M(\tau) - M(\tau)\right)$$

$$= M(0) + \sup_{\tau \in \mathcal{T}(0)} E^{Q^N}\left(\frac{U(\tau)}{N(\tau)} - M(\tau)\right).$$

In the second equality, we have relied on the *optional sampling theorem*, a result that states that the martingale property is satisfied up to a bounded random stopping time, i.e. that $\mathrm{E}^{Q^N}(M(\tau)) = M(0)$; see Karatzas and Shreve [1991] for details. We now turn the above result into an upper bound by forming a pathwise maximum at all possible future exercise dates $\mathcal{D}(0)$:

$$V(0) = M(0) + \sup_{\tau \in \mathcal{T}(0)} \mathrm{E}^{Q^N}\left(\frac{U(\tau)}{N(\tau)} - M(\tau)\right)$$

$$\leq M(0) + \mathrm{E}^{Q^N}\left(\max_{t \in \mathcal{D}(0)}\left(\frac{U(t)}{N(t)} - M(t)\right)\right). \quad (1.71)$$

With (1.70) and (1.71) we have, as desired, established upper and lower bounds for values of options with early exercise rights. Let us consider how to make these bounds tight. As mentioned earlier, to tighten the lower bound we need to pick exercise strategies close to the optimal one. Tightening the upper bound is a bit more involved and requires the following basic theorem, proven in Karatzas and Shreve [1991]:

Theorem 1.10.1 (Doob-Meyer Decomposition). *Let $\{Y(t),\ t \in [0,T]\}$ be a positive $\mathcal{F}_t$-adapted supermartingale process with right-continuous sample paths. Then we can write*

$$Y(t) = m(t) - A(t),$$

where $m(t)$ is a martingale process with $m(0) = Y(0)$ and $A(t)$ is an increasing predictable process with $A(0) = 0$.

Applying the Doob-Meyer decomposition on the supermartingale process $V(t)/N(t)$ under Q^N shows that

$$V(t)/N(t) = m(t) - A(t),$$

and $V(0) = m(0)$. Consider taking $M(t) = m(t)$ in equation (1.71), to get

$$V(0) \leq V(0) + \mathrm{E}^{Q^N}\left(\max_{t \in \mathcal{D}(0)}\left(\frac{U(t)}{N(t)} - m(t)\right)\right)$$

$$= V(0) + \mathrm{E}^{Q^N}\left(\max_{t \in \mathcal{D}(0)}\left(\frac{U(t)}{N(t)} - \frac{V(t)}{N(t)} - A(t)\right)\right)$$

$$\leq V(0).$$

The last inequality follows from the fact that $V(t) \geq U(t)$ and $A(t) \geq 0$. In conclusion, we have arrived at a *dual* formulation of the option price

$$V(0) = \inf_{M \in \mathcal{K}}\left\{M(0) + \mathrm{E}^{Q^N}\left(\max_{t \in \mathcal{D}(0)}\left(\frac{U(t)}{N(t)} - M(t)\right)\right)\right\}, \quad (1.72)$$

and have demonstrated that the infimum is attained when the martingale M is set equal to the martingale component of the deflated price process

$V(t)/N(t)$. In practice, we are obviously not privy to $V(t)/N(t)$ (which is a quantity that we are trying to estimate), but we are nevertheless provided with a strategy to make the upper bound (1.71) tight: use a martingale that is "close" to the martingale component of the true deflated option price process. In Chapter 18 we shall demonstrate how to make this strategy operational.

1.10.3 Early Exercise Premia

We finish our discussion of options with early exercise rights by listing some known results for puts and calls, including an interesting decomposition of American and Bermudan option prices into the sum of a European option price and an *early exercise premium*. For convenience, we work in a Markovian setting where the single state variable, denoted $S(t)$, follows one-dimensional GBMD. Specifically, we assume that

$$dS(t)/S(t) = (r - q)\,dt + \sigma\,dW^\beta(t), \qquad (1.73)$$

with $W^\beta(t)$ being a one-dimensional Brownian motion in the risk-neutral measure, i.e. the measure induced by the money market account $\beta(t) = e^{rt}$. For simplicity we assume that the interest rate r, the dividend yield q, and the volatility σ are all constants; the extension to time-dependent parameters is straightforward.

Let $c(t)$, $C_A(t)$, and $C_B(t)$ be the time t European, American, and Bermudan prices of the call option with terminal maturity T, conditional on no exercise prior to time t. While obviously $c(t) \leq C_B(t) \leq C_A(t)$, in some cases these inequalities are equalities, as the following straightforward lemma shows.

Lemma 1.10.2. *Suppose that $r \geq 0$ and $q \leq 0$ in (1.73). It is then never optimal to exercise a call option early, and*

$$c(t) = C_A(t) = C_B(t).$$

Proof. Notice that, by Jensen's inequality,

$$c(t) = e^{-r(T-t)} E_t^Q \left((S(T) - K)^+ \right)$$
$$\geq e^{-r(T-t)} \left(\left(E_t^Q (S(T)) - K \right)^+ \right) = \left(e^{-q(T-t)} S(t) - e^{-r(T-t)} K \right)^+.$$

It is therefore clear that if $r \geq 0$ and $q \leq 0$, then for any value of $T - t$,

$$c(t) \geq (S(t) - K)^+,$$

i.e. the European call option price dominates the exercise value. As the hold value of American and Bermudan options must be at least as large as the European option price, it follows that the option to exercise early is worthless. □

Remark 1.10.3. For the put option, early exercise is never optimal if $r \leq 0$ and $q \geq 0$. As this situation rarely happens in practice, American put options nearly always trade at a premium to their European counterparts.

Lemma 1.10.2 demonstrates the well-known fact that American or Bermudan call options on stocks that pay no dividends ($q = 0$) should never be exercised early. On the other hand, if the stock does pay dividends, for an American call option there will, at time t, be a critical value of the stock, $S_A(t)$, at which the value of the stream of dividends paid by the stock will compensate for the cost of accelerating the payment of the strike K. In other words, an American option should be exercised at time t, provided that $S(t) \geq S_A(t)$. The deterministic curve $S_A(t)$ is known as the *early exercise boundary* and marks the boundary between the exercise and continuation regions, $\mathcal{X}$ and $\mathcal{C}$. Writing $C_A(t) = C_A(t, S(t))$, we formally have

$$S_A(t) = \inf\left\{S : C_A(t, S) = (S - K)^+\right\}, \quad t \leq T.$$

For a Bermudan option, we may similarly define

$$S_B(t) = \inf\left\{S : C_B(t, S) = (S - K)^+\right\}, \quad t \in \mathcal{D}(0),$$

where we recall that $\mathcal{D}(0)$ is the (discrete) set of allowed exercise dates for the Bermudan option.

The following important result characterizes the exercise boundary of American call options.

Proposition 1.10.4. *For the American call option on a stock that follows (1.73), we have*

$$\frac{\partial S_A(t)}{\partial t} \leq 0, \quad t < T, \tag{1.74}$$

and

$$\left.\frac{\partial C_A(t, S)}{\partial S}\right|_{S=S_A(t)} = 1, \quad t < T. \tag{1.75}$$

Equation (1.74) states that the exercise boundary decreases as we approach maturity, a result that is easily understood. Statement (1.75) is more subtle, however, and amounts to a tangency condition that ensures that the American call option value transitions smoothly from hold value to exercise value across the early exercise boundary. As a consequence, (1.75) is often known as the *smooth pasting condition* or the *high contact condition*. A similar tangency condition does *not* hold for the Bermudan option value, which is not differentiable at the boundary but instead transitions into the exercise region at a "kink":

$$\lim_{\varepsilon \downarrow 0} \left.\frac{C_B(t, S) - C_B(t, S - \varepsilon)}{\varepsilon}\right|_{S=S_B(t)} < 1, \quad t \in \mathcal{D}(0).$$

38   1 Introduction to Arbitrage Pricing Theory

Fig. 1.1. Call Option Prices

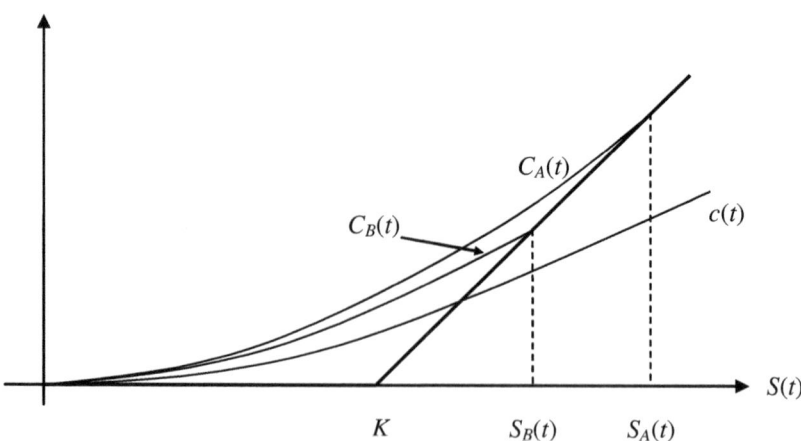

Notes: Time t prices of American, Bermudan, and European call options, as a function of the asset price. The Bermudan option is assumed to be exercisable at time t.

Figure 1.1 shows a typical value profile for a Bermudan call, along with the corresponding profiles for the European and American options.

Smooth pasting is essentially an optimality condition, which is how Proposition 1.10.4 is traditionally derived (see, e.g., Merton [1973] or the more recent Brekke and Øksendal [1991]). A more descriptive proof based on hedging arguments is given in Tavella and Randall [2000] and Wilmott et al. [1993]. Loosely speaking, the idea is here that a delta hedger should not be able to make riskless profits when the underlying asset crosses into the exercise region. This requires that the delta is continuous across the boundary, which is (1.75).

Remark 1.10.5. For the American put option, $\partial S_A(t)/\partial t \geq 0$ and the high contact condition states that the delta equals -1 at the exercise boundary.

Establishing the boundary $S_A(t)$ will virtually always require numerical methods, although asymptotic results are known for t close to T (see for instance Lipton [2001]). One simple result is listed below.

Lemma 1.10.6. *Assume that $r \geq 0$ and $q > 0$, such that the early exercise boundary exists for the American call option. The exercise boundary just prior to maturity is then*

$$\lim_{\varepsilon \downarrow 0} S_A(T - \varepsilon) = K \max\left(1, \frac{r}{q}\right).$$

1.10 Options with Early Exercise Rights

Proof. An informal proof of Lemma 1.10.6 proceeds as follows. At time $T - dt$, assume that $S(T - dt) > K$; otherwise it clearly makes no sense to exercise the option. If we exercise the option, we receive $S(T - dt) - K$ at time $T - dt$. On the other hand, if we postpone exercise, at time $T - dt$ our hold value is

$$e^{-r\,dt} \mathrm{E}^Q_{T-dt}\left(S(T) - K\right) = S(T-dt)e^{-q\,dt} - Ke^{-r\,dt}$$
$$= S(T-dt) - K - S(T-t)q\,dt + Kr\,dt.$$

Clearly, we should then only exercise if

$$S(T-dt) - K > S(T-dt) - K - S(T-dt)q\,dt + Kr\,dt$$

or if

$$S(T-dt)q > Kr.$$

□

Notice that since clearly $S_A(T) = K$, the call option exercise boundary will have a *discontinuity* at time T, if $q < r$.

One might guess that complete knowledge of the curve $S_A(t)$ should suffice to price the American option analytically. This intuition is confirmed by the following result due to Jamshidian [1992], Carr et al. [1992], Kim [1990], and Jacka [1991].

Proposition 1.10.7. *The American option price $C_A(t)$ satisfies*

$$C_A(t) = c(t) + E_A(t), \quad t \leq T, \tag{1.76}$$

where the (American) early exercise premium $E_A(t)$ is defined as

$$E_A(t) = \int_t^T e^{-r(u-t)} \mathrm{E}^Q_t \left(1_{\{S(u) \geq S_A(u)\}} (qS(u) - rK)\right) du \tag{1.77}$$

$$= \int_t^T \left(qS(t)e^{-q(u-t)}\Phi(d_+(u)) - rKe^{-r(u-t)}\Phi(d_-(u))\right) du, \tag{1.78}$$

where

$$d_\pm(u) = \frac{\ln(S(t)/S_A(u)) + \left(r - q \pm \tfrac{1}{2}\sigma^2\right)(u-t)}{\sigma\sqrt{u-t}}.$$

Proof. Due to the smooth pasting condition in Proposition 1.10.4, we are justified[17] in applying Ito's lemma. In informal notation,

$$dC_A(t) = 1_{\{S(t) \geq S_A(t)\}}\,dS(t)$$
$$+ 1_{\{S(t) < S_A(t)\}}\left\{\frac{\partial C_A(t)}{\partial t}\,dt + \frac{\partial C_A(t)}{\partial S}\,dS(t) + \frac{1}{2}\sigma^2 S^2 \frac{\partial^2 C_A(t)}{\partial S^2}\,dt\right\}, \tag{1.79}$$

---

[17] In particular, there is no local time contribution to $dC_A(t)$ at the boundary.

where we have used the fact that

$$1_{\{S(t)\geq S_A(t)\}}C_A(t) = 1_{\{S(t)\geq S_A(t)\}}(S(t) - K).$$

In the continuation region, $C_A(t, S)$ satisfies the PDE (1.47), i.e.

$$\frac{\partial C_A(t,S)}{\partial t} + (r-q)S\frac{\partial C_A(t,S)}{\partial S} + \frac{1}{2}\sigma^2 S^2 \frac{\partial^2 C_A(t,S)}{\partial S^2} = rC_A(t,S).$$

Inserting this into (1.79) we get, after a few rearrangements,

$$dC_A(t) = rC_A(t)\,dt + 1_{\{S(t)<S_A(t)\}}\sigma S(t)\frac{\partial C_A(t)}{\partial S}\,dW^\beta(t)$$
$$+ 1_{\{S(t)\geq S_A(t)\}}\left\{((r-q)S(t) - rC_A(t))\,dt + \sigma S(t)\,dW^\beta(t)\right\}$$
$$= rC_A(t)\,dt + 1_{\{S(t)<S_A(t)\}}\sigma S(t)\frac{\partial C_A(t)}{\partial S}\,dW^\beta(t)$$
$$+ 1_{\{S(t)\geq S_A(t)\}}\left\{(rK - qS(t))\,dt + \sigma S(t)\,dW^\beta(t)\right\}$$

Setting $y(t) = C_A(t)/\beta(t)$, it follows from Ito's lemma that

$$dy(t) = e^{-rt}1_{\{S(t)<S_A(t)\}}\sigma S(t)\frac{\partial C_A(t)}{\partial S}\,dW^\beta(t)$$
$$+ 1_{\{S(t)\geq S_A(t)\}}e^{-rt}\left\{(rK - qS(t))\,dt + \sigma S(t)\,dW^\beta(t)\right\}.$$

Integrating and taking expectations leads to

$$\mathrm{E}^Q_t(y(T)) = y(t) + \int_t^T e^{-ru}\mathrm{E}^Q_t\left(1_{\{S(u)\geq S_A(u)\}}(rK - qS(u))\right)\,du.$$

Applying the definition of $y(t)$ and the fact that $y(T) = e^{-rT}(S(T) - K)^+$ proves (1.76). The explicit form of the early exercise premium in (1.78) follows from the properties of GBMD. □

Remark 1.10.8. Combining results from Lemma 1.10.6 and Proposition 1.10.4, it follows that $E_A(t) \geq 0$, so $C_A(t) \geq c(t)$ as expected.

The integral representation of the American call option in Proposition 1.10.7 forms the basis for a number of proposed computational methods for American option pricing. Loosely speaking, these methods are based on the idea of iteratively estimating the exercise boundary $S_A(t)$, often working backwards from $t = T$, after which an application of Proposition 1.10.7 will yield the American option price. A representative example of these methods can be found in Ju [1998]. See Chiarella et al. [2004] for a survey of the literature, and Section 19.7.3 for an application in interest rate derivative pricing.

For a Bermudan option, an integral representation such as that in Proposition 1.10.7 is not possible. Nevertheless, it is still possible to break the

Bermudan call option into the sum of a European option and an early exercise premium. To show this, assume that the allowed exercise dates are $\mathcal{D}(0) = \{T_1, T_2, \ldots, T_B\}$, and let $S_B(T_i)$ be the exercise level above which the Bermudan option should be exercised at time T_i, $i = 1, \ldots, B$. Notice that if at time T_i we have $S(T_i) > S_B(T_i)$, then C_B will *jump down* in value when time progresses past time T_i, as a reflection of the missed exercise opportunity. Indeed, in the earlier notation of hold and exercise values, we have

$$C_B(T_i) = \max\left(U(T_i), H(T_i)\right),$$
$$C_B(T_i+) = H(T_i),$$

which makes the jump in value evident. Given the existence of these jumps, we may write

$$dC_B(t) = rC_B(t)\,dt + dM(t)$$
$$+ \sum_{i=1}^{B} 1_{\{S(T_i) > S_B(T_i)\}} \delta(T_i - t)\left(H(T_i) - U(T_i)\right)\,dt,$$

where $H(T_i) = C_B(T_i+)$ is the hold value at time T_i, $U(T_i) = S(T_i) - K$, and $M(t)$ is a martingale,

$$dM(t) = \frac{\partial C_B(t)}{\partial S}(r-q)S(t)\,dW^\beta(t).$$

Deflating C_B by the money market account and forming expectations, we get, since $c(t) = e^{-r(T-t)} E_t^Q(C_B(T))$,

$$C_B(t) = c(t) + \sum_{T_i \geq t} e^{-r(T_i - t)} E_t^Q \left(1_{\{S(T_i) > S_B(T_i)\}} \left(U(T_i) - H(T_i)\right)\right).$$

As $H(T_i)$ must be less than the exercise value $U(T_i)$ whenever $S(T_i) > S_B(T_i)$ we can simplify this expression to the following result that we, in Section 18.2.3, call the *marginal exercise value decomposition*.

Proposition 1.10.9. *The Bermudan option price $C_B(t)$ satisfies*

$$C_B(t) = c(t) + E_B(t), \quad t \leq T,$$

where the (Bermudan) early exercise premium $E_B(t)$ is defined as

$$E_B(t) = \sum_{T_i \geq t} e^{-r(T_i - t)} E_t^Q \left((U(T_i) - H(T_i))^+\right),$$

with $T_1 < T_2 < \ldots < T_B = T$ being the set of exercise dates.

As shown in Section 18.2.3, the result in Proposition 1.10.9 may be extended to more complicated processes and payouts than those considered here.

2
Finite Difference Methods

In Chapter 1 we described how the pricing of a derivative security typically requires either the solution of a parabolic partial differential equation (PDE) or the evaluation of an expectation of a random variable. In realistic applications, both of these price formulations often do not allow for closed-form solution, in which case we must resort to either analytical approximations or, more generally, numerical techniques. In the next two chapters we will describe a number of numerical algorithms useful in derivatives pricing. Analytical approximations will receive ample treatment later in this book, in the context of specific problems.

Our treatment of numerical methods is broken into two main subjects. In this chapter, we cover finite difference solutions of PDEs; and in Chapter 3 we turn to Monte Carlo evaluation of expectations. Many excellent specialist books exist on both topics, including Mitchell and Griffiths [1980], Tavella and Randall [2000], and Glasserman [2004]; our treatment only surveys the most important concepts, as required for our needs in this book. We do provide, however, a number of schemes rarely described in detail in the finance literature and also supplement our analysis with a number of "tricks of the trade", particularly in the application of finite difference grids.

The analysis of numerical PDE solutions in this chapter is arranged in two blocks. First, in Sections 2.1–2.8 we study the basic mechanics of the finite difference grid method for one-dimensional PDEs. Subsequently, Sections 2.9–2.12 then apply operator splitting techniques to extend the finite difference method to PDE of dimensions two and higher. The analysis culminates with a presentation of *ADI schemes* for multi-dimensional PDEs with mixed partial derivatives.

2.1 1-Dimensional PDEs: Problem Formulation

Initially, we will consider the numerical solution of the general one-dimensional terminal value PDE problem

$$\frac{\partial V}{\partial t} + \mathcal{L}V = 0, \tag{2.1}$$

where $\mathcal{L}$ is the operator

$$\mathcal{L} = \mu(t,x)\frac{\partial}{\partial x} + \frac{1}{2}\sigma(t,x)^2\frac{\partial^2}{\partial x^2} - r(t,x),$$

and where $V = V(t,x)$ satisfies a terminal condition $V(T,x) = g(x)$. We recognize the PDE as being an extension of the Black-Scholes PDE (1.47) to general time- and state-dependent drift (μ), volatility (σ), and interest rate (r). Underneath the PDE lies a physical model where a state variable process $x(\cdot)$ follows an SDE of the form

$$dx(t) = \mu(t,x(t))\,dt + \sigma(t,x(t))\,dW(t) \tag{2.2}$$

where $W(t)$ is a Brownian motion in the risk-neutral probability measure Q. Let the range of values attainable by $x(t)$ on $t \in [0,T]$ be denoted $\mathcal{B} \subseteq \mathbb{R}$, and assume that the functions $\mu, \sigma, r : [0,T) \times \mathcal{B} \to \mathbb{R}$ are sufficiently regular to make (2.1) and (2.2) meaningful (see Chapter 1).

The terminal value problem above is, as discussed earlier, a *Cauchy problem* to be solved for $V(t,x)$ on $(t,x) \in [0,T) \times \mathcal{B}$. In many cases of practical interest, further boundary conditions are applied in the spatial (x) domain. If such boundary conditions are expressed directly in terms of V (rather than its derivatives) we have a *Dirichlet boundary problem*. For instance, a so-called *up-and-out barrier option* will pay out $g(x(T))$ at time T if and only if $x(t)$ stays strictly below a contractually specified barrier level H at all times $t \leq T$. If, on the other hand, $x(t)$ touches H at any time during the life of the contract, it will expire worthless (or "knock out"). In this case, the PDE is only to be solved on $(t,x) \in [0,T) \times (\mathcal{B} \cap (-\infty, H))$ and is subject to the Dirichlet boundary condition

$$V(t,H) = 0, \quad t \in [0,T],$$

which expresses that the option has no value for $x \geq H$. We note that it is not uncommon to encounter options where the spatial domain boundaries are functions of time, a situation we shall deal with in Section 2.7.1. Also, as we shall see shortly, sometimes boundary conditions are conveniently expressed in terms of derivatives of V.

For numerical solution of the PDE (2.1), we often need to assume that the domain of the state variable x is finite, even in situations where (2.1) is supposed to hold for an infinite domain. Suitable truncation of the domain can often be done probabilistically, based on a confidence interval for $x(T)$. To illustrate the procedure, consider the Black-Scholes PDE (1.47) applied to a call option with strike K. A common first step is to use the transformation $x = \ln S$, such that the PDE has constant coefficients,

$$\frac{\partial V}{\partial t} + \left(r - \frac{1}{2}\sigma^2\right)\frac{\partial V}{\partial x} + \frac{1}{2}\sigma^2\frac{\partial^2 V}{\partial x^2} - rV = 0, \tag{2.3}$$

with terminal value (for a call option) $V(T,x) = (e^x - K)^+$. The domain of x is here the entire real line, $\mathcal{B} = \mathbb{R}$. We know (from (1.39)) that

$$x(T) = x(0) + \left(r - \frac{1}{2}\sigma^2\right)T + \sigma\left(W(T) - W(0)\right), \tag{2.4}$$

which is a Gaussian random variable with mean $\bar{x} = x(0) + (r - \frac{1}{2}\sigma^2)T$ and variance $\sigma^2 T$. Consider now replacing the domain $(-\infty, \infty)$ with the finite interval $[\bar{x} - \alpha\sigma\sqrt{T}, \bar{x} + \alpha\sigma\sqrt{T}]$ for some positive constant α. The likelihood of $x(T)$ falling outside of this interval is easily seen to be $2\Phi(-\alpha)$ (where, as always, $\Phi(z)$ is the standard Gaussian cumulative distribution function). If, say, we set α to 4, $2\Phi(-\alpha) = 6.3 \times 10^{-5}$, which is an insignificant probability for most applications. Larger (smaller) values of α will make the truncation error smaller (larger) and will ultimately require more (less) effort in a numerical scheme. We recommend values of α somewhere between 3 and 5 for most applications. For the Black-Scholes case, a rigorous estimate of the error imposed by domain truncation is given in Kangro and Nicolaides [2000].

In many cases of practical interest, it is not possible to write down an exact confidence interval for $x(T)$. In such cases, one instead may use an approximate confidence interval, found by, for instance, using "average" values for $\mu(t,x)$ and $\sigma(t,x)$. High precision in these estimates is typically not needed.

2.2 Finite Difference Discretization

In order to solve the PDE (2.1) numerically, we now wish to discretize it on the rectangular domain $(t,x) \in [0,T] \times [\underline{M}, \overline{M}]$, where $\overline{M}$ and $\underline{M}$ are finite constants, possibly found by a truncation procedure such as the one outlined above. We first introduce two equidistant[1] grids $\{t_i\}_{i=0}^{n}$ and $\{x_j\}_{j=0}^{m+1}$ where $t_i = iT/n \triangleq i\Delta_t$, $i = 0, 1, \ldots, n$, and $x_j = \underline{M} + j(\overline{M} - \underline{M})/(m+1) \triangleq \underline{M} + j\Delta_x$, $j = 0, 1, \ldots, m+1$. The terminal value $V(T,x) = g(x)$ is imposed at $t_n = T$, and spatial boundary conditions are imposed at x_0 and x_{m+1}.

2.2.1 Discretization in x-Direction. Dirichlet Boundary Conditions

We first focus on the spatial operator $\mathcal{L}$ and restrict x to take values in the interior of the spatial grid $x \in \{x_j\}_{j=1}^{m}$. Consider replacing the first- and second-order partial derivatives with first- and second-order difference operators:

---

[1] Non-equidistant grids are often required in practice and will be covered in Section 2.4.

$$\delta_x V(t, x_j) \triangleq \frac{V(t, x_{j+1}) - V(t, x_{j-1})}{2\Delta_x}, \tag{2.5}$$

$$\delta_{xx} V(t, x_j) \triangleq \frac{V(t, x_{j+1}) + V(t, x_{j-1}) - 2V(t, x_j)}{\Delta_x^2}. \tag{2.6}$$

These operators are accurate to second order. Formally[2],

Lemma 2.2.1.

$$\delta_x V(t, x_j) = \frac{\partial V(t, x_j)}{\partial x} + O\left(\Delta_x^2\right),$$

$$\delta_{xx} V(t, x_j) = \frac{\partial^2 V(t, x_j)}{\partial x^2} + O\left(\Delta_x^2\right).$$

Proof. A Taylor expansion of $V(t, x)$ around the point $x = x_j$ gives

$$V(t, x_{j+1}) = V(t, x_j) + \Delta_x \frac{\partial V(t, x_j)}{\partial x}$$
$$+ \frac{1}{2}\Delta_x^2 \frac{\partial^2 V(t, x_j)}{\partial x^2} + \frac{1}{6}\Delta_x^3 \frac{\partial^3 V(t, x_j)}{\partial x^3} + O\left(\Delta_x^4\right),$$

and

$$V(t, x_{j-1}) = V(t, x_j) - \Delta_x \frac{\partial V(t, x_j)}{\partial x}$$
$$+ \frac{1}{2}\Delta_x^2 \frac{\partial^2 V(t, x_j)}{\partial x^2} - \frac{1}{6}\Delta_x^3 \frac{\partial^3 V(t, x_j)}{\partial x^3} + O\left(\Delta_x^4\right).$$

Insertion of these expressions into (2.5) and (2.6) gives the desired result. □

In other words, if we introduce the discrete operator

$$\widehat{\mathcal{L}} = \mu(t, x)\delta_x + \frac{1}{2}\sigma(t, x)^2 \delta_{xx} - r(t, x),$$

we have, for $x \in \{x_j\}_{j=1}^m$,

$$\mathcal{L}V(t, x) = \widehat{\mathcal{L}}V(t, x) + O\left(\Delta_x^2\right).$$

With attention restricted to values on the grid $\{x_j\}_{j=1}^m$, we can view $\widehat{\mathcal{L}}$ as a matrix, once we specify the side boundary conditions at x_0 and x_{m+1}. For the Dirichlet case, assume for instance that

$$V(t, x_0) = \underline{f}(t, x_0), \quad V(t, x_{m+1}) = \overline{f}(t, x_{m+1}),$$

---

[2] Recall that a function $f(h)$ is of order $O(e(h))$ if $|f(h)|/|e(h)|$ is bounded from above by a positive constant in the limit $h \to 0$.

2.2 Finite Difference Discretization

for given functions $f, \overline{f} : [0,T] \times \mathbb{R} \to \mathbb{R}$. With[3] $\mathbf{V}(t) \triangleq (V(t,x_1), \ldots, V(t,x_m))^\top$ and, for $j = 1, \ldots, m$,

$$c_j(t) \triangleq -\sigma(t,x_j)^2 \Delta_x^{-2} - r(t,x_j), \tag{2.7}$$

$$u_j(t) \triangleq \frac{1}{2}\mu(t,x_j)\Delta_x^{-1} + \frac{1}{2}\sigma(t,x_j)^2 \Delta_x^{-2}, \tag{2.8}$$

$$l_j(t) \triangleq -\frac{1}{2}\mu(t,x_j)\Delta_x^{-1} + \frac{1}{2}\sigma(t,x_j)^2 \Delta_x^{-2}, \tag{2.9}$$

we can write

$$\widehat{\mathcal{L}}\mathbf{V}(t) = \mathbf{A}(t)\mathbf{V}(t) + \mathbf{\Omega}(t), \tag{2.10}$$

where $\mathbf{A}$ is a *tri-diagonal matrix*

$$\mathbf{A}(t) = \begin{pmatrix} c_1(t) & u_1(t) & 0 & 0 & 0 & \cdots & 0 \\ l_2(t) & c_2(t) & u_2(t) & 0 & 0 & \cdots & 0 \\ 0 & l_3(t) & c_3(t) & u_3(t) & 0 & \cdots & 0 \\ 0 & 0 & l_4(t) & c_4(t) & u_4(t) & \cdots & 0 \\ \vdots & \vdots & \vdots & \ddots & \ddots & \ddots & 0 \\ 0 & 0 & 0 & 0 & l_{m-1}(t) & c_{m-1}(t) & u_{m-1}(t) \\ 0 & 0 & 0 & 0 & 0 & l_m(t) & c_m(t) \end{pmatrix} \tag{2.11}$$

and $\mathbf{\Omega}(t)$ is a vector containing boundary values

$$\mathbf{\Omega}(t) = \begin{pmatrix} l_1(t)\underline{f}(t,x_0) \\ 0 \\ \vdots \\ 0 \\ u_m(t)\overline{f}(t,x_{m+1}) \end{pmatrix}.$$

As discussed earlier, sometimes one or both of the functions $\overline{f}$ and $\underline{f}$ are explicitly imposed as part of the option specification (as is the case for a knock-out options). In other cases, asymptotics may be necessary to establish these functions. For instance, for the case of a simple call option on a stock paying no dividends, we can set

$$\overline{f}(t,x) = e^x - Ke^{-r(T-t)},$$
$$\underline{f}(t,x) = 0,$$

where we, as before, have set $x = \ln S$ (S being the stock price) and assumed that the strike K is positive. The result for $\underline{f}$ is obvious; the result for $\overline{f}$ follows from the fact that a deep in-the-money call option will almost certainly pay at maturity the stock (the present value of which is just $S = e^x$) minus the strike (the present value of which is $Ke^{-r(T-t)}$).

---

[3] For clarity, this chapter uses boldface type for all vectors and matrices.

2.2.2 Other Boundary Conditions

Deriving asymptotic Dirichlet conditions can be quite involved for complicated option payouts and is often inconvenient in implementations. Rather than having to perform an asymptotic analysis for each and every type of option payout, it would be preferable to have a general-purpose mechanism for specifying the boundary condition. One common idea involves making assumptions on the form of the functional dependency between V and x at the grid boundaries, often from specification of relationships between spatial derivatives. For instance, if we impose the condition that the second derivative of V is zero at the upper boundary (x_{m+1}) — that is, V is a linear function of x — we can write (effectively using a downward discretization of the second derivative)

$$\frac{V(t, x_{m+1}) + V(t, x_{m-1}) - 2V(t, x_m)}{\Delta_x^2} = 0$$

$$\Rightarrow V(t, x_{m+1}) = 2V(t, x_m) - V(t, x_{m-1}).$$

A similar assumption at the lower spatial boundary yields

$$V(t, x_0) = 2V(t, x_1) - V(t, x_2).$$

For PDEs discretized in the logarithm of some asset, it may be more natural to assume that $V(t, x) \propto e^x$ at the boundaries; equivalently, we can assume that $\partial V/\partial x = \partial^2 V/\partial x^2$ at the boundary. When discretized in downward fashion at the upper boundary (x_{m+1}), this implies that

$$\frac{V(t, x_{m+1}) - V(t, x_m)}{\Delta_x} = \frac{V(t, x_{m+1}) + V(t, x_{m-1}) - 2V(t, x_m)}{\Delta_x^2}$$

or (assuming that $\Delta_x \neq 1$)

$$V(t, x_{m+1}) = V(t, x_{m-1}) \frac{1}{\Delta_x - 1} + V(t, x_m) \frac{\Delta_x - 2}{\Delta_x - 1}.$$

Similarly,

$$V(t, x_0) = V(t, x_1) \frac{2 + \Delta_x}{1 + \Delta_x} - V(t, x_2) \frac{1}{\Delta_x + 1}.$$

Common for both methods above — and for the Dirichlet specification discussed earlier — is that they give rise to boundary specifications through simple linear systems of the general form

$$V(t, x_{m+1}) = k_m(t) V(t, x_m) + k_{m-1}(t) V(t, x_{m-1}) + \overline{f}(t, x_{m+1}), \quad (2.12)$$
$$V(t, x_0) = k_1(t) V(t, x_1) + k_2(t) V(t, x_2) + \underline{f}(t, x_0). \quad (2.13)$$

This boundary specification can be captured in the matrix system (2.10) by simply rewriting a few components of $\mathbf{A}(t)$; specifically, we must set

$$c_m(t) = -\sigma(t, x_m)^2 \Delta_x^{-2} - r(t, x_m) + k_m(t) u_m(t),$$
$$l_m(t) = -\frac{1}{2}\mu(t, x_m)\Delta_x^{-1} + \frac{1}{2}\sigma(t, x_m)^2 \Delta_x^{-2} + k_{m-1}(t) u_m(t),$$
$$c_1(t) = -\sigma(t, x_1)^2 \Delta_x^{-2} - r(t, x_1) + k_1(t) l_1(t),$$
$$u_1(t) = \frac{1}{2}\mu(t, x_1)\Delta_x^{-1} + \frac{1}{2}\sigma(t, x_1)^2 \Delta_x^{-2} + k_2(t) l_1(t).$$

All other components of $\mathbf{A}$ remain as in (2.11); note that $\mathbf{A}$ remains tri-diagonal.

An alternative approach to specification of boundary conditions in the x-domain involves using the PDE itself to determine the boundary conditions, through replacement of all central difference operators with one-sided differences at the boundaries. Section 10.1.5.2 contains a detailed example of this idea; ultimately, this approach leads to boundary conditions that can also be written in the form (2.12)–(2.13).

2.2.3 Time-Discretization

To simplify notation, assume for now that $\Omega(t) = 0$ for all t, as will be the case if, say, we use the linear or linear-exponential boundary conditions outlined earlier. On the spatial grid, our original PDE can be written

$$\frac{\partial \mathbf{V}(t)}{\partial t} = -\mathbf{A}(t)\mathbf{V}(t) + O\left(\Delta_x^2\right)$$

which, ignoring the error term[4], defines a system of coupled ordinary differential equations (ODEs).

A number of methods are available for the numerical solution of coupled ODEs; see, e.g., Press et al. [1992]. We here only consider basic two-level time-stepping schemes, where grid computations at time t_i involve only PDE values at times t_i and t_{i+1}. Focusing the attention on a particular bucket $[t_i, t_{i+1}]$, the choice for the finite difference approximation of $\partial V/\partial t$ is obvious:

$$\frac{\partial \mathbf{V}}{\partial t} \approx \frac{\mathbf{V}(t_{i+1}) - \mathbf{V}(t_i)}{\Delta_t}.$$

Not so obvious, however, is to which time in the interval $[t_i, t_{i+1}]$ we should associate this derivative. To be general, consider picking a time $t_i^{i+1}(\theta) \in [t_i, t_{i+1}]$, given by

$$t_i^{i+1}(\theta) = (1-\theta)t_{i+1} + \theta t_i, \qquad (2.14)$$

where $\theta \in [0, 1]$ is a parameter. We then write

$$\frac{\partial \mathbf{V}\left(t_i^{i+1}(\theta)\right)}{\partial t} \approx \frac{\mathbf{V}(t_{i+1}) - \mathbf{V}(t_i)}{\Delta_t}.$$

---

[4]Note that the error term $O(\Delta_x^2)$ is here to be interpreted as an m-dimensional vector. We will use such short-hand notation throughout this chapter.

By a Taylor expansion, it is easy to see that this expression is first-order accurate in the time step when $\theta \neq \frac{1}{2}$, and second-order accurate when $\theta = \frac{1}{2}$. Written compactly,

$$\frac{\partial \mathbf{V}\left(t_i^{i+1}(\theta)\right)}{\partial t} = \frac{\mathbf{V}(t_{i+1}) - \mathbf{V}(t_i)}{\Delta_t} + 1_{\{\theta \neq \frac{1}{2}\}} O\left(\Delta_t\right) + O\left(\Delta_t^2\right). \quad (2.15)$$

This result on the convergence order is intuitive since only in the case $\theta = \frac{1}{2}$ is the difference coefficient precisely central; for all other cases, the difference coefficient is either predominantly backward in time or predominantly forward in time.

The time-discretization technique introduced above is known as a *theta scheme*. The special cases of $\theta = 1$, $\theta = 0$, and $\theta = \frac{1}{2}$ are known as the *fully implicit scheme*, the *fully explicit scheme*, and the *Crank-Nicolson scheme*, respectively. In light of the convergence result (2.15), one may wonder why anything other than the Crank-Nicolson scheme is ever used. The CN method is, indeed, often the method of choice, but there are situations where a straight application of the Crank-Nicolson scheme can lead to oscillations in the numerical solution or its spatial derivatives. Judicial application of the fully implicit method can often alleviate these problems, as we shall discuss later. The fully explicit method should never be used due to poor convergence and stability properties (see Section 2.3), but has nevertheless managed to survive in a surprisingly large number of finance texts and papers.

2.2.4 Finite Difference Scheme

We now proceed to combine the discretizations (2.10) and (2.15) into a complete finite difference scheme. First, we expand

$$\begin{aligned}\mathbf{A}\left(t_i^{i+1}(\theta)\right) \mathbf{V}\left(t_i^{i+1}(\theta)\right) &= \theta \mathbf{A}\left(t_i^{i+1}(\theta)\right) \mathbf{V}(t_i) \\ &\quad + (1-\theta) \mathbf{A}\left(t_i^{i+1}(\theta)\right) \mathbf{V}(t_{i+1}) + 1_{\{\theta \neq \frac{1}{2}\}} O(\Delta_t) + O\left(\Delta_t^2\right),\end{aligned}$$

such that our PDE can be represented as

$$\begin{aligned}\frac{\mathbf{V}(t_{i+1}) - \mathbf{V}(t_i)}{\Delta_t} &+ 1_{\{\theta \neq \frac{1}{2}\}} O(\Delta_t) + O\left(\Delta_t^2\right) \\ &= -\mathbf{A}\left(t_i^{i+1}(\theta)\right) \mathbf{V}\left(t_i^{i+1}(\theta)\right) + O\left(\Delta_x^2\right) \\ &= -\theta \mathbf{A}\left(t_i^{i+1}(\theta)\right) \mathbf{V}(t_i) - (1-\theta) \mathbf{A}\left(t_i^{i+1}(\theta)\right) \mathbf{V}(t_{i+1}) \\ &\quad + 1_{\{\theta \neq \frac{1}{2}\}} O(\Delta_t) + O\left(\Delta_t^2\right) + O\left(\Delta_x^2\right).\end{aligned}$$

Multiplying through with Δ_t gives rise to the complete finite difference representation of the PDE solution at times t_i and t_{i+1}:

Proposition 2.2.2. *On the grid $\{x_j\}_{j=1}^m$, the solution to (2.1) at times t_i and t_{i+1} is characterized by*

$$\left(\mathbf{I}-\theta\Delta_t\mathbf{A}\left(t_i^{i+1}(\theta)\right)\right)\mathbf{V}(t_i) = \left(\mathbf{I}+(1-\theta)\Delta_t\mathbf{A}\left(t_i^{i+1}(\theta)\right)\right)\mathbf{V}(t_{i+1}) + e_i^{i+1}, \quad (2.16)$$

where $\mathbf{I}$ is the $m \times m$ identity matrix, and e_i^{i+1} is an error term

$$e_i^{i+1} = \Delta_t O\left(\Delta_x^2\right) + 1_{\{\theta \neq \frac{1}{2}\}} O\left(\Delta_t^2\right) + O\left(\Delta_t^3\right). \quad (2.17)$$

Let $\widehat{V}(t_i, x_j)$ denote the approximation to the true solution $V(t_i, x_j)$ obtained by using (2.16) without the error term. Defining

$$\widehat{\mathbf{V}}(t) = \left(\widehat{V}(t, x_1), \ldots, \widehat{V}(t, x_m)\right)^\top,$$

we have

$$\left(\mathbf{I}-\theta\Delta_t\mathbf{A}\left(t_i^{i+1}(\theta)\right)\right)\widehat{\mathbf{V}}(t_i) = \left(\mathbf{I}+(1-\theta)\Delta_t\mathbf{A}\left(t_i^{i+1}(\theta)\right)\right)\widehat{\mathbf{V}}(t_{i+1}). \quad (2.18)$$

For a known value of $\widehat{\mathbf{V}}(t_{i+1})$, (2.18) defines a simple linear system of equations that can be solved for $\widehat{\mathbf{V}}(t_i)$ by standard methods. Simplifying matters is the fact that the matrix $(\mathbf{I}-\theta\Delta_t\mathbf{A}(t_i^{i+1}(\theta)))$ is tri-diagonal, allowing us to solve (2.18) in only $O(m)$ operations; see Press et al. [1992] for an algorithm[5].

Starting from the prescribed terminal condition $V(t_n, x_j) = g(x_j)$, $j = 1, \ldots, m$, we can now use (2.18) to iteratively step backward in time until we ultimately recover $\widehat{\mathbf{V}}(0)$. This procedure is known as *backward induction*.

Proposition 2.2.3. *The theta scheme (2.18) recovers $\widehat{\mathbf{V}}(0)$ in $O(mn)$ operations. If the scheme converges, the error on $\widehat{\mathbf{V}}(0)$ compared to the exact solution $\mathbf{V}(0)$ is of order*

$$O\left(\Delta_x^2\right) + 1_{\{\theta \neq \frac{1}{2}\}} O\left(\Delta_t\right) + O\left(\Delta_t^2\right).$$

Proof. The backward induction algorithm requires the solution of n tri-diagonal systems, one per time step, for a total computational cost of $O(mn)$. The local truncation error on $\widehat{\mathbf{V}}(t_i)$ is e_i^{i+1}, making the global truncation error after n time steps of order $n e_i^{i+1}$. Combining (2.17) with the fact that $n = T/\Delta_t = O(\Delta_t^{-1})$ gives the order result listed in the proposition. □

---

[5] The special case of an explicit scheme ($\theta = 0$) provides us with a direct expression for $V(t_i, x_j)$ in terms of $V(t_{i+1}, x_{j-1})$, $V(t_{i+1}, x_j)$, and $V(t_{i+1}, x_{j+1})$, a scheme that is easily visualized as a "trinomial tree". The intuitive nature of the explicit scheme coupled with the fact that no matrix equation must be solved may explain the popularity of this scheme in the finance literature, despite its poor numerical qualities (see Section 2.3). We stress that the workload of the explicit scheme is still $O(m)$ per time step, as is the case for all theta schemes.

It follows from Proposition 2.2.3 that the Crank-Nicolson scheme is second-order convergent in the time step, and all other theta schemes are first-order convergent in the time step. All theta-schemes are second-order convergent in the spatial step Δ_x.

In deriving (2.18), we assumed earlier that the boundary vector was zero, $\Omega(t) = 0$. Including a non-zero boundary vector into the scheme is, however, straightforward and results in a time-stepping scheme of the form

$$\left(\mathbf{I} - \theta \Delta_t \mathbf{A}\left(t_i^{i+1}(\theta)\right)\right) \widehat{\mathbf{V}}(t_i) = \left(\mathbf{I} + (1-\theta) \Delta_t \mathbf{A}\left(t_i^{i+1}(\theta)\right)\right) \widehat{\mathbf{V}}(t_{i+1})$$
$$+ (1-\theta)\Delta_t \Omega(t_{i+1}) + \theta \Delta_t \Omega(t_i). \quad (2.19)$$

As an alternative, we can use $\Delta_t \Omega(t_i^{i+1}(\theta))$ instead of $(1-\theta)\Delta_t \Omega(t_{i+1}) + \theta \Delta_t \Omega(t_i)$. Again, this system is easily solved for $\widehat{\mathbf{V}}(t_i)$ by a standard tri-diagonal equation solver.

As a final point, we stress that the finite difference scheme above ultimately yields a full vector of values $\widehat{\mathbf{V}}(0)$ at time 0, with one element per value of x_j, $j = 1, \ldots, m$. In general, we are mainly interested in $V(0, x(0))$, where $x(0)$ is the known value of x at time 0. There is no need to include $x(0)$ in the grid, as we can simply employ an interpolator (e.g., a cubic spline) on this vector $\widehat{\mathbf{V}}(0)$ to compute $V(0, x(0))$. Clearly, such an interpolator should be at least second-order accurate to avoid interfering with the overall $O(\Delta_x^2)$ convergence of the finite difference scheme. Assuming the interpolator is sufficiently smooth, we can also use it to compute various partial derivatives with respect to x that we may be interested in. Alternatively, these can be computed by the same type of finite difference coefficients discussed in Section 2.2.1. The derivative $\partial V(0, x(0))/\partial t$ — the *time decay* — can be picked up from the grid in the same fashion.

Remark 2.2.4. The scheme (2.18) may, without affecting convergence order, be replaced with

$$\left(\mathbf{I} - \theta \Delta_t \mathbf{A}(t_i)\right) \widehat{\mathbf{V}}(t_i) = \left(\mathbf{I} + (1-\theta)\Delta_t \mathbf{A}(t_{i+1})\right) \widehat{\mathbf{V}}(t_{i+1}).$$

2.3 Stability

2.3.1 Matrix Methods

Ignoring the contributions from boundary conditions, the finite difference scheme developed in the previous section can be rewritten

$$\widehat{\mathbf{V}}(t_i) = \mathbf{B}_i^{i+1} \widehat{\mathbf{V}}(t_{i+1}), \quad (2.20)$$

where

$$\mathbf{B}_i^{i+1} \triangleq \left(\mathbf{I} - \theta \Delta_t \mathbf{A}\left(t_i^{i+1}(\theta)\right)\right)^{-1} \left(\mathbf{I} + (1-\theta)\Delta_t \mathbf{A}\left(t_i^{i+1}(\theta)\right)\right).$$

2.3 Stability

That is, for any $0 \leq k < n$,

$$\widehat{\mathbf{V}}(t_k) = \mathbf{B}_k^n \widehat{\mathbf{V}}(t_n), \ \mathbf{B}_k^n \triangleq \mathbf{B}_k^{k+1} \mathbf{B}_{k+1}^{k+2} \cdots \mathbf{B}_{n-1}^n.$$

We say that the scheme is *stable* if $|\widehat{\mathbf{V}}(t_k)|$ is bounded for all $0 \leq k < n$. Assuming $|\widehat{\mathbf{V}}(T)| < \infty$, a necessary and sufficient condition for stability is that there exists a constant K such that for all $0 \leq k < n$

$$|\mathbf{B}_k^n| \leq K, \tag{2.21}$$

where $|\cdot|$ is any matrix norm, e.g. the spectral norm or the infinity norm[6]. See Mitchell and Griffiths [1980] for further details.

2.3.2 Von Neumann Analysis

For simple problems with time- and space-independent coefficients, it may be possible to establish the spectral norm of $\mathbf{B}_k^n$ by direct methods (see e.g. Mitchell and Griffiths [1980], Kraaijevanger et al. [1987], Lenferink and Spijker [1991], Spijker and Straetemans [1997]), but generally the stability criterion (2.21) is difficult to evaluate. While certain somewhat simpler matrix-based methods exist to establish necessary conditions for stability (again, see Mitchell and Griffiths [1980]), we shall here only consider a "local" method, known as the *von Neumann method*. In principle, the von Neumann method only holds for finite difference schemes where the underlying PDE has constant coefficients, but there is much numerical evidence to support wider application[7]. The von Neumann method does not directly consider the effect of boundary conditions on stability, but (for constant coefficient problems) provides a necessary condition for stability irrespective of the type of boundary condition.

The basis for the von Neumann analysis is the observation that a real function sampled on a finite number of points is uniquely defined by a complex Fourier series. For our PDE solution sampled on the spatial grid, the precise result is

$$V(t_k, x_j) = \sum_l H_l(t_k) e^{-i\omega_l j \Delta_x},$$

where $H_l(t_k)$ and ω_l are the amplification factor (discrete Fourier transform) and wave number for the l-th mode, respectively. Notice that i here denotes

---

[6]The spectral norm of a matrix $\mathbf{C}$ is defined as the largest absolute eigenvalue of $(\mathbf{C}^\top \mathbf{C})^{1/2}$. The infinity norm is defined as $\max_i \sum_j |C_{i,j}|$.

[7]In the application to PDEs with non-constant coefficients, it may help to think of the von Neumann analysis as being applied to the PDE locally with "frozen" coefficients, followed by an examination of the worst case among all frozen coefficients.

the imaginary unit, $i^2 = -1$, with k (momentarily) having taken the role of the time index in the finite difference grid. For the constant coefficient case, a key fact for our PDE problem is that

$$H_l(t_k) = H_l(t_{k+1}) \xi_l^{-1},$$

where ξ_l is a mode-specific *amplification factor* independent of time. To determine how a solution is propagated back through the finite difference grid, it thus suffices to consider a test function of the form

$$v(t_k, x_j) = \xi(\omega)^{n-k} e^{i\omega j \Delta_x}. \tag{2.22}$$

According to the Von Neumann criterion, stability of (2.20) requires that the modulus of the amplification factor $\xi(\omega)$ is less or equal to one, independent of the wave number:

$$\forall \omega : |\xi(\omega)| \leq 1. \tag{2.23}$$

This criterion is natural and merely expresses that all eigenmodes should be dampened, and not exponentially amplified, by the finite difference scheme.

Turning to our system (2.20), assume for simplicity that $r(t,x) = 0$. A positive interest rate (we will nearly always have $r(t,x) > 0$) introduces some extra dampening through discounting effects and will, if anything, lead to better stability properties than the case of zero interest rates. Writing $v(t_k, x_j) = v_{k,j}$, $\sigma(t_k^{k+1}(\theta), x_j) = \sigma_{k,j}$, and $\mu(t_k^{k+1}(\theta), x_j) = \mu_{k,j}$, the von Neumann analysis gives the following result:

Proposition 2.3.1. *Define $\alpha = \Delta_t/(\Delta_x)^2$. For (2.20) with $r(t,x) = 0$, the von Neumann stability criterion is*

$$1 \geq \theta \geq \frac{1}{2} - \frac{1}{\alpha} \left(\frac{\sigma_{k,j}^2}{\sigma_{k,j}^4 + \mu_{k,j}^2 \Delta_x^2 + \left|\mu_{k,j}^2 \Delta_x^2 - \sigma_{k,j}^4\right|} \right), \tag{2.24}$$

to hold for all $k = 0, 1, \ldots, n-1$, $j = 1, 2, \ldots, m$.

Proof. Define $\varsigma_{k,j}^\pm = \sigma_{k,j}^2 \pm \Delta_x \mu_{k,j}$. A local application of (2.20) gives

$$v_{k,j-1}\left(-\frac{\alpha\theta}{2}\varsigma_{k,j}^-\right) + v_{k,j}(1 + \alpha\theta\sigma_{k,j}^2) + v_{k,j+1}\left(-\frac{\alpha\theta}{2}\varsigma_{k,j}^+\right) =$$

$$v_{k+1,j-1}\left(\frac{\alpha(1-\theta)}{2}\varsigma_{k,j}^-\right) + v_{k+1,j}(1 - \alpha(1-\theta)\sigma_{k,j}^2) + v_{k+1,j+1}\left(\frac{\alpha(1-\theta)}{2}\varsigma_{k,j}^+\right)$$

with α defined above. Inserting (2.22) and rearranging (using Euler's formulas for sin and cos) yields

$$\xi(\omega) = \frac{1 - (1-\theta)\alpha\sigma_{k,j}^2(1 - \cos\omega\Delta_x) + i(1-\theta)\alpha\Delta_x \mu_{k,j} \sin\omega\Delta_x}{1 + \theta\alpha\sigma_{k,j}^2(1 - \cos\omega\Delta_x) - i\theta\alpha\Delta_x \mu_{k,j} \sin\omega\Delta_x}.$$

Note that ξ is a function of k and j, due to the non-constant PDE parameters. As discussed earlier (see also Mitchell and Griffiths [1980]), we expect the system to be stable if the criterion (2.23) holds for all k and j in the grid. Computing the modulus of ξ and requiring that it does not exceed one leads, after straightforward manipulations, to the stability criterion

$$\forall \omega : 2\alpha\sigma_{k,j}^2 + (2\theta - 1)\alpha^2 \left[\sigma_{k,j}^4 + \mu_{k,j}^2 \Delta_x^2 + \cos\omega\Delta_x \left(\mu_{k,j}^2 \Delta_x^2 - \sigma_{k,j}^4\right)\right] \geq 0.$$

As $\cos\omega\Delta_x \in [-1, 1]$, this expression can be simplified to (2.24). □

From (2.24) we can immediately conclude that the finite difference scheme is always stable if $\frac{1}{2} \leq \theta \leq 1$, irrespective of the magnitudes of Δ_x and Δ_t. For $\frac{1}{2} \leq \theta \leq 1$, we therefore say that the theta scheme is *absolutely stable*, or simply *A-stable*. Both the fully implicit ($\theta = 1$) and the Crank-Nicolson ($\theta = \frac{1}{2}$) finite difference schemes are thus *A*-stable. For the explicit scheme ($\theta = 0$), however, stability is *conditional*, requiring

$$\frac{2}{\alpha}\sigma_{k,j}^2 \geq \sigma_{k,j}^4 + \mu_{k,j}^2 \Delta_x^2 + \left|\mu_{k,j}^2 \Delta_x^2 - \sigma_{k,j}^4\right|.$$

For small drifts, this expression amounts to the restriction $\sigma_{k,j}^2 \leq \Delta_x^2/\Delta_t$ which can be quite onerous, often requiring the (laborious) use of thousands of time steps in the finite difference grid. We shall not consider fully explicit methods any further in this book.

Returning to the case $\frac{1}{2} \leq \theta \leq 1$, let us introduce a stronger definition of stability. A time-stepping method is said to be *strongly A-stable* if the modulus of the amplification factor ξ is strictly below 1 for any value of the time step, including the limit[8] $\Delta_t \to \infty$. From (2.24), we see that if $\Delta_t \to \infty$ (which implies $\alpha \to \infty$), then the modulus of the amplification factor could reach 1 in the special case of $\theta = 1/2$. In other words, the Crank-Nicolson scheme is *not* strongly *A*-stable. For large time steps, harmonics in the Crank-Nicolson finite difference solution will effectively not be dampened from one time step to the next, opening up the possibility that unwanted high-frequency oscillations can creep into the numerical solution. In practice, this is primarily a problem if high-frequency eigenmodes have high amplification factors, as can happen if there is an outright discontinuity in the terminal value function g. The problem is especially noticeable if the discontinuity in the value function is "close" in both time and space to $t = 0$ and $x = x(0)$ (as would be the case for a short-dated option with a discontinuity close to the starting value of x). Oscillations can be prevented by setting the time step smaller than twice the maximum stable explicit time step (see Tavella and Randall [2000]), but this can often be computationally expensive. We shall deal with other methods to suppress oscillations in Section 2.5.

We conclude this section by noting a deep connection between the stability of a finite difference scheme and its convergence to the true solution

---

[8] If further $|\xi|$ approaches zero for $\Delta_t \to 0$, the scheme is said to be *L-stable*.

of the PDE as $\Delta_t \to 0$ and $\Delta_x \to 0$. First, we define a finite difference scheme to be *consistent* if local (Taylor) truncation errors approach zero for $\Delta_t \to 0$ and $\Delta_x \to 0$. All the schemes we have encountered so far are consistent. Further, define a finite difference scheme to be *convergent* if the difference between the numerical solution and the exact PDE solution at a fixed point in the domain converges to zero uniformly as $\Delta_t \to 0$ and $\Delta_x \to 0$ (not necessarily independently of each other). We then have

Theorem 2.3.2 (Lax Equivalence Theorem). *For a well-posed[9] linear terminal value PDE, a consistent 2-level finite difference scheme is convergent if and only if it is stable.*

A more precise statement of the above result, as well as a proof, can be found in Mitchell and Griffiths [1980].

2.4 Non-Equidistant Discretization

In practice, we often wish to align the finite difference grid to particular dates (e.g., those on which a coupon or a dividend is paid) and particular values of x (e.g., those on which strikes and barriers are positioned). Also, for numerical reasons we may want to make certain important parts of the finite difference grid more densely spaced to concentrate computational effort on domains of particular importance to the solution of the PDE. To do so, we will now relax our earlier assumption of equidistant discretization in time and space. Doing so for the time domain is actually trivial and merely requires us to replace Δ_t in (2.18) with $\Delta_{t,i} \triangleq t_{i+1} - t_i$, where the spacing of the time grid $\{t_i\}_{i=0}^n$ is now no longer constant. The backward induction algorithm can proceed as before. We note that the ability to freely select the time grid will allow us to line up perfectly with dates that carry high significance for the product in question (e.g. dates on which cash flows take place, see Section 2.7.3) or to, say, use coarser time steps for the part of the finite difference grid that is far in the future. For an adaptive algorithm to automatically select the time-step, see d'Halluin et al. [2001].

For the spatial step, we have a number of options to induce non-equidistant spacing. One method involves a non-linear change of variables $y = h(x)$ in the PDE, followed by a regular equidistant discretization in the new variable y. This maps into a non-equidistant discretization in x which, provided that $h(\cdot)$ is chosen carefully, will have the desired geometry. Discussion of this method along with guidelines for choosing $h(\cdot)$ can be found in Chapter 5 of Tavella and Randall [2000]. We will here pursue a more direct alternative, where we simply introduce an irregular grid $\{x_j\}_{j=0}^{m+1}$

---

[9]Well-posed means that the PDE we are solving has a unique solution that depends continuously on the problem data (PDE coefficients, domain, boundary conditions, etc.)

and redefine the finite difference operators (2.5)–(2.6) to achieve maximum precision. For this, define

$$\Delta_{x,j}^+ \triangleq x_{j+1} - x_j, \quad \Delta_{x,j}^- \triangleq x_j - x_{j-1},$$

and set

$$\delta_x^+ V(t, x_j) = \frac{V(t, x_{j+1}) - V(t, x_j)}{\Delta_{x,j}^+}, \quad \delta_x^- V(t, x_j) = \frac{V(t, x_j) - V(t, x_{j-1})}{\Delta_{x,j}^-}.$$

By a Taylor expansion, we get

$$\delta_x^+ V(t, x_j) = \frac{\partial V(t, x_j)}{\partial x} + \frac{1}{2} \frac{\partial^2 V(t, x_j)}{\partial x^2} \Delta_{x,j}^+$$
$$+ \frac{1}{6} \frac{\partial^3 V(t, x_j)}{\partial x^3} \left(\Delta_{x,j}^+\right)^2 + O\left(\left(\Delta_{x,j}^+\right)^3\right), \quad (2.25)$$

$$\delta_x^- V(t, x_j) = \frac{\partial V(t, x_j)}{\partial x} - \frac{1}{2} \frac{\partial^2 V(t, x_j)}{\partial x^2} \Delta_{x,j}^-$$
$$+ \frac{1}{6} \frac{\partial^3 V(t, x_j)}{\partial x^3} \left(\Delta_{x,j}^-\right)^2 + O\left(\left(\Delta_{x,j}^-\right)^3\right). \quad (2.26)$$

Maximum accuracy on the first-order derivative approximation is achieved by selecting a weighted combination of (2.25)–(2.26) such that the terms of order $O(\Delta_{x,j}^+)$ and $O(\Delta_{x,j}^-)$ cancel. That is, we set

$$\delta_x V(t, x_j) = \frac{\Delta_{x,j}^-}{\Delta_{x,j}^+ + \Delta_{x,j}^-} \cdot \delta_x^+ V(t, x_j) + \frac{\Delta_{x,j}^+}{\Delta_{x,j}^+ + \Delta_{x,j}^-} \cdot \delta_x^- V(t, x_j) \quad (2.27)$$
$$= \frac{\partial V(t, x_j)}{\partial x} + O\left(\frac{\left(\Delta_{x,j}^+\right)^2 \Delta_{x,j}^- + \left(\Delta_{x,j}^-\right)^2 \Delta_{x,j}^+}{\Delta_{x,j}^+ + \Delta_{x,j}^-}\right)$$

which is second-order accurate, in the sense that reducing both $\Delta_{x,j}^+$ and $\Delta_{x,j}^-$ by a factor of k will reduce the error by a factor of k^2. To estimate the derivative $\partial^2 V(t, x_j)/\partial x^2$ we set

$$\delta_{xx} V(t, x_j) = \frac{\delta_x^+ V(t, x_j) - \delta_x^- V(t, x_j)}{\frac{1}{2}\left(\Delta_{x,j}^+ + \Delta_{x,j}^-\right)} \quad (2.28)$$
$$= \frac{\partial^2 V(t, x_j)}{\partial x^2} + O\left(\frac{\left(\Delta_{x,j}^+\right)^2 - \left(\Delta_{x,j}^-\right)^2}{\Delta_{x,j}^+ + \Delta_{x,j}^-} + \frac{\left(\Delta_{x,j}^+\right)^3 + \left(\Delta_{x,j}^-\right)^3}{\Delta_{x,j}^+ + \Delta_{x,j}^-}\right)$$

which is only first-order accurate, unless $\Delta_{x,j}^+ = \Delta_{x,j}^-$. Despite this, the global discretization error will typically remain second-order in the spatial step, even for a non-equidistant grid. A proof of this perhaps somewhat surprising result can be found in the monograph Axelsson and Barker [1991] on finite element methods.

Development of a theta scheme around the definitions (2.27) and (2.28) proceeds in the same way as in Section 2.2. The resulting time-stepping scheme is identical to (2.18), after a modification of the matrix $\mathbf{A}$. Specifically, we must simply redefine the c-, u-, and l-arrays in (2.7)–(2.9) as follows:

$$c_j(t) \triangleq \frac{\Delta^+_{x,j} - \Delta^-_{x,j}}{\Delta^+_{x,j}\Delta^-_{x,j}}\mu(t,x_j) - \frac{1}{\Delta^-_{x,j}\Delta^+_{x,j}}\sigma(t,x_j)^2 - r(t,x_j), \tag{2.29}$$

$$u_j(t) \triangleq \frac{\Delta^-_{x,j}}{\left(\Delta^+_{x,j}+\Delta^-_{x,j}\right)\Delta^+_{x,j}}\mu(t,x_j) + \frac{1}{\left(\Delta^+_{x,j}+\Delta^-_{x,j}\right)\Delta^+_{x,j}}\sigma(t,x_j)^2, \tag{2.30}$$

$$l_j(t) \triangleq -\frac{\Delta^+_{x,j}}{\left(\Delta^+_{x,j}+\Delta^-_{x,j}\right)\Delta^-_{x,j}}\mu(t,x_j) + \frac{1}{\left(\Delta^+_{x,j}+\Delta^-_{x,j}\right)\Delta^-_{x,j}}\sigma(t,x_j)^2. \tag{2.31}$$

For an example where having a non-equidistant grid is essential to the numerical performance of the scheme, see Section 9.4.3.

2.5 Smoothing and Continuity Correction

2.5.1 Crank-Nicolson Oscillation Remedies

As discussed earlier, for discontinuous terminal conditions, the Crank-Nicolson scheme may exhibit localized oscillations if the time step is too coarse relative to the spatial step. Depending on the timing and spatial position of the discontinuities, these spurious oscillations may negatively affect the computed option value or, more likely, its first ("delta") or second ("gamma") x-derivatives. Further, in the presence of discontinuous terminal conditions, the expected $O(\Delta_t^2)$ convergence order of the Crank-Nicolson scheme may not be realized. While $O(\Delta_t^2)$ convergence is possible without spurious oscillations in some multi-level time-stepping schemes, there is evidence that these schemes are less robust than the Crank-Nicolson scheme for many financially relevant problems, see, e.g., Windcliff et al. [2001]. Fortunately, it is relatively easy to remedy the problems in the Crank-Nicolson scheme. Specifically, a theoretical result by Rannacher [1984] shows that second-order convergence can be achieved for the Crank-Nicolson scheme, provided that two simple algorithm modifications are taken:

- The discontinuous terminal payout is least-squares (L^2) projected onto the space of linear Lagrange basis functions[10].

---

[10]Recall that the linear Lagrange basis functions (also called "hat" functions) are simply small triangles given by $l_j(x) = 1_{\{x_{j-1} < x \leq x_j\}} \cdot \frac{x - x_{j-1}}{x_j - x_{j-1}} + 1_{\{x_j < x \leq x_{j+1}\}} \cdot \frac{x_{j+1} - x}{x_{j+1} - x_j}$, $j = 1, \ldots, m$. For an algorithm to perform the L^2-projection, see Pooley et al. [2003].

- Two fully implicit time steps ($\theta = 1$) are taken before we switch to Crank-Nicolson ($\theta = \frac{1}{2}$) time stepping ("Rannacher stepping").

Both techniques effectively smoothen out the discontinuity before the Crank-Nicolson scheme is applied, dampening the problematic high-frequency modes of the numerical solution. As demonstrated in Pooley et al. [2003] (see also Giles and Carter [2006]), applying either technique in isolation will typically not suffice; both are jointly required to ensure smooth second-order convergence. That said, the application of Lagrange basis function projection may conveniently be substituted with simpler smoothing techniques, with no loss of convergence order. The usefulness of such payoff smoothing extends beyond the case of discontinuous boundary conditions, so we proceed to discuss a few common techniques next.

2.5.2 Continuity Correction

By the Shannon sampling theorem, (see Shannon [1949]) if the spectrum of $g(x)$ contains frequencies higher than $1/(2\Delta_x)$ (the *Nyquist frequency*), information is lost when we sample $g(x)$ on our mesh $\{x_j\}_{j=0}^{m+1}$. In other words, whenever $g(x)$ or its derivatives are non-smooth, we will incur a *quantization* error where important features of the payout (e.g., the discontinuity of the slope of a call option at the strike) will be lost between grid points. As the grid geometry is modified, and the location of critical points (strikes, barriers, etc.) relative to x-grid changes, the computed finite difference solution will jump back and forth in erratic fashion. This so-called *odd-even effect* will result in poor convergence and an undesirably strong dependence of the solution on the grid geometry.

One straightforward way to reduce the odd-even effect (and to smooth out the high-frequency components of the payoff) is to apply a common technique from probability theory known as a *continuity correction*. Here, we simply imagine that the value of g at a grid point x_j represents the average value of the function over the interval $[x_j - (x_j - x_{j-1})/2, x_j + (x_{j+1} - x_j)/2]$. In setting the terminal boundary value $V(T, x_j)$ we thus write

$$V(T, x_j) = \frac{1}{(x_{j+1} - x_{j-1})/2} \int_{x_j - (x_j - x_{j-1})/2}^{x_j + (x_{j+1} - x_j)/2} g(x)\, dx. \quad (2.32)$$

We note that this implies that $V(T, x_j) \neq g(x_j)$, unless g is linear in x and the x-grid is equidistant. The application of continuity correction to parabolic PDE solvers was first proposed in Kreiss et al. [1970].

2.5.3 Grid Shifting

Consider the effect of using (2.32) on a *digital call option*, $g(x) = 1_{\{x > H\}}$, where the level H (the digital strike) is located between nodes x_k and

x_{k+1}. For nodes x_j, $j > k+1$, clearly $V(T, x_j) = 1$; for nodes x_j, $j < k$, $V(T, x_j) = 0$. The smoothing algorithm will have effect only at x_k or x_{k+1}, and will set either $V(T, x_k)$ or $V(T, x_{k+1})$ to a value somewhere between 0 and 1, depending on which of x_k or x_{k+1} is closest to H. If H happens to be exactly midway between x_k or x_{k+1}, the continuity correction is seen to have no effect whatsoever.

The digital option example above gives rise to a method listed in Tavella and Randall [2000] (see also Cheuk and Vorst [1996]). Here, we simply arrange the spatial grid such that the x-values where the payoff (or its derivatives) is discontinuous are exactly midway between grid nodes. If necessary, we can use a scheme with non-equidistant grid spacing to accomplish this (see Section 2.4). Our example above shows that aligning the grid in this way will, in a loose sense, make the payoff smooth.

For digital options, the grid shifting technique can be very efficient, and such "locking" of the location of strikes and barriers relative to the spatial grid can often reduce odd-even effects even better than the continuity correction discussed earlier. To demonstrate, consider the concrete task of using a finite difference grid to price a digital call option on a stock S in the Black-Scholes model. In this case, we conveniently have a theoretical option price to compare against, since it is easily shown that the time 0 value $V(0)$ must be

$$V(0) = e^{-rT} Q\left(S(T) > H\right) = e^{-rT} \Phi\left(\frac{\ln(S(0)/H) + (r - \sigma^2/2)T}{\sigma\sqrt{T}}\right). \tag{2.33}$$

For our numerical work, we discretize the asset equidistantly in log-space (i.e., we work with the PDE (2.3)) and determine the spatial grid boundaries by probabilistic means using a multiplier of $\alpha = 4.5$, see Section 2.1. Spatial boundary conditions are $\partial V/\partial x = \partial^2 V/\partial x^2$, implemented as described in Section 2.2.2. In one experiment, we apply a straight Crank-Nicolson approach, with no attempt to regularize the payoff condition. In a second experiment, we combine Crank-Nicolson with Rannacher stepping and also nudge the entire spatial grid upwards until the log-barrier $\ln(H)$ is located exactly half-way between two spatial grid points. Numerical results are shown in Figure 2.1.

As Figure 2.1 shows, a naive Crank-Nicolson implementation is plagued by severe odd-even effects and very slow convergence — 100's of spatial steps appear to be necessary before acceptable levels of the option price are reached. On the other hand, grid shifting combined with Rannacher stepping results in a perfectly smooth[11] convergence profile, and 5-digit price precision is here reached in less than 30 steps.

---

[11] It can be verified that the convergence order in m is, as expected, close to 2 in this experiment.

Fig. 2.1. 3 Year Digital Option Price

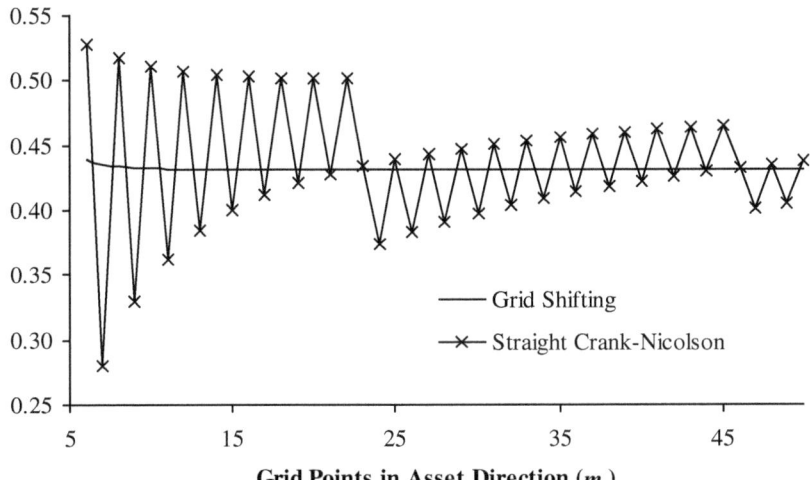

Notes: Finite difference estimates for the Black-Scholes price of a 3 year digital option with a strike of $H = 100$. The initial asset price is $S(0) = 100$, the interest rate is $r = 0$, and the volatility is $\sigma = 20\%$. Time stepping is performed with an equidistant grid containing $n = 50$ points. Spatial discretization in log-space is equidistant, as described in the main text; the number of grid points (m) is as listed on the x-axis of the figure. The "Straight Crank-Nicolson" graph shows the convergence profile for a pure Crank-Nicolson finite difference grid. The "Grid Shifting" graph shows the convergence profile for a Crank-Nicolson finite difference grid with Rannacher stepping and a shift of the spatial grid to center $\ln(H)$ midway between two grid points. From (2.33), the theoretical value of the option is 0.4312451.

2.6 Convection-Dominated PDEs

Recall from Section 2.3 that stability of the explicit finite difference scheme requires that (omitting grid subscripts on μ and σ)

$$\frac{2\Delta_x^2}{\Delta_t}\sigma^2 \geq \sigma^4 + \mu^2\Delta_x^2 + \left|\mu^2\Delta_x^2 - \sigma^4\right|.$$

As discussed, this condition can be violated if Δ_t is too large relative to Δ_x. However, for fixed Δ_t and Δ_x we notice that instability can also be triggered if the absolute value of the drift μ is raised to be sufficiently large relative to the diffusion coefficient σ.

While theta schemes with $\theta \geq 1/2$ are always stable, large drifts in the PDE can nevertheless cause spurious oscillations and an overall deterioration in numerical performance of these schemes. PDEs for which this effect occurs are said to be *convection-dominated*. To quantify matters, assume for simplicity that the finite difference grid is equidistant in the x-direction, and consider the matrix $\mathbf{A}$ in (2.11) with tri-diagonal coefficients c, u, and l given by (2.7)–(2.9). As discussed in e.g. d'Halluin et al. [2005], spurious oscillations can occur when, for some t and some j, either $u_j(t) < 0$ or $l_j(t) < 0$. From (2.8) and (2.9), to avoid spurious oscillations we would thus need

$$\sigma(t, x_j)^2 \geq |\mu(t, x_j)| \Delta_x. \tag{2.34}$$

Intuitively, in convection-dominated systems, the central difference coefficient δ_x and δ_{xx} used to discretize the PDE can no longer fully contain the large expected up- or downward trend of the underlying process for x; as a result, spurious oscillations can occur.

2.6.1 Upwinding

There are a number of well-established techniques to deal with convection-dominated PDEs. First, we can obviously attempt to lower Δ_x such that (2.34) is satisfied. This, however, may not be practical from a computational standpoint (and may require that Δ_t is lowered as well to avoid spurious oscillations originating from the time-stepping scheme). An alternative is to modify the first-order discrete operator δ_x such that it points in the direction of the large absolute drift. For instance, we can simply elect to use a suitably oriented one-sided difference, rather than a central difference, whenever (2.34) is violated. This procedure is known as *upstream differencing* or *upwinding*. To formalize the idea, introduce a new first-order difference operator δ_x^* given as

$$\delta_x^* V(t, x_j) = \begin{cases} \frac{1}{2}(V(t, x_{j+1}) - V(t, x_{j-1})) \Delta_x^{-1}, & |\mu(t, x_j)| \Delta_x \leq \sigma(t, x_j)^2, \\ (V(t, x_j) - V(t, x_{j-1})) \Delta_x^{-1}, & \mu(t, x_j) \Delta_x < -\sigma(t, x_j)^2, \\ (V(t, x_{j+1}) - V(t, x_j)) \Delta_x^{-1}, & \mu(t, x_j) \Delta_x > \sigma(t, x_j)^2. \end{cases}$$

Using δ_x^* instead of δ_x modifies the matrix $\mathbf{A}$ in (2.11). Specifically, if $\mu(t, x_j) \Delta_x < -\sigma(t, x_j)^2$ we replace (2.7)–(2.9) with:

$$c_j(t) = \mu(t, x_j) \Delta_x^{-1} - \sigma(t, x_j)^2 \Delta_x^{-2} - r(t, x_j), \tag{2.35}$$

$$u_j(t) = \frac{1}{2}\sigma(t, x_j)^2 \Delta_x^{-2}, \tag{2.36}$$

$$l_j(t) = -\mu(t, x_j) \Delta_x^{-1} + \frac{1}{2}\sigma(t, x_j)^2 \Delta_x^{-2}. \tag{2.37}$$

And when $\mu(t, x_j) \Delta_x > \sigma(t, x_j)^2$, we use

$$c_j(t) = -\mu(t, x_j) \Delta_x^{-1} - \sigma(t, x_j)^2 \Delta_x^{-2} - r(t, x_j), \tag{2.38}$$

$$u_j(t) = \mu(t, x_j) \Delta_x^{-1} + \frac{1}{2}\sigma(t, x_j)^2 \Delta_x^{-2}, \tag{2.39}$$

$$l_j(t) = \frac{1}{2}\sigma(t, x_j)^2 \Delta_x^{-2}. \tag{2.40}$$

For non-equidistant grids, a similar modification to (2.29)–(2.31) is required. We omit the straightforward details.

Let us try to gain some further understanding of the upwind algorithm. Comparison of (2.35)–(2.40) with (2.7)–(2.9), shows that upwinding amounts to using a regular central difference operator δ_x on a PDE with a diffusion coefficient modified to be $\sqrt{\sigma(t,x)^2 + |\mu(t,x)|\Delta_x}$. The numerical scheme in effect introduces enough artificial diffusion into the PDE to satisfy (2.34). Doing so, however, comes at a cost: the convergence order of the scheme will be reduced to $O(\Delta_x)$ if one-sided differencing ends up being activated in a significant part of the grid. We note that higher-order upwinding schemes are possible if the finite difference operator δ_x^* is allowed to act on more than three neighboring points. For such schemes, the matrix **A** will no longer be tri-diagonal.

2.6.2 Other Techniques

As discussed earlier, upwinding amounts to adding numerical diffusion at nodes where the scheme is convection dominated. Alternatively, we can increase $\sigma(t,x)$ directly, to $\sigma(t,x) + \varepsilon$ where ε is chosen to be large enough for the scheme to satisfy (2.34). By solving the resulting PDE for different values of ε, it may be possible to determine how the error associated with ε scales in ε. This, in turn, will allow us to extrapolate to the limit $\varepsilon = 0$. See p. 135 of Tavella and Randall [2000] for an example.

The upwinding scheme presented in Section 2.6.1 switches abruptly from central differencing to one-sided differencing when the condition (2.34) is violated. In some schemes, the switch from central to one-sided differencing is made smooth by using a weighted average of a one-sided and a central difference operator. The weight on the central difference is close to one when $\sigma(t,x)^2 \gg |\mu(t,x)|\Delta_x$, but decreases smoothly to zero as $\sigma(t,x)^2/|\mu(t,x)|$ tends to zero. While it is unclear whether a smooth transition to upwinding is truly important (the convergence order is typically not improved over straight upwinding), Duffy [2000] suggests that the class of exponentially fitted schemes (see Duffy [2000] and Stoyan [1979]) may be quite robust in derivatives pricing applications.

In some finance applications, multi-dimensional PDEs might arise where $\sigma(t,x) = 0$ for one of the underlying variables; see for instance Section 2.7.5. While upwinding techniques still apply here, we note that specialized methods exist with better ($O(\Delta_x^2)$) convergence, should they become necessary. See, for instance, d'Halluin et al. [2005] for details on the so-called *semi-Lagrangian* methods.

2.7 Option Examples

In our discussion so far, we have assumed that options are characterized by a single terminal payoff function $g(x)$ and a set of spatial boundary

64   2 Finite Difference Methods

conditions determining the option price at the boundaries of the x-domain. In reality, many options are more complicated than this and may involve early exercise decisions, pre-maturity cash flows, path dependency, and more. In this section, we provide some relatively straightforward examples of such complications and how to modify the basic finite difference algorithm to deal with them. More examples will be provided later, in the context of specific fixed income securities.

2.7.1 Continuous Barrier Options

We have already touched upon the concept of an up-and-out knock-out option, an option that expires worthless if the x-process ever rises above a critical level H. As we described, we here must simply solve the PDE (2.1) on a domain $[M, H]$, where M represents the lowest attainable value of the process $x(t)$ on $[0, T]$. The boundary condition at the upper boundary is then dictated to be $V(t, H) = 0$, i.e. of the Dirichlet type. We can generalize this to allow both "up" and "down" type barriers, and to perhaps give a non-zero payout (a "rebate") at the time the barrier(s) are hit (provided this happens before the option maturity). Specifically, if we have a lower barrier at $\underline{H}$, an upper barrier of $\overline{H}$, a time-dependent lower rebate function of $\underline{f}(t)$, and a time-dependent upper rebate function of $\overline{f}(t)$, we must dimension our spatial grid $\{x_j\}_{j=0}^{m+1}$ to have $x_0 = \underline{H}$, $x_{m+1} = \overline{H}$, and we then simply impose the Dirichlet boundary conditions $V(t, x_0) = \underline{f}(t)$ and $V(t, x_{m+1}) = \overline{f}(t)$. See (2.10) and the definition of Ω for the algorithm required to incorporate such Dirichlet boundary conditions into the finite difference scheme.

In practice, barrier options sometimes involve time-dependent barriers, possibly with discontinuities. For instance, *step-up* and *step-down* barrier options will have piecewise flat barriers that increase (step-up) or decrease (step-down) at discrete points in time. Extension of the finite difference algorithm to cover step-up and step-down options is relatively straightforward. As an illustration, consider a zero-rebate up-and-out single-barrier option where the (upper) barrier is flat, except for a discontinuous change at time $T^* < T$, at which point the barrier moves from a value of H^* to a value of H, with $H > H^*$. We set the x-domain of our finite difference grid to $x \in [M, H]$, with M a probabilistic lower limit, as defined above; accordingly, our spatial grid would be $\{x_j\}_{j=0}^{m+1}$, where $x_0 = M$ and $x_{m+1} = H$. In preparation for the shift in barrier levels at time T^*, we make sure that one level in the spatial grid — say x_{k+1}, $k < m$, — is set exactly at the level H^*. Similarly, we make sure that one level in the time grid is set exactly to T^*. Starting at time T, we then iterate backwards in time by repeated solution of m-dimensional tri-diagonal systems of equations, at each step integrating a prescribed rebate function by supplying the Dirichlet boundary condition $V(t, x_{m+1}) = 0$. The moment we hit T^*, the PDE now only applies to the smaller region $[M, H^*]$, covered by the reduced spatial grid $\{x_j\}_{j=0}^{k+1}$ with

$x_{k+1} = H^*$. From T^* back to time 0, the backward induction algorithm then involves only k-dimensional tri-diagonal systems of equations, with the Dirichlet boundary condition $V(t, x_{k+1}) = 0$. Spatial nodes above x_{k+1} correspond to zero option value and can be ignored[12]. Modification of the algorithm outlined above to handle more than two barrier discontinuities is straightforward.

We can extend our definition of barrier options even further by making the topology of "alive" and "dead" regions more complicated. At time t, assume for instance that the PDE applies in an "alive" region of $x \in L(t)$ and a rebate function $R(t, x)$ that applies in the "dead" region $D(t) = \mathcal{B}\backslash L(t)$. Assume that we discretize the problem on a single rectangular finite difference grid spanning the spatial domain $[\underline{M}, \overline{M}]$, where $\underline{M}$ and $\overline{M}$ are set such that the alive regions are covered, up to probabilistic limits (if necessary). Given option values at time t_{i+1}, we then only need to run the basic matrix equation (2.18) for values in our grid $\{x_j\}$ that lie inside $L(t_i)$. This requires scaling down the dimension of the matrix $\mathbf{A}$ as needed, and providing the relevant boundary conditions (given through $R(t_i, x_j)$) at the boundary (or boundaries) of $L(t_i)$. The parts of the spatial grid that lie outside of $L(t_i)$ can be directly filled in with values provided by the rebate function R. Notice that, if possible, the spatial grid should be set such that the boundaries of $L(t_i)$ are contained in the mesh; this will likely require us to use the techniques outlined in Section 2.4.

If the alive region has the simple form $L(t) = [\alpha(t), \beta(t)]$ for smooth deterministic functions α and β, an alternative to the scheme above is to introduce a time-dependent transformation that straightens out the barriers, allowing us to return to the standard finite difference setup where the PDE applies to a single rectangular (t, x) domain. One possible transformation involves using a spatial variable of

$$y = y(t, x) = \frac{x - \alpha(t)}{\beta(t) - \alpha(t)}, \qquad (2.41)$$

which transforms the curved x-barriers $\alpha(t)$ and $\beta(t)$ into flat y-barriers at $y = 0$ and $y = 1$, respectively. The linearity of the transformation (2.41) makes it easy to work with; see Tavella and Randall [2000] for details and a discussion of extensions to multi-dimensional PDEs and to barriers with discontinuities.

---

[12] An obvious twist to the algorithm involves using different spatial grids over $[0, T^*]$ and $[T^*, T]$, allowing for more flexibility in node placement. In this case, values computed by backward induction must, at time T^*, be interpolated from one x-grid to another. The interpolation rule should be at least third-order accurate; see the discussion in Section 2.7.3.

2.7.2 Discrete Barrier Options

The barrier options considered in Section 2.7.1 are continuously monitored, in the sense that the barrier condition is observed for all times in a given interval. In practice, monitoring the barrier condition continuously can be impractical, and it may instead only be imposed on a discrete set of dates $T_1 < T_2 < \ldots < T_K$, with $T_K \leq T$ and $T_1 > 0$. For the sake of concreteness, let us consider a discretely monitored up-and-out option with a constant barrier H. For a continuously monitored up-and-out barrier option it would suffice to solve the PDE on a domain $x \in [\underline{M}, H]$, where $\underline{M}$ is a probabilistic lower limit. This is, however, no longer the case for a discretely monitored option where we need to allow the value function to "diffuse" above the barrier levels between dates in the monitoring set $\{T_k\}_{k=1}^K$. To allow for this, we discretize the PDE on a larger domain $x \in [\underline{M}, \overline{M}]$, $\overline{M} > H$. We can determine $\overline{M}$ probabilistically by determining a confidence interval for how far above the barrier $x(t)$ can rise between monitoring dates. For instance, for the Black-Scholes PDE (2.3), assume that $\max_{k=2,\ldots,K}(T_k - T_{k-1}) = \Delta_T$. Conditioned on $x(t) = H$, the probability that $x(t + \Delta_T)$ exceeds

$$x_\alpha = H + \left(r - \frac{1}{2}\sigma^2\right)\Delta_T + \alpha\sigma\sqrt{\Delta_T}$$

is $\Phi(-\alpha)$. As in Section 2.1, we recommend setting $\overline{M} = x_\alpha$ for values of α somewhere between 3 and 5. To properly capture diffusion between barrier observation dates, we should also dimension the time grid of the finite difference scheme such that multiple time steps (at least two or three, say) are taken between observation dates. All observation dates $\{T_k\}_{k=1}^K$ should obviously be contained in the time grid.

Between barrier observation dates, we solve our PDE by the standard finite difference algorithm outlined in Section 2.2.4, as always imposing either an asymptotic Dirichlet condition at $x = \overline{M}$ or a condition on the x-derivatives of the value function. At each barrier observation time T_k, we must impose a *barrier jump condition*

$$V(T_k-, x) = V(T_k+, x)1_{\{x<H\}}, \quad k = 1, \ldots, K, \qquad (2.42)$$

where the notation $T_k\pm$ was introduced in Section 1.10.1 to denote the limit $T_k \pm \varepsilon$ for $\varepsilon \downarrow 0$. This merely states that all values $V(T_k, x)$ are zero for $x \geq H$, consistent with the definition of an up-and-out option. In our finite difference scheme, we incorporate this jump condition by simply interpreting the vector $\widehat{\mathbf{V}}(T_k)$ as found by regular backward induction as $\widehat{\mathbf{V}}(T_k+)$ and then replacing

$$\widehat{\mathbf{V}}(T_k+) = \left(\widehat{V}_1(T_k+), \ldots, \widehat{V}_m(T_k+)\right)^\top$$

with

2.7 Option Examples   67

$$\widehat{\mathbf{V}}(T_k-) = \left(\widehat{V}_1(T_k+)1_{\{x_1<H\}}, \ldots, \widehat{V}_m(T_k+)1_{\{x_m<H\}}\right)^\top$$

before continuing the algorithm backwards from T_k.

The jump condition (2.42) will generally produce a discontinuity in V as a function of x, around the barrier level H. If we use Crank-Nicolson time-stepping, it will then be prudent to employ a fully implicit scheme for the first few backwards time steps (Rannacher stepping) past each barrier observation date T_k. As discussed in Section 2.5, ideally this should be combined with a smoothing algorithm acting on $\widehat{\mathbf{V}}(T_k-)$ or, perhaps more conveniently, a shift of the spatial grid such that H lies exactly mid-way between two spatial nodes in the grid.

We round off by noting that the discussion above for an up-and-out option easily extends to more complicated discrete barrier options, including those with time-varying barrier levels and rebates. For instance, assume that an option involves upper and lower time-varying barriers of $\overline{H}(t)$ and $\underline{H}(t)$, respectively, as well as a time- and state-dependent rebate of $R(t,x)$. In this case, we simply replace the jump condition (2.42) with

$$V(T_k-, x) = V(T_k+, x)1_{\{\underline{H}(T_k)<x<\overline{H}(T_k)\}}$$
$$+ R(T_k, x)\left(1_{\{x\geq\overline{H}(T_k)\}} + 1_{\{x\leq\underline{H}(T_k)\}}\right),$$

and otherwise proceed as above. We note that time-dependent barriers will typically require flexibility in setting the spatial grid, as there are now multiple critical x-levels to consider. The discretization in Section 2.4 can obviously assist with this.

2.7.3 Coupon-Paying Securities and Dividends

Many fixed-income securities are coupon-bearing and involve periodic transfer of a cash amount between the buyer and the seller. This can easily be incorporated into a finite difference grid, through a jump condition. Specifically, consider a security that pays its owner a single cash amount of $p(T^*, x)$ at time $T^* < T$, where p is a deterministic function $p : [0,T] \times \mathcal{B} \to \mathbb{R}$. We dimension our time grid such that T^* is contained in the grid, and then apply at time T^* the condition

$$V(T^*-, x) = V(T^*+, x) + p(T^*, x). \qquad (2.43)$$

This simply expresses that V will decrease by an amount p immediately after p is paid (and thereby no longer contained in V). In a finite difference algorithm, (2.43) is incorporated by replacing $\widehat{\mathbf{V}}(T^*+)$, as found by regular backward induction, with

$$\widehat{\mathbf{V}}(T^*-) = \left(\widehat{V}_1(T^*+) + p(T^*, x_1), \ldots, \widehat{V}_m(T^*+) + p(T^*, x_m)\right)^\top$$

before continuing the algorithm backwards from T^*. Extensions to multiple coupons are trivial.

In some cases a derivative security does not itself pay coupons, but is written on a security that does. This involves no particular complications, except for the case where payments may affect the state variable underlying the PDE. For instance, consider the classical case of a stock paying a dividend: at the time of the dividend payment, the stock jumps down by an amount equal to the dividend payment. For a model that uses the stock price (or a transformation of the stock price) as the state variable x, a dividend payment at time T^* would thus be associated with a discontinuity in the state variable, $x(T^*+) = x(T^*-) - d(T^*, x(T^*-))$, where d is the magnitude of the jump[13]. As long as the dividend-payment does not come as a surprise (i.e., at a random time), it must already be contained into the option price at T^*-, and will have no price effect as we move forward from T^*- to T^*+. We can express this continuity restriction through yet another jump condition

$$V(T^*-, x) = V(T^*+, x - d(T^*, x)). \quad (2.44)$$

See Wilmott et al. [1993] for more discussion. Implementation of (2.44) in a finite difference grid proceeds as follows. First, we use regular backward induction to establish

$$\widehat{\mathbf{V}}(T^*+) = \left(\widehat{V}_1(T^*+), \ldots, \widehat{V}_m(T^*+)\right)^\top$$
$$= \left(\widehat{V}(T^*+, x_1), \ldots, \widehat{V}(T^*+, x_m)\right)^\top.$$

Then we write

$$\widehat{\mathbf{V}}(T^*-) = \left(\widehat{V}(T^*+, x_1 - d(T^*, x_1)), \ldots, \widehat{V}(T^*+, x_m - d(T^*, x_m))\right)^\top.$$

The values $\widehat{V}_j(T^*+, x_j - d)$ here can be found by interpolation in the x-direction on the $\widehat{\mathbf{V}}(T^*+)$-array. As shown in Tavella and Randall [2000], the order of the interpolator should be strictly higher than two, to avoid inducing spurious numerical diffusion into our θ-style finite difference schemes. We note that this rules out the piecewise linear interpolation rule proposed in Wilmott et al. [1993]. A common choice is to use cubic spline interpolation; see Chapter 6 for much information on cubic splines.

2.7.4 Securities with Early Exercise

In Section 1.10 we introduced the concept of Bermudan and American securities with early exercise features. Under the assumption that exercise

---

[13] To prevent negative stock prices, it may be necessary to truncate the size of d locally in the finite difference grid. For simplicity, we ignore this complication here.

values are determined by a deterministic function[14] $h(t,x)$, $h:[0,T]\times \mathcal{B} \to \mathbb{R}$, finite difference grids are ideal for pricing of such securities. Let us first consider a Bermudan option with exercise opportunities restricted to the finite set $\{T_k\}_{k=1}^K$. The Bellman principle (1.67) in Section 1.10 can, as shown there, be expressed as a simple jump condition

$$V(T_k-,x) = \max\left(V(T_k+,x), h(T_k,x)\right), \quad k=1,\ldots,K, \qquad (2.45)$$

which can be incorporated into a finite difference solver precisely the same way as in previous sections. The condition (2.45) will result in a kink in the value function around the level of x at which we shift from the hold region into the exercise region. If Crank-Nicolson time-stepping is used, one should ideally apply smoothing on the finite difference value vector $\widehat{\mathbf{V}}(T^*-)$, particularly around the kink.

If exercise can take place continuously (that is, American-style) on a given time interval, a crude way to incorporate this into a finite difference grid is by simply applying (2.45) to every point in the time grid of the finite difference scheme. By not specifically imposing the partial differential inequalities (see Section 1.10.1), this algorithm, however, will generally only be accurate to first order in the time step, even if a Crank-Nicolson scheme is used; see Carverhill and Clewlow [1990] for a proof. As American-style exercise is rarely used in fixed income markets, we shall not pursue this issue further but just point out that a number of schemes exist to restore second-order time convergence to finite difference pricing of American options, see, e.g., Forsyth and Vetzal [2002].

2.7.5 Path-Dependent Options

Finite difference methods are normally limited to Markovian problems where dynamics are characterized by SDEs and where payouts are simple deterministic functions of the underlying state variables. A number of options, however, have terminal time T payouts that depend not only on the state of x at time T, but on the entire path $\{x(t),\ t\in[0,T]\}$. In general, such options must be priced by Monte Carlo methods (see Chapter 3), but exceptions exist. Indeed, barrier and American options can be considered path-dependent options, yet, as we have seen, can still be priced in a finite difference grid. Even stronger path-dependence can sometimes be handled, through the introduction of new state variables to the PDE.

To give an example, consider a path-dependent contract where the terminal payout at time T can be written as

---

[14] If h represents the value of a derivative security that has no closed-form pricing formula, it may be necessary to estimate this function by backward induction in the finite difference grid itself. Such a "preprocessing" step is typically straightforward to execute.

$$V(T) = g(x(T), I(T)), \qquad (2.46)$$

where I is a path integral of the type

$$I(t) = \int_0^t h(x(s)) \, ds, \qquad (2.47)$$

for some deterministic function h. For instance, if $h(x) = x$, we say that the option is a continuously sampled *Asian option*.

For the payout (2.46) we have $V(t) = V(t, x(t), I(t))$ where $x(t)$ satisfies the SDE (2.2) and

$$dI(t) = h(x(t)) \, dt, \quad I(0) = 0.$$

From the backward Kolmogorov equation, it follows that $V(t, x, I)$ solves

$$\frac{\partial V}{\partial t} + \mu(t,x) \frac{\partial V}{\partial x} + \frac{1}{2} \sigma(t,x)^2 \frac{\partial^2 V}{\partial x^2} + h(x) \frac{\partial V}{\partial I} = r(t,x) V, \qquad (2.48)$$

subject to the terminal condition $V(T, x, I) = g(x, I)$. There are several complications with this PDE. First, it involves *two* spatial variables, x and I, requiring the use of a two-dimensional PDE solver. Second, the PDE contains no second-order derivative in the variable I, i.e. it is convection dominated in the I-direction. We have discussed methods to handle the latter issue in Section 2.6.1 and will turn to address the former in Section 2.9. Another complication is the fact that the term $h(x)$ multiplying $\partial V/\partial I$ may be of a different order of magnitude than the other coefficients in (2.48), increasing the difficulty of solving the equation numerically. We refer to Zvan et al. [1998] for a more detailed discussion of PDEs of the type (2.48).

In practice, it is rare that a continuous-time integral such as (2.47) is used in an option payout. Instead, one normally samples the function $h(x(t))$ only on a discrete set of dates, i.e. we replace $I(T)$ with

$$I(T) = \sum_{i=1}^{n} h(x(T_i)) (T_i - T_{i-1}),$$

where $T_0 < T_1 < \ldots < T_n$ is a discrete schedule, with $T_0 = 0$ and $T_n = T$. Informally, we now have

$$dI(t) = \delta(T_i - t) \cdot h(x(T_i)) (T_i - T_{i-1}), \quad I(0) = 0, \qquad (2.49)$$

where $\delta(\cdot)$ is the Dirac delta function. In a PDE setting, we incorporate a process such as (2.49) through appropriate jump conditions, writing

$$V(T_i-, x, I) = V(T_i+, x, I + h(x)(T_i - T_{i-1})). \qquad (2.50)$$

In the same fashion as for discrete dividends (Section 2.7.3), the jump condition enforces continuity of the option price across the dates where I

gets updated. The condition is applied at each date in the discrete schedule, $i = 1, \ldots, n$; in between schedule dates (where now $dI(t) = 0$), we solve the PDE

$$\frac{\partial V}{\partial t} + \mu(t,x)\frac{\partial V}{\partial x} + \frac{1}{2}\sigma(t,x)^2\frac{\partial^2 V}{\partial x^2} = r(t,x)V,$$

which has no term involving I. When the I-direction is discretized in, say, m_I different values, the solution scheme thus involves solving m_I different *one-dimensional* PDEs backward in time; the solutions of these m_I PDEs exchange information with each other at each date in the schedule, in accordance with (2.50). As was the case for cash dividends, implementation of (2.50) will normally require support from an interpolation scheme, to align the (x-dependent) jumps in I with the knots of the discretized I-grid used in the finite difference scheme. See, e.g., Zvan et al. [1999] or Wilmott et al. [1993] for further details. An application of this idea in the context of interest rate derivatives is given in Section 18.4.5.

On rare occasions — basically when the homogeneity condition $V(t, \eta x, \lambda I) = \lambda^\eta V(t, x, I)$, $\lambda, \eta > 0$, holds — it is possible to make a change of variables or a change of probability measure that will reduce (2.48) or its discrete-time version to a one-dimensional PDE; see e.g. Rogers and Shi [1995] or Andreasen [1998] for the case of various Asian options. Section 18.4.5 demonstrates one such method, sometimes called the method of *similarity reduction*, for pricing of "weakly path-dependent" securities, including certain callable interest rate derivatives where the notional accretes at a stochastic coupon rate (see Section 5.14.5 for definitions).

2.7.6 Multiple Exercise Rights

Certain financial products with early exercise rights allow the holder to exercise more than once. Such "multi-exercise" options are relatively rare, but the so-called *chooser cap* (also known as a *flexi-cap*) is occasionally traded and constitutes a good example for describing how to handle multi-exercise options in a PDE setting. Let there be given a set of L possible exercise dates, $T_1 < T_2 < \ldots < T_L$, and assume that we have the right to exercise no more than l times, with $l < L$. Provided that we exercise at time T_i, in a chooser cap we are paid[15] $(S(T_i) - K)^+$, where $S(\cdot)$ is some interest rate index and K is the strike. Clearly, we would never exercise at time T_i unless $S(T_i) > K$, but how much larger than K the rate $S(T_i)$ needs to be to trigger optimal exercise is not obvious, and must at least depend on i) how many of our l exercise opportunities we have already used up at time T_i; and ii) how much value is lost by using (rather than postponing) one of the remaining exercise opportunities.

---

[15] We have ignored a day count scaling constant in the payout. Also, in most cases payment takes place at time T_{i+1}, rather than at T_i; such a payment delay can be handled by a discount operation.

72   2 Finite Difference Methods

While the question of how to exercise optimally on a chooser cap may appear quite complex, it is surprisingly easy to implement in a finite difference setting by combining techniques from Sections 2.7.4 and 2.7.5 above. The key to the method is to introduce an additional state variable I to keep track of how many exercise opportunities are left. Assume that all interest rates are functions of a Markov state variable $x(\cdot)$, and let therefore $V(t,x,I)$ denote the value of the chooser cap at time t, given $x(t) = x$ and given that there are still I exercise opportunities left. Notice that the variable I can only take $l+1$ distinct values: $0, 1, \ldots, l$; notice also that $V(t,x,0) = 0$ for all t and x, since $I = 0$ corresponds to the situation where there are no exercise opportunities left. Additionally, at the terminal time T_L we clearly have
$$V(T_L, x, I) = (S(T_L, x) - K)^+, \quad I = 1, 2, \ldots, l, \qquad (2.51)$$
where we have written $S(T_L) = S(T_L, x)$ to emphasize the deterministic dependence of S on the state variable x.

For given dynamics of $x(t)$, starting with the terminal conditions in (2.51), we may roll the l different value functions $V(\cdot, x, I)$, $I = 1, 2, \ldots, l$, back through time in standard finite difference manner. At each time T_i, $i = 1, \ldots, L-1$, jump conditions similar to (2.45) must be applied, for all $I = 1, 2, \ldots, l$:
$$V(T_i-, x, I) = \max\left(V(T_i+, x, I), V(T_i+, x, I-1) + (S(T_i, x) - K)^+\right).$$
Notice that these conditions simply express that exercise is optimal only if the exercise value (the cap payout plus the value of a chooser cap with one less exercise opportunity) exceeds the hold value (the non-exercised chooser cap). Once we have rolled all the way back to $t = 0$, the chooser cap value at time $t = 0$ may be identified as $V(0) = V(0, x(0), l)$.

We should note that the "chooser" or "flexi" feature can be added to securities other than caps (and floors). For instance, in Section 19.5 we study the so-called *flexi-swap*, another security with multiple embedded exercise rights.

2.8 Special Issues

In this section, we briefly show a few techniques that may come in handy for certain applications.

2.8.1 Mesh Refinements for Multiple Events

As discussed in Section 2.1, the domain of the state variable x is often determined as an exact or approximate confidence interval for the random variable $x(T)$, where T is the final time of interest for a particular valuation problem we want to solve. Given the number of desired spatial steps in the scheme, the discretization step in x-direction is then obtained by dividing the size of the confidence interval by the number of steps. Similarly, the discretization step in t-direction is typically obtained by dividing T by the number of desired time steps. This is a standard procedure for building a simple rectangular mesh, and it works well if the derivative we wish to value does not have any "interesting" features between the valuation time 0 and the final time T (e.g., for a simple European option). However, as should be evident from the examples in Section 2.7, many real-life derivative securities are characterized by a multitude of events during their lifetimes, all of which must be adequately captured in the PDE scheme. It is not hard to see that a grid dimensioning scheme based solely on the last event date may yield inappropriate mesh resolution at earlier dates.

To make the discussion above concrete, let us consider the example of a Bermudan option (see Section 2.7.4) with two exercise dates, T_1 and T_2. Assume that $0 < T_1 \ll T_2$, i.e. that the first exercise date is much closer to the valuation date than the second (and last) one. Also assume that there is a decent chance that the option actually will be exercised at time T_1, making it important to capture to good precision the value of the option expiring at T_1. Now, if we build our mesh based only on the distribution of the state variable $x(T_2)$ at time T_2, there would typically be too few t-points in the interval $[0, T_1]$. Also, the x-direction discretization step would be too large compared to the range of possible values of the state variable $x(T_1)$ at time T_1, i.e. the x-grid would be too coarse for the process $x(\cdot)$ on the time interval $[0, T_1]$. Both issues would typically lead to a large discretization error in the finite difference stepping of the option over the time period $[0, T_1]$, leading to problems with accuracy in values and risk sensitivities.

The issue of the sparsity of the time grid is fairly easy to deal with, as we are free to add extra points to the time grid before time T_1. This by itself, however, will not solve precision problems, as the space step remains large. Any proper solution should, of course, come in the form of refining both the t- and x-grids at the same time.

One possible way of refining the x-discretization is to abandon the usage of a single rectangular (t, x)-domain, and instead link together different equidistant rectangular meshes for different periods in the life of the derivative. These mesh "blocks" would generally increase in spatial width with time and would connect to each other via an interpolation scheme. To be more specific, let us assume, as in Section 2.1, that the state variable $x(\cdot)$ is the logarithm of the stock in the Black-Scholes model and is given by (2.4), with the PDE to solve given by (2.3). We extend our simple two-period example above to a derivative with K times of interest, $0 < T_1 < \ldots < T_K$; these times could be specified as an additional input into valuation, or derived from the trade description (e.g. they could represent the exercise dates for a Bermudan option, or the knock-out dates for the discretely-monitored barrier option of Section 2.7.2). Suppose we are given values of m and n, and now wish to construct the mesh for the time period $[T_{k-1}, T_k]$, by using the same time and space steps Δ_t^k, Δ_x^k as would be used in the standard scheme of Section 2.1 for a derivative security with the terminal payoff at T_k. That is, having fixed the cutoff α we would set

$$\Delta_t^k = T_k/n, \quad \Delta_x^k = 2\alpha\sigma\sqrt{T_k}/(m+1). \tag{2.52}$$

Then the rectangular, equidistant mesh for the time period $[T_{k-1}, T_k]$ is given by

$$\{t_i^k\}_{i=0}^{\lfloor(1-T_{k-1}/T_k)n\rfloor} \times \{x_j^k\}_{j=0}^{m+1}, \quad t_i^k = T_{k-1} + i\Delta_t^k, \quad x_j^k = x_{\min}^k + j\Delta_x^k, \tag{2.53}$$

where $\lfloor \cdot \rfloor$ denotes the integer part of a real number and (see (2.4))

$$x_{\min}^k = x(0) + \left(r - \frac{1}{2}\sigma^2\right)T_k - \alpha\sigma\sqrt{T_k}. \tag{2.54}$$

Note that in reality we would want to make sure that the point T_k is also in the mesh for the time period $[T_{k-1}, T_k]$, even though for simplicity of notations we did not reflect it in (2.53). It is also useful to note that the total number of time points is not going to be n, but is actually equal to

$$\sum_{k=1}^{K} \lfloor (1 - T_{k-1}/T_k)\, n \rfloor,$$

which scales linearly with n. Clearly, if exactly n points were required, a simple adjustment to the definition of the time step in (2.52) could be applied.

With a mesh as defined above, when arriving at time T_k in a backward induction scheme the solution $V(T_k, \cdot)$ would be discretized on the x-grid $\{x_j^{k+1}\}_{j=0}^{m+1}$. To solve the PDE backwards over the time period $[T_{k-1}, T_k]$, we would need to resample it on the different x-grid $\{x_j^k\}_{j=0}^{m+1}$. As with interpolation across dividends (Section 2.7.3), simple cubic interpolation would be a good choice here. Specifically, one would fit a cubic spline to the values $V(T_k, x_j^{k+1})$, $j = 0, \ldots, m+1$, and then calculate $V(T_k, x_j^k)$, $j = 0, \ldots, m+1$, by valuing the spline at the required grid points.

The "interpolated mesh" scheme above is rather intuitive and straightforward, but it does suffer from the need to do interpolation work that could slow down the PDE (especially in dimensions higher than 1 and/or for a large number of interface points K). Also, it is not entirely clear how interpolation will affect stability and convergence properties of the PDE. Finally, linking the interface mesh geometry to the trade specifics (such as exercise dates) may not be ideal from the point of view of designing an efficient valuation flow in a risk management system. These considerations lead us to an alternative approach that relies on *non*-equidistant discretization as developed in Section 2.4. The idea of this method is to use non-uniform discretization to concentrate more points, both in time and space, around the initial point $t = 0$, $x = x(0)$. Clearly many ways of achieving this are possible — below we present a simple scheme we have used with good results.

We define K, the user input, to be the number of spatial refinement levels (with $K = 2$ or 3 typically used), and τ, another user input, to be a time scaling constant (typically $\tau = 4$). If T is the final horizon for valuation, we then introduce times

$$0 = T_0 < T_1 < \ldots < T_K = T$$

by

$$T_k = \frac{T}{\tau^{K-k}}, \quad k = 1, \ldots, K.$$

Then, the time grid for the time period $[T_{k-1}, T_k]$ is given by uniformly distributing $\widetilde{n} \triangleq \lfloor n/K \rfloor$ points[16] over $[T_{k-1}, T_k]$, i.e. is given by $\{t_i^k\}_{i=0}^{\widetilde{n}}$ with

$$t_i^k = T_{k-1} + i \frac{T_k - T_{k-1}}{\widetilde{n}} = T_{k-1} + i \frac{T_k - T_{k-1}}{\lfloor n/K \rfloor}.$$

(Note that we can use this specification with the interpolated mesh as well, instead of the time grid definition in (2.53)). The fact that the width of the intervals $[T_{k-1}, T_k]$ grow with k means that the time grid is more finely spaced in the beginning of the interval $[0, T]$ than at the end.

The x-grid we are going to define will be universal — i.e. the same for all time steps on the whole time interval $[0, T]$ — and non-uniform. To construct it, we first define a set of nested x-subdomains $[x_{\min}^k, x_{\max}^k]$, with $x_{\min}^k$ defined by (2.54) and $x_{\max}^k$ defined accordingly, i.e.

$$x_{\max}^k = x(0) + \left(r - \frac{1}{2}\sigma^2\right) T_k + \alpha \sigma \sqrt{T_k},$$

for $k = 0, \ldots, K$. Then we define step sizes by

$$\Delta_x^k = \frac{x_{\max}^k - x_{\min}^k}{\widetilde{m} + 1}, \quad \widetilde{m} = \left\lfloor \frac{m}{K} \right\rfloor.$$

The x-grid is then constructed by distributing grid points uniformly in subintervals $[x_{\min}^k, x_{\min}^{k-1}]$ and $[x_{\max}^{k-1}, x_{\max}^k]$ with the space step Δ_x^k, and is given by

$$\left(\bigcup_{k=1}^{K} \{x_j^{\min,k}\}_{j=0}^{m^k+1}\right) \cup \left(\bigcup_{k=1}^{K} \{x_j^{\max,k}\}_{j=0}^{m^k+1}\right),$$

where

$$x_j^{\min,k} = x_{\min}^k + j\Delta_x^k, \quad x_j^{\max,k} = x_{\max}^{k-1} + j\Delta_x^k,$$

---

[16] And, as advised earlier, adding trade event dates that fall into this period — although we do not reflect this in our notations for simplicity.

and
$$m^k = \left\lceil \frac{x_{min}^{k-1} - x_{min}^k}{\Delta_x^k} \right\rceil - 1 = \left\lceil \frac{x_{max}^k - x_{max}^{k-1}}{\Delta_x^k} \right\rceil - 1.$$

This distribution of space points results in an x-grid that is more dense around the point $x = x(0)$ than at the edges. It is worth noting that with only one refinement level $K = 1$, the standard rectangular uniform mesh sized by the terminal distribution of the state variable is recovered.

2.8.2 Analytics at the Last Time Step

In cases where the dynamics of underlying PDE variables are tractable, one naturally wonders whether finite difference methods could somehow be improved by incorporating analytical results into the scheme. Here, and in the next section, we discuss two simple ideas.

Suppose that we are faced with the problem of pricing a contingent claim with terminal boundary condition $g(x(T))$, where $x(t)$ is a Markovian process with known Arrow-Debreu state prices:

$$G(t, x; s, y) = \mathrm{E}^Q \left(\delta \left(x(s) - y \right) e^{-\int_t^s r(u, x(u))\, du} | x(t) = x \right), \quad s > t.$$

Assume also that the claim in question involves a jump condition at time $0 < T^* < T$ (but no jump conditions between T^* and T). If our finite difference grid is $\{x_j\}_{j=0}^{m+1}$, we can now use a series of $m+2$ outright convolutions to compute

$$V(T^*, x_j) = \int_{\mathbb{R}} G(T^*, x_j; T, y)\, g(y)\, dy, \quad j = 0, 1, \ldots, m+1. \quad (2.55)$$

If we are lucky (i.e., if both g and G are sufficiently simple), then the integral on the right-hand side may be known in closed form for all values of x_j. If not, we can always perform a series of numerical integrations, the total cost of which is typically[17] $O(m^2)$, i.e. more expensive than the typical $O(m)$ cost of a single time step in a finite difference method. There are several reasons why we may want to perform the numerical integrations nevertheless. First, the convolution expression (2.55) is exact, as it is based on the true transition density. Second, if the gap between T^* and T is large, an ordinary finite difference grid would need to roll back from T to T^* using multiple time steps n^*, at a total cost of $O(n^*m)$; if n^* is of the same magnitude as m, the computational effort of the convolution scheme would be comparable to that of a finite difference grid. Third, for discontinuous payouts, the integration in (2.55) will have a naturally smoothing effect, similar to (but often better than) the continuity correction method of Section 2.5.2. The smoothing

---

[17] There are exceptions. For instance, if fast Fourier transform (FFT) methods are applicable, the cost may be reduced to $O(m \ln(m))$. See Section 8.4 for details.

effect is discussed in more detail in Section 23.2.4 and is also demonstrated below, in Figure 2.2, where we have continued our investigation of the 3 year digital option considered earlier in Section 2.5.3. Since the model used in Figure 2.2 is ordinary Black-Scholes and $g(x) = 1_{\{x>H\}}$, the integrals in (2.55) can here be computed in closed form from (2.33).

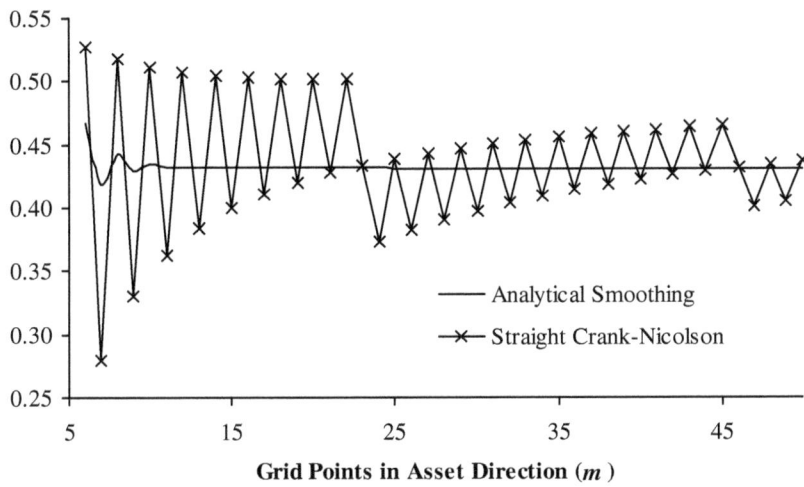

Fig. 2.2. 3 Year Digital Option Price

Notes: Finite difference estimates for the Black-Scholes price of a 3 year digital option. All contract and model parameters are as in Figure 2.1. Time stepping is performed with an equidistant grid containing $n = 50$ points. Spatial discretization in log-space is equidistant, as described in the main text; the number of grid points (m) is as listed on the x-axis of the figure. The "Straight Crank-Nicolson" graph shows the convergence profile for a pure Crank-Nicolson finite difference grid. The "Analytical Smoothing" graph shows the convergence profile for a Crank-Nicolson finite difference grid starting at $T^* = 2.5$ years, with the terminal boundary condition set equal to a 0.5 year digital option price (as in (2.55)). The theoretical value of the option is 0.4312451.

In principle, we could continue rolling back from T^* (through, possibly, jump conditions at earlier times) by performing convolutions, rather than solving finite difference grids. In practice, this rarely leads to improvements over a finite difference grid, unless the densities and payoffs are quite simple[18]. Moreover, in many cases we may not have *exact* Arrow-Debreu

---

[18]For simple densities (especially Gaussian), special-purpose methods exist to compute convolutions rapidly, typically involving payoff approximations through piecewise polynomials or other simple functions. We do not cover these methods in

prices, only approximate ones based on, say, a small-time expansion (see, e.g., Section 13.1.9.1). In this case, a one-time convolution may be safe — especially if $T - T^*$ is small — whereas repeated convolutions may lead to unacceptable biases.

2.8.3 Analytics at the First Time Step

The idea in Section 2.8.2 of replacing the finite difference stepping with analytical integration is even easier to apply over the *first*, rather than the *last*, time step. Suppose T^* is the first "interesting" time for a given derivative security, i.e. there might be a jump condition at time T^* but none over the time interval $[0, T^*)$. Then, rather than stepping the finite difference scheme from T^* to 0, we can perform a *single* integration to calculate the value $V(0, x(0))$ of the derivative at time zero from the discretized values $\{V(T^*, x_j)\}_{j=0}^{m+1}$ of the derivative at time T^* (using the same notations as in Section 2.8.2),

$$V(0, x(0)) = \int_{\mathbb{R}} G(0, x(0); T^*, y) \, \widetilde{V}(T^*, y) \, dy,$$

where $\widetilde{V}(T^*, y)$ is interpolated (using cubic splines, say) from the values $\{V(T^*, x_j)\}_{j=0}^{m+1}$ on the grid. If the integral is computed numerically — as is most often the case — the numerical cost is often comparable with that of the finite difference stepping because only one value $V(0, x(0))$ is required at time 0, not the whole slice.

While there are typically no numerical cost savings that arise from using integration over the first time step, there are accuracy and stability considerations that favor this approach. We have already seen in Section 2.8.1 that the standard discretization of a PDE often leads to insufficient fidelity in resolving any features of the payoff that are close to today, and numerical integration can be of considerable help in this regard. Moreover, as we discuss in much detail later in Chapter 23, an integration scheme typically allows us to treat discontinuities in the value $V(T^*, x)$ arising from the jump condition at time T^* explicitly. If the discontinuity is introduced at the value of the state variable x^*, then the integration scheme can (and should) explicitly take this information into account. For example we would write

$$V(0, x(0)) = \int_{-\infty}^{x^*} G(0, x(0); T^*, y) \, \widetilde{V}^-(T^*, y) \, dy$$
$$+ \int_{x^*}^{\infty} G(0, x(0); T^*, y) \, \widetilde{V}^+(T^*, y) \, dy$$

---

this book except for a brief mention in Section 11.A. For a representative example see Hu et al. [2006].

and calculate $\widetilde{V}^-(T^*, y)$ by interpolating the grid values in the time interval $(-\infty, x^*)$, and $\widetilde{V}^+(T^*, y)$ by interpolating the grid values in (x^*, ∞), separately[19].

The usefulness of the method is only limited by the availability of the closed-form expression for the time 0 Arrow-Debreu prices $G(0, x(0); T^*, \cdot)$. For some models this is not an issue; for most others, sufficiently close approximations could be obtained in a small-time limit (see e.g. Section 13.1.9.1 for a typical approach) that can be useful for times T^* that are not too large. By a change of measure, we see that

$$V(0, x(0)) = \mathrm{E}\left(e^{-\int_0^{T^*} r(s)\, ds} V(T^*, x(T^*))\right)$$
$$= P(0, T^*) \mathrm{E}^{T^*}\left(V(T^*, x(T^*))\right),$$

where E^{T^*} is the expected value operator under the T^*-forward measure Q^{T^*}; so we really only need the expression for the *density* (rather than Arrow-Debreu security prices) of $x(T^*)$ under Q^{T^*}, either exact or approximate.

Finally, we note that while the integration over the first time step can be seen to offer similar advantages to those of the methods in Section 2.8.1, the two approaches are not substitutes for each other, but are complementary. We typically recommend using direct integration over the time step $[0, T^*]$, where T^* is the smaller of the time of the first jump condition or the limit of applicability of the approximation to the density of $x(T^*)$, and then (if needed) use the methods in Section 2.8.1 over the time interval $[T^*, T]$, with T being the final maturity of the option in question.

2.9 Multi-Dimensional PDEs: Problem Formulation

We now turn our attention to the numerical solution of multi-dimensional terminal value problems. Let the spatial variable x be p-dimensional, $x = (x_1, \ldots, x_p)^\top$, and consider the PDE

$$\frac{\partial V}{\partial t} + \sum_{h=1}^p \mu_h(t, x) \frac{\partial V}{\partial x_h} + \frac{1}{2} \sum_{h=1}^p \sum_{l=1}^p s_{h,l}(t, x) \frac{\partial^2 V}{\partial x_h \partial x_l} - r(t, x) V = 0, \quad (2.56)$$

where $s_{h,h}(t, x) \geq 0$ and $s_{h,l}(t, x) = s_{l,h}(t, x)$ for $h, l = 1, \ldots, p$. The PDE is assumed subject to the terminal value condition $V(T, x) = g(x)$, $g : \mathbb{R}^p \to \mathbb{R}$.

From the results in Chapter 1, we recognize that the PDE provides the solution to the expectation

---

[19] One of the functions $\widetilde{V}^-(T^*, y)$, $\widetilde{V}^+(T^*, y)$ is often known analytically and for all values of y (rather than sampled on the grid); this is for instance the case for the Bermudan options of Section 2.7.4. The integration algorithm should obviously take advantage of this.

$$V(t,x) = \mathrm{E}_t\left(e^{-\int_t^T r(u,x)\,du} g\left(x(T)\right) | x(t) = x\right),$$

where the components of $x(t)$ satisfy risk-neutral SDEs of the type

$$dx_h(t) = \mu_h(t, x(t))\,dt + \sigma_h(t, x(t))\,dW(t), \quad h = 1, \ldots, p. \quad (2.57)$$

Here $W(t)$ is a d-dimensional Brownian motion, $\mu_h : [0,T] \times \mathbb{R}^p \to \mathbb{R}$, $h = 1, \ldots, p$, are (scalar) drifts, and $\sigma_h : [0,T] \times \mathbb{R}^p \to \mathbb{R}^{1 \times d}$, $h = 1, \ldots, p$, are d-dimensional (row vector) diffusion coefficients. The PDE coefficients $s_{h,l}$ in (2.56) represent the instantaneous covariance matrix for the components of $x(\cdot)$, i.e., $s_{h,l}(t,x) = \sigma_h(t,x)\sigma_l(t,x)^\top$. We assume enough regularity on μ_h, σ_h, r, and g to ensure that (2.56) has a unique solution.

For the purpose of solving (2.56) numerically, we assume that the PDE is to be solved on a (finite) spatial domain in x, $x \in [\underline{M}_1, \overline{M}_1] \times \ldots \times [\underline{M}_p, \overline{M}_p]$, where $\underline{M}_h, \overline{M}_h$, $h = 1, \ldots, p$, are constants either dictated by the contract at hand (barrier options) or found by a suitable probabilistic truncation (see Section 2.1).

2.10 Two-Dimensional PDE with No Mixed Derivatives

To illustrate the construction of finite difference discretization of (2.56), we start out with the simple case where $p = d = 2$ and there are no mixed partial derivatives in the PDE: $s_{1,2}(t,x) = s_{2,1}(t,x) = 0$ for all t and x. Probabilistically, the absence of mixed derivatives corresponds to the case where the stochastic process increments $dx_1(t)$ and $dx_2(t)$ are independent. Defining $\gamma_h(t,x)^2 = s_{h,h}(t,x)$, $h = 1, 2$, the PDE to be solved now becomes

$$\frac{\partial V}{\partial t} + (\mathcal{L}_1 + \mathcal{L}_2)V = 0, \quad (2.58)$$

where

$$\mathcal{L}_h \triangleq \mu_h(t,x)\frac{\partial}{\partial x_h} + \frac{1}{2}\gamma_h(t,x)^2 \frac{\partial^2}{\partial x_h^2} - \frac{1}{2}r(t,x), \quad h = 1, 2.$$

Notice that we have divided the term $r(t,x)$ into equal pieces in $\mathcal{L}_1$ and $\mathcal{L}_2$.

To discretize (2.58) in x, introduce grids $x_1 \in \{x_1^{j_1}\}_{j_1=0}^{m_1+1}$ and $x_2 \in \{x_2^{j_2}\}_{j_2=0}^{m_2+1}$. To simplify notation, assume these grids are equidistant such that $x_1^{j_1} = \underline{M}_1 + j_1 \Delta_1$ and $x_2^{j_2} = \underline{M}_2 + j_2 \Delta_2$. Let $V_{j_1,j_2}(t) \triangleq V(t, x_1^{j_1}, x_2^{j_2})$. We define discrete central difference operators as before

$$\delta_{x_1} V_{j_1,j_2}(t) = \frac{V_{j_1+1,j_2}(t) - V_{j_1-1,j_2}(t)}{2\Delta_1},$$

$$\delta_{x_2} V_{j_1,j_2}(t) = \frac{V_{j_1,j_2+1}(t) - V_{j_1,j_2-1}(t)}{2\Delta_2},$$

and

$$\delta_{x_1 x_1} V_{j_1,j_2}(t) = \frac{V_{j_1+1,j_2}(t) - 2V_{j_1,j_2}(t) + V_{j_1-1,j_2}(t)}{\Delta_1^2},$$

$$\delta_{x_2 x_2} V_{j_1,j_2}(t) = \frac{V_{j_1,j_2+1}(t) - 2V_{j_1,j_2}(t) + V_{j_1,j_2-1}(t)}{\Delta_2^2}.$$

These operators, in turn, give rise to the discrete operators

$$\widehat{\mathcal{L}}_h \triangleq \mu_h(t,x)\delta_{x_h} + \frac{1}{2}\gamma_h(t,x)^2 \delta_{x_h x_h} - \frac{1}{2}r(t,x), \quad h = 1,2,$$

where x is constrained to take values in the spatial grid. A Taylor expansion shows that this operator is second-order accurate (compare to Lemma 2.2.1),

$$(\mathcal{L}_1 + \mathcal{L}_2)V(t,x) = (\widehat{\mathcal{L}}_1 + \widehat{\mathcal{L}}_2)V(t,x) + O\left(\Delta_1^2 + \Delta_2^2\right).$$

2.10.1 Theta Method

Turning to a theta-style time discretization, consider first proceeding exactly as in Section 2.2.3. Assuming equidistant time spacing Δ_t, we get for the period $[t_i, t_{i+1}]$,

$$\left(1 - \theta \Delta_t \left(\widehat{\mathcal{L}}_1 + \widehat{\mathcal{L}}_2\right)\right) V_{j_1,j_2}(t_i)$$
$$= \left(1 + (1-\theta)\Delta_t \left(\widehat{\mathcal{L}}_1 + \widehat{\mathcal{L}}_2\right)\right) V_{j_1,j_2}(t_{i+1}) + e_i^{i+1},$$

where

$$e_i^{i+1} = O\left(\Delta_t \left(\Delta_1^2 + \Delta_2^2 + 1_{\{\theta \neq \frac{1}{2}\}}\Delta_t + \Delta_t^2\right)\right),$$

and where it is understood that $\widehat{\mathcal{L}}_1$ and $\widehat{\mathcal{L}}_2$ are to be evaluated at $(t,x) = (t_i^{i+1}(\theta), x_1^{j_1}, x_2^{j_2})$ with $t_i^{i+1}(\theta)$ defined as in (2.14). If $\widehat{V}_{j_1,j_2}(t) \triangleq \widehat{V}(t, x_1^{j_1}, x_2^{j_2})$ is a finite difference approximation to $V_{j_1,j_2}(t)$, we thus get the scheme

$$\left(1 - \theta \Delta_t \left(\widehat{\mathcal{L}}_1 + \widehat{\mathcal{L}}_2\right)\right) \widehat{V}_{j_1,j_2}(t_i) = \left(1 + (1-\theta)\Delta_t \left(\widehat{\mathcal{L}}_1 + \widehat{\mathcal{L}}_2\right)\right) \widehat{V}_{j_1,j_2}(t_{i+1}),$$
(2.59)

to be solved for the $m_1 m_2$ interior points $\widehat{V}_{j_1,j_2}(t_i)$, $j_1 = 1, \ldots, m_1$, $j_2 = 1, \ldots, m_2$, given the values of $\widehat{V}_{j_1,j_2}(t_{i+1})$, and given appropriate boundary conditions at $j_1 = 0$, $j_1 = m_1 + 1$, $j_2 = 0$, and $j_2 = m_2 + 1$.

The scheme (2.59) represents a system of linear equations in $m_1 m_2$ unknowns $\{\widehat{V}_{j_1,j_2}(t_i)\}$. When written out as a matrix equation (which requires us to arrange the various $\widehat{V}_{j_1,j_2}(t_i)$ in some order in a $(m_1 m_2)$-dimensional vector), the matrix to be inverted is sparse but, unfortunately, no longer tri-diagonal. Solution of the system of equations by standard methods (e.g., Gauss-Jordan elimination or LU decomposition) is out of the question due to the size of the matrix[20]. We can proceed in two ways: either we use a specialized sparse-matrix solver; or we attempt to redo the discretization (2.59) to make it computationally efficient. We personally prefer the second approach and shall outline one method in the next section. As for the first approach, we simply note that a good iterative sparse solver should be able to solve (2.59) in order $O((m_1 m_2)^{5/4})$ operations. See Saad [2003] for concrete algorithms.

2.10.2 The Alternating Direction Implicit (ADI) Method

The ADI method is an example of a so-called *operator splitting* method, where the simultaneous application of two operators (here $\widehat{\mathcal{L}}_1$ and $\widehat{\mathcal{L}}_2$) is split into two *sequential* operator applications. To illustrate the idea, set $\theta = \frac{1}{2}$ (Crank-Nicolson scheme) in (2.59) and approximate

$$\left(1 - \frac{1}{2}\Delta_t\left(\widehat{\mathcal{L}}_1 + \widehat{\mathcal{L}}_2\right)\right) \approx \left(1 - \frac{1}{2}\Delta_t\widehat{\mathcal{L}}_1\right)\left(1 - \frac{1}{2}\Delta_t\widehat{\mathcal{L}}_2\right), \quad (2.60)$$

$$\left(1 + \frac{1}{2}\Delta_t\left(\widehat{\mathcal{L}}_1 + \widehat{\mathcal{L}}_2\right)\right) \approx \left(1 + \frac{1}{2}\Delta_t\widehat{\mathcal{L}}_1\right)\left(1 + \frac{1}{2}\Delta_t\widehat{\mathcal{L}}_2\right). \quad (2.61)$$

It is easy to see[21] (and to verify, by a Taylor expansion) that the operators on the right-hand sides of these approximations have the same order truncation error as do the left-hand sides, namely $O(\Delta_t(\Delta_1^2 + \Delta_2^2 + \Delta_t^2))$. To the order of our original scheme, no accuracy is gained or lost in using the right-hand sides of (2.60)–(2.61). What is gained, however, is a considerable improvement in computational efficiency, originating in the fact that the resulting scheme

$$\left(1 - \frac{1}{2}\Delta_t\widehat{\mathcal{L}}_1\right)\left(1 - \frac{1}{2}\Delta_t\widehat{\mathcal{L}}_2\right)\widehat{V}_{j_1,j_2}(t_i)$$
$$= \left(1 + \frac{1}{2}\Delta_t\widehat{\mathcal{L}}_1\right)\left(1 + \frac{1}{2}\Delta_t\widehat{\mathcal{L}}_2\right)\widehat{V}_{j_1,j_2}(t_{i+1}) \quad (2.62)$$

can be *split* into the system

$$\left(1 - \frac{1}{2}\Delta_t\widehat{\mathcal{L}}_1\right)U_{j_1,j_2} = \left(1 + \frac{1}{2}\Delta_t\widehat{\mathcal{L}}_2\right)\widehat{V}_{j_1,j_2}(t_{i+1}), \quad (2.63)$$

$$\left(1 - \frac{1}{2}\Delta_t\widehat{\mathcal{L}}_2\right)\widehat{V}_{j_1,j_2}(t_i) = \left(1 + \frac{1}{2}\Delta_t\widehat{\mathcal{L}}_1\right)U_{j_1,j_2}, \quad (2.64)$$

---

[20] Recall that the solution of a general linear system with $m_1 m_2$ unknowns is an $O(m_1^2 m_2^2)$ operation. For, say, m_1 and m_2 in the order of 100, this would involve around 1,000,000 times more work than what is required for a one-dimensional (tri-diagonal) scheme ($O(m)$).

[21] To those versed in operator notation, we notice that the right- and left-hand sides both approximate, to identical order, $\exp(\pm 0.5\Delta_t(\widehat{\mathcal{L}}_1 + \widehat{\mathcal{L}}_2))$.

2.10 Two-Dimensional PDE with No Mixed Derivatives

where we have introduced an *intermediate value* U_{j_1,j_2}. The advantage of this decomposition is the fact that in each of (2.63) and (2.64), there is only one operator on the left-hand side, leading to simple tri-diagonal equation systems. To formalize this, first define

$$\mathbf{U}_1^{j_2} = (U_{1,j_2}, U_{2,j_2}, \ldots, U_{m_1,j_2})^\top.$$

Then, for a fixed value of j_2 we can write for the first step

$$\left(\mathbf{I} - \frac{1}{2}\Delta_t \mathbf{A}_1^{j_2}\left(\frac{t_{i+1}+t_i}{2}\right)\right)\mathbf{U}_1^{j_2} = \mathbf{M}_2^{j_2}\left(\frac{t_{i+1}+t_i}{2}\right), \quad (2.65)$$

where $\mathbf{A}_1^{j_2}$ is an $(m_1 \times m_1)$-dimensional tri-diagonal matrix of the same form as (2.11) (to get $\mathbf{A}_1^{j_2}$, basically freeze $x_2 = x_2^{j_2}$ and substitute μ_1 and γ_1 for μ and σ in the definition of the one-dimensional matrix $\mathbf{A}$). The m_1-dimensional vector $\mathbf{M}_2^{j_2}$ has components $M_{2,j_1}^{j_2}$, $j_1 = 1, \ldots, m_1$, given by

$$M_{2,j_1}^{j_2}\left(\frac{t_{i+1}+t_i}{2}\right) = \left(1 + \frac{1}{2}\Delta_t \widehat{\mathcal{L}}_2\right)\widehat{V}_{j_1,j_2}(t_{i+1})$$

$$= \frac{1}{2}\varsigma_{j_1,j_2}^{-}\widehat{V}_{j_1,j_2-1}(t_{i+1}) + \frac{1}{2}\varsigma_{j_1,j_2}^{+}\widehat{V}_{j_1,j_2+1}(t_{i+1})$$

$$+ \left(1 - \frac{1}{2}\varsigma_{j_1,j_2}\right)\widehat{V}_{j_1,j_2}(t_{i+1}), \quad (2.66)$$

where we have defined

$$\varsigma_{j_1,j_2}^{\pm} \triangleq \frac{\Delta_t}{2\Delta_2^2}\left(\gamma_2\left(\frac{t_{i+1}+t_i}{2}, x_1^{j_1}, x_2^{j_2}\right)^2 \pm \Delta_2 \mu_2\left(\frac{t_{i+1}+t_i}{2}, x_1^{j_1}, x_2^{j_2}\right)\right),$$

$$\varsigma_{j_1,j_2} \triangleq \frac{\Delta_t}{\Delta_2^2}\left(\gamma_2\left(\frac{t_{i+1}+t_i}{2}, x_1^{j_1}, x_2^{j_2}\right)^2 + \frac{1}{2}\Delta_2^2 r\left(\frac{t_{i+1}+t_i}{2}, x_1^{j_1}, x_2^{j_2}\right)\right).$$

For known values of $\widehat{V}(t_{i+1})$, (2.65) defines a simple tri-diagonal equation system which can be solved for $\mathbf{U}_1^{j_2}$ in $O(m_1)$ operations. Repeating the procedure above for $j_2 = 1, \ldots, m_2$ allows us to find U_{j_1,j_2} for all $j_1 = 1, \ldots, m_1$, $j_2 = 1, \ldots, m_2$, at a total computational cost of $O(m_1 m_2)$.

Turning to the second step of (2.63)–(2.64), we first fix j_1 and define

$$\widehat{\mathbf{V}}_2^{j_1}(t) = \left(\widehat{V}_{j_1,1}(t), \widehat{V}_{j_1,2}(t), \ldots, \widehat{V}_{j_1,m_2}(t)\right)^\top.$$

In the same fashion as earlier, we can then write

$$\left(\mathbf{I} - \frac{1}{2}\Delta_t \mathbf{A}_2^{j_1}\left(\frac{t_{i+1}+t_i}{2}\right)\right)\widehat{\mathbf{V}}_2^{j_1}(t_i) = \mathbf{M}_1^{j_1}\left(\frac{t_{i+1}+t_i}{2}\right), \quad (2.67)$$

where $\mathbf{A}_2^{j_1}$ is an $(m_2 \times m_2)$-dimensional tri-diagonal matrix and where the right-hand side vector now has components

$$M_{1,j_2}^{j_1}\left(\frac{t_{i+1}+t_i}{2}\right) = \left(1+\frac{1}{2}\Delta_t\widehat{\mathcal{L}}_1\right)U_{j_1,j_2}, \quad j_2 = 1,\ldots,m_2.$$

For brevity we omit writing out the $M_{1,j_2}^{j_1}$ (which will be similar to (2.66)), but just notice that the right-hand side of (2.67) is known after the first step of the ADI algorithm (above) is complete. For a given value of j_1, we can solve the tri-diagonal system (2.67) for $\widehat{\mathbf{V}}_2^{j_1}(t_i)$ in $O(m_2)$ operations. Looping over all m_1 different values of j_1, the full matrix of time t_i values $\widehat{V}_{j_1,j_2}(t_i)$, $j_1 = 1,\ldots,m_1$, $j_2 = 1,\ldots,m_2$, can then be found at a total computational cost of $O(m_1 m_2)$.

The scheme outlined above is known as the *Peaceman-Rachford* scheme. As is the case for all ADI schemes, the scheme works by alternating the directions that are treated fully implicitly in the finite difference grid: in the first step, the x_1-direction is fully implicit and the x_2-direction is fully explicit, and in the second step the order is reversed. In effect, both spatial variables end up being discretized "semi-implicitly", i.e. similar to a Crank-Nicolson scheme, resulting in convergence order is $O(\Delta_1^2 + \Delta_2^2 + \Delta_t^2)$. We emphasize, however, that whereas a direct application of the Crank-Nicolson scheme will involve (if an efficient sparse-matrix solver is used) a computational cost of $O((m_1 m_2)^{5/4})$ per time step, the computational cost of the Peaceman-Rachford ADI scheme is only $O(m_1 m_2)$. A (tedious) von Neumann analysis reveals that the scheme is A-stable, but, like the Crank-Nicolson scheme, not strongly A-stable.

While the Peaceman-Rachford scheme is a classical example of an ADI scheme, there are many others. For instance, consider a theta-version of the *Douglas-Rachford* scheme:

$$\left(1-\theta\Delta_t\widehat{\mathcal{L}}_1\right)U_{j_1,j_2} = \left(1+(1-\theta)\Delta_t\widehat{\mathcal{L}}_1 + \Delta_t\widehat{\mathcal{L}}_2\right)\widehat{V}_{j_1,j_2}(t_{i+1}), \qquad (2.68)$$

$$\left(1-\theta\Delta_t\widehat{\mathcal{L}}_2\right)\widehat{V}_{j_1,j_2}(t_i) = U_{j_1,j_2} - \theta\Delta_t\widehat{\mathcal{L}}_2\widehat{V}_{j_1,j_2}(t_{i+1}), \qquad (2.69)$$

where we understand that in $\widehat{\mathcal{L}}_1$ and $\widehat{\mathcal{L}}_2$ the PDE coefficients are to be evaluated at time $t_i^{i+1}(\theta)$. Again, notice how the scheme consists of two steps, each involving the solution of tri-diagonal sets of equations along only one of the x_1- or x_2-directions. The computational cost thus remains at $O(m_1 m_2)$. It can be shown that the convergence order of this scheme is $O(\Delta_1^2 + \Delta_2^2 + 1_{\{\theta\neq\frac{1}{2}\}}\Delta_t + \Delta_t^2)$ and it is A-stable for $\theta \geq \frac{1}{2}$, and strongly A-stable for $\theta > \frac{1}{2}$. By elimination of U_{j_1,j_2} we note that the unsplit version of the Douglas-Rachford scheme is

$$\left(1-\theta\Delta_t\widehat{\mathcal{L}}_1\right)\left(1-\theta\Delta_t\widehat{\mathcal{L}}_2\right)\widehat{V}_{j_1,j_2}(t_i)$$
$$= \left(\left(1-\theta\Delta_t\widehat{\mathcal{L}}_1\right)\left(1-\theta\Delta_t\widehat{\mathcal{L}}_2\right) + \Delta_t\widehat{\mathcal{L}}_1 + \Delta_t\widehat{\mathcal{L}}_2\right)\widehat{V}_{j_1,j_2}(t_{i+1}).$$

It is not difficult to see that this approximates (2.59) to second order.

2.10.3 Boundary Conditions and Other Issues

The fact that ADI schemes reduce to solving sequences of matrix systems identical to the ones arising in the one-dimensional case is convenient, in the sense that many of the issues we have encountered for one-dimensional finite difference grids (oscillations, stability, convection dominance, etc.) and their remedies (smoothing, non-equidistant discretization, upwinding, etc.) carry over to the ADI setting with only minor modifications. Consider for instance the issue of applying spatial boundary conditions along the edges of the (x_1, x_2) domain, which we have so far not discussed. As for the one-dimensional PDEs, the most convenient way to express such boundary conditions is typically by imposing conditions on derivatives, like $\partial^2 V(t, x_1^0, x_2^{j_2})/\partial x_1^2 = \partial V(t, x_1^0, x_2^{j_2})/\partial x_1$ and so forth. For the Peaceman-Rachford scheme, say, such conditions can be incorporated directly into (2.65) and (2.67) by altering the matrices $\mathbf{A}_1^{j_2}$ and $\mathbf{A}_2^{j_1}$, as well as the boundary elements of $\mathbf{M}_1^{j_1}$ and $\mathbf{M}_2^{j_2}$, in the manner outlined in Section 2.2.1. If instead we wish to impose Dirichlet boundary conditions, we need to add corrective terms to the tri-diagonal systems, as in (2.19). To complete the first part of the split scheme, this then requires us to establish what boundary terms are needed for the intermediate quantity U_{j_1, j_2}, i.e. we must define $U_{j_1, 0}$ and U_{j_1, m_2+1} for $j_1 = 1, \ldots, m_1$, as well as U_{0, j_2} and U_{m_1+1, j_2} for $j_2 = 1, \ldots, m_2$. While U_{j_1, j_2} is a purely mathematic construct, sometimes it is adequate to think of U_{j_1, j_2} as a proxy for V_{j_1, j_2} evaluated at $t_i^{i+1}(\theta)$, which obviously makes determination of boundary conditions straightforward. For maximum precision, however, we should use the ADI equations themselves to express the boundary conditions of U directly in terms of boundary conditions for $V(t_i)$ and $V(t_{i+1})$. Here, the Douglas-Rachford scheme is particularly easy to deal with, as a rearrangement of (2.69) directly relates U_{j_1, j_2} to $\widehat{V}_{j_1, j_2}(t_i)$ and $\widehat{V}_{j_1, j_2}(t_{i+1})$,

$$U_{j_1, j_2} = \left(1 - \theta \Delta_t \widehat{\mathcal{L}}_2\right) \widehat{V}_{j_1, j_2}(t_i) + \theta \Delta_t \widehat{\mathcal{L}}_2 \widehat{V}_{j_1, j_2}(t_{i+1}).$$

The Peaceman-Rachford scheme requires some further manipulations to express U in terms of $V(t_i)$ and $V(t_{i+1})$; see Mitchell and Griffiths [1980] for the details.

2.11 Two-Dimensional PDE with Mixed Derivatives

Consider now the case where the 2-dimensional PDE (2.58) has a mixed partial derivative,

$$\frac{\partial V}{\partial t} + (\mathcal{L}_1 + \mathcal{L}_2 + \mathcal{L}_{1,2}) V = 0, \qquad (2.70)$$

where $\mathcal{L}_1$ and $\mathcal{L}_2$ are as in (2.58), and where

$$\mathcal{L}_{1,2} = s_{1,2}(t,x)\frac{\partial^2}{\partial x_1 \partial x_2} \triangleq \rho(t,x)\gamma_1(t,x)\gamma_2(t,x)\frac{\partial^2}{\partial x_1 \partial x_2}. \qquad (2.71)$$

The quantity $\rho(t,x)$ is the instantaneous correlation between the processes $x_1(t)$ and $x_2(t)$ in (2.57), i.e. $\rho(t,x) \in [-1,1]$.

The presence of $\mathcal{L}_{1,2}$ prevents a direct application of the ADI methods in Section 2.10.2, since the mixed operator $\mathcal{L}_{1,2}$ is not amenable to operator splitting. We shall demonstrate two ways to overcome this problem: a) orthogonalization of the PDE; and b) predictor-corrector schemes.

2.11.1 Orthogonalization of the PDE

The idea here is to introduce new variables $y_1(t, x_1, x_2)$ and $y_2(t, x_1, x_2)$ such that the PDE loses its mixed derivative term when stated in terms of these variables. To demonstrate this idea, assume first that $\rho(t,x)$, $\gamma_1(t,x)$, and $\gamma_2(t,x)$ are all functions of time only and independent of x. Then define, say,

$$y_1(t, x_1, x_2) = x_1, \qquad (2.72)$$

$$y_2(t, x_1, x_2) = -\rho(t)\frac{\gamma_2(t)}{\gamma_1(t)} x_1 + x_2 \triangleq a(t)x_1 + x_2, \qquad (2.73)$$

where we must assume that $\gamma_1(t) \neq 0$ for all t.

Lemma 2.11.1. *Consider the PDE (2.70) subject to the terminal value condition $V(T,x) = g(x)$. Define $y = (y_1, y_2)^\top$ and $v(t,y) = V(t,x)$. With the variable change defined in (2.72)–(2.73), v satisfies*

$$\frac{\partial v}{\partial t} + \mu_1^y(t,y)\frac{\partial v}{\partial y_1} + \mu_2^y(t,y)\frac{\partial v}{\partial y_2} + \frac{1}{2}\gamma_1(t)^2 \frac{\partial^2 v}{\partial y_1^2}$$
$$+ \frac{1}{2}\left(1 - \rho(t)^2\right)\gamma_2(t)^2 \frac{\partial^2 v}{\partial y_2^2} - r(t, y_1, y_2 - a(t)y_1)v = 0, \qquad (2.74)$$

where

$$\mu_1^y(t,y) \triangleq \mu_1(t, x_1, x_2) = \mu_1(t, y_1, y_2 - a(t)y_1), \qquad (2.75)$$

$$\mu_2^y(t,y) \triangleq \frac{da(t)}{dt}x_1 + a(t)\mu_1(t, x_1, x_2) + \mu_2(t, x_1, x_2)$$
$$= \frac{da(t)}{dt}y_1 + a(t)\mu_1^y(t,y) + \mu_2(t, y_1, y_2 - a(t)y_1). \qquad (2.76)$$

The equation (2.74) is subject to the terminal value condition $v(T, y_1, y_2) = g(x_1, x_2) = g(y_1, y_2 - a(T)y_1)$.

Proof. While the result can be established by the usual mechanics of ordinary calculus, we will take the opportunity to show how stochastic calculus can also conveniently prove results of this type. Going back to the processes underlying the PDE (see (2.57)), we write

$$dx_1(t) = \mu_1(t,x)\,dt + \gamma_1(t)\,dW_1(t), \tag{2.77}$$

$$dx_2(t) = \mu_2(t,x)\,dt + \gamma_2(t)\left(\rho(t)\,dW_1(t) + \sqrt{1-\rho(t)^2}\,dW_2(t)\right), \tag{2.78}$$

for independent scalar Brownian motions $W_1(t)$ and $W_2(t)$; this is easily seen to generate the correct correlation $\rho(t)$ between x_1 and x_2. An application of Ito's lemma then shows that the processes for y_1 and y_2 are

$$dy_1(t) = dx_1(t) = \mu_1(t,x)\,dt + \gamma_1(t)\,dW_1(t),$$

$$dy_2(t) = \frac{da(t)}{dt}x_1(t)\,dt + a(t)\mu_1(t,x)\,dt + a(t)\gamma_1(t)\,dW_1(t)$$
$$+ \mu_2(t,x)\,dt + \gamma_2(t)\left(\rho(t)\,dW_1(t) + \sqrt{1-\rho(t)^2}\,dW_2(t)\right)$$
$$= \left(\frac{da(t)}{dt}x_1(t) + a(t)\mu_1(t,x) + \mu_2(t,x)\right)dt$$
$$+ \gamma_2(t)\sqrt{1-\rho(t)^2}\,dW_2(t).$$

With the definitions (2.75)–(2.76), this becomes simply

$$dy_1(t) = \mu_1^y(t,y(t))\,dt + \gamma_1(t)\,dW_1(t), \tag{2.79}$$

$$dy_2(t) = \mu_2^y(t,y(t))\,dt + \gamma_2(t)\sqrt{1-\rho(t)^2}\,dW_2(t). \tag{2.80}$$

Equations (2.79)–(2.80) define a Markov SDE in $y_1(t)$ and $y_2(t)$ where, importantly, the Brownian motions on $y_1(t)$ and $y_2(t)$ are now *independent*. Writing $V(t,x) = v(t,y)$, it then follows immediately from the backward Kolmogorov equation (see Section 1.8) that v satisfies the PDE (2.74). □

Through the chosen transformation (2.72)–(2.73), our original PDE has now been put into a form where we can immediately apply the ADI schemes outlined in Section 2.10.2.

In performing the orthogonalization of the PDE in Lemma 2.11.1 we relied on $\rho(t,x)$, $\gamma_1(t,x)$, and $\gamma_2(t,x)$ all being independent of x. This can often be relaxed. Consider for instance the case where $\rho(t,x) = \rho(t)$, $\gamma_1(t,x) = \gamma_1(t,x_1)$, and $\gamma_2(t,x) = \gamma_2(t,x_2)$; here the correlation ρ is still assumed deterministic, but we now allow for some (though not full) x-dependence in γ_1 and γ_2. Assuming that $\gamma_1(t,x_1) > 0$ and $\gamma_2(t,x_2) > 0$ we can introduce new variables

$$z_1(t,x_1) = \int \frac{1}{\gamma_1(t,x_1)}dx_1, \tag{2.81}$$

$$z_2(t,x_2) = \int \frac{1}{\gamma_2(t,x_2)}dx_2. \tag{2.82}$$

Applying Ito's lemma to (2.77)–(2.78) we see that

$$dz_1(t,x_1) = \left(-\int \frac{\partial \gamma_1(t,x_1)}{\partial t} \frac{1}{\gamma_1(t,x_1)^2} dx_1 + \frac{\mu_1(t,x)}{\gamma_1(t,x_1)}\right.$$
$$\left. - \frac{1}{2}\frac{\partial \gamma_1(t,x_1)}{\partial x_1}\right) dt + dW_1(t) \quad (2.83)$$

and

$$dz_2(t,x_2) = \left(-\int \frac{\partial \gamma_2(t,x_2)}{\partial t} \frac{1}{\gamma_2(t,x_2)^2} dx_2 + \frac{\mu_2(t,x)}{\gamma_2(t,x_2)}\right.$$
$$\left.-\frac{1}{2}\frac{\partial \gamma_2(t,x_2)}{\partial x_2}\right) dt + \rho(t)\, dW_1(t) + \sqrt{1-\rho(t)^2}\, dW_2(t). \quad (2.84)$$

As we assumed that $\gamma_1(t,x_1) > 0$ and $\gamma_2(t,x_2) > 0$, the functions z_1 and z_2 are increasing in x_1 and x_2, respectively, and are thereby invertible. As such, we can rewrite (2.83)–(2.84) in the more appealing form

$$dz_1(t,x_1) = \mu_1^z(t,z_1,z_2)\, dt + dW_1(t),$$
$$dz_2(t,x_1) = \mu_2^z(t,z_1,z_2)\, dt + \rho(t)\, dW_1(t) + \sqrt{1-\rho(t)^2}\, dW_2(t).$$

Through the transformation (2.81)–(2.82), we have reduced our original system to one where the coefficients on $W_1(t)$ and $W_2(t)$ are no longer state-dependent, similar to the case that lead to Lemma 2.11.1. We can now proceed with another variable transformation, as in (2.72)–(2.73), to orthogonalize the system and prepare it for an application of the ADI method.

While the orthogonalization method outlined here can be very effective on a range of practical problems, it suffers from a few drawbacks. Most obviously, the method is not completely general and requires a certain structure on the parameters of the PDE. Another drawback is that the introduction of a time-dependent transformation on one or more variables (Lemma 2.11.1) often makes the alignment of the finite difference grid along (time-independent) critical level points in x-space impossible. Also, the introduction of terms like $y_1 da(t)/dt$ in the drift of y_2 (see (2.76)) can be problematic, particularly if the functions $\gamma_1(t)$ and $\gamma_2(t)$ are not smooth. For instance, it is not unlikely that $y_1 da(t)/dt$ will locally be of such magnitude that upwinding will be necessary to prevent oscillations; see Section 2.6.1. Further, we note that inversion of the transformations (2.81)–(2.82) will not always be possible to perform analytically and may require numerical (root-search) work, complicating the scheme and potentially slowing it down. Finally, as we shall highlight in future chapters, maintaining the "continuity" of a numerical scheme with respect to input parameters is of critical importance for the smoothness of risk sensitivities. Such continuity is difficult to ensure if complicated transformations are applied to model variables. So, in the end, we recommend formulating the PDEs in terms of financially meaningful variables, avoiding excessive transformations, and relying on methods such as developed in the next section when dealing with mixed derivatives and other numerical complications.

2.11.2 Predictor-Corrector Scheme

In this section we shall consider a completely general method for handling mixed derivatives in two-dimensional PDEs. While a bit slower than the method outlined in Section 2.11.1, it does not involve any variable transformations and, by extension, does not suffer from the drawbacks associated with such transformations. As a first step, consider the discretization of the mixed derivative $\partial^2 V/\partial x_1 \partial x_2$. There are a few possibilities (see Mitchell and Griffiths [1980]), but we shall just use

$$\delta_{x_1 x_2} V_{j_1,j_2}(t) = \delta_{x_1}\delta_{x_2} V_{j_1,j_2}(t)$$
$$= \frac{V_{j_1+1,j_2+1}(t) - V_{j_1+1,j_2-1}(t) - V_{j_1-1,j_2+1}(t) + V_{j_1-1,j_2-1}(t)}{4\Delta_1 \Delta_2}. \quad (2.85)$$

Extensions to non-equidistant grids follow directly from (2.27) and the relation $\delta_{x_1 x_2} V_{j_1,j_2}(t) = \delta_{x_1}\delta_{x_2} V_{j_1,j_2}(t)$. As we have not encountered mixed difference operators before, for completeness we show the following lemma.

Lemma 2.11.2. *For the discrete operator (2.85) we have*

$$\delta_{x_1 x_2} V_{j_1,j_2}(t) = \frac{\partial^2 V(t, x_1^{j_1}, x_2^{j_2})}{\partial x_1 \partial x_2} + O\left(\Delta_1^2 + \Delta_2^2\right).$$

Proof. A Taylor expansion of $V(t,x)$ around the point $x = (x_1^{j_1}, x_2^{j_2})^\top$ gives

$$V_{j_1+1,j_2\pm 1}(t) = V_{j_1,j_2}(t) + \Delta_1 \frac{\partial V}{\partial x_1} \pm \Delta_2 \frac{\partial V}{\partial x_2} + \frac{1}{2}\Delta_1^2 \frac{\partial^2 V}{\partial x_1^2} + \frac{1}{2}\Delta_2^2 \frac{\partial^2 V}{\partial x_2^2}$$
$$\pm \Delta_1 \Delta_2 \frac{\partial^2 V}{\partial x_1 \partial x_2} + \frac{1}{6}\Delta_1^3 \frac{\partial^3 V}{\partial x_1^3} \pm \frac{1}{6}\Delta_2^3 \frac{\partial^3 V}{\partial x_2^3}$$
$$+ \frac{1}{2}\Delta_1 \Delta_2^2 \frac{\partial^3 V}{\partial x_1 \partial x_2^2} \pm \frac{1}{2}\Delta_1^2 \Delta_2 \frac{\partial^3 V}{\partial x_1^2 \partial x_2} + \cdots,$$

$$V_{j_1-1,j_2\pm 1}(t) = V_{j_1,j_2}(t) - \Delta_1 \frac{\partial V}{\partial x_1} \pm \Delta_2 \frac{\partial V}{\partial x_2} + \frac{1}{2}\Delta_1^2 \frac{\partial^2 V}{\partial x_1^2} + \frac{1}{2}\Delta_2^2 \frac{\partial^2 V}{\partial x_2^2}$$
$$\mp \Delta_1 \Delta_2 \frac{\partial^2 V}{\partial x_1 \partial x_2} - \frac{1}{6}\Delta_1^3 \frac{\partial^3 V}{\partial x_1^3} \pm \frac{1}{6}\Delta_2^3 \frac{\partial^3 V}{\partial x_2^3}$$
$$- \frac{1}{2}\Delta_1 \Delta_2^2 \frac{\partial^3 V}{\partial x_1 \partial x_2^2} \pm \frac{1}{2}\Delta_1^2 \Delta_2 \frac{\partial^3 V}{\partial x_1^2 \partial x_2} + \cdots.$$

A little thought then shows that

$$V_{j_1+1,j_2+1}(t) - V_{j_1+1,j_2-1}(t) - V_{j_1-1,j_2+1}(t) + V_{j_1-1,j_2-1}(t)$$
$$= 4\Delta_1 \Delta_2 \frac{\partial^2 V}{\partial x_1 \partial x_2} + O\left(\Delta_1^3 \Delta_2 + \Delta_1 \Delta_2^3\right),$$

as error terms of order Δ_1^4, Δ_2^4, and $\Delta_1^2\Delta_2^2$ will cancel. The result follows. □

Equipped with (2.85), we can approximate the operator $\mathcal{L}_{1,2}$ in (2.71) as

$$\widehat{\mathcal{L}}_{1,2}V_{j_1,j_2}(t) \triangleq \rho\left(t, x_1^{j_1}, x_2^{j_2}\right) \gamma_1\left(t, x_1^{j_1}, x_2^{j_2}\right) \gamma_2\left(t, x_1^{j_1}, x_2^{j_2}\right) \delta_{x_1 x_2} V_{j_1,j_2}(t),$$

which is accurate to order $O(\Delta_1^2 + \Delta_2^2)$. The first easy way to modify our ADI scheme to incorporate $\widehat{\mathcal{L}}_{1,2}$ is to treat the mixed derivative fully explicitly. In the Douglas-Rachford scheme (2.68)–(2.69), for instance, we thus modify the right-hand side of the first step as follows:

$$\left(1 - \theta\Delta_t\widehat{\mathcal{L}}_1\right) U_{j_1,j_2} = \left(1 + (1-\theta)\Delta_t\widehat{\mathcal{L}}_1 + \Delta_t\widehat{\mathcal{L}}_2 + \Delta_t\widehat{\mathcal{L}}_{1,2}\right) \widehat{V}_{j_1,j_2}(t_{i+1}), \quad (2.86)$$

$$\left(1 - \theta\Delta_t\widehat{\mathcal{L}}_2\right) \widehat{V}_{j_1,j_2}(t_i) = U_{j_1,j_2} - \theta\Delta_t\widehat{\mathcal{L}}_2\widehat{V}_{j_1,j_2}(t_{i+1}). \quad (2.87)$$

The addition of $\widehat{\mathcal{L}}_{1,2}$ this way clearly preserves the ADI structure of the scheme which will continue to involve only sequences of tri-diagonal linear equations. However, having, in effect, only a one-sided time-differencing of the mixed derivative term will lower the convergence order of the time step to $O(\Delta_t)$, irrespective of the choice of θ.

To change the time at which the mixed operator $\widehat{\mathcal{L}}_{1,2}$ is evaluated, consider using a *predictor-corrector* scheme, where the results of (2.86)–(2.87) are re-used in a one-time[22] iteration. Specifically, we write, for some $\lambda \in [0,1]$,

Predictor:

$$\left(1 - \theta\Delta_t\widehat{\mathcal{L}}_1\right) U^{(1)}_{j_1,j_2} = \left(1 + (1-\theta)\Delta_t\widehat{\mathcal{L}}_1 + \Delta_t\widehat{\mathcal{L}}_2 + \Delta_t\widehat{\mathcal{L}}_{1,2}\right) \widehat{V}_{j_1,j_2}(t_{i+1}), \quad (2.88)$$

$$\left(1 - \theta\Delta_t\widehat{\mathcal{L}}_2\right) U^{(2)}_{j_1,j_2} = U^{(1)}_{j_1,j_2} - \theta\Delta_t\widehat{\mathcal{L}}_2\widehat{V}_{j_1,j_2}(t_{i+1}). \quad (2.89)$$

Corrector:

$$\left(1 - \theta\Delta_t\widehat{\mathcal{L}}_1\right) Z^{(1)}_{j_1,j_2} = \left(1 + (1-\theta)\Delta_t\widehat{\mathcal{L}}_1 + \Delta_t\widehat{\mathcal{L}}_2 \right.$$
$$\left. + (1-\lambda)\Delta_t\widehat{\mathcal{L}}_{1,2}\right) \widehat{V}_{j_1,j_2}(t_{i+1}) + \lambda\Delta_t\widehat{\mathcal{L}}_{1,2}U^{(2)}_{j_1,j_2}, \quad (2.90)$$

$$\left(1 - \theta\Delta_t\widehat{\mathcal{L}}_2\right) \widehat{V}_{j_1,j_2}(t_i) = Z^{(1)}_{j_1,j_2} - \theta\Delta_t\widehat{\mathcal{L}}_2\widehat{V}_{j_1,j_2}(t_{i+1}). \quad (2.91)$$

---

[22] We can run the iteration more than once if desired, but a single iteration will normally suffice.

Notice how the Douglas-Rachford scheme is first run once, in (2.88)–(2.89), to yield a first guess (a "predictor"), $U^{(2)}_{j_1,j_2}$, for the time t_i value $V_{j_1,j_2}(t_i)$. In a second run of the Douglas-Rachford scheme, in (2.90)–(2.91), this guess is used as a "corrector" to affect the time at which $\widehat{\mathcal{L}}_{1,2}$ is evaluated, by applying this operator to $(1-\lambda)\widehat{V}_{j_1,j_2}(t_{i+1}) + \lambda U^{(2)}_{j_1,j_2}$; when $\lambda = \frac{1}{2}$ we effectively center the time-differencing of the mixed term. The scheme now relies on three intermediate variables, $U^{(1)}_{j_1,j_2}$, $U^{(2)}_{j_1,j_2}$, and $Z^{(1)}_{j_1,j_2}$.

The combined predictor-corrector scheme above (in a slightly less general form, with $\Delta_1 = \Delta_2$) was suggested by Craig and Sneyd [1988]. It can be shown that the scheme has convergence order

$$O\left((\Delta_1 + \Delta_2)^2 + 1_{\{\theta \neq \frac{1}{2}\}}\Delta_t + 1_{\{\lambda \neq \frac{1}{2}\}}\Delta_t + \Delta_t^2\right),$$

so second order convergence in the time domain is still achievable by setting $\theta = \lambda = \frac{1}{2}$. The scheme will be A-stable for $\theta \geq \frac{1}{2}$ and $\frac{1}{2} \leq \lambda \leq \theta$. The computational cost of the predictor-corrector is clearly still $O(m_1 m_2)$ per time step, as both the predictor and corrector schemes have $O(m_1 m_2)$ cost per time-step. Even though the standard Douglas-Rachford scheme is effectively run twice, we should point out that when intelligently implemented, (2.88)–(2.91) is typically only about 30-40% slower than the Douglas-Rachford scheme, as a number of results from the predictor step can be cached and reused in the corrector step.

As for the standard ADI grids, extensions to non-equidistant grids are straightforward using the techniques in Section 2.4. Boundary conditions in the x-domain are imposed along the lines outlined in Section 2.10.3.

2.12 PDEs of Arbitrary Order

We now turn our attention back to the general p-dimensional PDE (2.56). To prepare for a numerical scheme, let us rewrite the PDE as follows:

$$\frac{\partial V}{\partial t} + \sum_{h=1}^{p} \mathcal{L}_h V + \sum_{h=1}^{p} \sum_{l=h+1}^{p} \mathcal{L}_{h,l} V = 0, \qquad (2.92)$$

where

$$\mathcal{L}_h = \mu_h(t,x)\frac{\partial}{\partial x_h} + \frac{1}{2}s_{h,h}(t,x)\frac{\partial^2}{\partial x_h^2} - p^{-1}r(t,x),$$

$$\mathcal{L}_{h,l} = s_{h,l}(t,x)\frac{\partial^2}{\partial x_h \partial x_l}.$$

The method we present here for solution of (2.92) is a p-dimensional version of the predictor-corrector scheme outlined above. The extension is straightforward and we simply list it here without further discussion; see Craig and Sneyd [1988] for additional background. To simplify notation, we have omitted sub-indices everywhere (i.e., $\widehat{V}(t_i)$ is used instead of $\widehat{V}_{j_1,j_2,\ldots,j_p}(t_i)$).

Predictor:

$$\left(1 - \theta \Delta_t \widehat{\mathcal{L}}_1\right) U^{(1)}$$
$$= \Delta_t \left(\Delta_t^{-1} + (1-\theta)\widehat{\mathcal{L}}_1 + \sum_{h=2}^{p} \widehat{\mathcal{L}}_h + \sum_{h=1}^{p} \sum_{l=h+1}^{p} \widehat{\mathcal{L}}_{h,l} \right) \widehat{V}(t_{i+1}),$$

$$\left(1 - \theta \Delta_t \widehat{\mathcal{L}}_2\right) U^{(2)} = U^{(1)} - \theta \Delta_t \widehat{\mathcal{L}}_2 \widehat{V}(t_{i+1}),$$

$$\vdots$$

$$\left(1 - \theta \Delta_t \widehat{\mathcal{L}}_p\right) U^{(p)} = U^{(p-1)} - \theta \Delta_t \widehat{\mathcal{L}}_p \widehat{V}(t_{i+1}).$$

Corrector:

$$\left(1 - \theta \Delta_t \widehat{\mathcal{L}}_1\right) Z^{(1)}$$
$$= \Delta_t \left(\Delta_t^{-1} + (1-\theta)\widehat{\mathcal{L}}_1 + \sum_{h=2}^{p} \widehat{\mathcal{L}}_h \right.$$
$$\left. + (1-\lambda) \sum_{h=1}^{p} \sum_{l=h+1}^{p} \widehat{\mathcal{L}}_{h,l} \right) \widehat{V}(t_{i+1}) + \lambda \Delta_t \sum_{h=1}^{p} \sum_{l=h+1}^{p} \widehat{\mathcal{L}}_{h,l} U^{(p)},$$

$$\left(1 - \theta \Delta_t \widehat{\mathcal{L}}_2\right) Z^{(2)} = Z^{(1)} - \theta \Delta_t \widehat{\mathcal{L}}_2 \widehat{V}(t_{i+1}),$$

$$\vdots$$

$$\left(1 - \theta \Delta_t \widehat{\mathcal{L}}_p\right) \widehat{V}(t_i) = Z^{(p-1)} - \theta \Delta_t \widehat{\mathcal{L}}_p \widehat{V}(t_{i+1}).$$

With m_h points in the x_h-direction, $h = 1, \ldots, p$, the computational cost of the predictor-corrector scheme is $O(\prod_{h=1}^{p} m_h)$. For $p \leq 3$, sufficient conditions for A-stability are $\theta \geq \frac{1}{2}$ and $\frac{1}{2} \leq \lambda \leq \theta$. For $p \geq 4$, sufficient conditions are $\theta \geq \frac{1}{2}$ and

$$\frac{1}{2} \leq \lambda \leq \frac{p^{p-1}}{(p-1)^p} \theta.$$

See Craig and Sneyd [1988] for a proof. Convergence is similar to the two-dimensional case.

2.12 PDEs of Arbitrary Order

As a final comment, let us note that as dimensionality increases, the computational complexity of an iterative sparse solver will start approaching that of ADI. Specifically, for a p-dimensional problem, the complexity of the former is $O(m_{\text{total}})$ and for the latter $O(m_{\text{total}}^{(2p+1)/2p})$, with $m_{\text{total}} = m_1 \cdot m_2 \cdot \ldots \cdot m_p$.

3
Monte Carlo Methods

While the finite difference method is flexible and powerful, it has a number of limitations. First, its usage is restricted to problems where the state variable dynamics are Markovian. Second, for strongly path-dependent problems, the method often does not apply. And third, it is unsuited for problems where the dimension of the underlying vector of state variables is high. To expand on the last point, recall from Section 2.9 that the (ADI) finite difference method applied to a p-dimensional problem has computational complexity $O(m^p)$ per time step, where m is the average number of spatial points per dimension. The exponential growth in p — the "curse of dimensionality" — is typical of grid-based methods and prevents the practical usage of the method for p larger than about 4 or 5.

In this chapter, we study the *Monte Carlo method*, a numerical technique where the computational effort grows only linearly in the problem dimension p. While convergence of the Monte Carlo method is relatively slow, it is nearly always the method of choice for high-dimensional pricing problems. Compared to finite difference methods, Monte Carlo methods are easy to apply to problems with non-Markovian dynamics as well as strong path-dependency in the payout. On the other hand, as Monte Carlo methods inherently run forward in time, dynamic programming techniques are challenging to implement, making Monte Carlo pricing of American and Bermudan options significantly more involved than for the naturally backward-working finite difference method.

3.1 Fundamentals

Consider a European-style derivative V with time T payout $V(T) = g(T)$, where $g(T)$ is an $\mathcal{F}_T$-measurable (and integrable) random variable. Where finite difference methods start with a PDE representation of the price of a contingent claim at times $t < T$, the starting point for the Monte Carlo method is the basic martingale relation (see (1.15))

3 Monte Carlo Methods

$$V(t) = N(t) \mathrm{E}_t^{\mathrm{Q}^N}\left(g(T)/N(T)\right), \qquad (3.1)$$

where $N(\cdot)$ is a numeraire and Q^N is the measure induced by $N(\cdot)$. To evaluate this expression numerically, we need a numerical technique to compute expectations of a random variable. For this, we turn to the law of large numbers:

Theorem 3.1.1 (Strong Law of Large Numbers). *Let $Y_1, Y_2, \ldots$ be a sequence of independent identically distributed (i.i.d.) random variables with expectation $\mu < \infty$. Define the sample mean*

$$\overline{Y}_n = \frac{1}{n} \sum_{i=1}^n Y_i. \qquad (3.2)$$

Then

$$\lim_{n \to \infty} \overline{Y}_n = \mu, \quad a.s.$$

This result forms the basis for the *Monte Carlo method*, which computes the expectation in (3.1) by simply i) generating independent realizations of $g(T)/N(T)$ under Q^N; and ii) forming their average. Specifically, let $g_1/N_1, \ldots, g_n/N_n$ denote n independent samples from the distribution of $g(T)/N(T)$, conditional on $\mathcal{F}_t$. Then our Monte Carlo estimator for $V(t)$ is the sample mean

$$\overline{V}(t) = N(t) \frac{1}{n} \sum_{i=1}^n g_i/N_i. \qquad (3.3)$$

We shall delve into how to generate samples from the distribution of $g(T)/N(T)$ shortly, but before doing so let us consider the expected convergence rate of the Monte Carlo method as n is increased. The key result is here the central limit theorem:

Theorem 3.1.2 (Central Limit Theorem). *Let $Y_1, Y_2, \ldots$ be a sequence of i.i.d. random variables with expectation μ and standard deviation $\sigma < \infty$. Let the sample mean be defined as in (3.2). Then, for $n \to \infty$,*

$$\frac{\overline{Y}_n - \mu}{\sigma/\sqrt{n}} \xrightarrow{d} \mathcal{N}(0,1),$$

where $\mathcal{N}(0,1)$ is a standard Gaussian distribution and $\xrightarrow{d}$ denotes convergence in distribution[1]. *Further, if we define*

$$s_n = \sqrt{\frac{1}{n-1} \sum_{i=1}^n \left(Y_i - \overline{Y}_n\right)^2},$$

---

[1] Recall that a sequence of variables X_n with cumulative distribution functions F_n converge in distribution to a random variable X with distribution F if $\lim_{n \to \infty} F_n(x) = F(x)$ for all $x \in \mathbb{R}$ at which $F(x)$ is continuous.

then also
$$\frac{\overline{Y}_n - \mu}{s_n/\sqrt{n}} \xrightarrow{d} \mathcal{N}(0,1).$$

Define the Gaussian percentile u_γ as $\Phi(u_\gamma) = 1 - \gamma$, where Φ is the Gaussian cumulative distribution function. From Theorem 3.1.2, and from the definition of convergence in distribution (see footnote 1), the probability that the confidence interval

$$\left[\overline{V}(t) - u_{\gamma/2} \cdot s_n/\sqrt{n}, \overline{V}(t) + u_{\gamma/2} \cdot s_n/\sqrt{n} \right] \quad (3.4)$$

fails to include the true value $V(t)$ approaches γ for large n. Here

$$s_n \triangleq \sqrt{\frac{1}{n-1} \sum_{i=1}^{n} \left(\frac{g_i N(t)}{N_i} - \overline{V}(t) \right)^2 },$$

with the quantity $s_n/\sqrt{n}$ known as the *standard error*. For given γ, the rate at which the confidence interval for $V(t)$ contracts is $O(n^{-\frac{1}{2}})$. This is relatively slow: to reduce the width of the interval by a factor of 2, n must increase by a factor of 4. On the other hand, we notice that the (asymptotic) convergence rate only depends on n, not on the specifics of the g_i's. In particular, if $g(T) = g(X(T))$ where X is p-dimensional, the asymptotic convergence rate is independent of p. As we shall see shortly, in most applications the work required to generate samples of $g(X(T))$ is (at most) linear in p.

3.1.1 Generation of Random Samples

At the most basic level, the Monte Carlo method requires the ability to draw independent realizations of a scalar random variable Z with a specified cumulative distribution function $F(z) = P(Z \leq z)$, where P is a probability measure. On a computer, the starting point for this exercise is normally a *pseudo-random number generator*, a software program that will generate a sequence of numbers uniformly distributed on $[0,1]$ (i.e. from $\mathcal{U}(0,1)$). Press et al. [1992] list a number of generators producing sequences of uniform numbers $u_1, u_2, \ldots$ from iterative relationships of the form

$$I_{i+1} = (aI_i + c) \bmod(m),$$
$$u_{i+1} = I_{i+1}/m.$$

The externally specified starting point I_0 is the *seed* of the random number generator. In this so-called *general linear congruential generator*, the choice of the *multiplier* a, the *modulus* m, and the *increment* c must be done with great care to ensure that the period length of the generator is large[2] and that the resulting algorithm is efficient on a computer. The latter, for instance, can be accomplished by setting m to be a power of 2 such that the modulo operation can be done by bit-shifting. For detailed discussion and a number of concrete algorithms (including computer code), we refer to Press et al. [1992]. The algorithms in Press et al. [1992] should suffice for most fixed income applications, but we should note the existence of more sophisticated methods that (theoretically, at least) have better performance than linear congruential generators. For instance, the so-called *Mersenne twister* proposed in Matsumoto and Nishimura [1998] has become popular, especially the specific variant MT19937 which has a period of $2^{19937} - 1$. For an extensive survey of pseudo-random number generators, see L'Ecuyer [1994].

So far we have only discussed techniques to generate $\mathcal{U}(0, 1)$ numbers, but many methods exist to convert uniformly distributed numbers into draws from the distribution F of Z. We cover a few important techniques next.

3.1.1.1 Inverse Transform Method

The idea of the inverse transform method is straightforward. Let U be a random variable uniformly distributed on $[0, 1]$, and consider setting

$$Z = F^{-1}(U), \qquad (3.5)$$

where we assume that F^{-1} is well-defined, for all but a finite number of points[3]. As desired,

$$\mathrm{P}(Z \leq z) = \mathrm{P}\left(F^{-1}(U) \leq z\right) = \mathrm{P}(U \leq F(z)) = F(z),$$

where the last equality follows from the property of uniformly distributed random variables. The inverse transform method (3.5) is quite general, but its practical usefulness hinges on being able to compute F^{-1} fast. Many distributions allow for closed-form inversion; this includes the *exponential distribution* where $F(z) = 1 - e^{-z\lambda}$ for some positive constant λ, and the *Cauchy distribution* where $F(z) = 1/2 + (1/\pi)\arctan((z-t)/s)$ for constants t and $s > 0$.

For the important case of the Gaussian distribution, no closed-form expression for the inverse distribution exists. Nevertheless, the inverse transform

---

[2] Note that if a number $I_k = I_i$, the sequences starting from I_k and I_i are identical. In practice, we would want the generator to have *full period*, in the sense that the sequence would produce $m - 1$ distinct values before repeating the sequence.

[3] For discrete random variables, the distribution function is discontinuous around each of the possible (discrete) outcomes of Z. We can handle this by simply defining $F^{-1}(u) = \inf\{z : F(z) \geq u\}$.

method can still be applied as fast and extremely accurate approximations for Φ^{-1} exist. For instance, Beasley and Springer [1977] suggest the rational approximation

$$\Phi^{-1}(x) \approx \frac{\sum_{i=0}^{3} a_i \left(x - \frac{1}{2}\right)^{2i+1}}{1 + \sum_{i=0}^{3} b_i \left(x - \frac{1}{2}\right)^{2i}}, \quad 0.5 \leq x \leq 0.92, \tag{3.6}$$

for constants a_i, b_i, $i = 0, \ldots, 3$, listed in Appendix 3.A. For values of x greater than 0.92, Moro [1995] proposes the approximation

$$\Phi^{-1}(x) \approx \sum_{i=1}^{8} c_i \left[\ln\left(-\ln(1-x)\right)\right]^i, \quad 0.92 \leq x < 1, \tag{3.7}$$

for constants c_i, $i = 0, \ldots, 8$, given in Appendix 3.A. Taken together, (3.6) and (3.7) provide an approximation valid for $0.5 \leq x < 1$; when $0 < x < 0.5$ we can compute $\Phi^{-1}(x)$ by symmetry: $\Phi^{-1}(1-x) = -\Phi^{-1}(x)$. The precision of (3.6)–(3.7) is excellent[4], with the error less than 3×10^{-9} for x in the range $x \in [\Phi(-7), \Phi(7)]$. For alternative algorithms, see for instance Acklam [2003] and Wichura [1988].

Well-known alternative methods for sampling in the Gaussian distribution include the *Box-Muller method* and the related *Marsaglia polar method* (see Press et al. [1992]).

3.1.1.2 Acceptance-Rejection Method

In cases where F^{-1} is cumbersome to compute, the so-called *acceptance-rejection method* may be preferable. To describe the method, suppose that we want to sample from a density $f(z) = dF(z)/dz$, and further suppose that we have a good method to sample from a density $e(z)$, where

$$e(z)c \geq f(z), \quad z \in \mathbb{R}, \tag{3.8}$$

for some positive constant c. By necessity, $c \geq 1$ as both e and f integrate to 1. In the acceptance-rejection method, we

1. Draw a sample Z from $e(z)$.
2. Draw an independent uniform variable U, $U \sim \mathcal{U}(0,1)$.
3. Accept the sample Z if $U \leq f(Z)/(ce(Z))$; otherwise discard it.

---

[4] If even higher precision is required, we can use (3.6)–(3.7) as a guess for the root y in the equation $\Phi(y) = x$. Any number of numerical root search routines (e.g. Newton-Raphson) can then be applied to improve the precision of the solution further. Typically only one or two iterations will be required to get the solution to within machine precision on a PC.

The proof of why this algorithm works is straightforward and we omit it. Note that the third step of the acceptance-rejection method can be wasteful if too many samples need rejection. The key to the numerical efficiency of the acceptance-rejection method is thus evidently the ability to identify densities $e(z)$ that are "close" to $f(z)$, in the sense that c is close to 1 for all x. Indeed, it can easily be shown that the probability of accepting a sample is $1/c$. Press et al. [1992] list good choices for $e(z)$ for a number of standard densities $f(z)$.

To demonstrate the mechanics of setting up an acceptance-rejection scheme for a particular distribution, let us consider sampling of a variable χ_ν^2 from a *chi-square distribution* with ν degrees of freedom. This distribution arises in a number of interest rate applications and is characterized by the cumulative distribution function

$$P\left(\chi_\nu^2 \leq z\right) = \frac{1}{2^{\nu/2}\Gamma(\nu/2)} \int_0^z e^{-y/2} y^{(\nu/2)-1} \, dy, \quad \nu > 0, \ z \geq 0,$$

where Γ is the gamma function. For reasonably large degrees of freedom ν, the chi-square density is typically bell-shaped. The chi-square distribution is a special case of the *gamma distribution* with density

$$f(z; a, b) = \frac{a(az)^{b-1} e^{-az}}{\Gamma(b)}, \quad a, b > 0, \ z \geq 0. \tag{3.9}$$

The chi-square distribution corresponds to $a = \frac{1}{2}$ and $b = \frac{\nu}{2}$. Rather than considering how to simulate a chi-square distribution, we will consider the more general question of how to draw from (3.9). We note that if a variable X has gamma density $f(z; 1, b)$, then aX, $a > 0$, has gamma density $f(z; 1/a, b)$, so, in fact, it suffices to consider a simulation algorithm for the unit-scale density

$$f(z) = \frac{z^{b-1} e^{-z}}{\Gamma(b)},$$

where we assume that $b \geq 1$. One simple choice of "comparison" density for an acceptance-rejection algorithm is the exponential density

$$e(z) = \lambda e^{-\lambda z},$$

which, as mentioned earlier, can easily be simulated by inverse transform techniques. Note that

$$\frac{f(z)}{e(z)} = \frac{1}{\lambda \Gamma(b)} z^{b-1} e^{(\lambda-1)z},$$

which can be checked to have a maximum value of

$$\sup\left(\frac{f(z)}{e(z)}\right) = \frac{1}{\lambda \Gamma(b)} \left(\frac{b-1}{e(1-\lambda)}\right)^{b-1}, \tag{3.10}$$

where we must assume that $\lambda < 1$. To satisfy (3.8) we take $c = \sup(f(z)/e(z))$ and now search for the value of λ that minimizes c, thereby optimizing computational speed. It is easy to see that (3.10) is minimized for $\lambda = 1/b$, corresponding to $c = b^b e^{1-b}/\Gamma(b)$. Note that

$$\frac{f(z)}{ce(z)} = \frac{z^{b-1}}{\lambda} e^{b-1+(\lambda-1)z} b^{-b},$$

with the third step of the acceptance-rejection algorithm best done in logarithms.

The algorithm outlined above was proposed by Fishman [1976] and works best for moderate values of b. For larger values, the Gamma distribution starts looking like a bell-shaped Gaussian distribution and is no longer well-approximated by an exponential distribution. Indeed, we notice that the probability of acceptance $(1/c)$ is approximately $\sqrt{2\pi/(e^2 b)}$, so of order $O(1/\sqrt{b})$, for $b \gg 1$. Modifications to the basic Fishman algorithm to accelerate sampling can be found in Cheng and Feast [1980]. Another common idea is to set $e(z)$ to the *Cauchy density*

$$e(z) = \frac{1}{s\pi \left(1 + ((z-t)/s)^2\right)},$$

where $s > 0$ and t are constants. This distribution is bell-shaped and, as discussed earlier, can be simulated by the inverse transform method. Press et al. [1992] list computer code and references for this case. For values $b \in [0, 1]$, the acceptance-rejection technique of Ahrens and Dieter [1974] can also be used.

3.1.1.3 Composition

A third and final method to generate random variables from a given distribution function exploits known functional relationships that map variables sampled from one or more distributions to variables sampled from a target distribution. This technique is known as *composition*. A classical example of composition is the *log-normal distribution* $\mathcal{LN}(\mu, \sigma^2)$ which, as we saw earlier in Chapter 1, is defined through the relation

$$X \sim \mathcal{N}(\mu, \sigma^2) \Rightarrow e^X \sim \mathcal{LN}(\mu, \sigma^2),$$

where $\sim$ denotes "distributed as", and where $\mathcal{N}(\mu, \sigma^2)$ is the Gaussian distribution with mean μ and variance σ^2. In other words, a sample Z from $\mathcal{LN}(\mu, \sigma^2)$ can be generated by drawing (by the inverse transformation method, say) a $\mathcal{N}(0, 1)$ variable X, and then setting $Z = e^{\mu + \sigma X}$.

Another classical example of a functional map is the *Student's t-distribution*, where samples can be generated by multiplying independent samples from a standard Gaussian and a chi-square distribution; see Andersen et al. [2003] for a financial application of this. While we earlier demonstrated that the chi-square and gamma distributions can be generated by acceptance-rejection techniques, in fact we can also use composition for this. For instance, it is known that if $X_1, X_2, \ldots, X_\nu$ are independent standard Gaussian variables, then

$$Z = \sum_{i=1}^{\nu} X_i^2 \tag{3.11}$$

is distributed chi-square with ν degrees of freedom. Also, if $U_1, \ldots, U_b$ are independent uniformly distributed variables, then

$$Z = -a \sum_{i=1}^{b} \ln U_i \tag{3.12}$$

is gamma distributed with density (3.9). For small integer-valued distribution parameters b or ν, (3.11) or (3.12) often define a faster simulation scheme than acceptance-rejection methods.

For later use, we note that the relationship (3.11) can be generalized to

$$\tilde{\chi}_\nu^2(\lambda) = \sum_{i=1}^{\nu} (X_i + a_i)^2$$

for a series of constants a_i, $i = 1, \ldots, \nu$. The random variable $\tilde{\chi}_\nu^2(\lambda)$ follows a so-called *non-central chi-square distribution* with ν degrees of freedom and *non-centrality parameter* $\lambda = \sum_i a_i^2$. The distribution function is given by

$$P\left(\tilde{\chi}_\nu^2(\lambda) \leq z\right) = e^{-\lambda/2} \sum_{j=0}^{\infty} \frac{\left(\frac{1}{2}\lambda\right)^j}{j!\Gamma\left(\frac{\nu}{2}+j\right) 2^{(\nu/2)+j}} \int_0^z y^{\nu/2+j-1} e^{-y/2} \, dy, \tag{3.13}$$

an expression that also holds for non-integer ν. If $\nu > 1$, samples from a non-central chi-squared distribution can be generated by composition, using the relation

$$\tilde{\chi}_\nu^2(\lambda) = \left(Z + \sqrt{\lambda}\right)^2 + \chi_{\nu-1}^2,$$

where Z is a standard Gaussian random variable independent of $\chi_{\nu-1}^2$. To handle the case $\nu \leq 1$, one can observe from the expression (3.13) that a non-central chi-square variable can be expressed as a regular chi-square variable $\chi_{\nu+2N}^2$, where N is an independent *Poisson-distributed* discrete variable with intensity $\lambda/2$,

$$P(N = j) = e^{-\lambda/2} \frac{(\lambda/2)^j}{j!}, \quad j = 0, 1, \ldots.$$

This suggests a composition rule for arbitrary ν: draw Poisson variables N (by the inverse transformation method, say) and then draw $\chi_{\nu+2N}^2$ using the methods in Section 3.1.1.2.

3.1.2 Correlated Gaussian Samples

The previous section dealt with the generation of scalar random variables. In applications, however, we may face the task of generating *vectors* of random variables, drawn from a joint multi-variate distribution. Of primary importance in financial applications is the multi-variate Gaussian distribution, so we devote this section to issues surrounding the generation of correlated Gaussian samples.

Recall that a p-dimensional Gaussian distribution $\mathcal{N}(\mu, \Sigma)$ is characterized by a p-dimensional vector-valued mean μ and a $p \times p$ symmetric, positive semi-definite[5] covariance matrix Σ. The joint density is

$$\phi_p(z; \mu, \Sigma) = \frac{1}{(2\pi)^{p/2}(\det \Sigma)^{1/2}} \exp\left(-\frac{1}{2}(z-\mu)^\top \Sigma^{-1}(z-\mu)\right), \quad z \in \mathbb{R}^p.$$

The following result is useful:

Lemma 3.1.3 (Linear Transformation). *Let $Z \sim \mathcal{N}(\mu, \Sigma)$ be p-dimensional. Given a $d \times p$ matrix A and a d-dimensional vector B, then*

$$AZ + B \sim \mathcal{N}\left(A\mu + B, A\Sigma A^\top\right).$$

We can use this lemma as follows. Suppose that we generate p independent standard (that is, $\mathcal{N}(0,1)$) Gaussian samples and collect them in a p-dimensional vector X. This can be accomplished using the techniques in Section 3.1.1. Clearly $X \sim \mathcal{N}(0, I)$, where I is the p-dimensional identity matrix. Define a $(p \times p)$-dimensional matrix C satisfying

$$CC^\top = \Sigma. \tag{3.14}$$

Then

$$Z = \mu + CX$$

is distributed $\mathcal{N}(\mu, \Sigma)$.

It remains to determine a matrix C that satisfies (3.14). While there is generally an infinite number of such matrices, two particular choices are of primary importance. We discuss these below.

3.1.2.1 Cholesky Decomposition

In the Cholesky decomposition, we impose the constraint that the matrix C be lower triangular (that is, having all zeros above the diagonal), thereby conveniently reducing the number of multiplications required to compute CX to $p(1+(p-1)/2)$, rather than p^2. Assuming that the matrix is positive definite (not only positive semi-definite), the Cholesky decomposition is well-defined, and given by

---
[5] That is, all eigenvalues of Σ are non-negative.

$$C_{i,i} = \sqrt{\Sigma_{i,i} - \sum_{k=1}^{i-1} C_{i,k}^2}, \quad i = 1, \ldots, p,$$

$$C_{i,j} = \frac{1}{C_{j,j}} \left(\Sigma_{i,j} - \sum_{k=1}^{j-1} C_{i,k} C_{j,k} \right), \quad j = 1, \ldots, p-1, \quad j < i.$$

For instance, if

$$\Sigma = \begin{pmatrix} \sigma_1^2 & \rho \sigma_1 \sigma_2 \\ \rho \sigma_1 \sigma_2 & \sigma_2^2 \end{pmatrix},$$

where $\rho \in [-1, 1]$ and $\sigma_1, \sigma_2 > 0$, then

$$C = \begin{pmatrix} \sigma_1 & 0 \\ \sigma_2 \rho & \sigma_2 \sqrt{1-\rho^2} \end{pmatrix},$$

a result that we have already used in Section 2.11. Press et al. [1992], among others, list computer code implementing the relations above.

If the matrix Σ is only positive semi-definite (but not positive definite), the Cholesky decomposition will fail. In this case, linear algebra tells us that the matrix Σ is rank-deficient, with rank $r < p$. As such, we must be able to set $Z = \mu + MY$, where M is a $p \times r$ matrix and $Y \sim \mathcal{N}(0, \Sigma_Y)$ is r-dimensional, with the covariance matrix having full rank r. Using Cholesky decomposition instead on Σ_Y, we can find a lower diagonal matrix C_Y satisfying $C_Y C_Y^\top = \Sigma_Y$. Thus, in this case

$$Z = \mu + M C_Y X$$

where X is a vector of r (not p) independent standard Gaussian samples. The matrix M can be found by the singular value decomposition (SVD) algorithm, see Press et al. [1992], or the algorithm in the next section.

3.1.2.2 Eigenvalue Decomposition

As an alternative to Cholesky decomposition, we can also consider diagonalizing Σ through an eigenvalue decomposition. Here, we write

$$\Sigma = E \Lambda E^\top, \tag{3.15}$$

where Λ is a diagonal matrix of eigenvalues λ_i, $i = 1, \ldots, p$, and the columns of E contain the orthonormal eigenvectors of Σ. Some eigenvalues may be zero, if Σ is rank-deficient (positive semi-definite). Comparison with (3.14) implies that one choice of C is

$$C = E\sqrt{\Lambda} = E \begin{pmatrix} \sqrt{\lambda_1} & 0 & \cdots & 0 \\ 0 & \sqrt{\lambda_2} & \ddots & \vdots \\ \vdots & \ddots & \ddots & 0 \\ 0 & \cdots & 0 & \sqrt{\lambda_p} \end{pmatrix}. \tag{3.16}$$

The eigenvalue decomposition (3.15) is relatively straightforward, at least as eigenvalue problems go, due to the fact that Σ is symmetric and positive semi-definite; see Press et al. [1992] for an algorithm. While both Cholesky decomposition and eigenvalue decompositions have computational complexity $O(n^3)$, in practice the Cholesky method is often much faster than the eigenvalue method, making the Cholesky method preferable in practice. Nevertheless, decompositions of the type (3.16) have certain appealing theoretical properties that shall be useful later, so the next section explores (3.16) further.

3.1.3 Principal Components Analysis (PCA)

Consider a p-dimensional Gaussian variable Z with a given covariance matrix Σ. Assume, with no loss of generality, that the mean of Z is 0 and that Σ has full rank (positive definite). Consider now writing, as an approximation,

$$Z \approx DX, \qquad (3.17)$$

where X is an r-dimensional vector of independent standard Gaussian variables, $r \leq p$, and D is a $(p \times r)$-dimensional matrix. How should we choose D in an optimal way?

First, we obviously need to define what constitutes an "optimal" approximation in (3.17). We here have in mind L^2 closeness of the covariance matrix $DD^\top$ to Σ (see Lemma 3.1.3), so let us define the optimal D^* as the matrix that minimizes the norm $f(D)$, where

$$f(D)^2 = \mathrm{tr}\left(\left(\Sigma - DD^\top\right)\left(\Sigma - DD^\top\right)^\top\right).$$

This is just the matrix representation of the usual Frobenius norm on the squared differences between Σ and $DD^\top$. The value of D that minimizes $f(D)$ can be shown to be

$$D^* = E_r \sqrt{\Lambda_r}, \qquad (3.18)$$

where Λ_r is an $r \times r$ diagonal matrix containing the *largest* r eigenvalues of Σ, and E_r is a $p \times r$ matrix of r p-dimensional eigenvectors corresponding to the eigenvalues in Λ_r.

Equipped with the optimal D, we now go back to the approximation (3.17) and write

$$Z \approx \widetilde{Z} \triangleq E_r \sqrt{\Lambda_r} X = \sqrt{\lambda_1} e_1 X_1 + \sqrt{\lambda_2} e_2 X_2 + \ldots + \sqrt{\lambda_r} e_r X_r, \quad (3.19)$$

where e_i denotes the i-th column of E_r and the λ_i's are the eigenvalues, sorted in decreasing order of magnitude. The (deterministic) vector e_i is known as the i-th *principal component* of Z, and the (random) variable $\sqrt{\lambda_i} X_i$ as the i-th *principal factor*. With (3.19), we have $\mathrm{tr}(\mathrm{Cov}(Z,Z)) =$

$E(Z^\top Z) = \sum_{i=1}^{p} \lambda_i$ and $\text{tr}(\text{Cov}(\tilde{Z}, \tilde{Z})) = E(\tilde{Z}^\top \tilde{Z}) = \sum_{i=1}^{r} \lambda_i$, i.e. the first r terms in the decomposition (3.19) explain a fraction

$$\frac{\sum_{i=1}^{r} \lambda_i}{\sum_{i=1}^{p} \lambda_i}$$

of the sum of the diagonal elements of the covariance matrix of Z. Principal components decomposition will thus result in a loss of total variance, unless the covariance matrix is either rank-deficient (i.e. has eigenvalues that are strictly zero), or we use a full set of principal components ($p = r$). In many cases of interest to us here, the loss of variance can be small, even if r is a modest number, e.g. 2 or 3. We notice that the covariance matrix for Z, as approximated by (3.19), will be *rank-deficient*, as the number r of non-zero eigenvalues is less than p.

While we have used a setting with Gaussian variables to motivate our treatment of principal components analysis (PCA), it is, in fact, a generically useful tool for uncovering the structure of large-dimensional random vectors, and replacing them with more manageable, lower-dimensional variables; see, e.g., Theil [1971] for more details and an application to empirical non-Gaussian data. Also, PCA identifies which directions of a multi-dimensional random variable are "important", potentially allowing us to allocate computational resources in an intelligent manner. One example of this is shown later in this chapter, in Section 3.2.10.

3.2 Generation of Sample Paths

So far, we have assumed that random variables are characterized by a known distribution function. In most of our applications, however, the random variables $g(T)/N(T)$ used in the basic pricing equation (3.1) are specified through an SDE or, more generally, an Ito process. In this section, we shall discuss Monte Carlo simulation of such processes. We start out with a motivating example, set in the Black-Scholes-Merton economy.

3.2.1 Example: Asian Basket Options in Black-Scholes Economy

Consider a dividend-free stock S, with Black-Scholes dynamics

$$dS(t)/S(t) = r\, dt + \sigma\, dW(t), \tag{3.20}$$

where $W(t)$ is a Brownian motion in the risk-neutral measure Q, r is a constant interest rate, and σ is a constant volatility. Let there be given an increasing set of observation times $\{t_1, t_2, \ldots, t_m\}$, with $t_m = T$, and define the $\mathcal{F}_T$-measurable (discretely observed) stock average

$$A(T) = \frac{1}{m} \sum_{i=1}^{m} S(t_i). \tag{3.21}$$

An *Asian* (or *average rate*) call option with strike K is defined by the terminal payout
$$g(T) = (A(T) - K)^+ ; \qquad (3.22)$$
we wish to price this option by Monte Carlo simulation.

As discussed earlier (see (1.39)), the geometric Brownian motion process (3.20) allows us to express S directly in terms of the Brownian motion,
$$S(t) = S(0)e^{rt - \frac{1}{2}\sigma^2 t + \sigma W(t)}, \quad t > 0,$$
whereby, with $\Delta_i \triangleq t_i - t_{i-1}$ and $t_0 = 0$,
$$S(t_i) = S(t_{i-1}) \exp\left(\left[r - \frac{1}{2}\sigma^2\right]\Delta_i + \sigma\left[W(t_i) - W(t_{i-1})\right]\right),$$
$i = 1, \ldots, m$. By the properties of Brownian motion, the increments $W(t_i) - W(t_{i-1})$ are independent Gaussian variables distributed as $\mathcal{N}(0, \Delta_i)$. For the purposes of Monte Carlo simulation, we can therefore write
$$S(t_i) = S(t_{i-1}) \exp\left(\left(r - \frac{1}{2}\sigma^2\right)\Delta_i\right) \exp\left(\sigma\sqrt{\Delta_i} Z_i\right), \quad i = 1, \ldots, m, \qquad (3.23)$$
where the Z_i are independent standard $\mathcal{N}(0,1)$ Gaussian random variables. To produce a single sample draw of $g(T)$, we thus

1. Draw independent standard Gaussian samples Z_i, $i = 1, \ldots, m$ (see Section 3.1.1).
2. Starting from $S(0)$, generate $S(t_i)$, $i = 1, \ldots, m$, from the iteration (3.23).
3. Compute $g(T)$ from (3.21)–(3.22).

Repeating this procedure n times (with Gaussian samples independent from one path to the next), we can generate n random samples $g_1, g_2, \ldots, g_n$ of $g(T)$. Our estimate of the time 0 price of the Asian option is then, from (3.3) with $N(t) = e^{rt}$ and non-random,
$$\overline{V}(0) = e^{-rT} \frac{1}{n} \sum_{j=1}^{n} g_j.$$

Asymptotic confidence intervals can be computed from (3.4). The pricing algorithm involves drawing mn Gaussian variables, so the computational cost of the pricing algorithm is $O(mn)$.

Increasing the complexity, let us now consider an Asian option on a p-dimensional basket of stocks $S_1, S_2, \ldots, S_p$, each following geometric Brownian motion,
$$dS_k(t)/S_k(t) = r\, dt + \sigma_k\, dW_k(t), \quad k = 1, \ldots, p.$$

The Brownian motions W_k and W_j are assumed correlated with constant correlation coefficient $\rho_{k,j}$, $j, k = 1, \ldots, p$, $j \neq k$. Define a unit-weighted basket price as

$$B(t) = \sum_{k=1}^{p} S_k(t),$$

and set the terminal Asian option payout to be

$$g(T) = \left(\frac{1}{m} \sum_{i=1}^{m} B(t_i) - K \right)^+, \qquad (3.24)$$

where the time line $\{t_i\}$ is as before. Equivalent to (3.23), we draw sample paths for each asset according to the prescription

$$S_k(t_i) = S_k(t_{i-1}) \exp\left(\left(r - \frac{1}{2}\sigma_k^2 \right) \Delta_i + \sigma_k \sqrt{\Delta_i} Z_{k,i} \right), \qquad (3.25)$$
$$i = 1, \ldots, m, \quad k = 1, \ldots, p,$$

where the $Z_{k,i}$ are Gaussian samples, independently drawn at each time step but correlated across k's. Let C be the Cholesky decomposition of the correlation matrix $\{\rho_{k,j}\}$ (see Section 3.1.2.1), in which case we can generate the correlated sample vectors $Z_i = (Z_{1,i}, Z_{2,i}, \ldots, Z_{p,i})^\top$ as

$$Z_i = C X_i$$

for a p-dimensional vector X_i of independent Gaussian samples. Given joint sample paths of all basket component assets S_k, $k = 1, \ldots, p$, pricing of the Asian basket option proceeds as above, substituting (3.24) for (3.22).

Completion of (3.25) requires pm samples to complete a full path of all p assets, making the total computational effort of an n-sample Monte Carlo scheme $O(nmp)$, with the (probabilistic) convergence order $O(n^{-1/2})$ and dependent only on n. As mentioned earlier, the linearity of computational cost on the dimension of the asset vector p compares favorably to the exponential growth in p of finite difference schemes. Notice also the ease with which the Monte Carlo scheme is able to incorporate path-dependence.

3.2.2 Discretization Schemes, Convergence, and Stability

At the heart of the example in Section 3.2.1 was an iterative scheme for the production of a sample path for a vector-valued SDE; see (3.25). For the simple Black-Scholes model, SDE state variables (stock prices) could be expressed analytically in terms of independent increments of a Brownian motion, making path generation straightforward. In practice, however, we are often working with SDEs that do not permit closed-form solution. In such cases, we need to *time-discretize* the SDE, much the same way as we did for the numerical solution of PDEs.

In the next few sections, we shall consider a few important SDE discretization schemes. Before moving on to this, it is useful to discuss the sense in which we consider a discretization scheme to converge to the true SDE solution. For this, consider a vector-valued SDE

$$dX(t) = \mu(t, X(t)) \, dt + \sigma(t, X(t)) \, dW(t), \tag{3.26}$$

where $X(t)$ is p-dimensional, W is a d-dimensional vector of independent Brownian motions, and $\mu : [0, T] \times \mathbb{R}^p \to \mathbb{R}^p$ and $\sigma : [0, T] \times \mathbb{R}^p \to \mathbb{R}^{p \times d}$ satisfy the usual regularity conditions. Consider an equidistant[6] time grid $\{0, \Delta, 2\Delta, \ldots, m\Delta\}$, the number of references and let $\widehat{X}$ be an approximation to X, based on some kind of time-discretization scheme on the grid $\{i\Delta\}$. For simplicity of notation, set $\widehat{X}_i \triangleq \widehat{X}(i\Delta)$. We say that the underlying approximation is *weakly consistent* if there exists a function $c(\Delta)$ with

$$\lim_{\Delta \downarrow 0} c(\Delta) = 0$$

such that (dropping the measure superscript on the expectation operator)

$$\mathrm{E}\left(\left| \mathrm{E}\left(\Delta^{-1} \left(\widehat{X}_{i+1} - \widehat{X}_i \right) \middle| \mathcal{F}_{i\Delta} \right) - \mu\left(i\Delta, \widehat{X}_i\right) \right|^2 \right) \le c(\Delta), \tag{3.27}$$

and

$$\mathrm{E}\left(\left| \mathrm{E}\left(\Delta^{-1} \left(\widehat{X}_{i+1} - \widehat{X}_i \right) \left(\widehat{X}_{i+1} - \widehat{X}_i \right)^\top \middle| \mathcal{F}_{i\Delta} \right) \right. \right.$$
$$\left. \left. - \sigma\left(i\Delta, \widehat{X}_i\right) \sigma\left(i\Delta, \widehat{X}_i\right)^\top \right|^2 \right) \le c(\Delta), \tag{3.28}$$

for all $i = 0, \ldots, m-1$. The notion of weak consistency[7] thus amounts to requiring that the mean and variance of the increments of the approximating process be close to those of the true SDE solution.

A concept related to consistency is the notion of *weak convergence*. We say that an approximate solution converges weakly to X at time $T = m\Delta$ with respect to a class $\mathcal{C}$ of test functions $g : \mathbb{R}^p \to \mathbb{R}$ if

$$\lim_{\Delta \downarrow 0} \left| \mathrm{E}\left(g(X(T))\right) - \mathrm{E}\left(g(\widehat{X}(T))\right) \right| = 0, \tag{3.29}$$

for all $g \in \mathcal{C}$. Notice that the limit necessarily involves $m \to \infty$.

---

[6] To keep notation manageable, we use a constant time step Δ in most of this chapter. All results are, however, easily extendable to non-equidistant grids.

[7] *Strong consistency* (which is of little use to us in this book) requires that (3.27) is satisfied, and that the variance of the difference between increments of the true process and the approximation vanish. The second requirement is stronger than (3.28).

The class of test functions used in (3.29) is normally always in the set C_P^l of functions with polynomially bounded[8] derivatives of order $0, 1, \ldots, l$ with maximum power l. We say that a scheme converges with *weak order* β if, for all $g \in C_P^{2(\beta+1)}$, (3.29) can be strengthened to

$$\left| \mathrm{E}\left(g(X(T))\right) - \mathrm{E}\left(g(\widehat{X}(T))\right) \right| \leq c\Delta^\beta, \tag{3.30}$$

for all $\Delta \in (0, \Delta_0)$, where Δ_0 and c are constants and c does not depend on Δ (but may depend on g).

One would generally expect that a weakly consistent scheme is weakly convergent. Indeed, this can be established to be the case under certain additional regularity conditions. We will not list the exact result here, but refer to Kloeden and Platen [2000], Theorem 9.7.4.

Finally, a brief word on stability of a time-discretized SDE. A commonly used definition of *A*-stability focuses on the behavior of a discretized test SDE of the type

$$dX(t) = \lambda X(t)\,dt + dW(t), \tag{3.31}$$

where λ is a complex-valued constant with real part $\mathrm{Re}\,(\lambda) < 0$. We suppose that a discretization scheme can be represented as

$$\widehat{X}_{i+1} = \widehat{X}_i G(\lambda \Delta) + Z_i^\Delta, \quad i = 0, 1, \ldots, m-1, \tag{3.32}$$

where G is a mapping of the complex plane onto itself and the Z_i^Δ's are random variables independent of the $\widehat{X}_i$'s. In this case, the *region of stability* for a scheme is the set of $\lambda\Delta$ for which $\mathrm{Re}\,(\lambda) < 0$ and

$$|G(\lambda\Delta)| < 1. \tag{3.33}$$

Similar to the definition used for finite difference scheme discretizations, we say that an SDE time-discretization scheme is *A-stable*, if the region of stability includes all values of λ with $\mathrm{Re}\,(\lambda) < 0$ and all $\Delta > 0$.

3.2.3 The Euler Scheme

An obvious first scheme to discretize (3.26) treats both dt and $dW(t)$ fully explicitly, evaluating all SDE coefficients on time step $[i\Delta, i\Delta + \Delta]$ at the left interval point $i\Delta$. In other words, we write, starting from $\widehat{X}_0 = X(0)$,

$$\widehat{X}_{i+1} = \widehat{X}_i + \mu\left(i\Delta, \widehat{X}_i\right)\Delta + \sigma\left(i\Delta, \widehat{X}_i\right)\left(W(i\Delta + \Delta) - W(i\Delta)\right), \tag{3.34}$$
$$i = 0, 1, \ldots, m-1.$$

---

[8] A function $f : \mathbb{R}^p \to \mathbb{R}$ is polynomially bounded if $|f(x)| \leq k(1 + |x|^q)$, $x \in \mathbb{R}^p$, for constants k and q.

With this scheme, Monte Carlo generation of paths is straightforward and involves, as in Section 3.2.1, replacing the increments $W(i\Delta + \Delta) - W(i\Delta)$ with $Z_i\sqrt{\Delta}$, for a d-dimensional vector of independent standard Gaussian samples Z_i.

The discretization scheme (3.34) is known as the *Euler scheme*, sometimes also called the *Euler-Maruyama* scheme. The Euler scheme is easy to implement and is a true workhorse that we will often use in this book. We note that the scheme is weakly consistent, as

$$\mathrm{E}\left(\left|\mathrm{E}\left(\Delta^{-1}\left(\widehat{X}_{i+1} - \widehat{X}_i\right)\Big|\mathcal{F}_{i\Delta}\right) - \mu\left(i\Delta, \widehat{X}_i\right)\right|^2\right) = 0,$$

and

$$\mathrm{E}\left(\left|\mathrm{E}\left(\Delta^{-1}\left(\widehat{X}_{i+1} - \widehat{X}_i\right)\left(\widehat{X}_{i+1} - \widehat{X}_i\right)^\top\Big|\mathcal{F}_{i\Delta}\right)\right.\right.$$
$$\left.\left. - \sigma\left(i\Delta, \widehat{X}_i\right)\sigma\left(i\Delta, \widehat{X}_i\right)^\top\right|^2\right) = O\left(\Delta^2\right).$$

While one might believe that the explicit discretization of the diffusion term — which is only accurate to order $O(\sqrt{\Delta})$ — would give the scheme weak convergence order[9] $1/2$, in fact we typically have that the Euler scheme has *weak convergence order* $\beta = 1$. We note that for this result to hold, however, regularity conditions on μ and σ stronger than those of the existence and uniqueness results (Theorem 1.6.1) are needed. For instance, in the case where μ and σ are functions of X alone, Theorem 9.7.6 in Kloeden and Platen [2000] requires that μ and σ be four times continuously differentiable with polynomial growth and uniformly bounded derivatives. See also their Theorem 15.4.2 for a more general result.

Given that the Euler scheme is fully explicit, our experience from finite difference methods suggests that the scheme may have stability problems. To investigate, we follow Section 3.2.2 and consider the test SDE

$$dX(t) = \lambda X(t)dt + dW(t),$$

which is discretized as

$$\widehat{X}_{i+1} = \widehat{X}_i\left(1 + \lambda\Delta\right) + \sqrt{\Delta}Z_i, \qquad (3.35)$$

where Z_i's are standard Gaussian. Comparison to (3.32) and (3.33) shows that the region of stability for the Euler scheme is

$$|(1 + \lambda\Delta)| < 1, \quad \mathrm{Re}\left(\lambda\right) < 0,$$

which is the unit disc in the complex plane centered at $\lambda\Delta = -1$. For a given λ, there are thus restrictions on how big a time step Δ can be used.

---

[9] The so-called *strong convergence order* of the Euler scheme is in fact only $1/2$. The concept of strong convergence order is defined in Kloeden and Platen [2000] and is of little importance to applications in this book.

3.2.3.1 Linear-Drift SDEs

The restricted stability region of the Euler scheme can be a practical concern. For instance, SDEs of the important type

$$dX(t) = \varkappa\left(\theta(t) - X(t)\right)dt + \sigma\left(t, X(t)\right)dW(t) \qquad (3.36)$$

arise quite frequently in fixed income modeling, and in cases where $\varkappa$ is big (which is often the case for, say, stochastic volatility models such as those covered in Chapters 8, 9 and 13) the Euler scheme can become unstable and return meaningless results. One way to solve the problem is to switch to an implicit scheme (see next section), but in the case (3.36) we can use the fact that the drift term can be removed by a simple change of variable. For instance, for the case where $X(t)$ is scalar we can set

$$Y(t) = e^{\varkappa t}X(t) - \varkappa \int_0^t e^{\varkappa u}\theta(u)\,du,$$

such that, from Ito's lemma,

$$dY(t) = e^{\varkappa t}\sigma\left(t, X(t)\right)dW(t)$$
$$= e^{\varkappa t}\sigma\left(t, e^{-\varkappa t}\left(Y(t) + \varkappa \int_0^t e^{\varkappa u}\theta(u)\,du\right)\right)dW(t).$$

Euler simulation of the process for $Y(t)$, rather than for $X(t)$, will center X around its analytically known mean

$$\mathrm{E}\left(X\left((i+1)\Delta\right)|\,X\left(i\Delta\right)\right) = e^{-\varkappa\Delta}X\left(i\Delta\right) + \varkappa \int_{i\Delta}^{i\Delta+\Delta} e^{-\varkappa((i+1)\Delta-u)}\theta(u)\,du$$

and will often alleviate any stability problems.

3.2.3.2 Log-Euler Scheme

One potential problem with the pure Euler scheme (3.34) is the fact that all increments are locally Gaussian, thereby implying a non-zero probability of $\widehat{X}$ crossing zero and becoming negative. Many SDEs, however, are known to produce only non-negative solutions, and the functions μ and σ may not allow for negative arguments. This, for instance, is the case for the square-root process

$$dX(t) = \sqrt{X(t)}\,dW(t), \quad X(0) > 0,$$

where the Euler scheme cannot be directly applied. Some authors (e.g., Kloeden and Platen [2000]) suggest heuristic modifications of the Euler scheme, such as

$$\widehat{X}_{i+1} = \widehat{X}_i + \sqrt{|\widehat{X}_i|}\left(W\left(i\Delta + \Delta\right) - W\left(i\Delta\right)\right),$$

but ultimately this is not very satisfying and the resulting scheme will often have large errors[10]. An alternative is to introduce an invertible transformation $X(t) = f(Y(t))$, with $f : \mathbb{R} \to \mathbb{R}_+$, and then apply the Euler scheme to Y, at each step recovering X as $f(Y)$. In finance applications, where many processes are based on SDEs that bear some resemblance to geometric Brownian motion, an often-used choice for f is $f(y) = e^y$. The resulting scheme is known as the *log-Euler scheme*.

Consider the SDE (3.26) and assume for simplicity that X is scalar (if X is vector valued, the log-transform can be applied to all, or a few selected, components of X). Set $X(t) = \exp(Y(t))$, such that $Y(t) = \ln(X(t))$. The process for Y then follows from Ito's lemma:

$$dY(t) = \left(\frac{\mu(t, X(t))}{X(t)} - \frac{1}{2}\frac{\sigma(t, X(t))^2}{X(t)^2}\right)dt + \frac{\sigma(t, X(t))}{X(t)}dW(t), \ X(t) = e^{Y(t)}.$$

Writing out a standard Euler scheme for Y and making the transformation $\widehat{X}_i = \exp(\widehat{Y}_i)$ gives us the (scalar) log-Euler scheme for X:

$$\widehat{X}_{i+1} = \widehat{X}_i \exp\left(\left(\frac{\mu(i\Delta, \widehat{X}_i)}{\widehat{X}_i} - \frac{1}{2}\frac{\sigma(i\Delta, \widehat{X}_i)^2}{\widehat{X}_i^2}\right)\Delta + \frac{\sigma(i\Delta, \widehat{X}_i)}{\widehat{X}_i}Z_i\sqrt{\Delta}\right),$$

where $Z_i \sim \mathcal{N}(0, 1)$. Generalizations of the technique above to situations where the valid range of X is some general set $\mathcal{C}$ are obvious and involve identifying an invertible mapping function $f : \mathbb{R} \to \mathcal{C}$, preferably one that can be inverted analytically. For instance, if $\mathcal{C} = [a, \infty)$, we could use $f(y) = a + e^y$.

3.2.4 The Implicit Euler Scheme

The implicit Euler scheme for the vector-valued SDE (3.26) takes the form

$$\widehat{X}_{i+1} = \widehat{X}_i + \mu\left(i\Delta + \Delta, \widehat{X}_{i+1}\right)\Delta + \sigma\left(i\Delta, \widehat{X}_i\right)\left(W\left(i\Delta + \Delta\right) - W\left(i\Delta\right)\right), \tag{3.37}$$

for $i = 0, 1, \ldots, m - 1$. We highlight the fact that the drift coefficient μ is now evaluated at time $i\Delta + \Delta$, rather than at time $i\Delta$. It is easy to show that the implicit Euler scheme is consistent. Under regularity conditions, it can also be shown that the weak convergence order is $\beta = 1$, just as was the case for the explicit Euler scheme.

The main advantage of the implicit Euler scheme over the explicit Euler scheme is numerical stability. To examine the region of stability for the implicit Euler scheme, consider again the test SDE

---

[10] For a dedicated treatment of the rather delicate problem of simulating square-root process, see Chapter 9.

$$dX(t) = \lambda X(t)\, dt + dW(t).$$

It will now be discretized as (compare to (3.35))

$$\widehat{X}_{i+1} = \widehat{X}_i + \widehat{X}_{i+1}\lambda\Delta + \sqrt{\Delta}Z_i,$$

or

$$\widehat{X}_{i+1}\left(1 - \lambda\Delta\right) = \widehat{X}_i + \sqrt{\Delta}Z_i.$$

Comparison to (3.32) and (3.33) shows that now

$$G(\lambda\Delta) = \frac{1}{1 - \lambda\Delta}$$

such that the stability criterion $|G(\lambda\Delta)| < 1$ is satisfied for any value of $\lambda\Delta$ where $\operatorname{Re}(\lambda) < 0$. In other words, the implicit scheme is A-stable.

3.2.4.1 Implicit Diffusion Term

The reader may at this point wonder why the implicit scheme (3.37) only discretized the drift term (μ) implicitly, and not the diffusion term (σ). The answer lies in the differences between a regular Riemann integral and the stochastic integral. Recall in particular that the stochastic Ito integral is defined to be non-anticipative, in the sense that the integrand is always evaluated "to the left" on any partitions of the Brownian motion. As a consequence, if $\sigma(i\Delta, \widehat{X}_i)$ were replaced with $\sigma(i\Delta + \Delta, \widehat{X}_{i+1})$ in (3.37), the resulting scheme would not be weakly consistent, in the sense defined earlier. To illustrate this point, just consider the simple scalar process

$$dX(t) = \sigma X(t)\, dW(t),$$

which we contemplate discretizing as

$$\widehat{X}_{i+1} = \widehat{X}_i + \sigma \widehat{X}_{i+1}\left(W(i\Delta + \Delta) - W(i\Delta)\right),$$

or

$$\widehat{X}_{i+1}\left(1 - \sigma Z_i\sqrt{\Delta}\right) = \widehat{X}_i, \quad i = 0, \ldots, m-1. \tag{3.38}$$

Here, a first difficulty arises: the term $(1 - \sigma Z_i\sqrt{\Delta})$ may become 0 (or very close to zero) if Z_i is an (unbounded) Gaussian variable. For fully implicit discretization schemes, it becomes necessary to use a bounded approximation to the Brownian motion. As discussed in Kloeden and Platen [2000], weak convergence order is preserved if in (3.38) we set the Z_i to be independent binomial variables with

$$\mathrm{P}(Z_i = 1) = \mathrm{P}(Z_i = -1) = \frac{1}{2}.$$

We assume that $1 - \sigma\sqrt{\Delta} > 0$. Rearranging and Taylor-expanding, we get

$$\frac{\widehat{X}_{i+1} - \widehat{X}_i}{\Delta} = \frac{\widehat{X}_i}{\Delta}\left(\frac{1}{1 - \sigma Z_i \sqrt{\Delta}} - 1\right)$$
$$= \frac{\widehat{X}_i}{\Delta}\left(1 + \sigma Z_i \sqrt{\Delta} + \sigma^2 Z_i^2 \Delta + O\left(\sigma^3 Z_i^3 \Delta^{3/2}\right) - 1\right)$$
$$= \widehat{X}_i \left(\sigma Z_i \Delta^{-1/2} + \sigma^2 Z_i^2 + O\left(\sigma^3 Z_i^3 \Delta^{1/2}\right)\right)$$

such that
$$\mathrm{E}\left(\left.\frac{\widehat{X}_{i+1} - \widehat{X}_i}{\Delta}\right|\widehat{X}_i\right) = \widehat{X}_i\left(\sigma^2 + O(\Delta)\right).$$

Clearly, this will cause a violation of the consistency condition (3.27).

In the example above, we notice that consistency can be restored if the drift of the original SDE is changed from 0 to $-\sigma^2 X(t)$ before the "doubly" implicit Euler discretization is employed. More generally, it is not difficult to show that (3.37) can be modified to treat the diffusion term implicitly, provided that the drift of the original vector-valued SDE (3.26) is first changed from μ to

$$\overline{\mu} = \mu - \sum_{j=1}^{d}\sum_{k=1}^{p}(\sigma_{X_k})_{\cdot,j}\,\sigma_{k,j}$$

where the p-dimensional vector $(\sigma_{X_k})_{\cdot,j}$ is the j-th column of the $(p \times d)$-dimensional matrix $\sigma_{X_k} = \{\partial \sigma_{i,j}/\partial X_k\}$. Inspired by the theta methods of Chapter 2, we can, in fact, introduce a family of discretizations

$$\widehat{X}_{i+1} = \widehat{X}_i + \left[(1-\theta)\overline{\mu}_\eta\left(i\Delta, \widehat{X}_i\right) + \theta\overline{\mu}_\eta\left(i\Delta + \Delta, \widehat{X}_{i+1}\right)\right]\Delta$$
$$+ \left[(1-\eta)\sigma\left(i\Delta, \widehat{X}_i\right) + \eta\sigma\left(i\Delta + \Delta, \widehat{X}_{i+1}\right)\right]Z_i\sqrt{\Delta}, \quad (3.39)$$

where the Z_i are binomially distributed variables, $\theta, \eta \in [0,1]$ are parameters, and

$$\overline{\mu}_\eta = \mu - \eta\sum_{j=1}^{d}\sum_{k=1}^{p}(\sigma_{X_k})_{\cdot,j}\,\sigma_{k,j}. \qquad (3.40)$$

As it turns out, all these schemes theoretically have identical convergence order $\beta = 1$, but in practice some choices of θ, η may turn out to work better than others. We shall discuss methods to raise the theoretical convergence order in Section 3.2.6. The scheme (3.39) can be verified to be A-stable for $\theta \in [1/2, 1]$.

3.2.5 Predictor-Corrector Schemes

A closer examination of the implicit Euler scheme (3.37) demonstrates the need to recover $\widehat{X}(i\Delta + \Delta)$ as the vector-valued root of a possibly nonlinear equation. In general, this must be done numerically (using, say, the

Newton-Raphson algorithm), causing a severe deterioration of computational performance. An alternative is to use the explicit Euler scheme as a *predictor* and the implicit scheme as a *corrector*, much the same way we used explicit finite difference approximations as predictors in the Craig-Sneyd algorithm of Section 2.11. Moving straight to the general implicit discretization family (3.39), we write the predictor-corrector as

$$\overline{X}_{i+1} = \widehat{X}_i + \mu\left(i\varDelta, \widehat{X}_i\right)\varDelta + \sigma\left(i\varDelta, \widehat{X}_i\right)(W\left(i\varDelta + \varDelta\right) - W\left(i\varDelta\right)), \quad (3.41)$$

$$\widehat{X}_{i+1} = \widehat{X}_i + \left[(1-\theta)\overline{\mu}_\eta\left(i\varDelta, \widehat{X}_i\right) + \theta\overline{\mu}_\eta\left(i\varDelta + \varDelta, \overline{X}_{i+1}\right)\right]\varDelta$$
$$+ \left[(1-\eta)\sigma\left(i\varDelta, \widehat{X}_i\right) + \eta\sigma\left(i\varDelta + \varDelta, \overline{X}_{i+1}\right)\right](W\left(i\varDelta + \varDelta\right) - W\left(i\varDelta\right)),$$
$$(3.42)$$

where $\theta, \eta \in [0, 1]$, and $\overline{\mu}_\eta$ is as given in (3.40). It is understood that the Brownian motion increments in (3.41) and (3.42) are to be identical.

For sufficiently smooth coefficients, it can be shown that the predictor-corrector scheme (3.41)–(3.42) converges weakly with order $\beta = 1$, independent of the choice of θ and η. As for stability, discretization of (3.31) leads to

$$\overline{X}_{i+1} = \widehat{X}_i(1 + \lambda\varDelta) + W\left(i\varDelta + \varDelta\right) - W\left(i\varDelta\right),$$
$$\widehat{X}_{i+1} = \widehat{X}_i + \left[(1-\theta)\lambda\widehat{X}_i + \theta\lambda\overline{X}_{i+1}\right]\varDelta + W\left(i\varDelta + \varDelta\right) - W\left(i\varDelta\right)$$
$$= \widehat{X}_i(1 + \lambda\varDelta(1 + \theta\lambda\varDelta)) + (W\left(i\varDelta + \varDelta\right) - W\left(i\varDelta\right))(1 + \theta\lambda\varDelta).$$

The region of stability can be verified to be

$$|1 + \lambda\varDelta(1 + \theta\lambda\varDelta)| < 1, \quad \mathrm{Re}\left(\lambda\right) < 0.$$

For $\theta = \frac{1}{2}$, the stability criterion above is identical to that of the classical *Heun scheme* (or *modified trapezoidal scheme*) used for ordinary differential equations. Indeed, the predictor-corrector scheme above can be seen as an adaptation of this scheme for SDEs. We note that SDE adaptations of more sophisticated ODE solvers (such as Runge-Kutta) are also possible, but this goes beyond the scope of this text.

3.2.6 Ito-Taylor Expansions and Higher-Order Schemes

Despite our various efforts at centering derivatives, none of the schemes listed above theoretically attain second-order weak convergence. To develop such schemes, we need to delve further into adapting classical Taylor expansions to the rules of stochastic (Ito) calculus. As we shall ultimately not have much use for higher-order schemes, we keep the treatment informal and limit ourselves to the scalar case where $p = d = 1$ in (3.26).

3.2.6.1 Ordinary Taylor Expansion of ODEs

To gain intuition, start by setting $\sigma = 0$ in (3.26), such that we first deal with an ordinary ODE

$$dX(t) = \mu(t, X(t))\, dt. \tag{3.43}$$

For a given value of t, we can use Taylor's theorem to write

$$X(t + \Delta) = X(t) + \frac{dX(t)}{dt}\Delta + \frac{1}{2}\frac{d^2 X(t)}{dt^2}\Delta^2 + O(\Delta^3),$$

where we stop at order $O(\Delta^3)$. We notice that

$$\frac{dX(t)}{dt} = \mu(t, X(t))$$

and

$$\begin{aligned}\frac{d^2 X(t)}{dt^2} &= \frac{\partial}{\partial t}\mu(t, X(t)) + \frac{\partial}{\partial x}\mu(t, X(t)) \cdot \frac{dX(t)}{dt} \\ &= \left(\frac{\partial}{\partial t} + \mu(t, X(t))\frac{\partial}{\partial x}\right)\mu(t, X(t)).\end{aligned}$$

Setting

$$\mathcal{L} \triangleq \frac{\partial}{\partial t} + \mu\frac{\partial}{\partial x},$$

we thus have

$$X(t + \Delta) = X(t) + \mu(t, X(t))\Delta + \frac{1}{2}\mathcal{L}\mu(t, X(t))\Delta^2 + O(\Delta^3). \tag{3.44}$$

Another way to develop (3.44) proceeds by iteration on the integral representation

$$X(t + \Delta) = X(t) + \int_t^{t+\Delta} \mu(u, X(u))\, du. \tag{3.45}$$

First we recognize that (as seen above)

$$d\mu(t, X(t)) = \mathcal{L}\mu(t, X(t))\, dt$$

such that

$$\mu(u, X(u)) = \mu(t, X(t)) + \int_t^u \mathcal{L}\mu(s, X(s))\, ds, \quad u > t. \tag{3.46}$$

Inserting this into (3.45) gives

$$X(t + \Delta) = X(t) + \mu(t, X(t))\int_t^{t+\Delta} du + \int_t^{t+\Delta}\int_t^u \mathcal{L}\mu(s, X(s))\, ds\, du.$$

Applied to $\mathcal{L}\mu(s, X(s))$ the steps that lead to (3.46) yield

$$\mathcal{L}\mu(s, X(s)) = \mathcal{L}\mu(t, X(t)) + \int_t^s \mathcal{L}^2\mu(v, X(v))\, dv, \quad s > t,$$

such that

$$X(t+\Delta) = X(t) + \mu(t, X(t)) \int_t^{t+\Delta} du + \mathcal{L}\mu(t, X(t)) \int_t^{t+\Delta} \int_t^u ds\, du$$

$$+ \int_t^{t+\Delta} \int_t^u \int_t^s \mathcal{L}^2\mu(v, X(v))\, dv\, ds\, du$$

$$= X(t) + \mu(t, X(t))\Delta + \frac{1}{2}\mathcal{L}\mu(t, X(t))\Delta^2 + O(\Delta^3), \quad (3.47)$$

which is just (3.44). We can continue the iteration to arbitrary high order.

3.2.6.2 Ito-Taylor Expansions

One may wonder why in the previous section we bothered with the integral representation of Taylor's theorem when the usual (differential) Taylor expansion lead to the correct result. The reason is that the integral approach can be extended to SDEs, leading to stochastic *Ito-Taylor expansions*. To give a flavor of these, reintroduce a diffusion term to (3.43), and start out with the integral representation

$$X(t+\Delta) = X(t) + \int_t^{t+\Delta} \mu(u, X(u))\, du + \int_t^{t+\Delta} \sigma(u, X(u))\, dW(u). \quad (3.48)$$

Applying Ito's lemma to μ gives (compare to (3.46))

$$\mu(u, X(u)) = \mu(t, X(t)) + \int_t^u \mathcal{L}_0\mu(s, X(s))\, ds + \int_t^u \mathcal{L}_1\mu(s, X(s))\, dW(s), \quad (3.49)$$

where

$$\mathcal{L}_0 \triangleq \frac{\partial}{\partial t} + \mu\frac{\partial}{\partial x} + \frac{1}{2}\sigma^2\frac{\partial^2}{\partial x^2}, \quad \mathcal{L}_1 \triangleq \sigma\frac{\partial}{\partial x}.$$

Similarly,

$$\sigma(u, X(u)) = \sigma(t, X(t)) + \int_t^u \mathcal{L}_0\sigma(s, X(s))\, ds + \int_t^u \mathcal{L}_1\sigma(s, X(s))\, dW(s). \quad (3.50)$$

Plugging (3.49) and (3.50) into (3.48) yields

$$X(t+\Delta) = X(t) + \mu(t, X(t)) \int_t^{t+\Delta} du + \sigma(t, X(t)) \int_t^{t+\Delta} dW(u) + R_1$$

$$= X(t) + \mu(t, X(t))\Delta + \sigma(t, X(t))(W(t+\Delta) - W(t)) + R_1, \quad (3.51)$$

where the remainder R_1 is

$$R_1 = \int_t^{t+\Delta} \int_t^u \mathcal{L}_0\mu\left(s, X(s)\right) ds\, du$$

$$+ \int_t^{t+\Delta} \int_t^u \mathcal{L}_1\mu\left(s, X(s)\right) dW(s)\, du$$

$$+ \int_t^{t+\Delta} \int_t^u \mathcal{L}_0\sigma\left(s, X(s)\right) ds\, dW(u)$$

$$+ \int_t^{t+\Delta} \int_t^u \mathcal{L}_1\sigma\left(s, X(s)\right) dW(s)\, dW(u).$$

As for the ODE example above, we can repeat this procedure arbitrarily many times. Going just one step further, we arrive at

$$X(t+\Delta) = X(t) + \mu\left(t, X(t)\right) \Delta + \sigma\left(t, X(t)\right)\left(W(t+\Delta) - W(t)\right)$$

$$+ \mathcal{L}_0\mu\left(t, X(t)\right) \frac{1}{2}\Delta^2$$

$$+ \mathcal{L}_1\mu\left(t, X(t)\right) \int_t^{t+\Delta} \int_t^u dW(s)\, du$$

$$+ \mathcal{L}_0\sigma\left(t, X(t)\right) \int_t^{t+\Delta} \int_t^u ds\, dW(u)$$

$$+ \mathcal{L}_1\sigma\left(t, X(t)\right) \int_t^{t+\Delta} \int_t^u dW(s)\, dW(u) + R_2, \qquad (3.52)$$

where R_2 contains triple integrals over t and W.

Stochastic Taylor expansions can be continued to arbitrary order, but we shall not go any further.

3.2.6.3 Milstein Second-Order Discretization Scheme

Discarding the remainder R_1 in the one-step iteration (3.51) is seen to lead to the Euler scheme (see Section 3.2.3), known to have weak convergence order $\beta = 1$. Under additional regularity (see Talay [1984]) of μ and σ, discarding the remainder R_2 in the higher-order expansion (3.52) can form the basis of a discretization scheme with weak order $\beta = 2$. For us to implement such a scheme, however, we need to concern ourselves with the simulation of the three stochastic double integrals figuring in (3.52). We go through the integrals in order below.

First,

$$I_{(1,1)} \triangleq \int_t^{t+\Delta} \int_t^u dW(s)\, dW(u)$$

$$= \int_t^{t+\Delta} (W(u) - W(t))\, dW(u) = \frac{1}{2}\left(W(t+\Delta) - W(t)\right)^2 - \frac{1}{2}\Delta,$$

$$(3.53)$$

where we have used the fact that

$$\int_0^t W(u)\,dW(u) = \frac{1}{2}W(t)^2 - \frac{1}{2}t,$$

as can be verified by Ito's lemma. Second,

$$I_{(0,1)} \triangleq \int_t^{t+\Delta} \int_t^u ds\,dW(u) = \int_t^{t+\Delta} (u-t)\,dW(u)$$

$$= \Delta(W(t+\Delta) - W(t)) - \int_t^{t+\Delta} (W(u) - W(t))\,du$$

$$\triangleq \Delta(W(t+\Delta) - W(t)) - I_{(1,0)}, \tag{3.54}$$

where we have used the integration-by-parts formula

$$\int_0^t u\,dW(u) = tW(t) - \int_0^t W(u)\,du,$$

which follows from applying Ito's lemma to $tW(t)$. In (3.54), the remaining integral $I_{(1,0)}$ on the right-hand-side is the same as the final double integral in (3.52), namely

$$I_{(1,0)} \triangleq \int_t^{t+\Delta} \int_t^u dW(s)\,du = \int_t^{t+\Delta} (W(u) - W(t))\,du.$$

Reversing the order of integration, we get

$$I_{(1,0)} = \int_t^{t+\Delta} \int_t^u dW(s)\,du$$

$$= \int_t^{t+\Delta} \int_u^{t+\Delta} ds\,dW(u) = \int_t^{t+\Delta} (t+\Delta-u)\,dW(u)$$

so we see, from Theorem 1.1.3 and the discussion in Section 1.6 on linear SDEs, that $I_{(1,0)}$ is Gaussian with mean 0 and variance

$$\text{Var}\left(I_{(1,0)}\right) = \int_t^{t+\Delta} (t+\Delta-u)^2\,du = \frac{1}{3}\Delta^3.$$

The covariance between $I_{(1,0)}$ and $W(t+\Delta) - W(t)$ can be computed as

$$\text{Cov}\left(I_{(1,0)}, W(t+\Delta) - W(t)\right) = \int_t^{t+\Delta} (t+\Delta-u)\,du = \frac{1}{2}\Delta^2.$$

With the results above, we can cast the Taylor expansion (3.52) in the form of a simulation scheme (μ, σ, and their derivatives are to be evaluated at $t = i\Delta$ and $X = \widehat{X}(i\Delta)$),

$$\widehat{X}_{i+1} = \widehat{X}_i + \mu\Delta + \sigma Z_{i,1}\sqrt{\Delta} + \mathcal{L}_0\mu\frac{1}{2}\Delta^2 + \mathcal{L}_1\mu Z_{i,2}\sqrt{\frac{1}{3}\Delta^3}$$

$$+ \mathcal{L}_0\sigma\left[\Delta Z_{i,1}\sqrt{\Delta} - Z_{i,2}\sqrt{\frac{1}{3}\Delta^3}\right] + \mathcal{L}_1\sigma\left(\frac{1}{2}Z_{i,1}^2\Delta - \frac{1}{2}\Delta\right), \quad (3.55)$$

where $Z_{i,1}$ and $Z_{i,2}$ are sequences of $\mathcal{N}(0,1)$ Gaussian variables with pairwise correlation

$$\rho(Z_{i,1}, Z_{i,2}) = \frac{\frac{1}{2}\Delta^2}{\sqrt{\frac{1}{3}\Delta^3}\sqrt{\Delta}} = \sqrt{\frac{3}{4}}.$$

The scheme above is known as the *Milstein scheme*. As mentioned earlier, the scheme has weak convergence order 2 under fairly strong regularity assumptions on μ and σ. We note that in the literature on SDE simulation, the Milstein scheme is often presented in a simplified form with the integral $I_{(1,0)}$ simulated as

$$\mathrm{E}\left(\int_t^{t+\Delta}\int_t^u dW(s)\,du \bigg| W(t), W(t+\Delta)\right) = \frac{1}{2}\Delta\left(W(t+\Delta) - W(t)\right),$$

which corresponds to replacing $Z_{i,2}\sqrt{\Delta^3/3}$ with $\frac{1}{2}\Delta Z_{i,1}$ in (3.55). See Kloeden and Platen [2000] for a discussion of why this type of simplification does not affect the weak convergence order. The same source also contains a full discussion of how to extend the Milstein scheme to multi-dimensional SDEs.

3.2.7 Other Second-Order Schemes

The need to explicitly compute derivatives of the functions μ and σ often makes the Milstein scheme inconvenient to apply. High-order simulation schemes that substitute finite difference approximations for derivatives exist, and retain second (or higher) order weak convergence, are surveyed in Kloeden and Platen [2000]. To give an example of such a scheme, consider the scalar case $d = p = 1$ and assume that SDE coefficient functions μ and σ are function of x only. A derivative-free scheme that achieves second-order weak convergence is (from Kloeden and Platen [2000], Chapter 15)

$$\widehat{X}_{i+1} = \widehat{X}_i + \frac{1}{2}\left(\mu\left(\overline{X}\right) + \mu\left(\widehat{X}_i\right)\right)\Delta$$

$$+ \frac{1}{4}\left(\sigma\left(\overline{X}^+\right) + \sigma\left(\overline{X}^-\right) + 2\sigma\left(\widehat{X}_i\right)\right)Z_i\sqrt{\Delta}$$

$$+ \frac{1}{4}\left(\sigma\left(\overline{X}^+\right) - \sigma\left(\overline{X}^-\right)\right)\left(Z_i^2\Delta - \Delta\right)\Delta^{-1/2}, \quad (3.56)$$

where the Z_i's are a sequence of $\mathcal{N}(0,1)$ Gaussian variables, and

$$\overline{X} = \widehat{X}_i + \mu\left(\widehat{X}_i\right)\Delta + \sigma\left(\widehat{X}_i\right) Z_i\sqrt{\Delta},$$
$$\overline{X}^\pm = \widehat{X}_i + \mu\left(\widehat{X}_i\right)\Delta \pm \sigma\left(\widehat{X}_i\right)\sqrt{\Delta}.$$

Comparison of (3.56) with the simplified Milstein scheme in the previous section shows that (3.56) avoids derivatives by using additional supporting values $\overline{X}$ and $\overline{X}^\pm$.

Another, quite different, approach to avoid explicit derivatives applies the classical idea of *Richardson extrapolation* to the Euler scheme. This idea was proposed by Talay and Tubaro [1990] and takes advantage of the fact that, under additional regularity conditions, the error of the Euler scheme can be sharpened beyond (3.30) (with $\beta = 1$) to

$$\mathrm{E}\left(g\left(\widehat{X}(T)\right)\right) = \mathrm{E}\left(g\left(X(T)\right)\right) + c\Delta + O\left(\Delta^2\right), \qquad (3.57)$$

for a constant c. Let $\widehat{X}_\Delta$ and $\widehat{X}_{2\Delta}$ be estimates of X based on Euler discretizations with time steps of Δ and 2Δ, respectively. Provided that (3.57) holds, we can write

$$2\mathrm{E}\left(g\left(\widehat{X}_\Delta(T)\right)\right) - \mathrm{E}\left(g\left(\widehat{X}_{2\Delta}(T)\right)\right) = \mathrm{E}\left(g\left(X(T)\right)\right) + O\left(\Delta^2\right), \quad (3.58)$$

which is our second-order extrapolation formula. As the Euler scheme is simple to set up, the extrapolation scheme is an attractive alternative to other second-order techniques. In practice, however, the convergence of the Euler scheme may not always be smooth enough to make (3.58) work well. Numerical experiments will nearly always be necessary (as is also the case of the Ito-Taylor schemes, for that matter).

A final word about generation of $\widehat{X}_\Delta$ and $\widehat{X}_{2\Delta}$ in the Richardson extrapolation scheme. To avoid duplication of work, we discretize time in buckets of Δ and generate both $\widehat{X}_\Delta$ and $\widehat{X}_{2\Delta}$ simultaneously, combining time steps in pairs for the purpose of generating $\widehat{X}_{2\Delta}$. That is, if we use Gaussian increments of $Z_1\sqrt{\Delta}, Z_2\sqrt{\Delta},\ldots$ for $\widehat{X}_\Delta$, we use $(Z_1+Z_2)\sqrt{\Delta}$, $(Z_3+Z_4)\sqrt{\Delta},\ldots$ for $\widehat{X}_{2\Delta}$. Not only do we save work by re-using Gaussian draws, we most likely also reduce the statistical error of our Monte Carlo estimate of the difference $2\mathrm{E}(g(\widehat{X}_\Delta(T))) - \mathrm{E}(g(\widehat{X}_{2\Delta}(T)))$ by raising correlation between $g(\widehat{X}_\Delta(T))$ and $g(\widehat{X}_{2\Delta}(T))$. We shall return to this idea in Section 3.3.1.

3.2.8 Bias vs. Monte Carlo Error

When we use an m-step discretization scheme in an n-path Monte Carlo run, we are exposed to two types of errors on the expectation we are trying to evaluate: i) the statistical Monte Carlo error e_s (the standard error); and ii) a bias e_b, originating from the discretization scheme. Raising n will reduce the standard error, but will not affect the bias which can only be reduced by increasing the number of steps m in the time discretization scheme. Raising m and/or n obviously involves a computational cost, so let us briefly consider explicitly the trade-offs involved in simultaneously reducing bias and standard error.

Assume first that the discretization scheme has weak order β. Proceeding informally, we interpret this as implying

$$e_b = c_b \Delta^\beta,$$

for some constant c_b. Also, we know that the variance of e_s is

$$\mathrm{Var}(e_s) = \frac{c_s}{n},$$

for a constant c_s. The total computing time τ is reasonably assumed to be proportional to nm or, using the fact that $\Delta = T/m$,

$$\tau = n \frac{c_\tau}{\Delta} \qquad (3.59)$$

for some constant c_τ. For a given computing budget τ, consider minimizing the total mean-square error (MSE) $c_b^2 \Delta^{2\beta} + \frac{c_s}{n}$. Using (3.59) to eliminate a variable, the optimization problem is

$$\min_\Delta \left(c_b^2 \Delta^{2\beta} + \frac{c_s c_\tau}{\tau \Delta} \right).$$

Let Δ^* be the value of Δ at which the minimum MSE is attained. Δ^* is seen to satisfy

$$\Delta^* = C \tau^{-\frac{1}{2\beta+1}}, \quad C \triangleq \left(\frac{c_s c_\tau}{2\beta c_b^2} \right)^{\frac{1}{2\beta+1}}, \qquad (3.60)$$

such that the minimum MSE becomes

$$C' \tau^{-\frac{2\beta}{2\beta+1}} \qquad (3.61)$$

for yet another constant C'.

Equations (3.60) and (3.61) reveal a number of structural characteristics of Monte Carlo simulation of discretized SDEs. For instance, according to (3.61), the optimal root-mean-square (RMS) error behaves with the computing time τ as

$$\mathrm{RMS} \propto \tau^{-\frac{\beta}{2\beta+1}}. \qquad (3.62)$$

The computational cost of working with SDEs that are not explicitly solvable are quantified by (3.62). For an unbiased (that is, exact) SDE simulation scheme, $\beta = \infty$ and the optimal RMS error converges at the rate of $\tau^{-\frac{1}{2}}$, consistent with the results of Section 3.1. However, for an Euler scheme ($\beta = 1$) the RMS error convergence rate is lowered to $\tau^{-\frac{1}{3}}$.

Equation (3.60) in principle tells us how to optimally allocate resources between the competing objectives of a lower bias and a lower standard error. Let m^* be defined through $\Delta^* = T/m^*$ and let n^* be defined through $\tau = c_\tau n^* m^*/T$. After a few rearrangements, we find the intuitive result

$$\sqrt{n^*} = C''(m^*)^\beta,$$

where C'' is a constant independent of τ. When we increase or decrease our computing budget, it is thus reasonable to allocate resources in such a way that we keep the factor $n^{1/2} m^{-\beta}$ constant. More detailed discussion, as well as asymptotic limit results, can be found in Duffie and Glynn [1995].

3.2.9 Sampling of Continuous Process Extremes

We round off our discussion of path simulation schemes by considering the pricing of options that depend on continuously or high-frequency sampled extremes of an SDE. We focus on the scalar case, with our SDE given as

$$dX(t) = \mu(t, X(t)) \, dt + \sigma(t, X(t)) \, dW(t),$$

where both X and W are 1-dimensional (i.e., $p = d = 1$). We also assume that the SDE is Euler-discretized according to

$$\widehat{X}_{i+1} = \widehat{X}_i + \mu\left(i\Delta, \widehat{X}_i\right) \Delta + \sigma\left(i\Delta, \widehat{X}_i\right) \sqrt{\Delta} Z_i, \quad i = 0, 1, \ldots, m-1,$$

with $m\Delta = T$.

On the interval $[0, T]$, let the maximum and minimum values of $X(t)$ be denoted $M_{[0,T]}$ and $m_{[0,T]}$, respectively. That is,

$$M_{[0,T]} \triangleq \max_{0 \leq t \leq T} X(t); \quad m_{[0,T]} \triangleq \min_{0 \leq t \leq T} X(t).$$

To give examples of options that depend on $M_{[0,T]}$ and $m_{[0,T]}$, consider for instance the up-and-out call option we encountered in Section 2.7. With a knock-out barrier of H and a terminal strike of K, the terminal maturity payout can be written as[11]

$$g(T) = 1_{\{M_{[0,T]} < H\}} (X(T) - K)^+.$$

A *double-barrier knock-out call option* with an upper barrier of H and a lower barrier of h pays

$$g(T) = 1_{\{m_{[0,T]} > h\}} 1_{\{M_{[0,T]} < H\}} (X(T) - K)^+.$$

Finally, a so-called *lookback call option* (see Section 2.7) pays

$$g(T) = \left(M_{[0,T]} - K\right)^+.$$

---

[11] If $X(t)$ is the logarithm of the asset price, we replace this expression by $g(T) = 1_{\{M_{[0,T]} \leq e^H\}} (e^{X(T)} - K)^+$, and similarly for the other payouts considered.

To price options such as those above, we must provide pathwise estimates of $M_{[0,T]}$ and $m_{[0,T]}$. Given our discretization schemes, natural estimators are

$$\widehat{M}_{[0,T]} = \max\left(X(0), \widehat{X}_1, \ldots, \widehat{X}_m\right), \tag{3.63}$$

$$\widehat{m}_{[0,T]} = \min\left(X(0), \widehat{X}_1, \ldots, \widehat{X}_m\right). \tag{3.64}$$

Even in cases where the discretization scheme itself is perfectly unbiased, it is clear that these estimators will understate the range of the extremes of $X(t)$, by consistently failing to account for the movement (the "overshoot" and "undershoot") of X between sample points $i\Delta$, $i = 0, 1, \ldots, m$. As a consequence, for each simulated path in an otherwise unbiased discretization scheme, almost surely

$$\widehat{M}_{[0,T]} < M_{[0,T]}, \quad \widehat{m}_{[0,T]} > m_{[0,T]}.$$

As shown in Andersen and Brotherton-Ratcliffe [1996], the bias introduced can be very significant, even if Δ is quite small. For instance, for a 1 year lookback option, Andersen and Brotherton-Ratcliffe [1996] report that even daily sampling produces a 6% price error.

To improve the price estimates of options that depend on continuously sampled extremes, we should alter (3.63) and (3.64) to take into consideration movements between sample dates. This can be accomplished by the Brownian bridge technique introduced in Andersen and Brotherton-Ratcliffe [1996] (see also Broadie et al. [1997]). Let us focus on a particular bucket $[i\Delta, (i+1)\Delta]$ and assume, consistent with the Euler scheme, that $\widehat{X}(t)$, $t \in [i\Delta, (i+1)\Delta]$, is a Gaussian process with conditional moments

$$\mathrm{E}\left(\widehat{X}(t) - \widehat{X}_i \Big| \widehat{X}_i\right) = \mu\left(i\Delta, \widehat{X}_i\right)(t - i\Delta), \quad t \in [i\Delta, (i+1)\Delta];$$

$$\mathrm{Var}\left(\widehat{X}(t) - \widehat{X}_i \Big| \widehat{X}_i\right) = \sigma\left(i\Delta, \widehat{X}_i\right)^2 (t - i\Delta), \quad t \in [i\Delta, (i+1)\Delta].$$

Assume that we have already simulated $\widehat{X}_i$ and $\widehat{X}((i+1)\Delta)$ by the Euler scheme above. Conditional on *both* $\widehat{X}_i$ and $\widehat{X}_{i+1}$, the process for $\widehat{X}(t)$, $t \in [i\Delta, (i+1)\Delta]$ is a Gaussian process "pinned" at the levels $\widehat{X}_i$ and $\widehat{X}_{i+1}$. The resulting process is known as a *Brownian bridge* with diffusion coefficient $\sigma(i\Delta, \widehat{X}_i)$. Let $\widehat{M}_i^c$ ($\widehat{m}_i^c$) be defined as the continuously sampled maximum (minimum) of $\widehat{X}(t)$ on $[i\Delta, (i+1)\Delta]$. The following lemma is a special case of a result in Andersen and Brotherton-Ratcliffe [1996]:

Lemma 3.2.1.

$$\mathrm{P}\left(\widehat{M}_i^c \leq s \Big| \widehat{X}_i, \widehat{X}_{i+1}\right) = 1 - \xi_i(s), \quad s > \max\left(\widehat{X}_i, \widehat{X}_{i+1}\right),$$

$$\mathrm{P}\left(\widehat{m}_i^c \leq s | \widehat{X}_i, \widehat{X}_{i+1}\right) = \xi_i(s), \quad s < \min\left(\widehat{X}_i, \widehat{X}_{i+1}\right),$$

where
$$\xi_i(s) \triangleq \exp\left(\frac{2\left(s - \widehat{X}_i\right)\left(\widehat{X}_{i+1} - s\right)}{\sigma\left(i\Delta, \widehat{X}_i\right)^2 \Delta}\right).$$

We can use the result of the lemma in a number of ways. Most obviously, we can apply it to sample $\widehat{M}_i^c$ and $\widehat{m}_i^c$ directly, by the inverse transform method (see Section 3.1.1.1). To illustrate, consider for instance sampling $\widehat{M}_i^c$. Having first drawn $\widehat{X}_i$ and $\widehat{X}_{i+1}$ by usual means, we draw an additional independent $\mathcal{U}(0,1)$ uniform variable U_i, and set

$$1 - \xi_i(\widehat{M}_i^c) = U_i$$

or, after a few rearrangements,

$$\widehat{M}_i^c = \frac{1}{2}\left(\widehat{X}_{i+1} + \widehat{X}_i\right) + \frac{1}{2}\sqrt{\left(\widehat{X}_{i+1} - \widehat{X}_i\right)^2 - 2\sigma\left(i\Delta, \widehat{X}_i\right)^2 \Delta \ln(1 - U_i)}.$$

This procedure can be repeated for $i = 0, 1, \ldots, m-1$, giving us the improved estimator for the maximum of X over $[0, T]$,

$$\widehat{M}_{[0,T]}^c = \max\left(\widehat{M}_0^c, \ldots, \widehat{M}_{m-1}^c\right).$$

For options depending on both the minimum and maximum (such as double barrier options and the double lookback options in He et al. [1998]), the necessary extensions required for joint sampling of minimum and maximum are developed in Andersen [1998].

For barrier options, we note that locating $\widehat{M}_i^c$ and $\widehat{m}_i^c$ directly is typically not necessary, as it suffices to check locally whether the barrier is breached. For an up-and-out knock-out option with barrier H, for each interval it thus suffices to check whether $\widehat{M}_i^c > H$ which, conditional on $\widehat{X}_i$ and $\widehat{X}_{i+1}$, happens with likelihood $\xi_i(H)$. So, provided that $\widehat{X}_i$ and $\widehat{X}_{i+1}$ are both below H, determining whether the barrier was nevertheless breached in $[i\Delta, (i+1)\Delta]$ is a matter of drawing a uniform variable U_i and setting

$$1_{\{\widehat{M}_i^c \geq H\}} = 1_{\{U_i \leq \xi_i(H)\}}. \tag{3.65}$$

This scheme is easily extended to time-dependent barriers and to cases where there are rebates[12]. As pointed out in Glasserman and Staum [2001], for

---

[12] To get the timing of rebate payments right, the exact time that the barrier is breached must, in principle, be located. Andersen and Brotherton-Ratcliffe [1996] list analytical Brownian bridge hitting time results that can be used for this purpose. For reasonably fine discretizations, it will often suffice to set the hitting time to, say, the mid-point of the time bucket where the barrier is known to be breached.

Markov processes and the special case of barrier options with no rebates, one can in fact avoid drawing U_i's altogether, as

$$\mathrm{E}\left(g\left(\widehat{X}_m\right) 1_{\{\widehat{M}^c_{[0,T]} < H\}} \middle| \widehat{X}_0, \widehat{X}_1, \ldots, \widehat{X}_m\right)$$
$$= \mathrm{E}\left(g\left(\widehat{X}_m\right) \prod_{i=0}^{m-1} 1_{\{\widehat{M}^c_i < H\}} \middle| \widehat{X}_0, \widehat{X}_1, \ldots, \widehat{X}_m\right)$$
$$= g\left(\widehat{X}_m\right) \prod_{i=0}^{m-1} \mathrm{E}\left(1_{\{\widehat{M}^c_i < H\}} \middle| \widehat{X}_i, \widehat{X}_{i+1}\right).$$

Here,

$$\mathrm{E}\left(1_{\{\widehat{M}^c_i < H\}} \middle| \widehat{X}_i, \widehat{X}_{i+1}\right) = \begin{cases} 0, & \widehat{X}_i \geq H \text{ or } \widehat{X}_{i+1} \geq H, \\ 1 - \xi_i(H), & \widehat{X}_i < H \text{ and } \widehat{X}_{i+1} < H. \end{cases}$$

In other words, rather than explicitly simulating the indicator functions (3.65), it suffices to adjust the terminal payout by the product of conditional survival probabilities along the path. This scheme is an example of *conditional Monte Carlo*, a variance-reduction scheme discussed in more detail in Section 25.2 and in Boyle et al. [1997]. One potential drawback of the scheme is the fact that we typically need to continue the paths for a longer period of time before a barrier crossing is detected and the path can be stopped.

We round off this section with a few comments. First, we note that the schemes above assume that X is well approximated by a Gaussian process. In some applications the geometric Brownian motion, say, may be a more appropriate model. In Lemma 3.2.1, we can easily accommodate this by simply replacing $\widehat{M}^c_i$, $\widehat{m}^c_i$, s, $\widehat{X}_i$, and $\widehat{X}_{i+1}$ with $\ln \widehat{M}^c_i$, $\ln \widehat{m}^c_i$, $\ln s$, $\ln \widehat{X}_i$, and $\ln \widehat{X}_{i+1}$. Other transformations are handled the same way.

Secondly, it should be pointed out that many real options are, in fact, not sampled continuously but rather at some finite but high frequency, often daily. Running an Euler scheme with daily discretization is obviously computationally inefficient. Fortunately, we can often use our scheme above as part of a Richardson-type interpolation idea. Indeed, it can often be established (see Andersen and Brotherton-Ratcliffe [1996]) that options on process extremes converge as $O(\sqrt{\Delta})$. If we first compute a price estimate $\widehat{V}^\Delta$ based on a relatively coarse value of Δ, we can then write

$$\widehat{V}^{\Delta^*} \approx \widehat{V}^c + (\widehat{V}^\Delta - \widehat{V}^c)\Delta^*/\Delta, \quad \Delta^* < \Delta,$$

where $\widehat{V}^c$ is the continuously monitored price computed by the scheme outlined above. We have here implicitly made the assumption that the regular Euler bias is small relative to the bias induced by using the wrong sampling frequency. The idea above is developed further in Chapter V of Andersen [1996], where a number of numerical results can also be found.

And finally, one may wonder whether it is possible to deal with a continuously monitored barrier option in a discrete-time simulation by adjusting the *barrier*, rather than the underlying *process*. As it turns out, this is indeed possible. Specifically, in the Black-Scholes-Merton model with volatility σ, let $V^c(H)$, $V^\Delta(H)$ be the values of a continuously and discretely sampled barrier options with barrier H, respectively. Assuming that the discrete sampling happens on a time grid with spacing Δ, we have the following result from Broadie et al. [1997]:

Theorem 3.2.2. *The following holds,*

$$V^\Delta(H) = V^c\left(He^{\pm\beta\sigma\sqrt{\Delta}}\right) + o\left(\sqrt{\Delta}\right),$$

where $+$ applies if $H > X(0)$, $-$ applies if $H < X(0)$, and $\beta = \zeta(1/2)/\sqrt{2\pi} \approx 0.5826$, with $\zeta(\cdot)$ being the Riemann zeta function.

According to this result we can price a continuous barrier option by evaluating a discrete barrier option instead (e.g., one where the barrier monitoring takes place only on the simulation dates of the Monte Carlo scheme used), but with the discrete barrier level shifted according to the theorem. Theorem 3.2.2 can also be used to save computation time by, say, turning a barrier option with daily observations into an option with quarterly observations, as the theorem shows that

$$V^{\Delta^*}(H) \approx V^\Delta\left(He^{\pm\beta\sigma(\sqrt{\Delta^*}-\sqrt{\Delta})}\right).$$

While the result of Theorem 3.2.2 is only proved for the log-normal process, practical experience shows that it is robust across a wide variety of models. A similar approach exists for lookback options, see Broadie et al. [1999].

3.2.10 PCA and Bridge Construction of Brownian Motion Paths

3.2.10.1 Brownian Bridge and Quasi-Random Sequences

To close out the section on sample path simulation, let us address alternative ways of generating sample paths of Brownian motion. So far, to produce a sample of the vector $\mathbf{W} = (W(\Delta), W(2\Delta), \ldots, W(m\Delta))^\top$, we have relied exclusively on the forward recursion

$$W(i\Delta + \Delta) = W(i\Delta) + Z_i\sqrt{\Delta}, \quad W(0) = 0, \tag{3.66}$$

where $Z_0, Z_1, \ldots, Z_{m-1}$ is a sequence of independent standard Gaussian variables. Rather than filling out the elements of $\mathbf{W}$ in order, we may, for instance, rely on a *Brownian bridge (BB) construction* where we first sample the end-point $W(m\Delta)$, then sample the mid-point[13] $W(\lfloor m/2 \rfloor \Delta)$ *conditional* on $W(m\Delta)$, and so forth. In executing this scheme, we can use the easily proven result below.

Lemma 3.2.3. *Let $\underline{t} < t < \overline{t}$. Conditional on $W(\underline{t})$ and $W(\overline{t})$, $W(t)$ is Gaussian with moments*

$$\mathrm{E}\left(W(t)|W(\underline{t}) = \underline{w}, W(\overline{t}) = \overline{w}\right) = \underline{w} \times \frac{\overline{t} - t}{\overline{t} - \underline{t}} + \overline{w} \times \frac{t - \underline{t}}{\overline{t} - \underline{t}},$$

$$\mathrm{Var}\left(W(t)|W(\underline{t}) = \underline{w}, W(\overline{t}) = \overline{w}\right) = \frac{(\overline{t} - t)(t - \underline{t})}{\overline{t} - \underline{t}}.$$

The BB scheme for construction of **W** relies on repeated application of the result in Lemma 3.2.3 to progressively fill in **W** in the "bisection" manner described above; consult any Monte Carlo textbook (e.g. Jäckel [2002] or Glasserman [2004]) if further details are required. As is the case for the standard scheme (3.66), a total of m standard Gaussian random variables are needed to construct a single sample of **W** by the Brownian bridge scheme, so the latter offers no computational advantage over the former. Why then use the Brownian bridge construction?

One important distinction between the BB construction and (3.66) is the fact that the Brownian bridge assigns different importance to the random numbers used to produce **W**. For instance, the very first Gaussian number drawn in the BB technique *alone* determines the end-point $W(m\Delta)$ of the Brownian motion — and thereby establishes a significant part of the overall coarse structure of the path of W. Subsequent random number draws contribute by filling in the details of the W-path, with late draws adding only to the fine-structure of the path. In contrast, with (3.66) the end-point $W(m\Delta)$ is affected equally by the m random numbers $Z_0, Z_1, \ldots, Z_{m-1}$. In most financial problems the coarse shape of the path of W is more critical than finer details, so ultimately the BB technique allows us to identify and isolate the important features of the Brownian motion path. In some variance reduction techniques this can be important, as it allows us to focus computational effort on the random numbers that matter the most. Also, some variance reduction techniques that are known to work particularly well on low-dimensional problems can now be applied to the (low-dimensional) random numbers that contribute most to the sample path.

One relevant technique is *quasi-random sequences* (also known as *low-discrepancy sequences*), a method of generating points on the hypercube that are as "dispersed" as possible. A good survey of the underlying ideas and theory can be found in Jäckel [2002] or Glasserman [2004], with source code available (for the special case of *Sobol sequences*) in Press et al. [1992]; suffice to say that quasi-random sequences can, under some circumstances,

---

[13] $\lfloor x \rfloor$ denotes the integer part of a real variable x.

accelerate Monte Carlo convergence substantially[14]. It is well-known, however, that the efficacy of quasi-random sequences depend strongly on the problem dimension (here: m, the number of random numbers needed per path), and that the sequences deteriorate in higher dimensions. When quasi-random sequences are combined with BB simulation of the path, however, the (low-dimensional) points of the sequences that are well-distributed can be applied to generate — by the methods in Section 3.1.1 — the Gaussian samples that determine the coarse structure of the paths, whereas the poorly distributed (high-dimensional) parts of the sequence can be relegated to the generation of less important fine-structure details[15]. A full account of this idea can be found in Moskowitz and Caflisch [1996].

3.2.10.2 PC Construction

With the Brownian Bridge (BB) construction of Brownian motion, much of the variance of the sample paths $\mathbf{W}$ is explained by the values of the first few (Gaussian) random variables drawn in the path simulation. We recall from Section 3.1.3, however, that the *optimal* way to project the variation of a Gaussian vector onto a low-dimensional set of random variables is done through a principal component (PC) construction, rather than the Brownian bridge. To demonstrate how a PC construction of $\mathbf{W} = (W(\Delta), W(2\Delta), \ldots, W(m\Delta))^\top$ would proceed, we first notice that the $m \times m$ variance-covariance matrix Σ of $\mathbf{W}$ has elements

$$\Sigma_{i,j} = \mathrm{E}\left(W(i\Delta) W(j\Delta)\right) = \min(i\Delta, j\Delta), \quad i,j = 1, 2, \ldots, m.$$

As shown in Åkesson and Lehoczky [1998], the eigenvalues of Σ can be found analytically to be

$$\lambda_i = \frac{\Delta}{4} \sin\left(\frac{\pi}{2} \cdot \frac{2i-1}{2m+1}\right)^{-2}, \quad i = 1, \ldots, m,$$

where $\lambda_1 > \lambda_2 > \ldots > \lambda_m$. Let e_i be the eigenvector associated with λ_i, then it is also known that $e_i = (e_{i,1}, e_{i,2}, \ldots, e_{i,m})^\top$, where

$$e_{i,j} = \frac{2}{\sqrt{2m+1}} \sin\left(j\pi \cdot \frac{2i-1}{2m+1}\right), \quad j = 1, \ldots, m.$$

From the results in Section 3.1.3, we know that we can write

$$\mathbf{W} = \sum_{i=1}^{m} Z_i \sqrt{\lambda_i} e_i, \tag{3.67}$$

---

[14]Theoretically from $O(1/\sqrt{N})$ to (nearly) $O(1/N)$. Comparative tests on actual finance problems can be found in, for instance, Brotherton-Ratcliffe [1994], Paskov and Traub [1995], and Joy et al. [1996].

[15]Alternatively we can use regular *pseudo-random numbers* for this.

where $Z_1, Z_2, \ldots, Z_m$ is a sequence of independent standard Gaussian random variables. This equation constitutes the principal components construction of the Brownian path, and it is characterized by the fact that for any $k \leq m$, the first k terms of (3.67) (that is, $\sum_{i=1}^{k} Z_i \sqrt{\lambda_i} e_i$) explain as much of the variance of $\mathbf{W}$ as is possible with k Gaussian variables. Even more so than for the Brownian bridge, the PC construction of a Brownian motion thus connects the overall shape of the Brownian path to a few of the Gaussian random variables Z_i, with the remaining random variables contributing only high-frequency details. As explained above, this can be useful in certain variance reduction techniques by allowing us to focus our attention and resources on just a few of the m random variables needed to simulate $\mathbf{W}$. We note that the PC construction is more expensive to compute than the BB technique (the latter is $O(m)$ whereas the former can be seen from (3.67) to be $O(m^2)$), so the optimality of the PC approach may, in some applications, be outweighed by its lack of speed.

While we developed the BB and PC constructions exclusively on an equidistant time grid, they easily extend to non-equidistant grids. When the grid is non-equidistant, the variance-covariance matrix of $\mathbf{W}$ has elements

$$\Sigma_{i,j} = \mathrm{E}\left(W(t_i)W(t_j)\right) = \min\left(t_i, t_j\right), \quad i,j = 1, 2, \ldots, m,$$

and eigenvectors and eigenvalues must then be found numerically, rather than through the analytical results listed earlier. Also, both the BB and PC techniques can easily be extended to the case of multi-variate Brownian motions, see Jäckel [2002].

Finally, for those interested in such matters, we note that in the limit $m \to \infty$, the PC construction of Brownian motion is known as the *Karhunen-Loeve decomposition*. In the continuous-time limit, the BB representation is sometimes known as a *Haar function* decomposition of Brownian motion.

3.3 Sensitivity Computations

In most finance applications, the fact that options must be dynamically hedged and risk managed requires us not only to produce an estimate of an option price, but also to compute reliable estimates of the sensitivity of the price with respect to the underlying state variables, as well as various other model parameters. In this section, we will present a number of methods for sensitivity computations by Monte Carlo methods. For each method, we use the problem of estimating the stock price delta of options in the Black-Scholes economy as a motivating example. We shall spend much more time on sensitivity computations (by PDE and Monte Carlo methods) in Part V of this book, often in the context of particular interest rate products. Here we just give a flavor of things to come.

3.3.1 Finite Difference Estimates

3.3.1.1 Black-Scholes Delta

Consider a T-maturity European option on a dividend-free stock S in the Black-Scholes economy. Let the payout function be $g(S(T))$, and assume that the continuously compounded interest rate is a constant r. With $V(S_0)$ denoting the time 0 price of the option given $S(0) = S_0$, we are interested in computing

$$\frac{dV}{dS_0} = \lim_{h \to 0} \frac{V(S_0 + h) - V(S_0)}{h}. \tag{3.68}$$

In a Monte Carlo setting, we can approximate this derivative ("delta") by finite difference techniques as follows. First, for some fixed number ε draw random standard Gaussian variables Z and Z_ε, and set (see Section 3.2.1)

$$S(T) = S_0 \exp\left(\left(r - \frac{1}{2}\sigma^2\right)T + \sigma\sqrt{T}Z\right),$$

$$S_\varepsilon(T) = (S_0 + \varepsilon) \exp\left(\left(r - \frac{1}{2}\sigma^2\right)T + \sigma\sqrt{T}Z_\varepsilon\right),$$

where σ as always denotes the constant volatility of the stock. We then form the difference

$$\delta = e^{-rT}\varepsilon^{-1}\left(g\left(S_\varepsilon(T)\right) - g(S(T))\right),$$

such that δ constitutes a single-sample estimate for dV/dS_0. By generating n independent replications of δ and forming the sample average, we will obtain in the limit $n \to \infty$ the finite difference ratio

$$\frac{V(S_0 + \varepsilon) - V(S_0)}{\varepsilon}. \tag{3.69}$$

We know from Chapter 2 that this estimate will be biased relative to the true derivative dV/dS_0 by an amount of order $O(\varepsilon)$.

We have so far not mentioned whether the standard Gaussian variables Z and Z_ε should be independent or not. To analyze this, we need to consider the variance of the Monte Carlo estimator of (3.69). From Theorem 3.1.2, we know that for a finite number of trials n, the variance of our sample average will decrease as v_ε/n, where

$$v_\varepsilon = \varepsilon^{-2}e^{-2rT}\text{Var}\left(g\left(S_\varepsilon(T)\right) - g(S(T))\right)$$
$$= \varepsilon^{-2}e^{-2rT}\left[\text{Var}\left(g\left(S(T)\right)\right) + \text{Var}\left(g\left(S_\varepsilon(T)\right)\right)\right.$$
$$\left. - 2\text{Cov}\left(g\left(S_\varepsilon(T)\right), g\left(S(T)\right)\right)\right].$$

If the random numbers Z and Z_ε are independent, $\text{Cov}(g(S_\varepsilon(T)), g(S(T)))$ will be zero and

$$v_\varepsilon \approx 2\varepsilon^{-2}e^{-2rT}\text{Var}\left(g\left(S(T)\right)\right).$$

Making ε approach zero — as is needed to reduce the bias of the finite difference approximation (3.69) — will cause v_ε grow at a rate of $O(\varepsilon^{-2})$. This is obviously not ideal as our Monte Carlo estimate will be swamped by noise if ε is picked too small. On the other hand, if we set Z and Z_ε to be *identical*, we see that
$$S_\varepsilon(T) = (S_0 + \varepsilon)\, S(T)/S_0$$
and would expect
$$\text{Cov}\,(g\,(S_\varepsilon(T)), g(S(T))) > 0,$$
which would reduce v_ε relative to the independent case. For smooth g, a Taylor expansion in ε shows that
$$\begin{aligned} g\,(S_\varepsilon(T)) &= g\,(S(T)) + (S_\varepsilon(T) - S(T))\, g'\,(S(T)) + \ldots \\ &= g\,(S(T)) + \varepsilon S(T)/S_0 \cdot g'\,(S(T)) + \ldots. \end{aligned}$$
If derivatives of g are bounded
$$\text{Cov}\,(g\,(S_\varepsilon(T)), g\,(S(T))) = \text{Var}\,(g\,(S(T))) + O\,(\varepsilon),$$
and similarly for $\text{Var}(g(S_\varepsilon(T)))$. In other words,
$$v_\varepsilon = e^{-2rT} O(1) \tag{3.70}$$
which is a clear improvement over the earlier $O(\varepsilon^{-2})$ result.

The result (3.70) hinged on the payout function having bounded derivatives. We can, in fact, relax this considerably, to functions that are essentially just continuous in the stock price; see Section 3.3.2.2 for a discussion. For discontinuous payouts, however, (3.70) will not hold. To demonstrate, consider a digital option paying
$$g\,(S(T)) = 1_{\{S(T) > K\}},$$
for some strike K. With $Z = Z_\varepsilon$ we get (assuming $\varepsilon > 0$ and that the probability measure is P)
$$\begin{aligned} \mathrm{E}\left([g\,(S_\varepsilon(T)) - g\,(S(T))]^2\right) &= \mathrm{P}\,(S(T) \le K < S_\varepsilon(T)) \\ &= \mathrm{P}\,(S(T) \le K < (1 + \varepsilon/S_0)\, S(T)) = O(\varepsilon), \end{aligned}$$
compared with the $O(\varepsilon^2)$ result for smooth g.

3.3.1.2 General Case

To generalize the problem considered in the previous section, we consider a setting where a random variable Y depends on a parameter $\alpha \in \mathbb{R}$, in the sense that each value of α uniquely determines a scheme for the generation of Y. The random variable Y will typically represent a (discounted) option payout, and α is typically an initial value of an asset price (as in Section 3.3.1.1) or a parameter in the (vector) equations determining the dynamics of the underlying model. Let

$$V(\alpha) = E(Y(\alpha)),$$

and consider the problem of determining $dV/d\alpha$.

In the basic finite difference Monte Carlo approximation to $dV/d\alpha$, we use the sample average of one-sided difference coefficients,

$$\overline{\delta}_n = \frac{\overline{Y}_n(\alpha + \varepsilon) - \overline{Y}_n(\alpha)}{\varepsilon},$$

where $\overline{Y}_n(\alpha)$ is the sample average of n realizations of $Y(\alpha)$. In the limit,

$$\lim_{n \to \infty} \overline{\delta}_n = \frac{V(\alpha + \varepsilon) - V(\alpha)}{\varepsilon} = dV/d\alpha + O(\varepsilon).$$

If we instead wish to use a central estimator

$$\overline{\delta}_n^c = \frac{\overline{Y}_n(\alpha + \varepsilon) - \overline{Y}_n(\alpha - \varepsilon)}{2\varepsilon},$$

we get

$$\lim_{n \to \infty} \overline{\delta}_n^c = \frac{V(\alpha + \varepsilon) - V(\alpha - \varepsilon)}{2\varepsilon} = dV/d\alpha + O(\varepsilon^2),$$

but now need to simulate an extra random variable (that is, $Y(\alpha - \varepsilon)$), increasing the computational cost.

In the generation of $\overline{Y}_n(\alpha \pm \varepsilon)$ and $\overline{Y}_n(\alpha)$, the individual samples of $Y(\alpha + \varepsilon)$, $Y(\alpha - \varepsilon)$, and $Y(\alpha)$ would typically be based on a series of draws of vector-valued support variables Z, with $Y(\alpha) = Y(Z; \alpha)$, and so forth. For instance, in an m-step Euler simulation of an SDE with d Brownian motions, each SDE path (and each outcome of Y) would involve $d \cdot m$ i.i.d. standard Gaussian variables $Z_1, \ldots, Z_{d \cdot m}$. The observations in the previous section tell us that to minimize variance we should use the same Z for $Y(\alpha + \varepsilon)$, $Y(\alpha - \varepsilon)$, and $Y(\alpha)$. In practice, this is often easiest to accomplish by simply using the same random number seed (see Section 3.1.1) in otherwise separate computations of each of the quantities $\overline{Y}_n(\alpha + \varepsilon)$, $\overline{Y}_n(\alpha - \varepsilon)$, and $\overline{Y}_n(\alpha)$. Assuming this so-called *common random number scheme* is followed, the variance analysis in Section 3.3.1.1 can be generalized to our setting, and we would expect that either i) $\text{Var}(\overline{\delta}_n) = \text{Var}(\overline{\delta}_n^c) = O(\varepsilon^{-1}n^{-1})$; or ii) $\text{Var}(\overline{\delta}_n) = \text{Var}(\overline{\delta}_n^c) = O(n^{-1})$. Case ii) essentially requires a.s. continuity[16] of $Y(\alpha)$ with respect to α, as would be the case when Y represents a continuous

---

[16] More precisely, we need uniform integrability in the difference coefficients $[Y(\alpha + \varepsilon) - Y(\alpha)]\varepsilon^{-1}$ and $\frac{1}{2}[Y(\alpha + \varepsilon) - Y(\alpha - \varepsilon)]\varepsilon^{-1}$. See Section 3.3.2.2 for more precise conditions.

option payout function. Case i) generally applies when Y represents a discontinuous option payout, such as the digital option considered in Section 3.3.1.1.

If case ii) above applies, the estimator variance is independent of ε, and ε should be picked as small as possible (a matter of machine precision) to minimize the $O(\varepsilon)$ and $O(\varepsilon^2)$ biases of $\overline{\delta}_n$ and $\overline{\delta}_n^c$. If the overhead of evaluating $Y(Z;\alpha)$ for given Z is small relative to the cost of generating Z, the central estimator $\overline{\delta}_n^c$ will dominate. For complicated payout functions, however, there may be situations when $\overline{\delta}_n$ is preferable, despite its slower convergence rate in ε. For case i), we must weigh bias against variance in a manner quite similar to the discussion in Section 3.2.8: if ε is small the difference coefficient bias will be small, but the variance of the estimators $\overline{\delta}_n$ and $\overline{\delta}_n^c$ will be high. An RMS minimization similar to the one in Section 3.2.8 is possible; in the interest of brevity, we leave this as an exercise to the reader (see also Glasserman [2004], Chapter 7).

3.3.2 Pathwise Estimate

3.3.2.1 Black-Scholes Delta

Reverting back to the setting of Section 3.3.1.1, let us take another look at the delta definition (3.68):

$$\begin{aligned}\frac{dV}{dS_0} &= \lim_{h \to 0} \frac{V(S_0 + h) - V(S_0)}{h} \\ &= e^{-rT} \lim_{h \to 0} \mathrm{E}\left(\frac{g(S_h(T)) - g(S(T))}{h}\right),\end{aligned} \qquad (3.71)$$

where we have used the same notation as earlier:

$$S_h(T) = (S_0 + h)\, S(T)/S_0. \qquad (3.72)$$

Under sufficient regularity on g, we can interchange expectation and limit in (3.71), such that simply

$$\begin{aligned}\frac{dV}{dS_0} &= e^{-rT}\mathrm{E}\left(\lim_{h \to 0} \frac{g(S_h(T)) - g(S(T))}{h}\right) \\ &= e^{-rT}\mathrm{E}\left(g'(S(T)) \frac{dS(T)}{dS_0}\right) \\ &= e^{-rT}\mathrm{E}\left(g'(S(T)) \frac{S(T)}{S_0}\right),\end{aligned} \qquad (3.73)$$

where $g'(x) \triangleq dg/dx$, and the last equality follows by the linearity of (3.72).

We can implement the result (3.73) directly in a Monte Carlo trial, by generating samples of $S(T)$ and recording the sample averages of

$g'(S(T))S(T)/S_0$. The resulting estimate for dV/dS_0 is a direct and unbiased estimate of the true derivative; it is known as a *pathwise estimate*.

For (3.73) to hold, g should be continuous, but does not necessarily need to be differentiable everywhere (it suffices that g is Lipschitz continuous, as discussed below). A regular call option payout $g(x) = (x-K)^+$, for instance, is non-differentiable at $x = K$, but we simply write

$$g'(x) = 1_{\{x>K\}}$$

and proceed directly with (3.73). For discontinuous payouts[17], however, care must be taken as a direct application of (3.73) will introduce a bias. To demonstrate, consider the case $g(x) = 1_{\{x>K\}}$. Proceeding informally, a literal application of (3.73) results in

$$\frac{dV}{dS_0} = e^{-rT}\mathrm{E}\left(\delta\left(S(T)-K\right)\frac{S(T)}{S_0}\right),$$

where $\delta(x)$ is the Dirac delta function. While correct, this result is unsuited for Monte Carlo simulation: no matter how many samples n we draw of $S(T)$, the likelihood of $\delta(S(T)-K)$ being non-zero is zero, and the derivative would almost surely be estimated as 0. The correct result, however, is

$$e^{-rT}\mathrm{E}\left(\delta\left(S(T)-K\right)\frac{S(T)}{S_0}\right) = e^{-rT}\mathrm{E}\left(\delta\left(S(T)-K\right)\frac{K}{S_0}\right)$$

$$= e^{-rT}\frac{K}{S_0}\mathrm{E}\left(\delta\left(S(T)-K\right)\right)$$

$$= e^{-rT}\frac{K}{S_0}\varphi_S(K),$$

where $\varphi_S(\cdot)$ is the density of $S(T)$.

3.3.2.2 General Case

In a general setting, the technique employed in Section 3.3.2.1 above is known as *infinitesimal perturbation analysis*. A broad overview of the technique can be found in Glasserman [2004], with applications to finance covered in Broadie and Glasserman [1996]. Our treatment follows the latter closely.

Borrowing the notation of Section 3.3.1.2, we again consider estimating $dV/d\alpha$, where $V(\alpha) = \mathrm{E}(Y(\alpha))$. The basic idea of the pathwise derivative estimate is to write

$$\frac{dV}{d\alpha} = \frac{d}{d\alpha}\mathrm{E}\left(Y(\alpha)\right) = \mathrm{E}\left(\frac{d}{d\alpha}Y(\alpha)\right). \qquad (3.74)$$

---

[17] Or for the evaluation of, say, the second derivative (gamma) of a call payout.

The exchange of expectation and differentiation requires certain regularity conditions to be valid. In practice, the most interesting situation arises when Y represents a (discounted) payout function, such that

$$Y(\alpha) = g(X(\alpha)),$$

where $X(\alpha) = (X_1(\alpha), \ldots, X_q(\alpha))^\top$ is a q-dimensional random vector of observations (possibly at different dates) of asset prices. In this case, we have the following result, from Broadie and Glasserman [1996]:

Proposition 3.3.1. *For all α in some open interval $\mathcal{A}$ assume that $dX_i/d\alpha$ exists almost surely for all $i = 1, \ldots, q$. Suppose that the function g is almost surely differentiable[18] and is Lipschitz, such that*

$$|g(x) - g(y)| \leq k|x - y|$$

for some constant k. Finally, assume that there exists finite-mean random variables β_i, $i = 1, \ldots, q$, such that for all $\alpha_1, \alpha_2 \in \mathcal{A}$

$$|X_i(\alpha_2) - X_i(\alpha_1)| \leq \beta_i |\alpha_2 - \alpha_1|.$$

In this case, (3.74) holds.

The first two assumptions of Proposition ensure that the random variable $dY(\alpha)/d\alpha$ exists almost surely, with its value given by the chain rule

$$\frac{d}{d\alpha} Y(\alpha) = \sum_{i=1}^{q} \frac{\partial g}{\partial X_i} \frac{dX_i}{d\alpha}. \tag{3.75}$$

As we saw earlier, in Section 3.3.2.1, almost sure existence of $dY(\alpha)/d\alpha$ is not sufficient for the pathwise method to yield an unbiased estimator, we also need, roughly speaking, for g to be continuous at the points at which differentiability fails. The last two conditions ensure this, and together imply that Y is almost surely Lipschitz in α:

$$|Y(\alpha_2) - Y(\alpha_1)| \leq \beta_Y |\alpha_2 - \alpha_1|, \quad \beta_Y = k \sum_{i=1}^{q} \beta_i.$$

As $\alpha^{-1}|Y(h+\alpha) - Y(h)|$ is then bounded by β_Y, where $E(\beta_Y) < \infty$, the result of Proposition 3.3.1 follows from the dominated convergence theorem. See Broadie and Glasserman [1996] for further details.

Remark 3.3.2. If in Proposition 3.3.1 we further assume that $E(\beta_i^2) < \infty$, it follows that $E(Y(h+\varepsilon) - Y(h)) \leq \beta_Y^2 \varepsilon^2$, such that

---

[18] That is, differentiable everywhere except on some set $\mathcal{X}$ where $P(X(\alpha) \in \mathcal{X}) = 0$.

$$\mathrm{Var}\,(Y(h+\varepsilon) - Y(h)) = O(\varepsilon^2).$$

We recognize this as case ii) from Section 3.3.1.2 on finite difference estimates, for which we have now made the regularity conditions more precise. In practice, the Lipschitz continuity of g is the critical condition.

Remark 3.3.3. For discontinuous payouts, the pathwise method will yield a biased estimator. As we saw in Section 3.3.2.1, however, for a simple process where the transition density is known, the bias can often be accounted for. We shall see an example of this in Chapter 24. Notice that if the transition density is known, another method for sensitivity simulation — the likelihood ratio method — also applies. See Section 3.3.3 below for details about this method.

3.3.2.3 Sensitivity Path Generation

In Section 3.3.2.1, simulation of $dY(\alpha)/d\alpha$ was straightforward due to the simplicity of the Black-Scholes dynamics. In general, we see from (3.75) that generation of the random variables $dY(\alpha)/d\alpha$ will require us to compute the partial derivatives of the payout with respect to the underlying assets ($\partial g/\partial X_i$), as well as the sensitivities of the assets with respect to the perturbation parameter α ($dX_i/d\alpha$). The latter is normally the most difficult, and we shall outline a general approach here.

For illustration, consider a scalar SDE of the usual form

$$dX(t) = \mu(t, X(t))\,dt + \sigma(t, X(t))\,dW(t),$$

where X and W are one-dimensional. Let α be a parameter on which $X(t)$ depends (such as $X(0)$ or some parameter of μ or σ). Let $D_\alpha(t)$ denote $dX(t)/d\alpha$. Formally differentiating the SDE with respect to α, we get

$$dD_\alpha(t) = \mu'(t, X(t))\,D_\alpha(t)\,dt + \sigma'(t, X(t))\,D_\alpha(t)\,dW(t), \qquad (3.76)$$

where $\mu'(t,x) = \partial\mu(t,x)/\partial x$, and similar for σ'. This SDE can be discretized and simulated in parallel with the simulation of the SDE for $X(t)$ itself. In general, the work associated with this will obviously be more substantial than for the Black-Scholes delta, where we saw that $D_\alpha(t)$ could be recovered as the simple fraction $X(t)/X(0)$ (see equation (3.73)).

A few notes on the technique above. First, some regularity is obviously needed for (3.76) to be meaningful; see Kunita [1990] for some relevant results. Second, extensions to multi-dimensional SDEs are straightforward, although the dimension of the total scheme can be large. For instance, if X is p-dimensional and we wish to compute sensitivities with respect to $X_i(0)$, $i=1,\ldots,p$, a $p \times p$ system of SDEs for quantities $dX_i(t)/dX_j(0)$ will be required. We shall discuss approximative methods to improve efficiency of such high-dimensional matrix SDEs later, in Chapter 24 (see also Glasserman and Zhao [1999]).

3.3.3 Likelihood Ratio Method

As discussed above, the pathwise derivative method typically applies only to options with sufficiently smooth payouts[19] and can be cumbersome for multi-dimensional SDEs. For processes with explicitly known transition densities, the alternative *likelihood ratio method* can be used. This method applies to discontinuous payout functions and, unlike the pathwise method, requires little knowledge of the payout function and its derivatives, making it convenient for general implementation on a computer. When both methods apply, however, the pathwise derivative method generally is more efficient.

3.3.3.1 Black-Scholes Delta

In the notation of Sections 3.3.1.1 and 3.3.2.1, the Black-Scholes price of a call option can be written

$$V(S_0) = e^{-rT} \mathrm{E}\left((S(T) - K)^+ \big| S(0) = S_0 \right)$$
$$= e^{-rT} \mathrm{E}\left(\left(e^{\ln S_0 + (r - \frac{1}{2}\sigma^2)T + \sigma\sqrt{T}Z} - K\right)^+ \right)$$
$$= e^{-rT} \mathrm{E}\left(\left(e^{Y(T)} - K\right)^+ \right),$$

where $Z \sim \mathcal{N}(0,1)$ and

$$Y(T) \sim \mathcal{N}\left(\ln S_0 + \left(r - \frac{1}{2}\sigma^2\right) T, \sigma^2 T \right).$$

The density of Y is thereby a function of S_0:

$$P(Y \in dy) = \frac{dy}{\sqrt{2\pi}\sigma\sqrt{T}} \exp\left(-\frac{1}{2}\left(\frac{y - \ln S_0 - (r - \frac{1}{2}\sigma^2)T}{\sigma\sqrt{T}} \right)^2 \right)$$
$$\triangleq \varphi(y; S_0)\, dy.$$

Thereby

$$V(S_0) = e^{-rT} \int_{-\infty}^{\infty} (e^y - K)^+ \varphi(y; S_0)\, dy, \tag{3.77}$$

such that

$$\frac{dV(S_0)}{dS_0} = e^{-rT} \int_{-\infty}^{\infty} (e^y - K)^+ \frac{\partial \varphi(y; S_0)}{\partial S_0}\, dy$$
$$= e^{-rT} \int_{-\infty}^{\infty} (e^y - K)^+ \frac{\partial \varphi(y; S_0)}{\partial S_0} \frac{\varphi(y; S_0)}{\varphi(y; S_0)}\, dy$$
$$= e^{-rT} \int_{-\infty}^{\infty} (e^y - K)^+ \frac{\partial \ln \varphi(y; S_0)}{\partial S_0} \varphi(y; S_0)\, dy. \tag{3.78}$$

---

[19]But see Remark 3.3.3.

Comparison of (3.78) with (3.77) demonstrates that we can effectively compute the Black-Scholes delta as the price of a security that pays out at time T the amount

$$\left(e^{Y(T)} - K\right)^+ l\left(Y(T)\right), \tag{3.79}$$

where l is the so-called *log-likelihood ratio* (also known as the *score function*)

$$l(Y(T)) = \frac{\partial \ln \varphi(Y(T); S_0)}{\partial S_0} = \frac{Y(T) - \ln S_0 - (r - \frac{1}{2}\sigma^2)T}{S_0 \sigma^2 T} = \frac{Z}{S_0 \sigma \sqrt{T}}. \tag{3.80}$$

By differentiating the density, rather than the payout itself, the likelihood ratio technique applies to even discontinuous payouts, requiring only that the density is smooth (which is clearly the case here). Notice in particular that the log-likelihood ratio is independent of the payout, allowing us to use the same function $l(Y(T))$ for all European style payout functions $V(T) = g(S(T))$.

3.3.3.2 General Case

As in Section 3.3.2.2, consider now the general case where a random variable $Y(\alpha)$ represents a (deflated) payout function g applied to a vector of random variables $X(\alpha) = (X_1(\alpha), \ldots, X_q(\alpha))^\top$. Again, α is a parameter with respect to which we wish to compute sensitivities. Let the joint density of $X(\alpha)$ be denoted $f(x; \alpha)$, $x \in \mathbb{R}^q$. We then have

$$V(\alpha) = \int_{\mathbb{R}^q} g(x) f(x; \alpha)\, dx.$$

Making the reasonable assumption that density $f(x; \alpha)$ is a smooth function of α, we interchange integration and differentiation, such that

$$\frac{\partial V(\alpha)}{\partial \alpha} = \int_{\mathbb{R}^q} g(x) \frac{\partial f(x; \alpha)}{\partial \alpha}\, dx = \int_{\mathbb{R}^q} g(x) \frac{\partial \ln f(x; \alpha)}{\partial \alpha} f(x; \alpha)\, dx.$$

As for the Black-Scholes case above, the derivative $\partial V(\alpha)/\partial \alpha$ can thus be computed as the expectation of the payout modified by a log-likelihood ratio:

$$g(x)l(x), \quad l(x) = \frac{\partial \ln f(x; \alpha)}{\partial \alpha}.$$

3.3.3.3 Euler Schemes

In practice, the reliance on explicit knowledge of a transition density can be a considerable obstacle, and may rule out the application of the likelihood ratio method for many complex models. In cases where process dynamics are simulated through a simple time-discretization scheme, the situation is, however, salvageable, as we shall now demonstrate.

3.3 Sensitivity Computations

For illustration, consider an asset $X(t)$ that follows an SDE of the type

$$dX(t) = \mu(t, X(t); \alpha) \, dt + \sigma(t, X(t); \alpha) \, dW(t),$$

where μ and σ are smooth functions, and where α is a parameter. In general, we do not know the exact transition density for $X(t)$. However, suppose now that we use an Euler scheme to simulate on some grid $\{t_i\}_{i=1}^m$, i.e.

$$\widehat{X}(t_{i+1}) = \widehat{X}(t_i) + \mu\left(t_i, \widehat{X}(t_i); \alpha\right)(t_{i+1} - t_i) + \sigma\left(t_i, \widehat{X}(t_i); \alpha\right)\sqrt{t_{i+1} - t_i} Z_i,$$

where $t_0 = 0$ and $Z_0, Z_1, \ldots, Z_{m-1}$ is a sequence of i.i.d. standard Gaussian random variables. Clearly, the transition density for the $\widehat{X}(t_{i+1})$ is now Gaussian,

$$\widehat{X}(t_{i+1})|\widehat{X}(t_i) \sim \mathcal{N}\left(m_i\left(\widehat{X}(t_i); \alpha\right), s_i\left(\widehat{X}(t_i); \alpha\right)\right),$$

where, for $i = 0, \ldots, m-1$,

$$m_i\left(\widehat{X}(t_i); \alpha\right) = \widehat{X}(t_i) + \mu\left(t_i, \widehat{X}(t_i); \alpha\right)(t_{i+1} - t_i),$$

$$s_i\left(\widehat{X}(t_i); \alpha\right) = \sigma\left(t_i, \widehat{X}(t_i); \alpha\right)^2 (t_{i+1} - t_i).$$

Set $\widehat{X} = (\widehat{X}(t_1), \ldots, \widehat{X}(t_m))^\top$; the density of this vector is (where $x_0 = X(0)$)

$$f(x_1, \ldots x_m; \alpha) = \prod_{i=1}^m \frac{1}{\sqrt{2\pi}\sqrt{s_{i-1}(x_{i-1}; \alpha)}} \exp\left(-\frac{(x_i - m_{i-1}(x_{i-1}; \alpha))^2}{2 s_{i-1}(x_{i-1}; \alpha)}\right). \tag{3.81}$$

Consider some (potentially path-dependent) security V with payout function $g(\widehat{X})$ and time 0 price of $V(\alpha)$. Equipped with (3.81), we can estimate the parameter sensitivity by the likelihood ratio method as

$$\frac{\partial V(\alpha)}{\partial \alpha} = \mathrm{E}\left(g(\widehat{X}) \frac{d}{d\alpha} \ln\left(f(\widehat{X}; \alpha)\right)\right),$$

where

$$\frac{d}{d\alpha} \ln\left(f(\widehat{X}; \alpha)\right)$$
$$= \frac{d}{d\alpha} \sum_{i=1}^m \left(-\frac{1}{2} \ln(s_{i-1}(x_{i-1}; \alpha)) - \frac{(x_i - m_{i-1}(x_{i-1}; \alpha))^2}{2 s_{i-1}(x_{i-1}; \alpha)}\right). \tag{3.82}$$

This derivative can typically be computed in closed form; if not, one can estimate it by finite differences (as in Su and Randall [2008]). Notice that when $\alpha = X(0)$ (i.e. we are trying to estimate the delta), only s_0 and m_0 will depend on α, simplifying computations.

The idea used above extends easily to vector-valued $X(t)$. In principle, higher-order schemes (e.g. the Milstein scheme) can also be used, although the complexity increases considerably.

3.3.3.4 Some Remarks

The main advantage of the likelihood ratio is the fact that it applies to classes of payouts for which other methods (pathwise methods, finite difference method) do not work well. Moreover, the method is easy and efficient to implement, as the log-likelihood ratio l is independent of the payout and does not — unlike the general pathwise method — require simulation of any quantities other than the vector X itself. As discussed above, the primary drawback of the method is its reliance on explicit knowledge of the process density, ruling out many of the more advanced models (although, as Section 3.3.3.3 demonstrated, there may sometimes be ways around this). Further, the variance of the likelihood ratio method can often be quite big, particularly if the parameter α simultaneously affects multiple stochastic variables. A fuller discussion of this issue, as well as the related issue of absolute continuity, can be found in Glasserman [2004]. We note in passing that the likelihood ratio method is a special case of a body of methods that have emerged from the so-called *Malliavin calculus*; see Fournié et al. [1999] for a survey. Most Malliavin methods other than the basic likelihood ratio method are, however, not particularly attractive due to computational issues[20].

We round off this section by noting that the various methods for derivative estimates can often be successfully combined. For instance, while it is common to use either the pathwise method or the likelihood ratio method to compute first order sensitivities (such as delta), second-order sensitivities (such as gamma) are often done by the finite difference method applied to first-order sensitivities, often using fairly sizable shifts of the underlying variables. By combining the pathwise method with the likelihood ratio method, we can also address the fact that the first derivative of many kinked option payouts is discontinuous, allowing us to produce a bias-free estimate of the second derivative. See Fournié et al. [1999] for other examples of combining the pathwise method with the likelihood ratio method.

3.4 Variance Reduction Techniques

As discussed earlier, the convergence of the Monte Carlo method is quite slow, of order $O(n^{-1/2})$ where n is the number of Monte Carlo samples. While there is little that can be done to improve[21] the $n^{-1/2}$ order itself, the constant multiplying $n^{-1/2}$ can be affected by a careful choice of the Monte Carlo estimator. Methods to improve numerical efficiency this way are

---

[20] Besides, the Malliavin calculus itself, a very technical area of mathematics even for specialists, can be avoided altogether, as Chen and Glasserman [2007b] demonstrate.

[21] As we discussed earlier, the quasi-random Monte Carlo method can theoretically achieve better convergence order than $O(n^{-1/2})$.

known collectively as *variance reduction techniques*, and constitute a major area of research in the theory of Monte Carlo methods. Our introduction of the topic is limited to a few basic examples. More details are provided later in the book for concrete models and products (see e.g. Chapter 25), and more information can be found in the standard Monte Carlo literature, including Hammersley and Handscomb [1965] and the survey article Boyle et al. [1997].

3.4.1 Variance Reduction and Efficiency

We recall that the goal of the Monte Carlo method is to estimate some quantity μ (e.g., the price of a financial contract) as the sample mean of n i.i.d. random variables $Y_1, \ldots, Y_n$, where each Y_i has expectation $E(Y_i) = \mu$ and variance $\text{Var}(Y_i) = \sigma^2$. From Section 3.1, we know that for large n the standard error of the sample mean $n^{-1} \sum Y_i$ is $\sigma n^{-1/2}$, with the probabilistic error bounds on the estimate of μ being proportional to the standard error.

Suppose now that we have available two sets of i.i.d. sequences $Y_{1,i}$ and $Y_{2,i}$, $i = 1, \ldots, n$, where $E(Y_{1,i}) = E(Y_{2,i}) = \mu$, but $\text{Var}(Y_{1,i}) = \sigma_1^2$ and $\text{Var}(Y_{2,i}) = \sigma_2^2$, with $\sigma_1 \neq \sigma_2$. Also suppose that the time it takes on a computer to generate individual samples $Y_{1,i}$ and $Y_{2,i}$ is τ_1 and τ_2, respectively. Which of the two estimators $n^{-1} \sum Y_{1,i}$ and $n^{-1} \sum Y_{2,i}$ is preferable? To answer this question, assume that we have a large fixed computing time budget τ. The number of replications of $Y_{1,i}$ and $Y_{2,i}$ that can be executed are thus (the integer parts of) τ/τ_1 and τ/τ_2, respectively. To this correspond sample mean standard errors of

$$\frac{\sigma_1}{\sqrt{\tau/\tau_1}} \text{ and } \frac{\sigma_2}{\sqrt{\tau/\tau_2}},$$

respectively. It follows that, for large τ, the estimator based on the sequence $Y_{1,i}$, $i = 1, \ldots, n$, is preferable, if

$$\frac{\sigma_1}{\sqrt{\tau/\tau_1}} < \frac{\sigma_2}{\sqrt{\tau/\tau_2}},$$

or equivalently

$$\sigma_1^2 \tau_1 < \sigma_2^2 \tau_2.$$

For obvious reasons, the product of variance and per-sample computing time is known as the *efficiency* of a Monte Carlo estimator. In devising methods to improve Monte Carlo performance, efficiency should always constitute the measure of comparison. For instance, a high-variance estimator may, in fact, be preferable to a low-variance estimator, provided that the former takes less time to compute than the latter.

Duffie and Glynn [1995] discuss Monte Carlo efficiency in more depth, with additional analysis of the effects of bias (see also Section 3.2.8) and cases where τ_1 and τ_2 are random.

3.4.2 Antithetic Variates

3.4.2.1 The Gaussian Case

A simple and easily implemented variance reduction technique is the method of *antithetic variates*. Assume that we are interested in estimating the expected value of a random variable $Y = G(Z)$, where G is a real-valued function and Z is a q-dimensional vector of independent standard Gaussian random variables. This problem routinely arises in determining the expected value of a function of assets driven by a vector SDE; Z then represents the aggregation of all independent standard Gaussian variables used to produce Brownian motion increments; see for instance Section 3.2.3. For n independent realizations of Z, $Z_1, \ldots, Z_n$, rather than using the regular sample average estimator for $E(Y)$, consider instead using

$$\overline{Y}_n^a = n^{-1} \sum_{i=1}^n \frac{G(Z_i) + G(-Z_i)}{2}.$$

In other words, in addition to the set $Z_1, \ldots, Z_n$ of Gaussian samples, we also effectively include the set $-Z_1, \ldots, -Z_n$ in the Monte Carlo trial. As $-Z_1, \ldots, -Z_n$ itself is a sequence of n independent Gaussian samples, we still must have

$$E\left(\overline{Y}_n^a\right) = E(Y),$$

so the antithetic estimator is unbiased. Also, as $G(Z)$ and $G(-Z)$ have identical variance,

$$\text{Var}\left(\overline{Y}_n^a\right) = n^{-1}\left[\frac{1}{2}\text{Var}(Y) + \frac{1}{2}\text{Cov}\left(G(Z), G(-Z)\right)\right] = \frac{\text{Var}(Y)}{n}\frac{(1+\rho)}{2},$$

where ρ is the correlation between $G(Z)$ and $G(-Z)$. Recalling that the regular sample average has variance

$$\text{Var}\left(\overline{Y}_n\right) = \frac{\text{Var}(Y)}{n},$$

we conclude that $\text{Var}(\overline{Y}_n^a) < \text{Var}(\overline{Y}_n)$ as long as $\rho < 1$ (which is obviously likely).

While use of antithetic variates can always be expected to lower the standard error, it is not necessarily more efficient than regular Monte Carlo, in the sense defined in Section 3.4.1. For instance, if generation of the Z_i's is of negligible cost relative to the evaluation of $G(Z)$, computation of $\overline{Y}_n^a$ will take about twice as long as the regular sample average $\overline{Y}_n$. For this case, the results in Section 3.4.1 show that for antithetic variates to constitute an improvement in computational efficiency, we must require that

$$\text{Var}\left(\overline{Y}_n^a\right) \leq \frac{1}{2}\text{Var}\left(\overline{Y}_n\right),$$

or
$$\rho \leq 0.$$

A sufficient condition for $\rho < 0$ is that G be monotone in all q elements of Z. Given this, we would expect antithetic variates to be most suitable for option payouts that depend monotonically on prices.

3.4.2.2 General Case

While the method of antithetic variates is primarily associated with the idea of changing signs on Gaussian variables, the method can, in fact, be extended to other distributions. At the most basic level, most simulation trials involve a series of uniform draws that are translated to other random variables, using techniques described in Section 3.1.1. In this case, we can focus our attention on estimating the mean of a random variable $Y = H(U)$, where H is a function and U is a vector of independent uniformly distributed random variables. We notice that if $U = (U_1, \ldots, U_q)^\top$ is a vector of independent uniform random variables on $[0,1]$, then so is $\widetilde{U} = (1 - U_1, \ldots, 1 - U_q)^\top$. The pair $\{U, \widetilde{U}\}$ is thereby antithetic (negatively dependent) in the same way as the Gaussian pair $\{Z, -Z\}$ above, and we can estimate the mean of Y as the average of independent samples of the form

$$\frac{H(U) + H(\widetilde{U})}{2}.$$

From the discussion above, it follows that if H is monotonic in U, the resulting scheme will exhibit better computational efficiency than regular Monte Carlo.

As an aside, we note that the simple "reflection" of a vector of uniforms advocated above is, as should be obvious, not the only possible way of generating an antithetic sample — for instance, we could have chosen to reflect only select dimensions of the U-vector. A similar observation holds for the case of vector-valued Gaussian variables. The general idea of applying deterministic transformations to a vector-valued sample of random numbers as a way to reduce variance is sometimes known as *systematic sampling*.

3.4.3 Control Variates

3.4.3.1 Basic Idea

While we may need to use Monte Carlo simulation to estimate the unknown mean of a random variable Y, there may be random variables "close" to Y with means that can be computed analytically. It seems reasonable that the additional information about Y revealed by these random variables could be useful in improving our estimate of E(Y). While a number of strategies are possible[22], we shall here focus on the so-called *control variate* method.

Formally, let
$$Y^c = (Y_1^c, \ldots, Y_q^c)^\top$$
be a vector of *control variates* (or just *controls*), ideally with strong negative or positive correlation to a variable Y. The mean of Y^c is known to be
$$E(Y^c) = \mu^c = (\mu_1^c, \ldots, \mu_q^c)^\top.$$

Now, introduce an arbitrary constant vector
$$\beta = (\beta_1, \ldots, \beta_q)^\top$$
and consider forming the linear combination
$$X = Y - \beta^\top (Y^c - \mu^c). \tag{3.83}$$

Clearly
$$E(X) = E(Y) - \beta^\top (E(Y^c) - \mu^c) = E(Y),$$
so using Monte Carlo sampling to estimate the mean of X will provide an unbiased estimate of E(Y), regardless of the choice of β.

To analyze the variance of the new variable X, let Σ_{Y^c} be the $q \times q$ covariance matrix of the vector Y^c, and let Σ_{Y,Y^c} be the q-dimensional vector of covariances between Y and the components of Y^c. The variance of X can then be shown to be
$$\text{Var}(X) = \text{Var}(Y) - 2\beta^\top \Sigma_{Y,Y^c} + \beta^\top \Sigma_{Y^c}\beta. \tag{3.84}$$

Whether or not this constitutes an improvement (in the sense that $\text{Var}(X) < \text{Var}(Y)$) is largely a matter of what β is chosen to be. We have the following easily proven lemma.

Lemma 3.4.1. *The function* $\text{Var}(X) = \text{Var}(Y) - 2\beta^\top \Sigma_{Y,Y^c} + \beta^\top \Sigma_{Y^c}\beta$ *is minimized at*
$$\beta^* = \Sigma_{Y^c}^{-1} \Sigma_{Y,Y^c}$$
with minimum value
$$\min_\beta \text{Var}(X) = (1 - R^2)\text{Var}(Y), \quad R^2 \triangleq \frac{\Sigma_{Y,Y^c}^\top \Sigma_{Y^c}^{-1} \Sigma_{Y,Y^c}}{\text{Var}(Y)} \geq 0. \tag{3.85}$$

---

[22] Other methods include *moment matching* and *importance sampling*. We shall cover the latter strategy shortly; the former is discussed in Boyle et al. [1997], where it is concluded that control variates are superior, at least asymptotically.

In the lemma, we recognize the scalar R^2 as the R-squared of a multidimensional regression of Y against Y^c. Similarly, the components of the optimal vector β^* are the regression coefficients (the slopes) on the vector Y^c. In practice, we may not know Σ_{Y^c} and Σ_{Y,Y^c} explicitly, in which case we simply replace these with empirical estimates, as obtained by an n-sample Monte Carlo trial. We note that if the random samples used to estimate β^* are the same as those used to estimate $E(X)$, a small bias is typically introduced. This can be circumvented by using separate random numbers for the estimates of β^* and $E(X)$, but in practice this is rarely worth the effort. Nelson [1990], among others, analyzes this issue in more detail.

While the usage of control variates will always lower variance (unless Y and Y^c are perfectly uncorrelated), an improvement of computational efficiency over standard Monte Carlo is, of course, not guaranteed. Consider, for instance, the case where the computational effort involved in generating a single sample of X is $q+1$ times that of generating Y itself. This will be the case, if i) the effort of drawing random numbers is small relative to computing Y itself; and ii) each of the components of Y^q take about the same time to compute as Y. According to the result in Section 3.4.1, for this special case the control variate method will only entail an increase in efficiency, if

$$(1 - R^2)\text{Var}(Y)(q+1) < \text{Var}(Y)$$

or

$$1 - R^2 < \frac{1}{q+1}.$$

As q grows large, this requirement obviously becomes increasingly difficult to satisfy. Rather than indiscriminately adding multiple controls, it is therefore normally best to properly analyze a given problem and use only a few well-chosen variables with strong (negative or positive) correlation to the variable in question.

3.4.3.2 Non-Linear Controls

Our discussion of the control variate method has so far only considered linear controls (3.83), where the modified estimator involves a linear combination of control variates. The resulting estimate of $E(Y)$ are n-point sample averages of the type

$$\overline{Y}_n - \beta^\top \left(\overline{Y}_n^c - \mu^c\right).$$

A more general formulation than (3.83) approximates $E(Y)$ with

$$f\left(\overline{Y}_n^c, \overline{Y}_n\right) \tag{3.86}$$

for some function f satisfying

$$f(\mu^c, y) = y. \tag{3.87}$$

The requirement (3.87) and smoothness of f ensure that $f(\overline{Y}_n^c, \overline{Y}_n)$ approaches $\mathrm{E}(Y)$ in the large-sample limit; unlike the regular control variate formulation, however, (3.86) may involve a bias for finite sample sizes.

If f is smooth, a result by Glynn and Whitt [1989] demonstrates that for sufficiently large samples, any non-linear control variate estimator of the type (3.86) is equivalent to an ordinary linear control variate estimator. Still, there may be situations where a non-linear control variate estimator is appropriate, either because i) the sample size is not large enough to justify the result in Glynn and Whitt [1989]; or ii) because the "effective" β weighting of $\overline{Y}_n^c$ implied by f is close to optimal, allowing us to skip the estimation of β^*.

To give an example of non-linear control variates, let us consider the "delta" method of Clewlow and Carverhill [1994]. To state the basic idea, consider the estimate of

$$V(0) = \mathrm{E}\left(g\left(X(T)\right)\right),$$

where $X(t)$ is a p-dimensional vector process and $g : \mathbb{R}^p \to \mathbb{R}$ is a smooth function. Assume that all components of X are martingales, as is the case when X represents assets deflated by a numeraire. We recall from Section 1.7 that, under certain regularity conditions, we have

$$V(T) = V(0) + \int_0^T \sum_{i=1}^p V_{x_i}(t)\, dX_i(t),$$

where we use the notation $V_{x_i}(t)$ from Section 1.7 to denote, informally, $V_{x_i}(t) = \partial V(t)/\partial X_i(t)$. On a simulation time line $\{t_j\}_{j=1}^m$, we can write, in the style of an Euler scheme,

$$V(T) \approx V(0) + \sum_{j=1}^m \sum_{i=1}^p V_{x_i}(t_{j-1})\left(X_i(t_j) - X_i(t_{j-1})\right).$$

As the zero-mean quantity

$$\sum_{j=1}^m \sum_{i=1}^p V_{x_i}(t_{j-1})\left(X_i(t_j) - X_i(t_{j-1})\right)$$

is likely to have high correlation to $V(T)$, we can consider using it as a control variate. One obstacle is the fact that the derivatives $V_{x_i}(t)$ are likely to be unknown (as the function $V(t)$ is unknown). Often, however, we can provide an inspired guess for these derivatives, based on perhaps a simpler model or on regression information. The former idea is outlined in Clewlow and Carverhill [1994], and the latter shall be discussed further in Chapter 25. In any case, the resulting scheme ends up effectively using the increments $X_i(t_j) - X_i(t_{j-1})$ as controls, with non-constant weights $V_{x_i}(t_{j-1})$ being functions of the X_i themselves.

3.4.4 Importance Sampling

3.4.4.1 Basic Idea

The basic idea of the *importance sampling method* is to use a measure shift to reduce variance. For a given measure P, consider estimating

$$\mu = \mathrm{E}^{\mathrm{P}}(Y), \tag{3.88}$$

where Y is a scalar random variable. Let $\widehat{\mathrm{P}}$ be a measure equivalent to P. From the Radon-Nikodym theorem in Chapter 1, we have

$$\mu = \mathrm{E}^{\widehat{\mathrm{P}}}(Y/R), \tag{3.89}$$

where R is the Radon-Nikodym derivative

$$R = d\widehat{\mathrm{P}}/d\mathrm{P}, \quad \mathrm{E}^{\mathrm{P}}(R) = 1.$$

While (3.88) and (3.89) are both valid expressions for μ, it is possible that the variance of Y/R under measure $\widehat{\mathrm{P}}$ is lower than the variance of Y under P, making (3.89) potentially more efficient for Monte Carlo purposes. As an extreme case, consider setting (assuming $Y > 0$ a.s.)

$$R = \frac{Y}{\mathrm{E}^{\mathrm{P}}(Y)} = \frac{Y}{\mu}. \tag{3.90}$$

In this case

$$Y/R = \mu$$

and non-random, implying that the measure shift from P to $\widehat{\mathrm{P}}$ has removed *all* variance. The problem with the "perfect" choice (3.90) is obviously that we do not know μ — if we did, there would be no need to estimate it by Monte Carlo methods. Nevertheless, we may be able to provide a good guess for μ, allowing us to use (3.90) in an approximate sense.

3.4.4.2 Density Formulation

Importance sampling methods are often most conveniently (and most intuitively) treated in terms of probability densities, so let us cast the description of Section 3.4.4.1 in such terms. Specifically, let us assume that Y can be represented as $g(X)$, where $g : \mathbb{R}^p \to \mathbb{R}$ is a well-behaved function and X is p-dimensional with probability density $f : \mathbb{R}^p \to \mathbb{R}$. We then write

$$\mu = \mathrm{E}^{\mathrm{P}}(g(X)) = \int_{\mathbb{R}^p} g(x) f(x) \, dx,$$

to which corresponds a regular Monte Carlo estimator

$$\bar{\mu}_n = \frac{1}{n}\sum_{i=1}^{n} g(X_i),$$

where the X_i are independent samples of X, drawn from the density f. Let $h: \mathbb{R}^p \to \mathbb{R}$ be another density, satisfying the continuity requirement that $h(x) > 0$ whenever $f(x) > 0$. We can then also represent μ as

$$\mu = \int_{\mathbb{R}^p} g(x)\frac{f(x)}{h(x)}h(x)\,dx,$$

which we can interpret as

$$\mu = \mathrm{E}^{\widehat{\mathrm{P}}}\left(g(X)\frac{f(X)}{h(X)}\right),$$

where $\widehat{\mathrm{P}}$ is a measure under which X has density $h(x)$. Comparison to the results above identifies the so-called *likelihood ratio* $l(x) = f(x)/h(x)$ as the Radon-Nikodym derivative $d\mathrm{P}/d\widehat{\mathrm{P}}$ (or $1/R$) governing the shift from P to $\widehat{\mathrm{P}}$. If now $X_1, \ldots, X_n$ are independent draws from h (and *not* f), the importance sampling Monte Carlo estimator for μ takes the form

$$\bar{\mu}_n^h = \frac{1}{n}\sum_{i=1}^{n} g(X_i)\frac{f(X_i)}{h(X_i)}.$$

Let us investigate under which circumstances importance sampling will lead to an improvement in variance. We have

$$\mathrm{Var}\left(\bar{\mu}_n^h\right) = \frac{1}{n}\left[\mathrm{E}^{\widehat{\mathrm{P}}}\left(g(X)^2\frac{f(X)^2}{h(X)^2}\right) - \mu^2\right]$$
$$= \frac{1}{n}\left[\mathrm{E}^{\mathrm{P}}\left(g(X)^2\frac{f(X)}{h(X)}\right) - \mu^2\right],$$

and

$$\mathrm{Var}\left(\bar{\mu}_n\right) = \frac{1}{n}\left[\mathrm{E}^{\mathrm{P}}\left(g(X)^2\right) - \mu^2\right].$$

Hence, importance sampling will lower variance, provided that

$$\mathrm{E}^{\mathrm{P}}\left(g(X)^2\frac{f(X)}{h(X)}\right) < \mathrm{E}^{\mathrm{P}}\left(g(X)^2\right).$$

Choosing the importance sampling density $h(x)$ wisely is key to the efficiency of the importance sampling. As an extreme, suppose we could set

$$h(x) = Cf(x)g(x), \qquad (3.91)$$

where the constant C is dictated by the need for $h(x)$ to integrate to 1:

$$C^{-1} = \int_{\mathbb{R}^p} g(x) f(x)\, dx = \mu.$$

In this case,

$$\mathrm{E}^{\mathrm{P}} \left(g(X)^2 \frac{f(X)}{h(X)} \right) = C^{-1} \mathrm{E}^{\mathrm{P}}(g(X)) = \mu^2$$

and

$$\mathrm{Var}\left(\overline{\mu}_n^h\right) = 0.$$

This replicates a similar argument in Section 3.4.4.1 (see equation (3.90)), and is equally useless in practice: to compute (3.91) we need to normalize by the constant $1/\mu$, where μ is the quantity that we are trying to estimate in the first place. Nevertheless, (3.91) provides some useful practical guidance: a good choice of likelihood density will sample in proportion to f and g. That is, values of X where both the density $f(X)$ and the payout $g(X)$ are high should be assigned a high value of $h(X)$ (high "importance"), and values of X where either $f(X)$ or $g(X)$ (or both) are low should be assigned a low value of $h(X)$ (low "importance"). This rule is often particularly easy and efficient to apply to situations where $g(X)$ is significant only for a set $X \in \mathcal{A}$, where $P(X \in \mathcal{A})$ is small. Such rare-event problems are a classical application of importance sampling; we give a simple example in Section 3.4.4.5. Related applications to barrier options can be found later, in Chapter 25, with more such examples in Boyle et al. [1997].

3.4.4.3 Importance Sampling and SDEs

Consider now a dynamic setting where we are given a P-measure SDE

$$dX(t) = \mu(t, X(t))\, dt + \sigma(t, X(t))\, dW(t), \tag{3.92}$$

where X is p-dimensional and W is d-dimensional. We wish to evaluate

$$\mathrm{E}^{\mathrm{P}}(g(X(T)))$$

for a real-valued function g. To shift measure, we introduce the density process

$$d\varsigma(t) = -\varsigma(t)\theta(t)^\top dW(t), \quad \varsigma(0) = 1, \tag{3.93}$$

for some adapted d-dimensional process $\theta(t)$, sufficiently regular to make $\varsigma(\cdot)$ a martingale (see Chapter 1). Let

$$\varsigma(t) = \mathrm{E}^{\mathrm{P}}_t \left(\frac{d\widehat{\mathrm{P}}}{d\mathrm{P}} \right),$$

for a new measure $\widehat{\mathrm{P}}$. By the Girsanov theorem, under $\widehat{\mathrm{P}}$,

$$dX(t) = [\mu(t, X(t)) - \sigma(t, X(t))\theta(t)]\, dt + \sigma(t, X(t))\, d\widehat{W}(t), \quad (3.94)$$

where $\widehat{W}$ is a Brownian motion in $\widehat{P}$. Also, by the Radon-Nikodym theorem,

$$E^P(g(X(T))) = E^{\widehat{P}}\left(\frac{g(X(T))}{\varsigma(T)}\right). \quad (3.95)$$

In a Monte Carlo setting, rather than simulating (3.92) (using methods from Section 3.2) and computing the sample mean of $g(X(T))$, we can instead jointly simulate (3.93) and (3.94) and compute the sample mean of $g(X(T))/\varsigma(T)$. The validity of this approach is independent of the choice of θ in (3.93), and we can use θ as a parameter to minimize the variance of $g(X(T))/\varsigma(T)$ under $\widehat{P}$.

To find the optimal choice for θ, define

$$u(t, X(t)) = E_t^P(g(X(T))), \quad t \le T,$$

and consider setting

$$\varsigma(t) u(0, X(0)) = u(t, X(t)).$$

By Ito's lemma,

$$d\varsigma(t) = -\varsigma(t)\theta(t)^\top dW(t), \quad \varsigma(0) = 1,$$

where

$$\theta(t) = -u(t, X(t))^{-1} \sigma(t, X(t))^\top \frac{\partial u(t, X(t))}{\partial x}, \quad (3.96)$$

with $\partial u(t, X(t))/\partial x$ being a p-dimensional vector of partial derivatives $\{\partial u(t, X(t))/\partial x_i\}$. The choice for θ in (3.96) is optimal as we have

$$g(X(T))/\varsigma(T) = u(0, X(0)) = E^P(g(X(T))),$$

which is non-random with zero variance. As in earlier examples, the optimal choice for $\theta(t)$ cannot be applied directly as it requires knowledge of $E_t^P(g(X(T)))$ for all t, knowledge which we never possess in practice. In many applications, however, we can often make an educated guess for u, based perhaps on either a simpler SDE than (3.92) or on a simpler payout function than g. We shall see an example of this in Chapter 25; another application can be found in Schoenmakers and Heemink [1997].

3.4.4.4 More on SDE Path Simulation

Let us consider an alternative point of view about SDE simulations, where we assume that the SDE (3.92) is simulated by an m–dimensional Euler scheme (or similar), such that we can write (see also Section 3.4.2.1) for some function $G : \mathbb{R}^{p \times m} \to \mathbb{R}$,

3.4 Variance Reduction Techniques 153

$$g\left(X(T)\right) = G\left(Z_1, \ldots, Z_m\right),$$

where the Z_i are independent p-dimensional Gaussian vectors. With the Gaussian density of Z_i being denoted $\phi(z)$, $z \in \mathbb{R}^p$, the independence of the Z_i's allows us to write

$$\mathrm{E}^{\mathrm{P}}\left(g\left(X(T)\right)\right) = \int_{\mathbb{R}^{p \times m}} G\left(z_1, \ldots, z_m\right) \prod_{i=1}^{m} \phi(z_i)\, dz, \quad z \triangleq (z_1, \ldots, z_m).$$

If we apply a change of measure that preserves independence of Z_i but alters the common marginal density from $\phi(z)$ to $h(z)$, the likelihood ratio is easily seen to be

$$l(z) = \prod_{i=1}^{m} \frac{\phi(z_i)}{h(z_i)},$$

such that

$$\mathrm{E}^{\mathrm{P}}\left(g\left(X(T)\right)\right) = \mathrm{E}^{\widehat{\mathrm{P}}}\left(G(Z_1, \ldots, Z_m) \prod_{i=1}^{m} \frac{\phi(Z_i)}{h(Z_i)}\right).$$

It is understood that the Z_i used to advance the SDE simulation under $\widehat{\mathrm{P}}$ are drawn from the density h, rather than ϕ.

To give a concrete example of a measure shift, assume for simplicity that $p = 1$ and consider shifting the means of the Z_i from zero to some scalar[23] μ, but retaining unit variance. For this, we must set

$$h(z_i) = \frac{1}{\sqrt{2\pi}} \exp\left(-\frac{1}{2}(z_i - \mu)^2\right),$$

whereby

$$l(z) = l(z; \mu) = \exp\left(-\mu \sum_{i=1}^{m} z_i + \frac{m}{2}\mu^2\right). \tag{3.97}$$

Here μ is a free variable, which can be set to minimize the variance of the term

$$G(Z)l(Z;\mu), \quad Z \triangleq (Z_1, \ldots, Z_m)$$

under $\widehat{\mathrm{P}}$. Sometimes this minimization problem can be handled analytically (see Section 3.4.4.5), but most often numerical methods are required. Examples of how to perform this minimization by Monte Carlo simulation can be found in, for example, Su and Fu [2002] and Capriotti [2007]. The approach in Capriotti [2007] (called *least-squares importance sampling*) is particularly straightforward, as the optimization problem is here cast as a least-squares regression problem for which well-known numerical schemes exist such as,

---

[23] It is also straightforward to introduce a measure shift that moves the means of the Z_i to *different* means μ_i, $i = 1, \ldots, m$.

e.g., the Levenberg-Marquardt routine in Press et al. [1992]. Both Su and Fu [2002] and Capriotti [2007] point out that, when computing variance, it is advantageous to cast the problem back into the original probability measure P by using

$$E^{\widehat{P}}\left(G(Z)^2 l(Z;\mu)^2\right) = E^P\left(G(Z)^2 l(Z;\mu)\right).$$

Let us finally note that the measure transformation employed above is a special case of so-called *exponential twisting* (also known as *Esscher transform*), under which a density $f(x)$, $x \in \mathbb{R}$, is transformed into

$$f_\theta(x) = e^{\theta x - \gamma(\theta)} f(x),$$

where θ is a twisting parameter and γ is the *cumulant-generating function*

$$\gamma(\theta) = \ln\left(\int_\mathbb{R} e^{\theta x}\, dx\right).$$

For a standard Gaussian variable, $\gamma(\theta) = \theta^2/2$, demonstrating that the shift of mean employed above is indeed a special case of exponential twisting. We notice that exponential twisting is often a very convenient starting point when working with parametric families of Radon-Nikodym derivatives.

3.4.4.5 Rare Event Simulation and Linearization

For illustrative purposes, consider finally the problem of estimating by Monte Carlo

$$P(Z > c), \tag{3.98}$$

where $Z \sim \mathcal{N}(0,1)$ is standard Gaussian under the measure P, and c is a big number. In ordinary Monte Carlo, we write

$$P(Z > c) = E^P\left(1_{\{Z>c\}}\right)$$

and use the sample mean estimator

$$P(Z > c) \approx \frac{1}{n}\sum_{i=1}^n 1_{\{Z_i > c\}},$$

where $Z_1, \ldots, Z_n$ are independent standard Gaussian samples. We notice that

$$\begin{aligned}
\mathrm{Var}^P\left(1_{\{Z>c\}}\right) &= E^P\left(\left(1_{\{Z>c\}}\right)^2\right) - E^P\left(1_{\{Z>c\}}\right)^2 \\
&= E^P\left(1_{\{Z>c\}}\right) - E^P\left(1_{\{Z>c\}}\right)^2 \\
&= P(Z>c)(1 - P(Z>c)),
\end{aligned}$$

with sample mean estimator variance being n times smaller. Consider now introducing a probability measure that shifts the mean of Z from 0 to μ. The likelihood ratio is seen from (3.97) to be

$$l(z) = e^{-\mu z + \mu^2/2},$$

such that

$$\mathrm{P}(Z > c) = \mathrm{E}^{\widehat{\mathrm{P}}}\left(e^{-\mu Z + \mu^2/2} 1_{\{Z > c\}}\right),$$

where $Z \sim \mathcal{N}(\mu, 1)$ in the measure $\widehat{\mathrm{P}}$. A Monte Carlo estimator for this is then

$$\frac{1}{n} \sum_{i=1}^{n} e^{-\mu(Z_i + \mu) + \mu^2/2} 1_{\{Z_i + \mu > c\}},$$

where $Z_1, \ldots, Z_n$ are again independent standard Gaussian samples. Notice that we have added to the Z_i the mean μ to reflect the shift of measure from P to $\widehat{\mathrm{P}}$. As for variance, we have

$$\begin{aligned}
\mathrm{Var}^{\widehat{\mathrm{P}}}\left(e^{-\mu Z + \mu^2/2} 1_{\{Z > c\}}\right) &= \mathrm{E}^{\widehat{\mathrm{P}}}\left(e^{-2\mu Z + \mu^2} (1_{\{Z > c\}})^2\right) - \mathrm{P}(Z > c)^2 \\
&= \mathrm{E}^{\widehat{\mathrm{P}}}\left(e^{-2\mu Z + \mu^2} 1_{\{Z > c\}}\right) - \mathrm{P}(Z > c)^2 \\
&= \mathrm{E}^{\mathrm{P}}\left(e^{-\mu Z + \mu^2/2} 1_{\{Z > c\}}\right) - \mathrm{P}(Z > c)^2 \\
&= e^{\mu^2} \mathrm{P}(Z > c + \mu) - \mathrm{P}(Z > c)^2, \quad (3.99)
\end{aligned}$$

where the last equation follows from the properties of the standard Gaussian density. The choice of μ that minimizes the variance under $\widehat{\mathrm{P}}$ is the solution to

$$\min_{\mu} e^{\mu^2} \mathrm{P}(Z > c + \mu).$$

Differentiating with respect to μ and setting the resulting expression to zero shows that the variance is minimized at μ^*, where

$$2\mu^* [1 - \Phi(c + \mu^*)] - \phi(c + \mu^*) = 0, \quad (3.100)$$

with Φ and ϕ being the standard Gaussian distribution function and density, respectively. This expression can be solved for μ^* with the aid of a numerical root solver. Alternatively, we can use the fact that c is large to rely on the asymptotic approximation

$$1 - \Phi(c + \mu^*) \approx \frac{\phi(c + \mu^*)}{c + \mu^*},$$

which leads to

$$2\mu^* \frac{\phi(c + \mu^*)}{c + \mu^*} \approx \phi(c + \mu^*) \Rightarrow \mu^* \approx c. \quad (3.101)$$

Note that this implies that the probability of Z exceeding c in measure $\widehat{P}$ is approximately $\frac{1}{2}$. This is an intuitive result[24], consistent with the discussion at the end of Section 3.4.4.2.

To measure the efficacy of importance sampling, we can use (3.99) to define a variance efficiency ratio as

$$\frac{P(Z>c)(1-P(Z>c))}{e^{\mu^2}P(Z>c+\mu)-P(Z>c)^2}. \quad (3.102)$$

Figure 3.1 graphs this ratio when μ is set to c, as prescribed in (3.101). For large c, the improvements to variance associated with using importance sampling can be seen to be extremely significant.

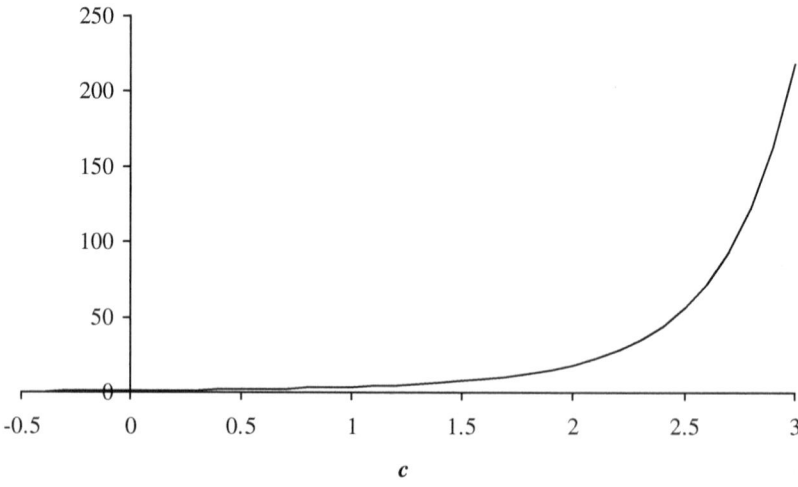

Fig. 3.1. Variance Ratio

Notes: The figure graphs the ratio (3.102), with μ set according to (3.101).

It is also illustrative to consider the multi-variate extension to the problem above. Here, we are interested in estimating

$$E^P\left(1_{\{X>c\}}\right),$$

where c is a p-dimensional constant and X is a p-dimensional vector of Gaussian random variables with mean 0 and covariance matrix Σ. Let C be the Cholesky decomposition of Σ, such that

$$E^P\left(1_{\{X>c\}}\right) = E^P\left(1_{\{CZ>c\}}\right) = E^P\left(1_{\{Z>c'\}}\right), \quad c' \triangleq C^{-1}c,$$

---

[24] For a somewhat more accurate approximation to μ^*, see Jäckel [2004].

where Z is a p-dimensional vector of independent standard Gaussian variables. Let us introduce a measure $\widehat{P}$ where the mean of Z has been shifted to μ, a p-dimensional vector. Following the same steps as for the univariate case, we have

$$E^P\left(1_{\{Z>c'\}}\right) = E^{\widehat{P}}\left(\exp\left(-\mu^\top Z + \frac{1}{2}\mu^\top \mu\right) 1_{\{Z>c'\}}\right),$$

with variance

$$\mathrm{Var}^{\widehat{P}}\left(\exp\left(-\mu^\top Z + \frac{1}{2}\mu^\top \mu\right) 1_{\{Z>c'\}}\right)$$
$$= e^{\mu^\top \mu} E^P\left(1_{\{Z>c'+\mu\}}\right) - P(X>c)^2.$$

A direct optimization of this expression in μ involves multi-dimensional Gaussian integrals, so we wish to resort to approximations. We can use the arguments of Section 3.4.4.2 to argue that the optimal importance sampling density should be proportional to

$$1_{\{z>c'\}} \exp\left(-\frac{1}{2}z^\top z\right), \tag{3.103}$$

since $\exp(-z^\top z/2)$ is proportional to the normal density. Following the idea in Glasserman et al. [1999], we can choose μ such that the location of the peak of an $\mathcal{N}(\mu, I)$ distribution coincides with the peak of (3.103). In other words, we approximate the optimal μ as the value μ^* of z that solves

$$\max_z \left\{1_{\{z>c'\}} \exp\left(-\frac{1}{2}z^\top z\right)\right\} = \min_{z>c'}\left\{z^\top z\right\}. \tag{3.104}$$

If we assume, say, that all components of c' are larger than 0, then obviously

$$\mu^* = c' = C^{-1}c,$$

consistent with the approximative univariate result (3.101).

We note that the idea behind (3.104) is not limited to situations where we evaluate expectations of an indicator function. For instance, suppose we, as in Section 3.4.4.4, wish to estimate

$$E^P\left(G(Z)\right),$$

for a smooth function $G : \mathbb{R}^p \to \mathbb{R}$. Restricting ourselves again to the class of measure shifts that only move the mean of Z, the approximately optimal mean shift μ solves

$$\max_z \left\{G(z) \exp\left(-\frac{1}{2}z^\top z\right)\right\}$$

158   3 Monte Carlo Methods

or, if $G(Z)$ is strictly positive,

$$\max_z \left\{ w(z) - \frac{1}{2} z^\top z \right\}, \quad w(z) \triangleq \ln(G(z)).$$

The first-order condition for the optimum is

$$\nabla w(\mu^*) = (\mu^*)^\top, \qquad (3.105)$$

where ∇ is the gradient operator, $\nabla = (\partial/\partial z_1, \ldots, \partial/\partial z_p)$ (row vector). This is a fixed-point condition that can be solved by numerical methods. The result (3.105) is exact if w is linear in its argument; the method above can thus be seen as a linearization through a first-order Taylor approximation. Glasserman et al. [1999] demonstrate that, under some conditions, (3.105) satisfies a certain asymptotic optimality property.

3.5 Some Notes on Bermudan Security Pricing

As alluded to in the beginning of this chapter, one drawback of Monte Carlo methods is the difficulty associated with the pricing of securities with early exercise rights. We demonstrated earlier in the chapter that Monte Carlo path generation runs *forward* in time, making direct application of dynamic programming and backward induction (see Chapters 1 and 2) impossible. Indeed, until the early 1990's, it was generally believed that Monte Carlo techniques were inherently incompatible with the pricing of early exercise rights. In the last decade, however, this belief has been overturned, with the advent of several different techniques for Monte Carlo pricing of options with early exercise rights. Most of these techniques are rather advanced and a detailed description will be postponed until later in this book, when the interest rate modeling foundation has been properly laid and the details of callable interest rate securities have been covered. For now, we only provide a brief discussion of certain generic principles, with additional details to be filled in later, in Chapters 18 and 19, among others. We start by establishing some notation and reminding the reader of some basic results from Chapter 1.

3.5.1 Basic Idea

For the remainder of this section, we consider the pricing of a Bermudan security C, with a payout function[25] $U(t) = U(t, x(t))$, where $x(t)$ is a p-dimensional vector of Markovian state variables. The allowed discrete set of exercise dates is denoted $\mathcal{D} = \{T_1, T_2, \ldots, T_B\}$, with $T_B = T$ being the

---

[25] For many exotic interest rate options, the function $U(t, x(t))$ may actually not be known in closed form. We deal with this complication in Chapter 18.

3.5 Some Notes on Bermudan Security Pricing

terminal maturity of C. We fix a numeraire N, assumed to be a function of $x(t)$, $N(t) = N(t, x(t))$. From Section 1.10, we recall that

$$C(0) = N(0) \sup_{\tau \in \mathcal{T}} \mathrm{E}^N \left(\frac{U(\tau)}{N(\tau)} \right),$$

where E^N denotes expectation in the measure Q^N induced by the numeraire N, and $\mathcal{T}$ is the set of stopping time strategies taking values in $\mathcal{D}$. More generally, we write

$$C(t) = N(t) \sup_{\tau \in \mathcal{T}(t)} \mathrm{E}^N_t \left(\frac{U(\tau)}{N(\tau)} \right), \tag{3.106}$$

where $\mathcal{T}(t)$ is the set of stopping time strategies in $\mathcal{D}$ for which $\tau \geq t$. We also recall that when $t \in (T_{i-1}, T_i]$, we have

$$C(t) = N(t) \mathrm{E}^N_t \left(N(T_i)^{-1} \max \left(H_i(T_i), U(T_i) \right) \right), \tag{3.107}$$

where the hold value (see Section 1.10) $H_i(T_i)$ is defined as

$$H_i(T_i) = N(T_i) \mathrm{E}^N_{T_i} \left(N(T_{i+1})^{-1} C(T_{i+1}) \right).$$

Notice that (3.107) establishes that the optimal exercise strategy, as seen from time t, is

$$\tau^* = \inf \left\{ T_i \geq t : U(T_i) \geq H_i(T_i) \right\}. \tag{3.108}$$

3.5.2 Parametric Lower Bound Methods

Assuming that we are able to simulate the p-dimensional vector $x(t)$ through time, it follows from (3.106) that a lower bound for $C(0)$ can be computed by Monte Carlo method through any exogenous guess for the optimal exercise strategy τ^*. One fairly intuitive approach involves a user-supplied specification of a parametric stopping rule, $\tau(\alpha) \in \mathcal{T}$, where $\alpha \in A \subset \mathbb{R}^m$ is an m-dimensional parameter vector. Defining $x = (x(T_1), \ldots, x(T_B))$, for a given value of α, we have the following algorithm.

1. Generate n independent paths $x^{(k)}$, $k = 1, \ldots, n$. For path k, let $\tau^{(k)}(\alpha)$ be the exercise date suggested by the parametric stopping rule.
2. For each path k, set $U^{(k)} = U(\tau^{(k)}(\alpha), x(\tau^{(k)}(\alpha)))$ and $C^{(k)}_\alpha = N(\tau^{(k)}(\alpha), x(\tau^{(k)}(\alpha)))^{-1} U^{(k)}$.
3. Return $\overline{C}_\alpha(0) = N(0) n^{-1} \sum_{k=1}^{n} C^{(k)}_\alpha$ as our estimate for the Bermudan option value $C(0)$.

Let $C_\alpha(0) \triangleq \mathrm{E}^N(\overline{C}_\alpha(0))$. As $\tau(\alpha)$ in general will be sub-optimal, it is clear that

$$C_\alpha(0) \leq C(0). \tag{3.109}$$

To get as close as possible to $C(0)$, it is, of course, preferable to use the value $\alpha^* \in A$ for which $C_\alpha(0)$ is optimized, i.e.

$$\alpha^* = \underset{\alpha \in A}{\mathrm{argsup}}\, C_\alpha(0).$$

A tempting way of estimating $C_{\alpha^*}(0)$ would be to modify step 3 in the algorithm above to

3a. Return $\overline{C}_{\alpha^*}(0) = \sup_{\alpha \in A} \overline{C}_\alpha(0)$.

Leaving aside the question of how one might execute the optimization in Step 3a, we notice that the estimator $\overline{C}_{\alpha^*}(0)$ is biased high relative to $C_{\alpha^*}(0)$:

$$\mathrm{E}^N\left(\overline{C}_{\alpha^*}(0)\right) \geq \sup_{\alpha \in A} C_\alpha(0). \tag{3.110}$$

This inequality states that the expected value of the maximum over α must be at least as large as the maximum over α of expected values, a consequence of Jensen's inequality. We may interpret the bias of $\mathrm{E}^N(\overline{C}_{\alpha^*}(0))$ as a *perfect foresight bias*: by using in-sample information to estimate α^*, we effectively "cheat" by making the optimum specific to the same n samples that are also used to determine the option value.

The combination of inequalities (3.109) and (3.110) shows that the quantity $\overline{C}_{\alpha^*}(0)$ from Step 3a has an *indeterminate* bias relative to the true option price. As a bias is generally inevitable when using parametric exercise strategies, in practice it is preferable to at least know its sign. To accomplish this, we can retain the estimated value of α^* found as described above, but draw a *separate* set of Monte Carlo paths when pricing the option. That is, we replace Step 3a with the following two steps:

3b. Set $\widehat{\alpha}^* = \mathrm{argsup}_{\alpha \in A}\, \overline{C}_\alpha(0)$.
4b. Draw a fresh set of n_2 independent paths for x and N, with α locked at the value $\widehat{\alpha}^*$. Return $\overline{C}_{\widehat{\alpha}^*}(0) = N(0)/n_2 \sum_{k=1}^{n_2} C_{\widehat{\alpha}^*}^{(k)}$, where the $C_{\widehat{\alpha}^*}^{(k)}$ are computed on the new set of paths.

As the parameter $\widehat{\alpha}^*$ will a.s. never equal α^*, it follows that

$$\mathrm{E}^N\left(\overline{C}_{\widehat{\alpha}^*}(0)\right) \leq C_{\alpha^*}(0) \leq C(0),$$

i.e. we are now assured that $\overline{C}_{\widehat{\alpha}^*}(0)$ is low-bound estimator.

3.5.3 Parametric Lower Bound: An Example

What constitutes a good parametric exercise rule is strongly instrument-specific and typically requires case-by-case analysis. Even for simple Markov models and standard option payouts, the topology of exercise and continuation regions can be highly complicated (see e.g. Broadie and Detemple

[1997]), so this exercise is by no means straightforward. As a first approximation, however, one can always attempt to use a simple rule based on outright "moneyness" of the underlying option payout, as in Andersen [2000a]. According to this rule, one sets $\alpha = (h_1, h_2, \ldots, h_B)$ (i.e. $m = B$) and writes

$$\tau(\alpha) = \inf\{T_i : U(T_i, x(T_i)) > h_i\}. \qquad (3.111)$$

That is, exercise of the option takes place when it is sufficiently deep in the money, with the term "sufficiently deep" quantified through unknown trigger thresholds $h_i \geq 0$, $i = 1, \ldots, B$.

While α is B-dimensional, finding its optimal value is not truly a B-dimensional optimization problem. Rather, due to the Markov assumption on $x(t)$, we may decompose it into a series of $B-1$ *one-dimensional* optimization problems. Specifically, working backwards in time, suppose that the optimal values of $h_{j+1}, h_{j+2}, \ldots, h_B$ are known. We then find the optimal value of h_j, by optimizing on $\overline{C}_\alpha(0)$, but subject to the constraint that exercise is *not* allowed to take place before time j. As $h_{j+1}, h_{j+2}, \ldots, h_B$ are assumed known — and $h_1, \ldots, h_{j-1}$ do not come into play — the only variable involved in this optimization is h_j. The algorithm starts with $j = B - 1$ and the known[26] boundary condition $h_B = 0$.

A couple of comments on the algorithm above are in order. First, we notice that $U(T_i, x(T_i)) > h_i$ can be replaced with any one-parameter boolean function $g(T_i, x(T_i); h_i)$ without affecting the basic algorithm — see Andersen [2000a] for some examples. Second, if in such a boolean function $g(T_i, x(T_i); h_i)$ each h_i is allowed to be q-dimensional, the optimization problem reduces to $B - 1$ q-dimensional optimization problems. And third, for a finite-path simulation, the objective functions in each of the $B - 1$ optimization problems will not be smooth; consequently, the optimization is best performed by an iterative search rather than a derivative-based method. Andersen [2000a] uses the golden section search (see Press et al. [1992]), but simpler strategies based on, say, outright sorting are also possible.

3.5.4 Regression-Based Lower Bound

According to (3.108), an approximation for the optimal exercise strategy can always be constructed through an estimate for hold values $H_i(T_i)$ at all $i = 1, 2, \ldots, B - 1$. In our Markov setting, we know that

$$H_i(T_i) = q_i(x(T_i)) = q_i(x_1(T_i), \ldots, x_p(T_i))$$

for a set of $B - 1$ functions $q_i : \mathbb{R}^p \to \mathbb{R}$, $i = 1, 2, \ldots, B - 1$; the problem of estimating hold values is equivalent to the problem of estimating the functions q_i.

---

[26] At the last possible exercise date, we would, of course, always exercise the option if it is in-the-money.

From Section 3.5.1, we know that

$$q_i(x) = N(T_i, x)\mathrm{E}^N\left(C(T_{i+1})/N\left(T_{i+1}, x(T_{i+1})\right)|x(T_i) = x\right), \qquad (3.112)$$

which can be interpreted as the *regression* of $C(T_{i+1})/N(T_{i+1})$ on the Markov state variables $x(T_i)$. Several authors — including Carrière [1996], Longstaff and Schwartz [2001], and Tsitsiklis and Roy [2001] — have used this observation to suggest that $q_i(x)$ be estimated by a linear combination of exogenously specified (basis) functions of $x(T_i)$, with least-squares regression on Monte Carlo paths used to determine the best weights for these functions. That is, we fundamentally assume that

$$q_i(x) = \sum_{j=1}^{d} \beta_{i,j} \psi_j(x), \qquad (3.113)$$

for a set of d basis-functions $\psi_j : \mathbb{R}^p \to \mathbb{R}$, $j = 1, 2, \ldots, d$. Setting $\beta_i = (\beta_{i,1}, \ldots, \beta_{i,d})^\top$ and $\psi(x) = (\psi_1(x), \ldots, \psi_d(x))^\top$, we can rewrite (3.113) as $q_i(x) = \psi(x)^\top \beta_i$ or, from (3.112),

$$\mathrm{E}^N\left(N(T_i, x(T_i))\, C(T_{i+1})/N\left(T_{i+1}, x(T_{i+1})\right)|x(T_i) = x\right)$$
$$= \mathrm{E}^N\left(\psi\left(x(T_i)\right)^\top |x(T_i) = x\right)\beta_i.$$

This, in turn, implies that

$$\Omega_i = \Psi_i \beta_i \quad \Rightarrow \quad \beta_i = \Psi_i^{-1} \Omega_i, \qquad (3.114)$$

where Ω_i is the d-dimensional vector

$$\Omega_i = \mathrm{E}^N\left(\psi\left(x(T_i)\right) \frac{N(T_i, x(T_i))}{N(T_{i+1}, x(T_{i+1}))} C(T_{i+1})\right),$$

and Ψ_i is the $d \times d$ matrix

$$\Psi_i = \mathrm{E}^N\left(\psi\left(x(T_i)\right) \psi\left(x(T_i)\right)^\top\right).$$

The rationale for rewriting (3.113) into the seemingly more convoluted representation (3.114) is that the latter leads naturally to the algorithm for a least-squares estimation of β_i: one simply replaces the expectations in Ψ_i and Ω_i with sample averages $\overline{\Psi}_i$ and $\overline{\Omega}_i$ computed on a set of Monte Carlo paths. That is, one uses[27]

$$\widehat{\beta}_i = \overline{\Psi}_i^{-1} \overline{\Omega}_i \qquad (3.115)$$

---

[27]In practice, a direct solution of linear equations in this fashion can be suboptimal if the matrix $\overline{\Psi}_i$ is ill-conditioned. Instead, one would use either truncated singular value decomposition (TSVD) or Tikhonov regularization to find $\widehat{\beta}_i$. We return to this issue in Chapter 18.

as the sample estimate.

We shall discuss the details of (and many variations on) the regression approach later, in Chapter 18. For now, let us just notice that computation of $\overline{\Omega}_i$ requires estimation of $C(T_{i+1})$, which naturally encourages running the estimation of the $\widehat{\beta}_i$ backwards in i, starting from $i = B - 1$. We also notice that the success of the regression approach depends critically on the choice and number of basis functions ψ_j. We give specific advice on this topic in Chapters 18 and 19.

3.5.5 Upper Bound Methods

Given a martingale M in measure Q^N, we recall from Section 1.10.2 that an M-specific upper bound $C_M(0)$ for a Bermudan option can always be constructed as

$$C_M(0) = N(0)\left\{M(0) + \mathrm{E}^N\left(\max_{t \in \mathcal{D}}\left(\frac{U(t,x(t))}{N(t,x(t))} - M(t)\right)\right)\right\} \geq C(0). \tag{3.116}$$

Let $\mathbf{M} = (M(T_1), \ldots, M(T_B))^\top$. As long as $M(\cdot)$ can be simulated along with the vector $x(\cdot)$, (3.116) suggests the following Monte Carlo algorithm:

1. Generate n independent paths $x^{(k)}$, $\mathbf{M}^{(k)}$, $k = 1, \ldots, n$. For path k, let $\gamma^{(k)}$ be the maximum value of $U(T_i, x(T_i))/N(T_i, x(T_i)) - M(T_i)$ over $i = 1, 2, \ldots, B$.
2. Return $\overline{C}_M(0) = N(0)\{M(0) + n^{-1}\sum_{k=1}^n \gamma^{(k)}\}$.

For the upper bound method to be practically useful, we would want the gap $\mathrm{E}^N(\overline{C}_M(0)) - C(0)$ to be small. This, in turn, requires that we specify the martingale $M(t)$ to be "reasonable". More specifically, from results in Section 1.10.2, we would like $M(t)$ to represent the martingale component of a good approximation to the supermartingale $C(t)$. For instance, if we happen to think that $C(t)$ is well-approximated by some known function v of time and x,

$$C(t) = v(t, x(t)), \tag{3.117}$$

we would set

$$dM(t) = \sum_{j=1}^p \frac{\partial v(t, x(t))}{\partial x_j}\left(dx_j(t) - \mathrm{E}_t^N(dx_j(t))\right). \tag{3.118}$$

As an example, suppose that W^N is a (possibly vector-valued) Brownian motion in Q^N and $dx_j(t) = \mu_j(t, x(t))\,dt + \sigma_j(t, x(t))\,dW^N(t)$, in which case we could use

$$dM(t) = \sum_{j=1}^p \frac{\partial v(t, x(t))}{\partial x_j} \sigma_j(t, x(t))\,dW^N(t). \tag{3.119}$$

Occasionally, a natural analytical guess for the function v may exist, but most often the only estimate for v is given only implicitly, through a low-bound estimator based on an approximation of the optimal exercise strategy. A completely generic algorithm to turn a guess for the optimal exercise strategy into a proxy for $M(t)$ is developed in Andersen and Broadie [2004]; we shall discuss this algorithm in detail in Chapter 18. If the evolution of $x(t)$ is described by an SDE, a regression approach to estimation of the terms multiplying $dW^N(t)$ in (3.119) can be found in Belomestny et al. [2007].

3.5.6 Confidence Intervals

Suppose that we simultaneously apply a lower bound and an upper bound method to provide two sample estimates $\overline{C}_{\text{lo}}$ and $\overline{C}_{\text{up}}$, with

$$\mathrm{E}^N\left(\overline{C}_{\text{lo}}\right) \leq C(0) \leq \mathrm{E}^N\left(\overline{C}_{\text{up}}\right). \tag{3.120}$$

Let us assume that the sample standard errors on $\overline{C}_{\text{lo}}$ and $\overline{C}_{\text{up}}$ have been computed as s_{lo} and s_{up}, respectively. For a sufficiently large number of Monte Carlo trials, we can then use the central limit theorem from Section 3.1 to set up a confidence interval

$$\left[\overline{C}_{\text{lo}} - u_{\gamma/2} \cdot s_{\text{lo}}, \overline{C}_{\text{up}} + u_{\gamma/2} \cdot s_{\text{up}}\right],$$

where $\Phi(u_{\gamma/2}) = 1 - \gamma/2$. It is clear from (3.120) that the likelihood of this interval bracketing the true price $C(0)$ is *at least* $1 - \gamma$. It is also clear that this confidence interval will not shrink to zero — even in the limit of an infinite number of samples where $s_{\text{lo}} \to 0$ and $s_{\text{up}} \to 0$ — unless $\overline{C}_{\text{lo}}$ and $\overline{C}_{\text{up}}$ simultaneously achieve the unlikely feat of being perfectly unbiased estimators for $C(0)$.

Finally, let us note that any number inside the interval $[\overline{C}_{\text{lo}}, \overline{C}_{\text{up}}]$ can reasonably be used as an estimator for $C(0)$. To the extent that we have reason to believe[28] that $\overline{C}_{\text{lo}}$ and $\overline{C}_{\text{up}}$ have roughly opposite biases, a natural estimator is $(\overline{C}_{\text{lo}} + \overline{C}_{\text{up}})/2$.

3.5.7 Other Methods

The methods described so far are those that, in our opinion, are most useful for practical Monte Carlo pricing of interest rate options with early exercise rights. Several other methods, however, have been proposed in the literature, some of which have interesting theoretical properties. We highlight the *random tree* method (Broadie and Glasserman [1997]) which builds a random

---

[28] There is some evidence that the upper-bound method in Andersen and Broadie [2004] produces a bias that is often roughly opposite of that of the low-bound method from which the martingale M in (3.116) is extracted.

non-recombining lattice by Monte Carlo methods; backward induction arguments are then used to construct high- and low-biased estimators, both of which are convergent to the true price as the number of Monte Carlo paths are increased. The drawback of the method is its computational complexity which increases exponentially in the number of exercise dates (B), ruling out its practical usage for many realistic applications. Broadie and Glasserman [2004] suggest a recombining *stochastic mesh* method that grows only linearly in the number of exercise weights; in its basic form, this method requires explicit knowledge of transition densities as it relies on likelihood ratios to set weights on nodes in the mesh (see Section 3.3.3). As discussed in Glasserman [2004], the concept of stochastic meshes can, however, be broadened to include several other methods, include the regression approach in Section 3.5.4.

3.A Appendix: Constants for Φ^{-1} Algorithm

a_0 2.50662823884
a_1 -18.61500062529
a_2 41.39119773534
a_3 -25.44106049637
b_0 -8.47351093090
b_1 23.08336743743
b_2 -21.06224101826
b_3 3.13082909833

c_0 0.3374754822726147
c_1 0.9761690190917186
c_2 0.1607979714918209
c_3 0.0276438810333863
c_4 0.0038405729373609
c_5 0.0003951896511919
c_6 0.0000321767881768
c_7 0.0000002888167364
c_8 0.0000003960315187

4
Fundamentals of Interest Rate Modeling

The purpose of this brief chapter is twofold. First, we introduce notations to characterize prices and yields of basic fixed income market securities. In addition to providing the foundation for a more expansive discussion of fixed income markets (which we shall undertake in Chapter 5), this part of the chapter serves to identify and characterize a number of probability measures that are of fundamental importance in models for the term structure of interest rates. A brief discussion of measures used in a two-currency setting is also provided.

In the second part of the chapter, we discuss general characteristics of models with dynamics driven by vector-valued Brownian motions. This analysis leads to the fundamental class of Heath-Jarrow-Morton (HJM) (see Heath et al. [1992]) models of continuously compounded forward rates. Among other special cases, we discuss in some detail tractable HJM models with Gaussian volatility structure, and provide some results for the case where such models are Markovian. These discussions continue in Chapters 10 through 12 where we consider one- and multi-factor short rate models, and in the Chapter 13 where we introduce the important class of quasi-Gaussian HJM models with local and stochastic volatility.

4.1 Fixed Income Notations

4.1.1 Bonds and Forward Rates

As in earlier chapters, let $P(t, T)$ denote the time t price of a zero-coupon bond (also known as a *discount bond*) delivering for certain \$1 at maturity $T \geq t$. Suppose we are interested in purchasing at some future time T a zero-coupon bond maturing at $T + \tau$, $\tau > 0$. At time $t < T$, the price of such a bond can be locked in by i) purchasing at time t one $(T + \tau)$-maturity zero-coupon bond; and ii) selling short ("shorting") $P(t, T + \tau)/P(t, T)$

168   4 Fundamentals of Interest Rate Modeling

T-maturity zero-coupon bonds. The time t cost of executing this strategy is zero,
$$-1 \cdot P(t, T+\tau) + P(t, T+\tau)/P(t, T) \cdot P(t, T) = 0,$$
but a flow of
$$-P(t, T+\tau)/P(t, T)$$
will take place at time T as the T-maturity short position matures. This is compensated by an inflow of $1 at time $T + \tau$. In other words, our trading strategy effectively fixes the time T purchase price of the $(T + \tau)$-maturity bond at
$$P(t, T, T+\tau) \triangleq P(t, T+\tau)/P(t, T), \quad \tau > 0,$$
a quantity known as the time t *forward price* for the zero-coupon bond spanning $[T, T + \tau]$.

It is often convenient to characterize a forward bond price by a discount rate. One such rate is the *continuously compounded forward yield* $y(t, T, T+\tau)$, defined by
$$e^{-y(t,T,T+\tau)\tau} = P(t, T, T+\tau). \tag{4.1}$$
The time between the maturity of the forward bond and the expiry of the forward contract, i.e. τ, is often called the *tenor* of the forward bond or the forward yield. In the definition of the continuously compounded yield lies an implicit, and idealized, assumption of continuous reinvestment of investment proceeds. Most actual market quotes, however, are based on discrete-time compounding of proceeds. Accordingly, we define a *simple forward rate* $L(t, T, T+\tau)$ as
$$1 + \tau L(t, T, T+\tau) = 1/P(t, T, T+\tau). \tag{4.2}$$
Again, τ is the *tenor* of the forward rate. For an arbitrary set of dates $T = T_0 < T_1 < T_2 < \ldots < T_n$, notice that forward bond prices can be recovered from forward rates by simple compounding,
$$P(t, T_n)/P(t, T) = \prod_{i=1}^{n} \frac{1}{1 + (T_i - T_{i-1}) L(t, T_{i-1}, T_i)}.$$

Unless we state otherwise, throughout this book we shall typically make the assumption that spot rates $L(T, T, T+\tau)$ are the *Libor (London Interbank Offered Rate) rates* quoted in the interbank market. Libor rates are quoted on values of τ ranging from one week $(\tau = 1/52)$[1] to 12 months $(\tau = 1)$, and form the basis for a number of floating-rate derivative contracts, such as interest rate swaps and Eurodollar futures. We shall examine these securities in more detail in Chapter 5.

---
[1] Note that in reality the calculation of year fractions τ are governed by fairly complicated market conventions. A brief discussion of this topic can be found in Appendix 5.A.

In the limit $\tau \downarrow 0$,

$$L(t, T, T + \tau) \to f(t, T),$$

where the quantity $f(t, T)$ is the time t *instantaneous forward rate* to time T. We think of $f(t, T)$ as the forward rate spanning $[T, T + dT]$, observed at time t. The relation between instantaneous forward rates and forward bond prices is given by the continuous compounding formula

$$P(t, T, T + \tau) = \exp\left(-\int_T^{T+\tau} f(t, u)\, du\right), \tag{4.3}$$

such that

$$f(t, T) = -\frac{\partial \ln P(t, T)}{\partial T}, \tag{4.4}$$

and, from (4.1),

$$y(t, T, T + \tau) = \tau^{-1} \int_T^{T+\tau} f(t, u)\, du, \quad f(t, T) = \lim_{\tau \downarrow 0} y(t, T, T + \tau).$$

We also notice the relationship

$$f(t, T) = \frac{\partial\, (y(t, t, T)(T - t))}{\partial T} = y(t, t, T) + (T - t)\frac{\partial y(t, t, T)}{\partial T}.$$

The quantity

$$r(t) \triangleq f(t, t) \tag{4.5}$$

is an $\mathcal{F}_t$-measurable random variable known as the *short rate* or sometimes the *spot rate*. Loosely speaking, we can think of $r(t)$ as the overnight rate in effect at time t.

Finally, let us note that interest rates of various flavors and related quantities are typically quoted in percentage points or sometimes in *basis points*, where 1 basis point = $1/100$ of one percent.

4.1.2 Futures Rates

Through the market for Eurodollar futures (see Chapter 5), investors can enter into standardized contracts that will pay at time T an amount of

$$1 - L(T, T, T + \tau). \tag{4.6}$$

At time 0, a Eurodollar futures contract can be entered into at no upfront cost, but with an implicit obligation of the holder to pay at time T per unit of notional

$$1 - F(0, T, T + \tau)$$

in return for the payout (4.6). Here, $F(t, T, T+\tau)$ is the time t *simple futures rate* for the period $[T, T+\tau]$. Importantly, the futures rate is *marked to market* (or *resettled*) each day, with the day's change in the futures rate immediately credited to or debited from the contract holder's account with the futures exchange. Specifically, after holding the contract for a period of $\Delta = 1$ day, the futures contract holder would thus experience a cash flow of

$$(1 - F(\Delta, T, T+\tau)) - (1 - F(0, T, T+\tau))$$
$$= -(F(\Delta, T, T+\tau) - F(0, T, T+\tau)).$$

Continuing the mark-to-market process to maturity shows that the total amount of cash flow received by the holder on $[0, T]$ is

$$-(F(T, T, T+\tau) - F(0, T, T+\tau)) = -(L(T, T, T+\tau) - F(0, T, T+\tau)) \quad (4.7)$$

where we have used the fact that $F(T, T, T+\tau)$ must equal $L(T, T, T+\tau)$ to avoid a *delivery arbitrage*.

The fact that the net cash flow payment (4.7) on a Eurodollar futures contract has been made incrementally on a daily basis has important valuation consequences, and causes the futures rate to differ from the forward rate defined earlier. For instance, under a scenario of rising interest rates, the holder of a Eurodollar futures contract must make payments to the futures exchange. As rates are rising, the contract holder will be faced with a high-rate — and thus unfavorable — borrowing environment for funding these payments. Conversely, when interest rates fall, the reinvestment of received funds will take place at increasingly low rates. Due to the adverse behavior of funding costs and reinvestment gains, we would expect the purchaser of a Eurodollar futures contract to pay less for these instruments than for a comparable instrument without daily mark-to-market. Consequently, we would expect the futures rate to be *above* the corresponding forward rate. We shall quantify this effect in Section 4.5.1 and, with more advanced models, in Section 16.8.

We notice that we can define *instantaneous futures rates* $q(t, T)$ in the same fashion as we defined instantaneous forward rates:

$$q(t, T) = \lim_{\tau \downarrow 0} F(t, T, T+\tau).$$

4.1.3 Annuity Factors and Par Rates

Most fixed income securities involve multiple cash flows taking place on a pre-set schedule of dates, often referred to as a *tenor structure*,

$$0 \leq T_0 < T_1 < \ldots < T_N.$$

Given a tenor structure, for any two integers k, m satisfying $0 \leq k < N$, $m > 0$, and $k + m \leq N$, we can define an *annuity factor* $A_{k,m}$ by

4.2 Fixed Income Probability Measures

$$A_{k,m}(t) = \sum_{n=k}^{k+m-1} P(t, T_{n+1})\tau_n, \quad \tau_n = T_{n+1} - T_n. \tag{4.8}$$

Annuity factors provide for compact notation when pricing coupon-bearing securities. For instance, a security making m coupon payments of $c\tau_n$ at all $T_{n+1}, n = k, \ldots, k+m-1$, is easily seen to have time t value of

$$cA_{k,m}(t), \quad t \leq T_k.$$

If the security also involves a back-end return of notional at time T_{k+m} (as is the case for a regular coupon-bearing bond), the pricing expression is

$$cA_{k,m}(t) + P(t, T_{k+m}), \tag{4.9}$$

where we assume that the bond has been normalized to have a unit notional. The time t forward price to T_k of the security (4.9) is

$$cA_{k,m}(t)/P(t, T_k) + P(t, T_{k+m})/P(t, T_k);$$

the value of the coupon c for which this expression equals 1 is known as the *forward par rate* or, when used in the context of swap pricing, as the *forward swap rate*. With $S_{k,m}(t)$ denoting the time t swap rate, we apparently have

$$S_{k,m}(t) = \frac{P(t, T_k) - P(t, T_{k+m})}{A_{k,m}(t)}, \quad t \leq T_k. \tag{4.10}$$

From the definition of $L(t, T_n, T_{n+1})$ in (4.2), a little thought shows that the numerator of the expression for $S_{k,m}(t)$ can be expanded into a weighted sum of forward rates, leading to the alternative expression

$$S_{k,m}(t) = \frac{\sum_{n=k}^{k+m-1} \tau_n P(t, T_{n+1}) L_n(t)}{A_{k,m}(t)}, \quad t \leq T_k, \tag{4.11}$$

where we have introduced the useful shorthand

$$L_n(t) \triangleq L(t, T_n, T_{n+1}).$$

It follows that the forward swap rate can be loosely interpreted as a weighted average of simple forward rates on the specified tenor structure. We note for the future that the time T_k is sometimes referred to as the *fixing date*, or *expiry*, of the swap rate $S_{k,m}$, while the length of the corresponding swap, $T_{k+m} - T_k$, is sometimes called the *tenor* of the swap rate.

4.2 Fixed Income Probability Measures

As discussed in Chapter 1, selecting an equivalent martingale measure is largely a matter of choosing a *numeraire*, an asset price process used to re-normalize the prices of other traded securities. For later reference, this section lists and names a number of important numeraires and measures used in fixed income pricing. Throughout the section, we assume that the market is complete, and we use $V(t)$ to denote the time t price of a derivative security making an $\mathcal{F}_T$-measurable payment of $V(T)$.

4.2.1 Risk Neutral Measure

The numeraire defining the risk-neutral measure Q is the continuously compounded money market account $\beta(t)$, satisfying the locally deterministic SDE
$$d\beta(t) = r(t)\beta(t)\,dt, \quad \beta(0) = 1, \tag{4.12}$$
where $r(t)$ is the short rate, $r(t) = f(t,t)$. Solving this equation yields
$$\beta(t) = e^{\int_0^t r(u)\,du}.$$
From the results of Chapter 1, in the absence of arbitrage the numeraire-deflated process $V(t)/\beta(t)$ must be a martingale, implying the derivative security valuation formula
$$V(t)/\beta(t) = \mathrm{E}_t^Q\left(V(T)/\beta(T)\right), \quad t \leq T, \tag{4.13}$$
or equivalently
$$V(t) = \mathrm{E}_t^Q\left(e^{-\int_t^T r(u)\,du} V(T)\right). \tag{4.14}$$

If we apply (4.14) to the special case of $V(T) = 1$, we obtain a fundamental bond pricing formula. We highlight the importance of this result by listing it in a lemma.

Lemma 4.2.1. *In the absence of arbitrage, the time t price $P(t,T)$ of a T-maturity zero-coupon bond is*
$$P(t,T) = \mathrm{E}_t^Q\left(e^{-\int_t^T r(u)\,du}\right). \tag{4.15}$$

It follows from Lemma 4.2.1 that specification of the dynamics of $r(t)$ under Q suffices to determine the prices of discount bonds at all times and maturities. Models that are based on such a direct specification of $r(t)$ dynamics are known as *short rate models* and are the subject of Chapters 10 through 12. Notice the resemblance between expressions (4.3) and (4.15); if $r(t)$ is deterministic, the two expressions will agree as $r(u) = f(t,u)$, $u \geq t$. If $r(t)$ is random, one may wonder whether this result will hold in expectation. The answer to this is negative, i.e.
$$f(t,u) \neq \mathrm{E}_t^Q\left(r(u)\right), \tag{4.16}$$
provided r is random. We prove this in the section below. Under certain idealized conditions, however, equality holds in (4.16) provided $f(t,u)$ is replaced by the *futures* rate $q(t,u)$. The exact result is as follows.

4.2 Fixed Income Probability Measures

Lemma 4.2.2. *Assume that mark-to-market takes place continuously. Under regularity conditions on the short rate $r(\cdot)$ — it suffices that $r(\cdot)$ is positive and bounded — the futures rate $F(\cdot, T, T+\tau)$ is a Q-martingale, and*

$$F(t, T, T+\tau) = \mathrm{E}_t^Q \left(L(T, T, T+\tau) \right). \tag{4.17}$$

Proof. Over a small interval $[t, t+dt]$, we have earlier shown that the cash proceeds from a futures contract are proportional to

$$dF(t, T, T+\tau) = F(t+dt, T, T+\tau) - F(t, T, T+\tau).$$

Suppose that we hold the futures contract up to some arbitrary horizon $t < \mathcal{T} \leq T$ at which point we exit (e.g., by selling the futures contract). Deflating all cash proceeds from this strategy with the numeraire $\beta(t)$ and integrating provides us with the time t value of the futures contract as

$$V_{\mathrm{fut}}(t) = \beta(t) \mathrm{E}_t^Q \left(\int_t^{\mathcal{T}} \beta(s)^{-1} \, dF(s, T, T+\tau) + \beta(\mathcal{T})^{-1} V_{\mathrm{fut}}(\mathcal{T}) \right),$$

where we have used the fact that $\beta(s)$ is locally deterministic. As it is always costless to enter into a futures contract, $V_{\mathrm{fut}}(t) = V_{\mathrm{fut}}(\mathcal{T}) = 0$ by definition, so for arbitrary $t < \mathcal{T} \leq T$ we must have (since $\beta(t)$ is positive)

$$\mathrm{E}_t^Q \left(\int_t^{\mathcal{T}} \beta(s)^{-1} \, dF(s, T, T+\tau) \right) = 0. \tag{4.18}$$

Under the stated assumptions on r, $\beta(s)^{-1}$ is bounded from above and below by positive constants, so (4.18) holds for arbitrary horizons $t < \mathcal{T} \leq T$ and it follows that

$$\mathrm{E}_t^Q \left(dF(s, T, T+\tau) \right) = 0, \quad t \leq s \leq T,$$

which demonstrates that F is a Q-martingale. The result (4.17) then immediately follows. □

Equation (4.17) states that the futures rate is the Q-expectation of the time T spot rate $L(T, T, T+\tau)$. A similar relation must then hold for the instantaneous futures rate, i.e.

$$q(t, u) = \mathrm{E}_t^Q \left(r(u) \right) \tag{4.19}$$

as stated earlier[2].

---

[2] This result should not be confused with the classical *expectations hypothesis* which states that futures (or sometimes forward) rates are unbiased estimators of future spot rates, in the *real-life* probability measure P: $\mathrm{E}_t^P(r(T)) = q(t, T)$. The expectations hypothesis amounts to a strong assumption about the market price of risk (see Chapter 1), whereas equation (4.19) is a preference-free arbitrage relationship.

Lemma 4.2.2 was first proven by non-probabilistic methods in Cox et al. [1981], who employed a direct, and quite instructive, hedging argument to show the result. The assumption of continuous resettlement in the lemma may appear idealized, but the difference between daily and continuous settlement is quite small, as shall be demonstrated in Chapter 16. Explicit modeling of discrete resettlement is nevertheless quite straightforward, and basically involves shifting measure, from the risk-neutral measure to the so-called spot measure, defined below. We return to this issue in Chapter 16.

4.2.2 T-Forward Measure

The T-forward measure Q^T was introduced in Jamshidian [1991b] (see also Geman et al. [1995]), and uses a T-maturity zero-coupon bond as the numeraire asset. As is customary, we let $E^T(\cdot)$ denote expectations in measure Q^T, such that

$$V(t)/P(t,T) = E^T_t \left(V(T)/P(T,T) \right), \quad t \leq T.$$

As obviously $P(T,T) = 1$, this expression simplifies to the convenient form

$$V(t) = P(t,T) E^T_t \left(V(T) \right). \tag{4.20}$$

Comparison of (4.20) and (4.14) shows that shifting to the T-forward measure in a sense decouples the expectation of the terminal payout $V(T)$ from that of the numeraire. As we shall see, this is often very convenient when we attempt to construct analytical formulas for prices of certain simple interest rate derivatives. From the results of Section 1.3, we note that the explicit connection between the risk-neutral and T-forward measures is given by the density

$$E^Q_t \left(\frac{dQ^T}{dQ} \right) = \frac{P(t,T)/P(0,T)}{\beta(t)}. \tag{4.21}$$

As $P(t, T+\tau)$ is the price of a traded asset, from the definition of the T-forward measure it follows that forward bond prices

$$P(t, T, T+\tau) = P(t, T+\tau)/P(t,T)$$

are martingales in the T-forward measure. We highlight a related result for forward rates below.

Lemma 4.2.3. *In the absence of arbitrage the forward Libor rate $L(t, T, T+\tau)$ is a martingale under $Q^{T+\tau}$, such that*

$$L(t, T, T+\tau) = E^{T+\tau}_t \left(L(T, T, T+\tau) \right), \quad t \leq T. \tag{4.22}$$

Proof. By definition (see (4.2))

$$L(t, T, T+\tau) = \tau^{-1} \left(P(t,T)/P(t, T+\tau) - 1 \right).$$

As $P(t,T)/P(t,T+\tau)$ is a martingale under $Q^{T+\tau}$, so is $L(t,T,T+\tau)$. The result follows. □

Taking the limit $\tau \downarrow 0$ and setting $T = u$ yields

$$f(t,u) = \mathrm{E}_t^u\left(f(u,u)\right) = \mathrm{E}_t^u\left(r(u)\right), \qquad (4.23)$$

which should be compared to the result (4.19).

4.2.3 Spot Measure

When working with a multitude of forward rates on a tenor structure $0 = T_0 < T_1 < \ldots < T_N$, it is often convenient to introduce a numeraire that can be extended to arbitrary horizons by compounding. While the continuously compounded money market account β would accomplish this, working with a continuously compounded numeraire is inherently awkward in a setting with a discrete tenor structure. As an alternative, we can introduce a discrete-time equivalent of the continuously compounded money market account to be the value of the following trading strategy. At time 0, \$1 is invested in $1/P(0,T_1)$ T_1-maturity discount bonds, returning the amount

$$1/P(0,T_1) = 1 + \tau_0 L(0,0,T_1)$$

at time T_1. This amount is then reinvested ("rolled") at time T_1 in T_2-maturity bonds, returning

$$1/P(0,T_1) \cdot 1/P(T_1,T_2) = (1+\tau_0 L(0,0,T_1))(1+\tau_1 L(T_1,T_1,T_2))$$

at time T_2. Repeating this re-investment strategy at each date in the tenor structure gives rise to an asset price process $B(t)$, where $B(0) = 1$ and

$$B(t) = \prod_{n=0}^{i}\left(1+\tau_n L_n(T_n)\right) P(t, T_{i+1}), \quad T_i < t \leq T_{i+1}, \qquad (4.24)$$

where we used the already introduced short-hand notation $L_n(t) = L(t, T_n, T_{n+1})$. The process $B(t)$ is effectively a rolling certificate of deposit, and can be interpreted as a discrete-time equivalent of $\beta(t)$. $B(t)$ will approach $\beta(t)$ as the time spacing of the tenor structure is made increasingly fine.

The measure induced by $B(t)$ is known as the *spot measure* (or sometimes *spot Libor measure*), denoted Q^B. With $\mathrm{E}^B(\cdot)$ denoting expectations in this measure we have

$$V(t) = \mathrm{E}_t^B\left(V(T)\frac{B(t)}{B(T)}\right),$$

where

$$\frac{B(t)}{B(T)} = \prod_{n=i+1}^{j} (1 + \tau_n L_n(T_n))^{-1} \frac{P(t, T_{i+1})}{P(T, T_{j+1})},$$
$$T_i < t \leq T_{i+1}, \quad T_j < T \leq T_{j+1}.$$

The similarity between the discrete and continuous money market accounts makes the spot Libor measure resemble the risk-neutral measure in many ways. For example, as we recall from Lemma 4.2.2, the risk-neutral measure is characterized by the fact that a continuously resettled futures rate is a martingale. In close parallel, the futures rate that is marked to market (resettled) discretely on dates $T_0, \ldots, T_N$ turns out to be a martingale in the spot Libor measure. We show this in Section 16.8.

4.2.4 Terminal and Hybrid Measures

One advantage of the spot measure over an arbitrary T-forward measure is the fact that the numeraire asset $B(t)$ will remain alive throughout the span of the tenor structure $\{T_n\}_{n=0}^{N}$. This property of $B(t)$ is necessary for the valuation of securities which may mature (randomly) at any date in the tenor structure. Securities of this type include for instance barrier options (such as range accruals) and options with early exercise rights (such as Bermudan swaptions). On the other hand, if we pick the T-forward measure corresponding to the *last* maturity in the tenor structure, $T = T_N$, this also yields a numeraire asset — the T_N-maturity zero-coupon bond — that is certain to remain alive at all dates in the tenor structure. The measure induced by $P(t, T_N)$ (Q^{T_N}) is often referred to as the *terminal measure*. For a security V maturing at a date $T \leq T_N$ we get, from the usual martingale pricing formula,

$$V(t) = P(t, T_N) \mathrm{E}_t^{T_N} \left(V(T)/P(T, T_N) \right), \quad t \leq T \leq T_N. \tag{4.25}$$

In (4.25) it is useful to notice that $V(T)/P(T, T_N)$ is the time T_N proceeds of rolling at time T the security payout $V(T)$ into a zero-coupon bond maturing at time T_N, effectively aligning the maturity of the numeraire and the cash flow date of the underlying asset. As an alternative, $V(T)$ could be rolled into the spot numeraire asset $B(T)$, leading to a T_N payout of $V(T)/B(T) \cdot B(T_N)$. This gives rise to the equivalent formula

$$V(t) = P(t, T_N) \mathrm{E}_t^{T_N} \left(V(T) B(T_N)/B(T) \right), \quad t \leq T \leq T_N. \tag{4.26}$$

We note that this formula can also be derived from the basic relationship between the measures Q^B and Q^{T_N} by simply noting that

$$P(T, T_N) \mathrm{E}_T^{T_N} \left(B(T_N)/B(T) \right) = B(T) \mathrm{E}_T^{B} \left(B(T_N)/B(T)/B(T_N) \right) = 1,$$

such that, by iterated conditional expectations[3],

---

[3] The *law of iterated conditional expectations*, sometimes known as the *tower rule*, states that for an $\mathcal{F}_T$-measurable random variable X, $E(E(X|\mathcal{F}_s)|\mathcal{F}_t) = E(X|\mathcal{F}_t)$, where $t \leq s \leq T$.

$$V(t) = P(t,T_N)\mathrm{E}_t^{T_N}\left(V(T)/P(T,T_N)\right)$$
$$= P(t,T_N)\mathrm{E}_t^{T_N}\left(V(T)\mathrm{E}_T^{T_N}\left(B(T_N)/B(T)\right)\right)$$
$$= P(t,T_N)\mathrm{E}_t^{T_N}\left(V(T)B(T_N)/B(T)\right),$$

as before.

As mentioned, equations (4.25) and (4.26) effectively involve reinvestment of the proceeds $V(T)$ to align cash payment with the numeraire $P(t,T_N)$. If the numeraire expires before the derivative security, we can apply the same reinvestment idea, but this time to the numeraire asset. Consider for instance a derivative security maturing at time T_N (paying $V(T_N)$), and suppose we wish to extend the T-forward measure to price this option. For instance, we can define a numeraire asset as follows:

$$\widetilde{P}(t,T) = \begin{cases} P(t,T), & t \leq T, \\ B(t)/B(T), & t > T. \end{cases}$$

This asset corresponds to an investment strategy where we i) at time 0 purchase the T-maturity zero-coupon bond; and ii) at time T invest the proceeds from the zero-coupon bond ($1) in the spot measure numeraire asset (4.24). Letting $\widetilde{\mathrm{Q}}^T$ denote the measure induced by $\widetilde{P}(t,T)$, we can write

$$V(t) = \widetilde{P}(t,T)\widetilde{\mathrm{E}}_t^T\left(V(T_N)/\widetilde{P}(T_N,T)\right)$$
$$= \widetilde{P}(t,T)\widetilde{\mathrm{E}}_t^T\left(V(T_N)B(T)/B(T_N)\right), \quad T < T_N,\ t < T_N,$$

where $\widetilde{\mathrm{E}}^T$ is the expectations operator for the measure $\widetilde{\mathrm{Q}}^T$. If also $t \leq T$, this expression becomes (compare to (4.26))

$$V(t) = P(t,T)\widetilde{\mathrm{E}}_t^T\left(V(T_N)B(T)/B(T_N)\right), \quad t \leq T < T_N,$$

which, in effect, uses $B(T)/B(T_N)$ to discount $V(T_N)$ back to time T. The equivalent result in the T-forward measure is

$$V(t) = P(t,T)\mathrm{E}_t^T\left(B(T)\mathrm{E}_T^B\left(V(T_N)/B(T_N)\right)\right).$$

Notice that if V matures at time T, rather than at T_N, we simply have

$$V(t) = P(t,T)\widetilde{\mathrm{E}}_t^T\left(V(T)\right) = P(t,T)\mathrm{E}_t^T\left(V(T)\right),$$

which is obvious from the definition of the numeraire $\widetilde{P}(t,T)$.

The measure $\widetilde{\mathrm{Q}}^T$ is by construction a hybrid between the spot measure and the T-forward measure. Obviously, many other such measures exist, corresponding to different reinvestment strategies of expiring numeraire assets.

4.2.5 Swap Measures

Being a linear combination of zero-coupon bonds (see (4.8)), an annuity factor $A_{k,m}(t)$ on a tenor structure qualifies as a numeraire asset. The measure $Q^{k,m}$ induced by this numeraire is known as a *swap measure* or an *annuity measure*. In the absence of arbitrage we have

$$V(t) = A_{k,m}(t) \mathrm{E}_t^{k,m} \left(V(T)/A_{k,m}(T) \right),$$

where $\mathrm{E}^{k,m}(\cdot)$ denotes expectation under $Q^{k,m}$.

Lemma 4.2.4. *In the absence of arbitrage, the forward swap rate $S_{k,m}(t)$ is a martingale in measure $Q^{k,m}$.*

Proof. By definition

$$S_{k,m}(t) = \frac{P(t,T_k) - P(t,T_{k+m})}{A_{k,m}(t)}.$$

As the numeraire deflated assets $P(t,T_k)/A_{k,m}(t)$ and $P(t,T_{k+m})/A_{k,m}(t)$ must both be martingales, so must be their difference. □

As we shall see later, swap measures are very useful for analytical manipulations of price formulas for options on swaps.

4.3 Multi-Currency Markets

While this book is primarily dedicated to the study of single-currency interest rate derivatives, occasionally it will be necessary to consider certain effects associated with trading in a multi-currency economy. For instance, in Chapter 6 we touch upon issues of yield curve constructions in non-domestic currencies, and in Chapter 16 we discuss the important practical case where a derivative pays out in a foreign currency, but has a payout function that depends on one or more domestic interest rate variables. This brief section provides background material and notation required for these and other cross-currency applications.

4.3.1 Notations and FX Forwards

We consider two economies, a "domestic" economy and a "foreign" economy. Let $P_d(t,T)$ and $P_f(t,T)$ denote time t zero-coupon bond prices in the domestic and foreign economies, respectively. So, $P_f(t,T)$ (say) is the time t price, in foreign currency, of one unit of foreign currency delivered for certain at time T. Translation of values in foreign currency to domestic currency takes place at a foreign exchange (FX) rate of $X(t)$, measured in units of domestic currency per unit of foreign currency. In other words, the value $\widetilde{P}_d$ to a domestic investor of one foreign zero-coupon bond is

$$\widetilde{P}_d(t,T) = X(t)P_f(t,T).$$

The quantity
$$X_T(t) = \frac{\widetilde{P}_d(t,T)}{P_d(t,T)} = X(t)\frac{P_f(t,T)}{P_d(t,T)}$$
is known as the *forward FX rate* to time T. The name is motivated by the following arbitrage strategy:

- Buy one foreign zero-coupon bond, at a cost of $\widetilde{P}_d(t,T)$ in domestic currency.
- Finance the purchase by selling short domestic zero-coupon bonds on a notional of $\widetilde{P}_d(t,T)/P_d(t,T)$.

With no outlay at time t, the strategy will generate a net cash flow at time T of one unit of foreign currency and $-X_T(t)$ units of domestic currency, such that the trading strategy in effect has locked in a time t a future time T exchange rate of $X_T(t)$.

4.3.2 Risk Neutral Measures

Let $\beta_d(t)$ and $\beta_f(t)$ be the continuously compounded money market accounts in the domestic and foreign economies, respectively. $\beta_d(t)$ and $\beta_f(t)$ induce two separate risk-neutral measures, denoted Q^d and Q^f; let us investigate how these measures are related. If $g(T)$ is a random payout at time T made in foreign currency, in a complete market the value (in units of foreign currency) of this payout to a foreign investor is, from standard principles,

$$V_f(t) = \beta_f(t)\mathrm{E}^f_t\left(g(T)\beta_f(T)^{-1}\right), \tag{4.27}$$

where E^f_t denotes expectations in the foreign risk-neutral measure Q^f. For a domestic investor, the payout of $g(T)$ must be translated to domestic currency units at a rate of $X(T)$, making the effective domestic payout function $g(T)X(T)$. Thereby,

$$V_d(t) = \beta_d(t)\mathrm{E}^d_t\left(g(T)X(T)\beta_d(T)^{-1}\right), \tag{4.28}$$

where E^d_t denotes expectations in measure Q^d. Importantly, the expressions in (4.27) and (4.28) are linked by the spot exchange rate, as the absence of a cross-currency arbitrage dictates that

$$V_d(t) = X(t)V_f(t),$$

or

$$\beta_d(t)\mathrm{E}^d_t\left(g(T)X(T)\beta_d(T)^{-1}\right) = X(t)\beta_f(t)\mathrm{E}^f_t\left(g(T)\beta_f(T)^{-1}\right). \tag{4.29}$$

We use this result to establish the following lemma:

Lemma 4.3.1. *The domestic and foreign risk-neutral probability measures Q^d and Q^f are related by the density process*

$$\mathrm{E}_t^d \left(\frac{dQ^f}{dQ^d} \right) = \frac{(\beta_f(t)/\beta_f(0))X(t)}{(\beta_d(t)/\beta_d(0))X(0)}, \quad t \geq 0.$$

Proof. For an $\mathcal{F}_T$-measurable variable $Y(T) = g(T)X(T)\beta_d(T)^{-1}$ satisfying regularity conditions, a rearrangement of the basic relation (4.29) yields

$$\mathrm{E}_t^d (Y(T)) = X(t) \frac{\beta_f(t)}{\beta_d(t)} \mathrm{E}_t^f \left(\frac{Y(T)}{X(T)} \frac{\beta_d(T)}{\beta_f(T)} \right),$$

From the results of Section 1.3, the density relating measures Q^d and Q^f is then as given in the lemma. □

With $\beta_f(t)X(t)/\beta_d(t)$ being a martingale in the domestic risk-neutral measure, we note that if $X(t)$ is an Ito process, it must take the form

$$dX(t) = X(t)\left(r_d(t) - r_f(t)\right)dt + \sigma_X(t)^\top dW(t),$$

where $r_d(t) - r_f(t)$ is the spread between domestic and foreign short rates, $W(t)$ is a (vector-valued) Q^d-Brownian motion, and $\sigma_X(t)$ is some adapted stochastic process satisfying regularity conditions.

4.3.3 Other Measures

Having established the Radon-Nikodym derivative relating the domestic and foreign risk-neutral measures, relations between various other domestic and foreign probability measures are easy to establish. For instance, the following result is easily proven the same way as Lemma 4.3.1.

Lemma 4.3.2. *Let $\mathrm{E}_t^{T,d}$ denote expectations in the domestic T-forward probability measure. The domestic and foreign T-forward probability measures $Q^{T,d}$ and $Q^{T,f}$ are related by the density process*

$$\mathrm{E}_t^{T,d} \left(\frac{dQ^{T,f}}{dQ^{T,d}} \right) = \frac{P_f(t,T)P_d(0,T)X(t)}{P_d(t,T)P_f(0,T)X(0)} = \frac{X_T(t)}{X_T(0)}, \quad t \geq 0.$$

We highlight the fact that the forward FX rate $X_T(t)$ is a $Q^{T,d}$-martingale satisfying $X_T(T) = X(T)$. For an $\mathcal{F}_T$-measurable variable $Y(T)$, we thereby have the convenient expression

$$\mathrm{E}_t^{T,d} (Y(T)) = X_T(t)\mathrm{E}_t^{T,f} \left(\frac{Y(T)}{X(T)} \right),$$

where $\mathrm{E}_t^{T,f}$ denotes expectations in the foreign T-forward probability measure.

4.4 The HJM Analysis

Having defined notations and established basic arbitrage relationships, let us turn to assigning dynamics to the many quantities we have introduced so far. We shall here follow the Heath et al. [1992] (Heath-Jarrow-Morton, or HJM) approach, where all information in the economy is assumed to originate with a finite number of Brownian motions. The resulting class of models is quite broad, and in much of the rest of this book we shall deal with ways to reduce the general HJM model to specific, and tractable, special cases. For now, however, we concentrate on a general analysis, although we keep our treatment fairly informal.

4.4.1 Bond Price Dynamics

In the HJM framework, we concern ourselves with the modeling of how an entire continuum of T-indexed bond prices $P(\cdot, T)$ jointly evolves over time, starting from a known condition $P(0,T)$. We consider models of finite horizon, i.e. with $T \in [0, \mathcal{T}]$, $\mathcal{T} < \infty$, and specialize to a filtration generated by a d-dimensional Brownian motion. We assume that a risk-neutral measure Q exists and is unique. Let $W(t)$ be an adapted d-dimensional Q-Brownian motion, and define deflated bond values as $P_\beta(t,T) = P(t,T)/\beta(t)$, where $\beta(t)$ as always is the continuously rolled money market account. In the absence of arbitrage, $P_\beta(t,T)$ is a martingale in the risk-neutral measure, and the martingale representation theorem then implies that

$$dP_\beta(t,T) = -P_\beta(t,T)\sigma_P(t,T)^\top dW(t), \quad t \leq T, \qquad (4.30)$$

where $\sigma_P(t,T) = \sigma_P(t,T,\omega)$ is a d-dimensional stochastic process adapted to the filtration generated by W. We assume that $\sigma_P(t,T)$ is regular enough for $P_\beta(t,T)$ to be a square-integrable martingale. Also, as the bond $P(t,T)$ must equal \$1 at $t=T$ ("pull to par"), we impose the consistency condition

$$\sigma_P(T,T) = 0.$$

Using (4.12) and Ito's lemma, it follows from (4.30) that

$$dP(t,T)/P(t,T) = r(t)\,dt - \sigma_P(t,T)^\top dW(t), \qquad (4.31)$$

where $r(t)$ is the short rate process. Equation (4.31) defines the class of d-dimensional HJM models.

Another application of Ito's lemma shows that forward bond prices $P(t,T,T+\tau) = P(t,T+\tau)/P(t,T)$ must satisfy

$$dP(t,T,T+\tau)/P(t,T,T+\tau) = -\left[\sigma_P(t,T+\tau) - \sigma_P(t,T)\right]^\top \sigma_P(t,T)\,dt$$
$$- \left[\sigma_P(t,T+\tau) - \sigma_P(t,T)\right]^\top dW(t). \quad (4.32)$$

In the T-forward measure Q^T, $P(t, T, T+\tau)$ is a martingale (see Section 4.2.2), and
$$dP(t, T, T+\tau)/P(t, T, T+\tau) = -\left[\sigma_P(t, T+\tau) - \sigma_P(t, T)\right]^\top dW^T(t), \tag{4.33}$$
where $W^T(t)$ is a Q^T-Brownian motion. Comparison of (4.32) and (4.33) shows that
$$dW^T(t) = dW(t) + \sigma_P(t, T)\, dt \tag{4.34}$$
which by Girsanov's theorem identifies the density process for the measure shift between Q^T and Q in the HJM setting:
$$\varsigma(t) = E_t^Q\left(\frac{dQ^T}{dQ}\right) = \mathcal{E}\left(-\int_0^t \sigma_P(u, T)^\top dW(u)\right), \tag{4.35}$$
or
$$d\varsigma(t)/\varsigma(t) = -\sigma_P(t, T)^\top dW(t).$$
This result could, of course, have been established from the first principles as well — see equation (4.21).

4.4.2 Forward Rate Dynamics

Traditionally, HJM models are stated in terms of instantaneous forward rates, rather than bond prices. Besides eliminating the need to consider the short rate r, this also reveals a number of fundamental properties of the class of HJM models. By Ito's lemma, in measure Q,
$$d\ln P(t, T) = O(dt) - \sigma_P(t, T)^\top dW(t),$$
where for convenience we have omitted writing out the drift term. Differentiating the right- and left-hand sides of this equation with respect to T, we get from equation (4.4),
$$df(t, T) = \mu_f(t, T)\, dt + \sigma_f(t, T)^\top dW(t),$$
where
$$\sigma_f(t, T) = \frac{\partial}{\partial T}\sigma_P(t, T), \tag{4.36}$$
and $\mu_f(t, T)$ is listed below.

Lemma 4.4.1. *The process for $f(t, T)$ in the T-forward measure is*
$$df(t, T) = \sigma_f(t, T)^\top dW^T(t). \tag{4.37}$$
In the risk-neutral measure, the process is
$$df(t, T) = \sigma_f(t, T)^\top \sigma_P(t, T)\, dt + \sigma_f(t, T)^\top dW(t)$$
$$= \sigma_f(t, T)^\top \int_t^T \sigma_f(t, u)\, du\, dt + \sigma_f(t, T)^\top dW(t). \tag{4.38}$$

Proof. The SDE (4.37) follows directly from the martingale relation (4.23). The risk-neutral process (4.38) then can be derived from the relations (4.34) and (4.35), with the second equality following from (4.36). □

The equation (4.38) is often considered to be the main result of Heath et al. [1992]. It demonstrates that an HJM model is fully specified once the forward rate diffusion coefficients $\sigma_f(t,T)$ have been specified for all t and T. Note that HJM models take initial forward rates $f(0,T)$ as exogenous inputs, ensuring that these models are automatically consistent with discount bond prices at time 0. This is true irrespective of the choice of $\sigma_f(t,T)$, which can be set freely (subject to regularity conditions) from either empirical analysis, or from a calibration to market prices of fixed income derivatives.

While it is convenient that HJM models are automatically calibrated to initial bond prices, a number of other features of the general HJM model are less attractive. Particularly problematic is the sheer dimensionality of the model: to describe the time t state of a discount bond curve spanning $[t,T]$, we need to keep track of a continuum of forward rates $\{f(t,u), t \leq u \leq T\}$. By Lemma 4.4.1 the forward rate curve follows an infinite-dimensional diffusion process, leaving us with an infinite number of state variables to diffuse. In practice, the implementation of an HJM model will require either making special assumptions about the σ_f process that permit a finite-dimensional Markovian representation of the forward rate curve; or moving from infinitesimal forward rates to continuously compounded forward rates that span time-buckets of finite length. Chapters 10 through 13 and Section 4.5.2 below give examples of the former idea, and Andersen [1995] discusses the latter approach in a Monte Carlo setting. An idea closely related to the discussion in Andersen [1995] is to built a model around a finite set of simple (Libor) forward rates on a fixed tenor structure. This approach has a number of computational and theoretical advantages, and is the subject of Chapter 14. For now, we note that *any* arbitrage-free interest rate model set in a filtration generated exclusively by Brownian motions must be a special case of an HJM model. In particular, any such model must correspond to a particular choice of $\sigma_f(t,T)$.

4.4.3 Short Rate Process

As discussed earlier, specification of a short rate process is, in principle, sufficient to completely specify a full yield curve model. In the HJM framework, it follows from (4.38) that the short rate $r(t)$ in measure Q is

$$r(t) = f(t,t) = f(0,t) + \int_0^t \sigma_f(u,t)^\top \int_u^t \sigma_f(u,s)\, ds\, du + \int_0^t \sigma_f(u,t)^\top dW(u).$$

The process for $r(t)$ is generally not Markovian, as can be seen by focusing on the path-dependent term

$$D(t) = \int_0^t \sigma_f(u,t)^\top dW(u)$$

for which we must have

$$D(T) = D(t) + \int_t^T \sigma_f(u,T)^\top dW(u)$$
$$+ \left\{ \int_0^t \sigma_f(u,T)^\top dW(u) - \int_0^t \sigma_f(u,t)^\top dW(u) \right\}. \quad (4.39)$$

Thereby
$$\mathrm{E}^Q\left(D(T)|\, D(t)\right) \neq \mathrm{E}_t^Q\left(D(T)\right)$$
unless the bracketed term in (4.39) is either non-random, or a deterministic function of $D(t)$ (which is generally not the case).

An interesting area of investigation concerns the conditions under which either $r(t)$ is outright Markov[4] or, less restrictively, can be written as

$$r(t) = h\left(t, x(t)\right),$$

for a deterministic function h and a finite-dimensional Markovian vector of state variables $x(t)$. Definitive results are given in Björk [2001], building on earlier (and considerably less abstract) work by Jamshidian [1991b], Cheyette [1991], and Ritchken and Sankarasubramanian [1995]. Section 4.5.2 and Chapter 13 list some of the results of these papers.

4.5 Examples of HJM Models

4.5.1 The Gaussian Model

In the HJM bond price dynamics (4.31), we now assume that $\sigma_P(t,T)$ is a bounded (d-dimensional) deterministic function of t and T. It follows from (4.32) and (4.33) that forward bond prices are then log-normally distributed in both Q and Q^T. The forward rate process in Q is

$$df(t,T) = \sigma_f(t,T)^\top \sigma_P(t,T)\, dt + \sigma_f(t,T)^\top dW(t), \quad \sigma_f(t,T) = \frac{\partial}{\partial T}\sigma_P(t,T), \tag{4.40}$$

which implies that $r(T) = f(T,T)$ is Gaussian with Q-moments

$$\mathrm{E}_t^Q\left(f(T,T)\right) = f(t,T) + \int_t^T \sigma_f(u,T)^\top \sigma_P(u,T)\, du,$$
$$\mathrm{Var}_t^Q\left(f(T,T)\right) = \int_t^T \sigma_f(u,T)^\top \sigma_f(u,T)\, du.$$

---

[4] In the sense that the time t expectations of functionals of $r(T)$ only require knowledge of $r(t)$ itself. In this case the process for $r(t)$ must be a diffusion characterized by an SDE $dr(t) = \mu_r(t, r(t))dt + \sigma_r(t, r(t))^\top dW(t)$.

The simple form of the Gaussian HJM model makes it quite tractable, permitting analytical price formulas for a number of European options and futures contracts[5]. While the Gaussian HJM model suffers from the drawback of allowing negative forward and spot rates, analytical results derived in the model are often very useful in gaining a deeper understanding of a given contract, even if ultimately a more realistic model will be required for serious pricing purposes. Indeed, results derived for the Gaussian HJM model can often be used as a starting point for development of closed-form approximations in other models; we shall see many examples of this later in the book.

For illustration, we list a few select analytical results below. More formulas can be found in numerous sources, including Chapter II in Andersen [1996], Jamshidian [1991b], and Jamshidian [1993], to name a few.

Proposition 4.5.1 (Option on Zero-Coupon Bond). *Consider a European call option paying at maturity T the amount*

$$V(T) = (P(T, T^*) - K)^+, \quad T^* > T.$$

In the Gaussian HJM model (4.40), we have

$$V(t) = P(t, T^*)\Phi(d_+) - P(t, T)K\Phi(d_-), \tag{4.41}$$

where

$$d_\pm = \frac{\ln(P(t, T^*)/(KP(t, T))) \pm v/2}{\sqrt{v}},$$

$$v = \int_t^T |\sigma_P(u, T^*) - \sigma_P(u, T)|^2 \, du.$$

Proof. In the T-forward measure Q^T we have, from (4.20),

$$V(t) = P(t, T)\mathrm{E}_t^T\left((P(T, T^*) - K)^+\right)$$
$$= P(t, T)\mathrm{E}_t^T\left((P(T, T, T^*) - K)^+\right).$$

From the discussion in Section 4.4.1 we know that $P(t, T, T^*)$ is a Q^T-martingale characterized by the SDE

$$dP(t, T, T^*)/P(t, T, T^*) = -\left[\sigma_P(t, T^*) - \sigma_P(t, T)\right]^\top dW^T(t),$$

where W^T is a d-dimensional Q^T-Brownian motion. As this is just a GBM process with time-dependent coefficients, the Black-Scholes-Merton results in Section 1.9 apply and lead to (4.41). □

---

[5] Indeed, we have already used this model in an equity context — see Section 1.9.3.2.

Proposition 4.5.2 (Caplet). *Consider a European call option paying at $T+\tau$ the amount (a caplet)*

$$V(T+\tau) = \tau \left(L(T,T,T+\tau) - K \right)^+, \quad \tau > 0.$$

In the Gaussian HJM model (4.40), we have

$$V(t) = P(t,T)\Phi(-d_-) - (1+K\tau)P(t,T+\tau)\Phi(-d_+),$$

where

$$d_\pm = \frac{\ln\left((1+K\tau)P(t,T+\tau)/P(t,T)\right) \pm v/2}{\sqrt{v}},$$

$$v = \int_t^T |\sigma_P(u,T+\tau) - \sigma_P(u,T)|^2 \, du.$$

Proof. By definition (4.2) of the Libor rate,

$$V(T+\tau) = \left(\frac{1}{P(T,T+\tau)} - 1 - K\tau \right)^+.$$

As the caplet payoff is $\mathcal{F}_T$-measurable we have

$$V(T) = P(T,T+\tau)V(T+\tau) = \left(1 - (1+K\tau)\,P(T,T+\tau)\right)^+,$$

and we see that the value of the caplet at time T can be written as a scaled payoff of a put option on the zero-coupon bond $P(T,T+\tau)$. Applying Proposition 4.5.1 and call-put parity immediately yields the result. □

Proposition 4.5.3 (Futures Rate). *In the Gaussian HJM model (4.40), futures rates are given by*

$$F(t,T,T+\tau) = \tau^{-1}\left((1/P(t,T,T+\tau))\, e^{\Omega(t,T)} - 1\right), \tag{4.42}$$

where

$$\Omega(t,T) = \int_t^T \left[\sigma_P(u,T+\tau) - \sigma_P(u,T)\right]^\top \sigma_P(u,T+\tau) \, du.$$

Proof. From Lemma 4.2.2,

$$\begin{aligned} F(t,T,T+\tau) &= \mathrm{E}_t^Q \left(L(T,T,T+\tau)\right) \\ &= \tau^{-1}\mathrm{E}_t^Q \left(1/P(T,T+\tau) - 1\right) \\ &= \tau^{-1}\mathrm{E}_t^Q \left(G(T) - 1\right), \end{aligned} \tag{4.43}$$

where we have introduced an auxiliary variable

$$G(t) \triangleq P(t,T)/P(t,T+\tau) = 1/P(t,T,T+\tau).$$

Ito's lemma shows that (see also (4.32)) in measure Q

$$dG(t)/G(t) = [\sigma_P(t,T+\tau) - \sigma_P(t,T)]^\top \sigma_P(t,T+\tau)\, dt$$
$$+ [\sigma_P(t,T+\tau) - \sigma_P(t,T)]^\top dW(t),$$

such that

$$E_t^Q(G(T)) = G(t) e^{\Omega(t,T)} = (1/P(t,T,T+\tau))\, e^{\Omega(t,T)},$$

where Ω is as given above. The result of Proposition 4.5.3 then follows directly from (4.43). □

In any rational model $\Omega(t,T) \geq 0$, such that $F(t,T,T+\tau) \geq L(t,T,T+\tau)$, consistent with the qualitative discussion in Section 4.1.2. As shown in Chapter II of Andersen [1996], the spread (also known as futures *convexity*) between futures and forward rates can be decomposed into two components: i) a term originating from the mark-to-market mechanism of a futures contract; and ii) a term originating from the fact that a futures contract — unlike a regular forward rate agreement — pays out the rate at the date it settles (at time T) rather than one period ahead (at time $T+\tau$). Andersen [1996], Chapter II, additionally contains a number of numerical examples examining typical futures-forward spreads, and also investigates the pricing of options on futures rates.

Section 16.8 looks in detail into pricing interest rate futures under more advanced models.

4.5.2 Gaussian HJM Models with Markovian Short Rate

Although quite tractable, the Gaussian HJM model generally does not allow for a finite-dimensional Markovian representation, and typically does not imply Markov-diffusive behavior of the short rate. As shown in Carverhill [1994], the short rate can be made Markovian, however, by imposing certain conditions on the deterministic forward rate volatility function $\sigma_f(t,T)$. To explore this, first recall from Section 4.4.3 the relation

$$r(t) = f(0,t) + \int_0^t \sigma_f(u,t)^\top \int_u^t \sigma_f(u,s)\, ds\, du + \int_0^t \sigma_f(u,t)^\top dW(u),$$

where now σ_f is deterministic. Consider imposing the special choice

$$\sigma_f(t,T) = g(t) h(T), \qquad (4.44)$$

where h is a positive real function and $g : \mathbb{R} \to \mathbb{R}^{d \times 1}$ can take any sign. For this case we have

$$\sigma_P(t,T) = \int_t^T \sigma_f(t,u)\, du = g(t) \int_t^T h(u)\, du,$$

and

$$r(t) = f(0,t) + h(t) \int_0^t g(u)^\top g(u) \left(\int_u^t h(s) \, ds \right) du + h(t) \int_0^t g(u)^\top dW(u)$$
(4.45)
$$\triangleq f(0,t) + h(t) \int_0^t m_f(t,u) \, du + h(t) \int_0^t g(u)^\top dW(u).$$

Importantly, the term

$$D(t) = \int_0^t \sigma_f(u,t)^\top dW(u) = h(t) \int_0^t g(u)^\top dW(u)$$

is now Markov, since

$$D(T) = h(T) \int_0^T g(u)^\top dW(u) = \frac{h(T)}{h(t)} D(t) + h(T) \int_t^T g(u)^\top dW(u),$$

which should be compared to the general (non-Markov) expression (4.39).

To show that the short rate is Markovian, we differentiate (4.45) with respect to t, yielding

$$dr(t) = \frac{\partial f(0,t)}{\partial t} dt + h'(t) \left(\int_0^t m_f(t,u) \, du + \frac{D(t)}{h(t)} \right) dt$$
$$+ h(t) \frac{\partial}{\partial t} \int_0^t m_f(t,u) \, du \, dt + h(t) g(t)^\top dW(t)$$
$$= \frac{\partial f(0,t)}{\partial t} dt + h'(t) \left(\frac{r(t) - f(0,t)}{h(t)} \right) dt$$
$$+ h(t)^2 \int_0^t g(u)^\top g(u) \, du \, dt + h(t) g(t)^\top dW(t)$$
$$= \left(\frac{\partial f(0,t)}{\partial t} - \frac{h'(t)}{h(t)} f(0,t) + h(t)^2 \int_0^t g(u)^\top g(u) \, du + \frac{h'(t)}{h(t)} r(t) \right) dt$$
$$+ h(t) g(t)^\top dW(t),$$
(4.46)

where the second equality follows from rearrangement of (4.45). This leads to the following result.

Proposition 4.5.4. *In the d-dimensional Gaussian HJM model, when (4.44) holds the short rate satisfies an SDE of the type*

$$dr(t) = (a(t) - \varkappa(t) r(t)) \, dt + \sigma_r(t)^\top dW(t),$$

where $\varkappa : \mathbb{R} \to \mathbb{R}$ and $\sigma_r : \mathbb{R} \to \mathbb{R}^{d \times 1}$ are deterministic functions of time, and

$$a(t) = \frac{\partial f(0,t)}{\partial t} + \varkappa(t)f(0,t) + \int_0^t e^{-2\int_u^t \varkappa(s)\,ds} \sigma_r(u)^\top \sigma_r(u)\,du$$

$$= \frac{\partial f(0,t)}{\partial t} + \varkappa(t)f(0,t) + \int_0^t \sigma_f(u,t)^\top \sigma_f(u,t)\,du.$$

Proof. First, by way of defining $\varkappa$ and σ_r, we set

$$h(T) = e^{-\int_0^T \varkappa(s)\,ds}; \quad g(t) = e^{\int_0^t \varkappa(s)\,ds} \sigma_r(t),$$

such that $h'(t)/h(t) = -\varkappa(t)$ and

$$\sigma_f(t,T) = e^{-\int_t^T \varkappa(s)\,ds} \sigma_r(t).$$

The result of Proposition 4.5.4 then follows directly by insertion into (4.46).
□

4.5.3 Log-Normal HJM Models

To avoid the negative forward rates inherent in Gaussian HJM models, it is tempting to consider forward rate specifications of the type

$$\sigma_f(t,T) = f(t,T)\sigma(t,T), \tag{4.47}$$

where $\sigma(t,T)$ is deterministic and bounded. In the T-forward measure

$$df(t,T) = f(t,T)\sigma(t,T)^\top\,dW^T(t)$$

such that $f(t,T)$ is log-normally distributed. While avoiding negative rates, the specification (4.47) has severe technical problems: in Q, forward rates will explode to infinity with non-zero probability. Attempts to apply the valuation formula (4.15) will thereby result in all zero-coupon bond prices being zero, implying obvious arbitrage opportunities. To suggest a rationale for the exploding rates, consider the Q-dynamics

$$df(t,T) = \left(f(t,T)\sigma(t,T)^\top \int_t^T f(t,u)\sigma(t,u)\,du\right)dt + f(t,T)\sigma(t,T)^\top dW(t).$$

Loosely speaking, the drift-term is proportional to forward rates *squared*, which, in the light of the linear growth condition in Theorem 1.6.1, may cause us to suspect problems with the existence of a non-exploding solution. Morton [1988] confirms this rigorously.

One solution to the explosion problem involves enforcing a strict upper bound on $\sigma_f(t,T)$, as in

$$\sigma_f(t,T) = \min\left(f(t,T), M\right)\sigma(t,T),$$

where M is a large positive constant. For the one-factor case ($d = 1$) Heath et al. [1992] demonstrate that this specification will ensure non-negative forward rates[6] and will prevent rate explosions. Nevertheless, the model is clearly awkward in its dependence on the arbitrary constant M. A more satisfying solution is discussed in Chapter 14, where we show that the explosion problem can be circumvented by working with simply — rather than continuously — compounded forward rates. A related issue in short rate models is also discussed in Chapter 11.

---

[6]To see this, notice that $M > 0$ guarantees that $df(t,T) = 0$ if $f(t,T)$ should ever reach 0.

5
Fixed Income Instruments

At this point, we have established the mathematical and numerical prerequisites needed for the remaining part of the book, much of which is devoted to the development of models for fixed income derivatives. Before delving into the modeling exercise, this final foundational chapter provides a tour of actual fixed income markets as well as an overview of the types of products traded. The simpler (and more liquid) of these products will typically serve as calibration targets to parameterize the models we develop; others (the more complicated and illiquid ones) will constitute the contracts that our model are ultimately meant to price and hedge. Throughout the chapter — and, indeed, this book — our focus is on the securities tied to the so-called *Libor rate*; this will include essentially all high-end exotic securities as well as more basic instruments such as FRAs, caps, and swaptions. Our priorities dictate that we leave out government, corporate, and mortgage bonds, as well as the derivatives associated with these types of securities. A discussion of these classes of securities, along with many more details on the organization and workings of fixed income markets, can be found in specialist literature, such as Fabozzi and Modigliani [1996], Fabozzi [1985], Fabozzi [2001], and Fabozzi and Fabozzi [1989].

5.1 Fixed Income Markets and Participants

At the most fundamental level, interest rates determine the economic cost of borrowing and lending, and as such define present values of future cash flows. In general, cash flows occurring at different times are discounted at different rates, reflecting market fluctuations in demand for money and risk preferences of market participants. The dependence of interest rates on time is described by the so-called *term structure of interest rates*, easily visualized as a curve that assigns a particular interest rate (or, equivalently, a discount factor) to each future date.

For a given entity, the cost of borrowing money will depend on its credit quality. Governments of developed countries, perceived to have virtually no possibility of default, issue bonds at comparatively low interest rates that reflect this perception. While the market in government debt is vast, corporations typically find it more convenient to use and originate fixed income instruments linked to rates that are more reflective of their own financing costs (i.e., credit quality). By far the most common of such reference rates is the London Interbank Offered rate, commonly known as the *Libor rate*. The Libor rate is a filtered average of bank estimates of rates at which they can borrow for a given term in the *interbank money market*, i.e. the wholesale market in which banks provide unsecured short-term credit to each other. Libor rates are quoted for multiple deposit maturities ranging from one day to one year, and are set every business day by averaging polling results from a number of large banks. Libor rates are available for deposits in different currencies, so that there is a USD-Libor rate, a EUR-Libor rate, and so on.

While Libor rates are probably the most used reference rates for interest rate contracts, there are other important rates to be aware of. For example, in the United States, banks are required to hold certain balances ("Federal funds") with the Federal Reserve, the central bank of the US. If a bank does not have sufficient balances, it can borrow them from another bank that has an excess on its account. The overnight interest rate charged in this case is called the *(effective) Federal funds rate*[1], or sometimes simply the Fed funds rate. This rate is often considered the best available proxy for a risk-free USD rate, in part because the Fed funds rate is normally the contractual rate used to accrue interest on posted collateral[2], as explained in Piterbarg [2010]. It is worth noting that the Fed funds rate used to be closely linked to the overnight Libor rate, with the spread between the two in the single basis points. However, in the subprime crisis of 2007–2009 the two have diverged significantly; the implications of this for interest rate curve construction are discussed in Section 6.5.3. Instruments linked to averages of the (effective) Fed fund rate over different terms are actively traded, giving rise to a term structure of Fed funds linked rates.

A special feature of the US public debt markets gives rise to another set of rates. In particular, interest on bonds issued by states and other local governments of the US is often free of the federal tax. The Bond Market Association, a trade association of the bond industry, publishes the *BMA rate* (or *BMA index*) which is the estimate of borrowing by such municipalities.

---

[1] The target rate, set by the Federal reserve, is aptly called the *target* Fed funds rate.

[2] To mitigate credit risk, many derivatives transactions require posting of collateral (normally cash or Treasury bonds) in the amount of the current mark-to-market. ISDA [2005] contains a detailed description of collateral agreements; according to ISDA [2009], in 2009 about 65% of all OTC derivatives transactions involved such agreements.

5.1 Fixed Income Markets and Participants 193

There is a well-developed market in interest rate derivatives that are linked to the BMA rate.

The Euro and GBP markets do not have the same mechanism as the US does for Federal funds, but overnight rates that are proxies for risk-free borrowing in those currencies do exist. They are called *Eonia* (Euro OverNight Index Average) in the Eurozone and *Sonia* (Sterling OverNight Index Average) in Great Britain, and are computed as averages of all *actual* overnight lending/borrowing transactions by qualifying banks weighted by the size of the transactions. We emphasize that these rates reflect the actual transactions that have happened, in contrast to Libor which reflect banks' estimates of rates at which borrowing (for a given term) might take place. In the crisis of 2007–2009 there have been serious concerns about the integrity of the Libor rate and whether it really reflected the actual cost of funding for banks, and even some calls to scrap the Libor rate altogether. While the Libor rate has survived the crisis, the importance of overnight rates has increased dramatically, with the market in FedFunds/Eonia/Sonia linked derivatives, most importantly in *overnight index swaps*, or OIS, of various maturities growing dramatically. As with the Fed funds rate, Eonia and Sonia have diverged significantly from the corresponding Libor rates during market turbulence, and the decoupling continues to persist. As with the Fed funds rate, the implications of these developments on interest rate curve construction are discussed in Section 6.5.3.

Interest rates change day-to-day in response to changing macroeconomic and market conditions. With the cost of borrowing and lending money affecting all aspects of the economy, it is no surprise that a vast market in derivatives on interest rates has developed. Motivations of participants are diverse, ranging from locking in the cost of financing to pure speculation.

The fixed income market can be broadly split into two (overlapping) segments: the *exchange* market and the *over-the-counter*, or *OTC*, market. Contracts linked to the level of interest rates are traded on many securities exchanges. The exchanges attract all types of investors, including market makers, hedgers and speculators; see Hull [2006] for details on all. As of March 2008, notional amounts outstanding were $26 trillion in exchange traded interest rate futures, and $45 trillion in exchange traded interest rate options. While these are impressive numbers, far more fixed income derivatives trade in OTC markets than in exchange markets: as of December 2007, the notional amounts outstanding of OTC interest rate derivatives amounted to $393 trillion[3]. The OTC market can loosely be visualized as a network of banks that trade with each other under terms governed by agreements spelled out by the trade organization International Swaps and Derivatives Association (ISDA). Central to OTC markets are the *interest*

---

[3] All figures from the report "Semiannual OTC derivatives statistics at end-December 2007" by Bank for International Settlements, available from www.bis.org.

rate dealers, banks with trading desks specializing in fixed income trading. The dealers provide liquidity in various types of securities, and are typically the most sophisticated players in the market. The dealers trade either on their own account or on behalf of customers such as *financial institutions* and *corporates*.

Financial institutions include mortgage companies (organizations that originate, package or service residential and commercial mortgage loans), pension funds, mutual funds, insurance companies, hedge funds, and other entities whose primary activities are related to financial markets. Financial institutions seek to either make money directly by engaging in trading activities (hedge funds), or to hedge their exposures (mortgage originators or servicers), or to achieve superior returns on their investments (pension funds, insurance companies). Among financial institutions, an important role is played by *issuers*, companies that issue structured notes for private and public placement. Structured notes deliver appealing return profiles to investors, returns that are essentially financed by selling options back to issuers. Issuance of increasingly complicated structured notes drives the exotic end of the fixed income markets.

Corporates are companies with primary activities not directly linked to fixed income markets, but whose operational results may be affected by the interest rate environment. For instance, many companies raise funds by borrowing from banks or by selling bonds, and are therefore affected by the prevailing levels of interest rates. Corporates often seek to lock in favorable interest rates for borrowing money, to hedge their interest rate exposures, to transform their liabilities from one type (e.g., a fixed rate liability) to another (a floating rate liability), or to design custom borrowing schemes around their expected future borrowing needs.

5.2 Certificates of Deposit and Libor Rates

Having identified the main types of market participants, we now proceed to define the universe of securities that this book will cover. For technical precision, we shall occasionally need to refer to the risk-neutral measure Q, as well as its associated expectation operator $\mathrm{E} = \mathrm{E}^Q$ and its numeraire $\beta(t)$.

We start with the *certificate of deposit* (or *CD*), a deposit of money for a pre-specified term at a pre-specified interest rate. Terms may range from one week to one year or more, with the most popular being a 3 month or a 6 month term, depending on the currency of the deposit. If 1 (dollar) is deposited at time T for a period of τ years, then the amount of capital to be returned at time $T + \tau$ is given by[4]

---

[4] As was mentioned earlier, the computation of τ from given start- and end-dates will involve certain formal day counting rules, see Appendix 5.A.

$$1 + \tau L,$$

where L is, by definition, the interest rate for the CD. The rate is quoted as a simple rate, i.e. a rate with the compounding frequency equal to the term of the deposit. Notice that the average value of L for CDs quoted in the interbank market will, by definition, be equal to the (spot) Libor rate for tenor τ. Spot Libor rates for various tenors are calculated daily and are published by major news services such as Bloomberg or Reuters. As mentioned above, Libor serves as the primary reference rate in fixed income markets.

If $P(T, T+\tau)$ is the (Libor-based) discount factor to date $T+\tau$ as observed at T, then the discounted value of receiving $1 + \tau L$ at time $T+\tau$ should be equal to 1 at time T, i.e.

$$1 = P(T, T+\tau)(1+\tau L).$$

In particular, recalling the definition (4.2) of $L(t, T, T+\tau)$, the rate L paid on the CD is a simple spot rate

$$L = L(T, T, T+\tau) = \frac{1}{\tau}\left(\frac{1}{P(T, T+\tau)} - 1\right). \tag{5.1}$$

5.3 Forward Rate Agreements (FRA)

A certificate of deposit allows a market participant to lock in an interest rate for a given period of time, effective immediately. Many market participants, however, find it convenient to lock in interest rates for a given period of time that starts in the future. Contracts that provide such a rate guarantee are known as *forward contracts* or, in a fixed income context, *forward rate agreements* (FRAs). An FRA for the period $[T, T+\tau]$ is a contract to exchange fixed rate payment (agreed at the initiation of the contract) against a payment based on the time T spot Libor rate of tenor τ. While all payments on an FRA are exchanged at, or near[5], time T, the contract is structured so that the payments are made in $T+\tau$ dollars.

Formally, consider the origination at time t, $t \leq T$, of a unit notional FRA contract with a rate of k. Ignoring payment delays, from the perspective of the fixed rate payer the net payment at time T will be

$$V_{\text{FRA}}(T) = \tau\left(L(T, T, T+\tau) - k\right)/(1 + \tau L(T, T, T+\tau)),$$

with the (contractually specified) factor $1/(1 + \tau L(T, T, T+\tau))$ applied to roll the payment to the future date $T+\tau$. We note that

---

[5] Typical market conventions call for a two business day payment delay, see Appendix 5.A for more details.

$$1/\left(1+\tau L(T,T,T+\tau)\right) = P(T,T+\tau)$$

so, by the fundamental pricing result (4.13), the value of this contract at time t is equal to

$$V_{\text{FRA}}(t) = \beta(t)\text{E}_t\left(\beta(T)^{-1}\tau\left(L\left(T,T,T+\tau\right)-k\right)P\left(T,T+\tau\right)\right)$$

(recall that $\beta(\cdot)$ is the money market account). Substituting (5.1) we obtain

$$V_{\text{FRA}}(t) = \beta(t)\text{E}_t\left(\beta(T)^{-1}\left(1-P\left(T,T+\tau\right)-\tau k P\left(T,T+\tau\right)\right)\right).$$

Since $P(\cdot, T+\tau)$ is a traded asset, its price deflated by the numeraire $\beta(\cdot)$ is a martingale. Thus

$$V_{\text{FRA}}(t) = P(t,T) - P(t,T+\tau) - \tau k P(t,T+\tau)$$
$$= \tau P(t,T+\tau)\left(\frac{P(t,T)-P(t,T+\tau)}{\tau P(t,T+\tau)} - k\right). \quad (5.2)$$

Most often, FRAs are issued at no cost to either party at the time of origination. The value of k that makes the FRA contract have value 0 at the contract initiation time t is given by the *forward Libor rate* (see (4.2)),

$$k = L(t,T,T+\tau) = \frac{P(t,T)-P(t,T+\tau)}{\tau P(t,T+\tau)}.$$

Thus, a forward Libor rate has the financial interpretation of being a break-even rate on an FRA contract in interbank markets.

5.4 Eurodollar Futures

FRAs, being forward contracts on Libor rates, allow market participants to either lock in favorable rates for future periods, or to speculate on the future direction of rates. FRAs trade in the OTC market, and are open only to institutions that participate in this market. Alternatively, *futures contracts* on Libor rates are available on a number of international exchanges, including the Chicago Mercantile Exchange (CME), London International Financial Futures and Options Exchange (LIFFE), and Marché à Terme International de France (MATIF). The CME interest rate futures contract on a three-month spot Libor rate on US dollar denominated deposits is called the *Eurodollar futures* or, simply, *ED futures* contract.

At maturity T, an ED futures contract is settled at

$$100 \times (1 - L(T,T,T+\tau)).$$

The *futures rate* $F(t,T,T+\tau)$ at time t (see (4.1.2)) is defined to be the rate such that the *quoted futures price* at time t is equal to[6]

---

[6] So, if the futures rate is 5%, the quoted futures price is 95.

$$100 \times (1 - F(t, T, T+\tau)).$$

As is the case for all futures contracts, ED futures are settled (marked to market) daily. Confusing matters somewhat, the actual amount of money that is settled between holders of the long and the short positions in an ED future is determined by the daily change in the *actual futures price* defined by

$$N_{\text{ED}} \times \left[1 - \frac{1}{4} F(t, T, T+\tau)\right],$$

where N_{ED} is the notional principal of the contract ($\$1,000,000$ for the CME's ED futures). In particular, for 1 basis point (0.01%) increase in the rate $F(t, T, T+\tau)$, the CME contract buyer pays $1,000,000 \times 0.25 \times 0.0001 = 25$ dollars to the seller.

As explained in Chapter 4, futures rates $F(t, T, T+\tau)$ are generally different from forward Libor rates $L(t, T, T+\tau)$. The problem of computing the difference, the *ED convexity adjustment*, is considered in Section 16.8.

Unlike FRAs, for which the deposit period is negotiated between two parties, ED futures are standardized. Available contracts expire on four specific dates, one each in March, June, September and December, over the next ten years. Such standardization increases liquidity in each particular contract.

5.5 Fixed-for-Floating Swaps

A *swap* is a generic term for an OTC derivative in which two counterparties agree to exchange one stream of cash flows against another stream. These streams are called the *legs* of the swap. A *plain vanilla fixed-for-floating interest rate swap* (a *plain vanilla swap*, or just a *swap* if there is no confusion) is a swap in which one leg is a stream of fixed rate payments and the other a stream of payments based on a floating rate, most often Libor. The legs are denominated in the same currency, have the same notional, and expire on the same date. Payment streams are made on a pre-defined schedule of contiguous time intervals, known as *periods*. Typically, the floating rate is observed (or *fixed*) at the beginning of each period, with both fixed and floating rate coupons being paid out at the end of the period. A plain-vanilla swap is economically equivalent[7] to a multi-period FRA, and serves the same purpose in the market as regular FRAs. Between interest rate dealers

---

[7]This is true up to subtle but potentially important discounting issues. As we have pointed out in Section 5.3, the net payment of an FRA is *contractually* discounted using Libor rate from $T+\tau$ to T, whereas in a swap, the net payment for a given period is discounted at the money market account rate from the end to the beginning of the accrual period. The two types of discounting can in fact be different in the presence of discounting-index *basis*, see Sections 6.5.2 and 6.5.3.

and financial institutions, swaps of different maturities are often traded to adjust interest risk positions of the parties involved, or to simply make bets on future direction of interest rates. Swaps are also used by corporates, often in conjunction with bond or note issuance, to transform fixed rate obligations into floating ones, or vice versa.

To formally define a fixed-floating swap, one specifies a tenor structure, i.e. an increasing sequence of maturity times, normally spaced roughly equidistantly (see Section 4.1.3)

$$0 \leq T_0 < T_1 < T_2 < \ldots < T_N, \quad \tau_n = T_{n+1} - T_n. \tag{5.3}$$

In a fixed-floating swap with fixed rate k, one party (the fixed rate payer) pays simple interest based on the rate k in return for simple interest payments computed from the Libor rate fixing on date T_n, for each period $[T_n, T_{n+1}]$, $n = 0, \ldots, N-1$. The payments are exchanged at the end of each period, i.e. at time T_{n+1}. In practice, the payments are netted, and only their difference changes hands. From the perspective of the fixed rate payer, the net cash flow of the swap at time T_{n+1} is therefore given by (on a unit notional)

$$\tau_n \left(L_n(T_n) - k \right), \quad L_n(t) = L\left(t, T_n, T_{n+1}\right),$$

for $n = 0, \ldots, N-1$. Dates when the Libor rates are observed are typically called *fixing dates*; dates when payments occur are called *payment dates*.

By the fundamental valuation result (4.13), the value of a swap is equal to the expected discounted value of its (netted) payments. Specifically, the value to the fixed rate payer of a unit notional fixed-floating swap at time t, $0 \leq t \leq T_0$, is given by[8]

$$V_{\text{swap}}(t) = \beta(t) \sum_{n=0}^{N-1} \tau_n \mathrm{E}_t \left(\beta\left(T_{n+1}\right)^{-1} \left(L_n(T_n) - k\right) \right)$$

$$= \beta(t) \sum_{n=0}^{N-1} \tau_n \mathrm{E}_t \left(\beta(T_n)^{-1} \left(L_n(T_n) - k\right) P\left(T_n, T_{n+1}\right) \right).$$

Using the definition of Libor rates $L_n(T_n)$,

$$V_{\text{swap}}(t) = \beta(t) \sum_{n=0}^{N-1} \mathrm{E}_t(\beta(T_n)^{-1}(1 - P(T_n, T_{n+1}) - \tau_n k P(T_n, T_{n+1}))).$$

For each n, $P(\cdot, T_n)$ is a traded asset, so its price deflated by the numeraire $\beta(\cdot)$ is a martingale. Hence

$$V_{\text{swap}}(t) = \sum_{n=0}^{N-1} \left(P(t, T_n) - P(t, T_{n+1}) - \tau_n k P(t, T_{n+1}) \right).$$

---

[8]This is a somewhat idealized expression. See Appendix 5.A for more details on market day counting conventions and related topics.

Recalling the definition of $L_n(t)$, this can be rewritten as

$$V_{\text{swap}}(t) = \sum_{n=0}^{N-1} \tau_n P(t, T_{n+1}) (L_n(t) - k).$$

An important observation is that a vanilla fixed-floating swap can be valued on date t using only the term structure of interest rates observed on that date. In particular, swap values are not affected by the *dynamics* of interest rates, only their current levels.

The swap valuation formula above can be rewritten as follows,

$$V_{\text{swap}}(t) = \left(\sum_{n=0}^{N-1} \tau_n P(t, T_{n+1})\right) \left(\frac{\sum_{n=0}^{N-1} \tau_n P(t, T_{n+1}) L_n(t)}{\sum_{n=0}^{N-1} \tau_n P(t, T_{n+1})} - k\right).$$

Using the definitions (4.8), (4.10) and (4.11) from Chapter 4:

$$A(t) \triangleq A_{0,N}(t) = \sum_{n=0}^{N-1} \tau_n P(t, T_{n+1}), \tag{5.4}$$

$$S(t) \triangleq S_{0,N}(t) = \frac{\sum_{n=0}^{N-1} \tau_n P(t, T_{n+1}) L_n(t)}{\sum_{n=0}^{N-1} \tau_n P(t, T_{n+1})}, \tag{5.5}$$

we obtain the convenient formula

$$V_{\text{swap}}(t) = A(t) (S(t) - k). \tag{5.6}$$

The quantity $A(\cdot)$ is the *annuity* of the swap (or its *PVBP*, for Present Value of a Basis Point), and the quantity $S(t)$ is the *forward swap rate*. Clearly, $S(t)$ is the value of the fixed rate that makes the swap have value 0 to both parties at time t; S is consequently often referred to as a *par* or *break-even* rate.

For plain-vanilla swaps, the fixed rate and the swap notional are constant through time. More general swaps are, however, not bound by such restrictions and both the fixed rate and the notional may vary from period to period. A non-standard swap with a notional schedule $\{q_n\}_{n=0}^{N-1}$ (non-constant but deterministic) and a fixed rate schedule $\{k_n\}_{n=0}^{N-1}$ has the value

$$V_{\text{genswap}}(t) = \beta(t) \sum_{n=0}^{N-1} \tau_n q_n E_t \left(\beta(T_{n+1})^{-1} (L_n(T_n) - k_n)\right)$$

$$= \sum_{n=0}^{N-1} \tau_n q_n P(t, T_{n+1}) (L_n(t) - k_n).$$

Certain general swaps have dedicated names, such as *amortizing swaps* (notional decreases with time) and *accreting swaps* (notional increases with time).

As we mentioned in Section 5.1, swaps linked to overnight rates (Fed-Funds/Eonia/Sonia) have recently become more popular. Among them the overnight index swap (OIS) is probably the most liquid, and is defined as a swap that pays a *compounded* overnight rate against fixed rate payments. To write down its definition, let us assume that a tenor structure (5.3) is given, and denote by $\{t_{n,i}\}_{i=0}^{K_n}$ the collection of all business days in the period $[T_n, T_{n+1}]$, so that $T_n = t_{n,0} < \ldots < t_{n,K_n} = T_{n+1}$. Then the net payment of the OIS with fixed rate k at time T_{n+1} is given by

$$\tau_n \left(\overline{L}_n - k \right),$$

where the floating rate $\overline{L}_n$ for the n-th period of OIS is given by

$$\overline{L}_n = \frac{1}{\tau_n} \left(\prod_{i=0}^{K_n-1} (1 + (t_{n,i+1} - t_{n,i}) L(t_{n,i}, t_{n,i}, t_{n,i+1})) - 1 \right). \quad (5.7)$$

Here we used the notation $L(t_{n,i}, t_{n,i}, t_{n,i+1})$ to denote the overnight rate. Equating the overnight rate with the short rate, we can use a more mathematically convenient (although not exactly correct) expression

$$\overline{L}_n = \frac{1}{\tau_n} \left(e^{\int_{T_n}^{T_{n+1}} r(t)\, dt} - 1 \right). \quad (5.8)$$

5.6 Libor-in-Arrears Swaps

Allowing the fixed rate and the notional to vary through time is not the only way to generalize a swap. For a *Libor-in-arrears swap*, Libor rates are observed (fixed) at the end of each period rather than at the beginning. Thus, a value of a Libor-in-arrears payer swap is equal to

$$V_{\text{LIA}}(t) = \beta(t) \sum_{n=0}^{N-1} \tau_n \mathrm{E}_t \left(\beta(T_{n+1})^{-1} (L_{n+1}(T_{n+1}) - k) \right).$$

Interestingly, this seemingly innocuous modification makes the value of a swap model-dependent, in contrast to the standard fixed-floating swap. We will discuss pricing of in-arrears swaps in Chapter 16.

Libor-in-arrears swaps are popular in upward-sloping interest rate curve environments, i.e. when long-tenor rates are higher than shorter-tenor ones. In such a scenario, the break-even fixed rate on the Libor-in-arrears swap tends to look more "attractive" than that of a standard fixed-floating swap, thus increasing the desirability of the swap to those seeking to receive fixed rate payments.

5.7 Averaging Swaps

Libor rates are not restricted to being observed on either the start date or the end of the pay period. A popular example is the *averaging swap*, i.e. a swap where the floating rate is determined as an average of Libor rate observations taken at regular intervals over each coupon period. For example, let $\{(t_{n,i}^f, t_{n,i}^s, t_{n,i}^e)\}_{i=1}^{K_n}$ be a collection of date triplets (fixing, start and end date) that define the rates to be used in calculating the payment in period n. Defining a set of weights $w_{n,i}$, $i = 1, \ldots, K_n$, the floating rate $\overline{L}_n$ for the period $[T_n, T_{n+1}]$ may be defined as

$$\overline{L}_n = \sum_{i=1}^{K_n} w_{n,i} L\left(t_{n,i}^f, t_{n,i}^s, t_{n,i}^e\right).$$

For the fixed rate swap payer, the averaging swap value is therefore

$$V_{\text{average}}(t) = \beta(t) \sum_{n=0}^{N-1} \tau_n E_t \left(\beta(T_{n+1})^{-1} \left(\overline{L}_n - k\right)\right). \quad (5.9)$$

As a rule, the weights $w_{n,i}$ sum up to 1, $\sum_{i=1}^{K_n} w_{n,i} = 1$; the weights usually reflect the number of days (using the appropriate day counting conventions) that a given rate $L(t_{n,i}^f, t_{n,i}^s, t_{n,i}^e)$ is supposed to be in effect. Computation of the valuation expression (5.9) can be done using techniques similar to those required for in-arrears swaps; see Chapter 16 for details.

Swaps linked to the average of the Federal funds rate are common examples of an averaging swap. Particularly noteworthy is the *Fed funds/Libor basis swap* which pays the average of the Fed funds rate (over a given period) against a payment based on a Libor rate for that period. This instrument is an example of a *floating-floating single-currency basis swap*, i.e., a swap that exchanges payments based on two different floating rates in the same currency. Closely related to Fed funds basis swaps are the *Fed funds futures* contracts traded on the Chicago Board of Trade (CBOT) exchange. These contract uses the 30 day running average of the Federal funds rate for settlement.

Remark 5.7.1. Going forward, in our product descriptions we shall normally assume that all cash flows pay at the end of the periods in which they fix. While this is common practice, as we have just seen the "pay-in-arrears" rule can be broken at will depending on the client's needs — the only (self-evident) restriction is that payments should be fixed by the time they are made.

5.8 Caps and Floors

A firm with liabilities funded at a floating (i.e., Libor) rate is naturally concerned with the possibility that interest rates, and thus its interest rate payments, may increase in the future. One way to immunize against this risk is to pay fixed on a fixed-floating interest rate swap, in effect turning floating rate payments into fixed ones. While this will guarantee a fixed rate for funding payments for the duration of the swap, it will also mean forgoing the possibility of benefiting from a potential future drop in rates. An *interest rate cap* is a security that allows one to benefit from low floating rates yet be protected from high rates. Similarly, for an investor with assets earning a floating rate, a low-rate scenario is unfavorable. An *interest rate floor* is an instrument designed to protect against low interest rates yet allow the holder to benefit from high rates.

Formally, a cap is a strip of *caplets*, call options on successive Libor rates, and a floor is a strip of *floorlets*, put options on successive Libor rates. We encountered caplets already in Section 4.5.1 and recall that this instrument pays

$$\tau_n \left(L_n(T_n) - k\right)^+$$

per unit notional at time T_{n+1}. Similarly, a floorlet pays

$$\tau_n \left(k - L_n(T_n)\right)^+$$

per unit notional at time T_{n+1}. Then, N-period caps and floors have values at time t of

$$V_{\text{cap}}(t) = \beta(t) \sum_{n=0}^{N-1} \tau_n \mathrm{E}_t \left(\beta \left(T_{n+1}\right)^{-1} \left(L_n(T_n) - k\right)^+\right),$$

$$V_{\text{floor}}(t) = \beta(t) \sum_{n=0}^{N-1} \tau_n \mathrm{E}_t \left(\beta \left(T_{n+1}\right)^{-1} \left(k - L_n(T_n)\right)^+\right).$$

By switching to the T_{n+1}-forward measure (see Section 4.2.2) for the n-th caplet/floorlet, the valuation formulas can be written in a more convenient form

$$V_{\text{cap}}(t) = \sum_{n=0}^{N-1} \tau_n P\left(t, T_{n+1}\right) \mathrm{E}_t^{T_{n+1}} \left(\left(L_n(T_n) - k\right)^+\right),$$

$$V_{\text{floor}}(t) = \sum_{n=0}^{N-1} \tau_n P\left(t, T_{n+1}\right) \mathrm{E}_t^{T_{n+1}} \left(\left(k - L_n(T_n)\right)^+\right).$$

By Lemma 4.2.3, the Libor rate $L_n(\cdot)$ is a martingale under the T_{n+1}-forward measure. Hence, caplets/floorlets can be priced using "vanilla" models[9], such as the log-normal Black model (see Remark 1.9.4).

---
[9] By a *vanilla* model we mean a model that specifies the dynamics (or just the terminal distribution) of only a single rate, or at most a few rates, in contrast to term structure models that specify consistent dynamics for the entire term structure of interest rates. Often vanilla models are borrowed from equity or FX modeling; having the underlying rate a martingale makes such borrowing painless. We discuss vanilla models in Chapters 7, 8, 9 and 17.

The OTC market in caps/floors is very liquid. While individual caplets/floorlets are not traded, caps/floors are available in a number of maturities. This allows the volatility information for individual forward Libor rates to be extracted from market quotes for caps/floors of different maturities, at least in principle[10]. Once extracted, these volatilities may be combined with the volatilities observed from European swaption quotes (see below), to form a set of market inputs to which interest rate models for exotics are calibrated.

5.9 Digital Caps and Floors

Digital caps and floors work like regular caps and floors, except that the n-th digital caplet pays
$$\tau_n \times 1_{\{L_n(T_n)>k\}}.$$
Similarly, the n-th digital floorlet pays
$$\tau_n \times 1_{\{L_n(T_n)<k\}}.$$
Digital caps and floors provide a leveraged way to bet on the future direction of interest rates, more so than through standard caps and floors.

5.10 European Swaptions

Caps and floors have an asymmetric exposure to interest rates, a characteristic used by both hedgers and speculators. A similar exposure profile is provided by options on swaps, the so-called *European swaptions*. A European swaption gives the holder a right, but not an obligation, to enter a swap at a future date at a given fixed rate. A *payer* swaption is an option to pay the fixed leg on a fixed-floating swap; a *receiver* swaption is an option to receive the fixed leg.

Assuming the underlying swap starts on the expiry date T_0 of the option (a typical situation), the payoff for a payer swaption at time T_0 then equals

$$V_{\text{swaption}}(T_0) = (V_{\text{swap}}(T_0))^+ = \left(\sum_{n=0}^{N-1} \tau_n P(T_0, T_{n+1})(L_n(T_0) - k)\right)^+. \tag{5.10}$$

The value at an intermediate time t, $t < T_0$, must then equal

$$V_{\text{swaption}}(t) = \beta(t)\mathrm{E}_t\left(\beta(T_0)^{-1}V_{\text{swaption}}(T_0)\right)$$
$$= \beta(t)\mathrm{E}_t\left(\beta(T_0)^{-1}\sum_{n=0}^{N-1}\tau_n P(T_0, T_{n+1})(L_n(T_0) - k)\right)^+,$$

---

[10]This "volatility bootstrap" is by no means trivial; we discuss it in Section 16.2.

which, using (5.6), can be rewritten in the more compact form

$$V_{\text{swaption}}(t) = \beta(t)\mathrm{E}_t\left(\beta(T_0)^{-1}A(T_0)\left(S(T_0) - k\right)^+\right). \tag{5.11}$$

Moreover, switching to the annuity measure, also known as the *swap measure*, Q^A from Section 4.2.5, the swaption value can be expressed as

$$V_{\text{swaption}}(t) = A(t)\mathrm{E}_t^A\left(S(T_0) - k\right)^+, \tag{5.12}$$

with the forward swap rate $S(\cdot)$ being a martingale in the swap measure Q^A; see Lemma 4.2.4.

It is evident from (5.12) that a payer European swaption is a call option — and a receiver European swaption is a put option — on the forward swap rate, struck at the fixed rate of the swap. Hence, swaptions could be priced using a vanilla model (see footnote 9), such as the Black model or similar. Conversely, values of European swaptions can be translated into market-implied distributional characteristics of forward swap rates, a topic discussed at length in Section 7.1.2. In particular, it is universal practice to quote swaption prices in terms of *implied* Black volatilities, i.e. volatilities that recover market price when used in the Black formula. In some markets (e.g., the US), it is also common to quote implied *Gaussian* volatilities, defined in the same way with regard to a Gaussian (rather than log-normal) model for the distribution of interest rates, see (7.16).

The market in swaptions is very liquid, with many different option maturities and swap underlyings actively traded. To characterize the full universe of traded instruments, given a tenor structure (5.3) we consider swaptions of different expiries $\{T_n\}_{n=0}^{N-1}$ that can be exercised into swaps that start at T_n and cover m periods[11], i.e. their last payment date is T_{n+m}. For a convenient way to denote the various swaptions, recall definitions (4.8), (4.10) and (4.11) and introduce

$$A_{n,m}(t) = \sum_{i=n}^{n+m-1} \tau_i P(t, T_{i+1}), \tag{5.13}$$

$$S_{n,m}(t) = \frac{\sum_{i=n}^{n+m-1} \tau_i P(t, T_{i+1}) L_i(t)}{\sum_{i=n}^{n+m-1} \tau_i P(t, T_{i+1})}, \tag{5.14}$$

for $n = 0, \ldots, N-1$, $m = 1, \ldots, N-n$. Then the value of the (n, m)-swaption (a short-hand for an "m-period swaption with expiry T_n") is equal to

$$A_{n,m}(t)\mathrm{E}_t^{n,m}\left(\left(S_{n,m}(T_n) - k\right)^+\right),$$

where $\mathrm{E}_t^{n,m}$ denotes time t expectation in the appropriate swap measure, $Q^{n,m}$. Note that in trader parlance, a (vanilla) T_n-maturity European swaption on a swap that runs from T_n to T_m is said to be a "T_n into $T_{m+n} - T_n$"

---

[11] A bit confusingly, such a swaption is often said to have *tenor* $T_{n+m} - T_n$, a characterization it inherits from the underlying swap rate.

swaption. For instance, a 5 year option on a 10 year swap would be a "5-into-10" (or "5y-into-10y", or simply "5y10y") swaption.

Clearly, when $m = 1$, the (n, m)-swaption reduces to a caplet (or floorlet) on the Libor rate $L_n(\cdot)$, so caplets and floorlets can be thought of as one-period swaptions. Whenever in this book swaptions are discussed or used, caplets and floorlets are thereby implicitly included. Collectively, all (n, m)-swaptions constitute *the swaption grid*.

Market quotes on swaptions, typically in terms of implied volatilities, in the swaption grid provide the most readily-available information on the volatility structure of interest rates. As swaptions in the grid cover overlapping sections of the term structure of interest rates, extracting clean volatility information from market quotes is a non-trivial exercise that forms the foundation for calibration of models used for exotic interest rate derivative pricing. We will have much to say about such *volatility calibration* later on.

While options on plain-vanilla swaps comprise the bulk of the liquid ("vanilla") interest rate market, options on general swaps (i.e. on swaps with non-constant notionals and fixed rates) also trade and are properly treated as exotic derivatives. Often, general swaps can be decomposed into baskets of standard swaps, in which case options on general swaps become *basket options*. Valuation of basket options requires information on the co-dependence structure of securities in the basket, information that is not readily available from the vanilla options markets. We demonstrate how to handle this complication in Section 19.4.

5.10.1 Cash-Settled Swaptions

The swaption contract discussed in the previous section involves *physical settlement*, in the sense that an actual interest rate swap is entered into, should the option be exercised at its expiry. Physically-settled swaptions are also known as *swap-settled swaptions*. An economically equivalent swaption contract is one that instead settles into a cash payment equal to the PV (present value) of the swap as observed at time T_0. Indeed, for both types, the swaption payoff (for a payer) is given by

$$A(T_0)\left(S(T_0) - k\right)^+, \qquad (5.15)$$

see (5.10) and (5.11). In the European markets, a third variety of swaptions is common, the so-called *cash-settled* swaptions. For this type of option, rather than entering into a swap, the option holder will receive a cash payout upon exercise. The settlement amount is calculated by a formula similar to (5.15), except the annuity $A(\cdot)$ is not calculated by (5.4), but instead by discounting fixed rate payments at the swap rate $S(T_0)$. Specifically,

$$V_{\text{css}}(T_0) = a\left(S(T_0)\right)\left(S(T_0) - k\right)^+,$$

where
$$a(x) = \sum_{n=0}^{N-1} \frac{\tau_n}{\prod_{i=0}^{n}(1+\tau_i x)}.$$

Notice that the cash settlement mechanism ensures a well-defined present value of the option payout, as long as the swap rate $S(T_0)$ is observable. In contrast, the value of exercise of a physically settled swaption — the computation of which requires knowledge of a strip of discount factors — may be estimated differently by different dealers, due to bid-ask spread effects and differences in curve building technology (see Chapter 6). Technically, however, the cash settlement mechanism induces certain valuation complications, and cash-settled swaptions cannot, strictly speaking, be considered vanilla options that can be priced using, e.g., a Black-type formula[12]. This follows from the fact that in the measure associated with the deflator $X(t) = a(S(t))$, the swap rate $S(\cdot)$ is *not* a martingale, and certain drift adjustments are required. We discuss valuation of cash-settled swaptions in Section 16.6.12. As they are the most liquidly-traded OTC interest rate options in the European market, cash-settled swaptions still could (and should) be used to extract information on the volatility structure of interest rates; the procedure, however, is necessarily more involved.

5.11 CMS Swaps, Caps and Floors

As the market in plain vanilla swaps is both deep and very active, market quotes of corresponding swap rates can be used as "indexes", i.e. market variables that can themselves be used in defining payoffs of other securities. The demand for such products is often driven by particular segments of fixed income markets. For example, mortgage lenders are primarily concerned with hedging interest rate risk arising from holding residential loans, some of which may have maturities as long as thirty years. Because of potential prepayments, the interest rate risk of a pool of such mortgages is often assumed to be closely connected to movements in the 10 year swap rate; hence, mortgage lenders are natural consumers of interest rate securities linked to the 10 year swap rate.

A constant-maturity swap (CMS) rate is defined as a break-even swap rate (see (5.5)) on a standard swap of a fixed maturity, e.g. 10 years or 30 years. A *CMS swap* works just like a standard fixed-floating (Libor) swap, except for the fact that floating leg payments are based on CMS, rather than Libor, rates. Formally, let $S_{n,m}(\cdot)$ be the m-period swap rate with the first fixing date T_n, as defined by (5.14). Then an N-period (payer) CMS swap (linked to m-period rate) value is given by

---

[12] Nevertheless, this practice has been widespread until recently, and may still be in use in some institutions.

$$V_{\text{cmsswap}}(t) = \beta(t) \sum_{n=0}^{N-1} \tau_n \mathrm{E}_t \left(\beta(T_{n+1})^{-1} (S_{n,m}(T_n) - k) \right)$$

or, using the T_{n+1}-forward measure for each period,

$$V_{\text{cmsswap}}(t) = \sum_{n=0}^{N-1} \tau_n P(t, T_{n+1}) \mathrm{E}_t^{T_{n+1}} \left((S_{n,m}(T_n) - k) \right).$$

While standard swaps can be valued solely from knowledge of the term structure of interest rates, CMS swaps require an interest rate model for valuation; we return to a complete discussion in Chapter 16.

CMS caps and floors are defined as strips of European options on CMS rates, just like regular caps and floors are strips of European options on Libor rates:

$$V_{\text{cmscap}}(t) = \sum_{n=0}^{N-1} \tau_n P(t, T_{n+1}) \mathrm{E}_t^{T_{n+1}} \left((S_{n,m}(T_n) - k)^+ \right),$$

$$V_{\text{cmsfloor}}(t) = \sum_{n=0}^{N-1} \tau_n P(t, T_{n+1}) \mathrm{E}_t^{T_{n+1}} \left((k - S_{n,m}(T_n))^+ \right).$$

CMS caplets are related to European swaptions, as both are European-style options on swap rates. The connection between the two types of securities is, however, subtle, as we shall discuss later in this book.

5.12 Bermudan Swaptions

A Bermudan swaption is an option to enter into a fixed-floating swap on any (or any from a given subset) of its fixing dates. For a given tenor structure (5.3), the holder of a standard Bermudan swaption has the right to exercise it on any of the dates $\{T_n\}_{n=0}^{N-1}$. Once exercised on date T_n, say, the option goes away, and the holder enters the swap with the first fixing date T_n and the final payment date T_N. The period up to $T_0 > 0$ is known as the *lockout* or *no-call* period. In common jargon, a Bermudan swaption on, say, a 10 year swap with a 2 year lockout period (at inception) is known as a "10 no-call 2", or "10nc2", Bermudan swaption.

Formally, at time T_n, the value of a payer[13], if exercised, is therefore

$$U_n(T_n) = \beta(t) \sum_{i=n}^{N-1} \tau_i \mathrm{E}_{T_n} \left(\beta(T_{i+1})^{-1} (L_i(T_i) - k) \right)$$

$$= \sum_{i=n}^{N-1} \tau_i P(T_n, T_{i+1}) (L_i(T_n) - k).$$

---

[13] Upon exercise, the holder of a payer (receiver) Bermudan swaption will pay the fixed (floating) leg of the swap.

Here, $U_n(T_n)$ here denotes the *exercise* value of the Bermudan swaption; loosely speaking, a Bermudan swaption contract is an option to chose between $U_n(T_n)$ for different $n = 0, \ldots, N-1$. More succinctly, we recall from Section 1.10 that the Bermudan option value at time T_n will be the maximum of $U_n(T_n)$ and the *hold value* $H_n(T_n)$, the latter defined as the value of a Bermudan swaption with the exercise dates $\{T_i\}_{i=n+1}^{N-1}$ only (compare to Sections 1.10 and 3.5).

Demand for Bermudan swaptions comes from different segments of fixed income markets. Mortgage companies use them to hedge pools of mortgages, with the flexibility of Bermudan exercise convenient in matching the uncertain timing of prepayments in mortgage pools. Investors seeking higher current income sell Bermudan-style options on swaps to increase the coupons they receive, as explained later in the context of callable Libor exotics. Bermudan swaptions are also used as hedges for callable coupon bonds.

While it may be tempting to think of Bermudan swaptions as straightforward generalizations of European swaptions, they are substantially more difficult to model and price. Indeed, it is fair to say that many valuation methods and techniques covered in this book were developed in response to the need to value and risk manage Bermudan swaptions. Bermudan swaptions are, by far, the most liquid exotic fixed income securities, with all interest rate dealers holding large inventories.

5.13 Exotic Swaps and Structured Notes

With market sophistication ever on the rise, clients demand increasingly complicated payouts, often in a familiar swap or bond format (although the appetite has waned somewhat post-crisis). In an *exotic swap*, a regular floating Libor leg is swapped against structured coupons that are allowed to be arbitrary functions of observed interest rates (such as Libor or CMS rates). A standard fixed-floating vanilla swap is an obvious and trivial example where the structured coupon simply is a fixed rate. A cap (or a floor) can be seen as another, less trivial, example. In particular, note that

$$(k - L_n(T_n))^+ = \Big((k - L_n(T_n))^+ + L_n(T_n)\Big) - L_n(T_n)$$
$$= \max\left(k, L_n(T_n)\right) - L_n(T_n),$$

which demonstrates that a floor can be represented as an exotic swap in which a Libor rate is exchanged for a *floored payoff* $\max(k, L_n(T_n))$.

Exotic swaps often start their life as bonds, or notes, sold by banks to investors. In a structured note, the investor pays an up-front *principal amount* (e.g., $10,000,000) to the issuer of the note, who in turn pays the investor a structured coupon, and repays the principal at the maturity of the note. The principal amount is invested by the issuer (or the trading

5.13 Exotic Swaps and Structured Notes

desk to which the issuer passes the note for risk management), and pays the Libor rate plus or minus a spread. From the perspective of the issuer (or the trading desk), the net cash flows of the note are those of an exotic swap.

In terms of valuation, if C_n is the structured coupon for the n-th period, the value of the exotic swap is equal to (from the perspective of structured leg buyer)

$$V_{\text{exotic}}(t) = \beta(t) \sum_{n=0}^{N-1} \tau_n \mathrm{E}_t \left(\beta \left(T_{n+1} \right)^{-1} \left(C_n - L_n(T_n) \right) \right),$$

where we for brevity have assumed that both legs of the swap pay at the end date of each coupon period (see Remark 5.7.1). As discussed earlier, in this valuation equation, the coupon C_n can be a complicated function of interest rates, structured to reflect investors' views on the market, or to take advantage of current interest rate market conditions. For example, a floored payoff can be offered to an investor who believes that interest rates are poised for a fall in the future.

There is no universally agreed "taxonomy" for exotic swaps, but for our purposes we can distinguish between exotic swaps that are i) Libor-based, ii) CMS-based, iii) multi-rate, iv) range accruals, and v) generally path dependent. We proceed to described each type of swap in more details.

5.13.1 Libor-Based Exotic Swaps

In a Libor-based exotic swap, the structured coupon is a function of a Libor rate:

$$C_n = C_n \left(L_n(T_n) \right).$$

A large variety of structured coupons $C_n(\cdot)$ can be used. For example:

- A standard swap,
$$C_n(x) = k.$$

- Capped and floored floaters. For strike s, gearing g, cap c and floor f,
$$C_n(x) = \max \left(\min \left(g \times x - s, c \right), f \right). \tag{5.16}$$

- Capped and floored inverse floaters. For spread s, gearing g, cap c and floor f,
$$C_n(x) = \max \left(\min \left(s - g \times x, c \right), f \right). \tag{5.17}$$

- Digitals. For strike s and coupon k,
$$C_n(x) = k \times 1_{\{x > s\}}$$

  or

$$C_n(x) = k \times 1_{\{x < s\}}.$$

- "Flip-flops" or "tip-tops". For strike s and two coupons, k_1 and k_2,

$$C_n(x) = \begin{cases} k_1, \ x \leq s, \\ k_2, \ x > s. \end{cases}$$

Different coupon types can be combined together to create new types of structured coupons.

A Libor-based exotic swap can usually be decomposed[14] into a sum of simpler instruments such as ordinary swap floating legs, fixed legs, caps and floors, and digital caps and floors. Therefore, if the prices of these simple contracts are available in the market (as is typically the case), Libor-based exotic swaps can be perfectly replicated by a one-time transaction in market-available instruments, a strategy referred to as *static replication*. Hence, by themselves, these instruments rarely present major valuation challenges. They do, however, serve as building blocks for more complicated securities.

5.13.2 CMS-Based Exotic Swaps

The payoffs from the previous section can be applied to CMS, rather than Libor, rates. Structured coupons are then deterministic functions of CMS rates. If an m-period rate is used, then a structured coupon for period n can be defined by

$$C_n = C_n\left(S_{n,m}(T_n)\right),$$

with $C_n(x)$ as defined in the previous section.

CMS-based exotic swaps can be decomposed into linear combinations of CMS swaps and CMS caps/floors and rarely present any extra modeling difficulties beyond those already present in CMS swaps and caps.

5.13.3 Multi-Rate Exotic Swaps

Multi-rate exotic swaps differ from the structures in Sections 5.13.1 and 5.13.2 by referencing multiple market rates (Libor or CMS) for the calculation of structured coupons. The most common example is a *CMS spread coupon*. To describe this contract, let $S_{n,a}(\cdot)$ and $S_{n,b}(\cdot)$ be two collections of CMS rates, fixing on T_n, $n = 0, \ldots, N-1$, and covering a and b periods, respectively. A CMS spread coupon with gearing g, spread s, cap c and floor f is then defined by

$$C_n = \max\left(\min\left(g \times (S_{n,a}(T_n) - S_{n,b}(T_n)) + s, c\right), f\right).$$

A typical example would be a 10 year/2 year (often abbreviated as 10y2y) CMS call spread option where a is 40 (40 quarterly periods to cover 10 years) and b is 8, with the quarterly coupon given by

---

[14] Indeed, this is the case for all the payouts listed above, a fact that we invite the reader to verify.

5.13 Exotic Swaps and Structured Notes

$$C_n = \max\left(S_{10y}(T_n) - S_{2y}(T_n), 0\right)$$

(using somewhat loose notation). A relatively liquid broker market exists for spread options on Euro and US dollar CMS rates.

A more general example is obtained by using one of the payoff functions $C_n(x)$ defined in Section 5.13.1, applied to the spread $x = S_{n,a}(T_n) - S_{n,b}(T_n)$. In particular, digital and flip-flop CMS spread swaps are quite popular.

Multi-rate exotic swaps typically cannot be decomposed into "standard" instruments (such as vanilla swaps, caps, etc.). Therefore, they, as a rule, cannot be valued by replication arguments, and a valuation model is required. Such a model, however, does not always need to be a full-blown term structure model: we shall show later that some types of spread-linked payoffs can be efficiently valued and risk managed by vanilla models (see footnote 9).

It should be noted that more than two rates can be used in the definition of a coupon. For example, in the so-called *curve cap* one takes a standard capped and floored payoff on a Libor or CMS (or CMS spread!) rate — see (5.16) — and makes the cap c and the floor f functions of, potentially different, CMS spreads:

$$C_n(x) = \max\left(\min\left(g \times x - s, c\right), f\right), \tag{5.18}$$
$$c = \max\left(\min\left(g_1 \times (S_{n,a_1}(T_n) - S_{n,b_1}(T_n)) + s_1, c_1\right), f_1\right),$$
$$f = \max\left(\min\left(g_2 \times (S_{n,a_2}(T_n) - S_{n,b_2}(T_n)) + s_2, c_2\right), f_2\right).$$

5.13.4 Range Accruals

A *range accrual* structured coupon is defined as a given rate — fixed in the simplest case, but potentially a Libor, CMS or a CMS spread rate — that only "accrues" when a different *reference* rate is inside (or, sometimes, outside) a given range. So, let $R_n(t)$ be the payment rate and $X_n(t)$ be the reference rate, and let l be the low bound and u be the upper bound. A range accrual coupon then pays

$$C_n = R_n(T_n) \times \frac{\sharp\{t \in [T_n, T_{n+1}] : X_n(t) \in [l, u]\}}{\sharp\{t \in [T_n, T_{n+1}]\}}, \tag{5.19}$$

where $\sharp\{\cdot\}$ is used to denote the number of days that a given criteria is satisfied.

The most common choice of the payment rate $R_n(t)$ is either a constant or Libor, but a CMS rate or any other structured coupon rate are also occasionally used. The reference rate $X_n(t)$ can be any market-observable rate such as a Libor rate fixing at t, a CMS rate fixing at t, or even a CMS spread rate.

We note that a range accrual coupon can always be decomposed into simpler digital payoffs, because

$$\#\{t \in [T_n, T_{n+1}] : X_n(t) \in [l, u]\} = \sum_{t \in [T_n, T_{n+1}]} 1_{\{X_n(t) \in [l, u]\}}, \tag{5.20}$$

where the sum on the right-hand side is over all business days in the period. This decomposition is particularly useful for fixed rate $(R_n(T_n) \equiv k)$ range accruals, as simple digital options can be priced directly from the market information on European options (see Section 7.1.2). For floating, or more complicated, range accruals the decomposition is useful but requires further work to turn it into valuation formulas — see Section 17.5 for further details.

The basic payout (5.19) can be extended to include more than one range condition. In a *dual range accrual*, the position of two different reference rates relative to the range are monitored, and (5.19) is generalized to

$$C_n = R_n(T_n) \frac{\#\{\{t \in I_n : X_{n,1}(t) \in [l_1, u_1]\} \diamond \{t \in I_n : X_{n,2}(t) \in [l_2, u_2]\}\}}{\#\{t \in I_n\}}$$

with $I_n = [T_n, T_{n+1}]$ and $\diamond$ denoting either intersection $\cap$ or union $\cup$. In the former case, one counts the number of days when *both* reference rates $X_{n,1}$ and $X_{n,2}$ are within their ranges; in the latter case, one counts the number of days when *either* of the two reference rates are within their ranges.

In a *curve cap range accrual*, the lower and upper bounds become functions of CMS spreads themselves, similar to (5.18).

A *product-of-ranges* range accrual multiplies up all range accrual factors to date to define the multiplier that is used for the current coupon, e.g.

$$C_n = R_n(T_n) \times Y_n,$$

where

$$Y_n = Y_{n-1} \times \frac{\#\{t \in [T_n, T_{n+1}] : X_n(t) \in [l, u]\}}{\#\{t \in [T_n, T_{n+1}]\}}, \tag{5.21}$$

$$Y_{-1} = 1.$$

5.13.5 Path-Dependent Swaps

The payoff (5.21) is an example of *path-dependence* in the payoff, where a coupon depends on rate observations from previous coupon periods. More commonly, path-dependence in exotic swaps is introduced by linking a structured coupon not only to interest rates observed during the coupon period, but to previous coupon(s) as well. This is often referred to as a *snowball* feature. The "original" snowball structure involved a coupon of an inverse floating type, with the n-th coupon C_n given by

$$C_n = (C_{n-1} + s_n - g_n \times L_n(T_n))^+. \tag{5.22}$$

Here $\{s_n\}$ and $\{g_n\}$ are contractually specified deterministic sequences of spreads and gearings. This type of a swap is sometimes also called a *ratchet* or a *ladder* swap.

The term snowball originates with the tendency of high initial coupons to spill into subsequent coupons, in a compounding or "snowballing" fashion". Indeed, a little reflection reveals that a snowball has a highly leveraged exposure to its first few coupons, a feature that makes it more attractive to some investors, but also quite difficult to risk manage.

A large number of snowball-like payoffs have been created, often with "snow"-themed — and rather nonsensical — names, such as "snowrange", "snowbear", and "snowstorm". For example, a "snowrange" combines a range-accrual feature and a snowball feature, in the following way

$$C_n = C_{n-1} \times Y_n + s_n + g_n \times X_{n,1}(T_n), \tag{5.23}$$

where $X_{n,1}(T_n)$ is some reference rate, and Y_n is a range-accrual factor depending on a second rate $X_{n,2}$,

$$Y_n = \frac{\sharp\{t \in [T_n, T_{n+1}] : X_{n,2}(t) \in [l, u]\}}{\sharp\{t \in [T_n, T_{n+1}]\}}.$$

The range accrual factor Y_n may be a product-of-ranges accrual factor, as in (5.21). Also, additional caps and/or floors are often added to the coupon (5.23).

Path-dependent swaps typically require a term structure model for valuation; for obvious reasons, Monte Carlo methods are often mandatory.

5.14 Callable Libor Exotics

5.14.1 Definitions

As described in Section 5.12, Bermudan swaptions are Bermudan-style options to enter into a regular fixed-floating swap. If we alter the swap underlying the Bermudan swaption from a regular swap to an exotic swap (see previous section), then a so-called *callable Libor exotic* (CLE) is created. CLEs most often emerge as part of callable structured notes in which an issuer receives the principal from an investor and pays a structured coupon in return. In addition, the issuer has the right to cancel — or *call* — the note on a schedule of dates; typically, this call schedule will coincide with coupon fixing dates, after some initial lock-out (or *no-call*) period. Should a note be called by its issuer[15], the principal is returned to the investor and no future coupons are paid.

A callable structured note is typically passed through by the issuer to an exotics trading desk (which could, but does not have to, be internal to the issuing bank) to deal with its risk management. Also, the principal is

---

[15]The call decision is most often made by the issuer's swap counterparty who is actually managing the risk, see next paragraph.

invested and pays a Libor rate, plus or minus a spread depending on the cost of financing. From a trading desk perspective what is left is an exotic swap paying structured coupons and receiving Libor, plus a Bermudan-style right to cancel the swap. For clarity, Figures 5.1–5.3 list the cash flow diagrams of a callable structured note[16].

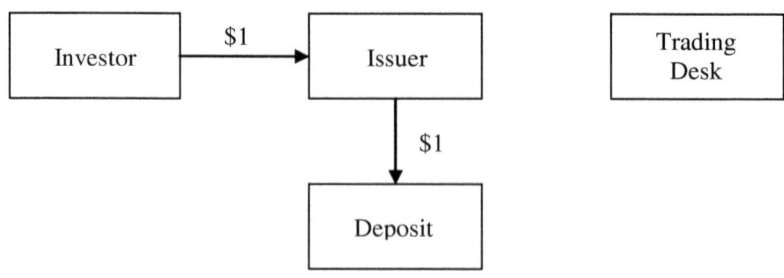

Fig. 5.1. Callable Note: Flows at Inception

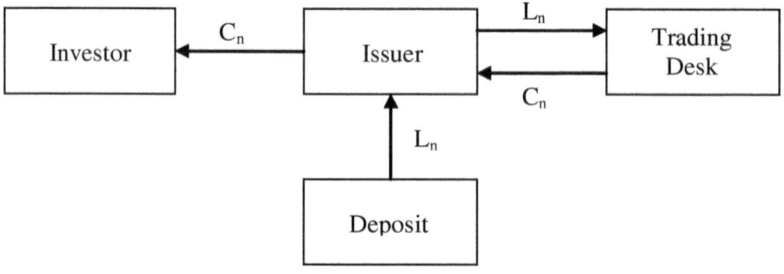

Fig. 5.2. Callable Note: Flows at Payment Times

Sometimes it is convenient to represent a cancelable exotic swap as a straight exotic swap, plus a Bermudan-style option to enter a reverse swap, i.e. a swap where legs are reversed relative to the original one. Beyond providing a break-down that is convenient for valuation purposes, this representation emphasizes the fact that the cancelability feature of a CLE benefits the party that owns it (typically a structured note issuer). Indeed, the feature is

---

[16] While, conceptually, the principal is deposited into a Libor-paying account, in practice it is used as part of cash management activities by the issuer. A structured note issuance program often provides cheaper funding to a bank than would be attainable by other means.

Fig. 5.3. Callable Note: Flows at Termination (Maturity or early Cancellation)

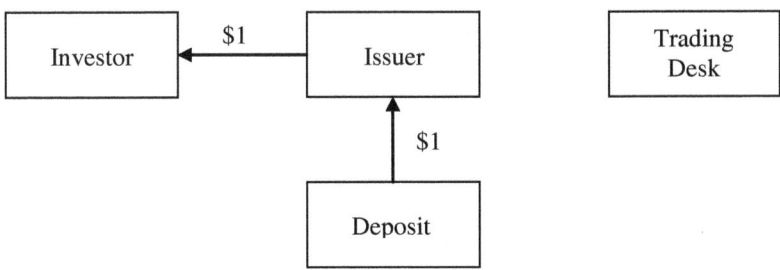

often added to a structured note as a way to offer a more attractive coupon to the investor, in return for the Bermudan-style option. Often, the coupons inside the non-call period are fixed rate coupons, and a typical way for the issuer to "pay" for the Bermudan option is to make these first coupons high, often much higher than the return available elsewhere. This "optical illusion" of high rate of return on investment is, at least in part, what drives investor interest in structured callable notes.

Consider a CLE on an exotic swap with structured coupons $\{C_n\}_{n=0}^{N-1}$. As for regular Bermudan swaptions, we denote the value of the exotic swap that one can exercise on date T_n by

$$U_n(T_n) = \beta(T_n) \sum_{i=n}^{N-1} \tau_i E_{T_n} \left(\beta(T_{i+1})^{-1} \times (C_i - L_i(T_i)) \right). \qquad (5.24)$$

Here, we recall that from a trading desk prospective, the cancelability feature of CLE involves an option to enter a reverse swap, with receipt of structured coupons and payment of Libor. Hold values are also defined analogously to the Bermudan swaption case: the n-th hold value $H_n(T_n)$ is defined as the time T_n value of the CLE on the same exotic swap, but with exercise dates $\{T_i\}_{i=n+1}^{N-1}$ only. That is, $H_n(T_n)$ is the time T_n value of the CLE provided it has not been exercised on or before T_n.

5.14.2 Pricing Callable Libor Exotics

A significant part of this book is dedicated to efficient methods for pricing and risk-managing callable Libor exotics. To provide a brief preview of the difficulties involved, we notice that the call feature embedded in CLEs may suggest application of PDE methods, using backward induction arguments outlined in previous chapters. However, often a CLE has explicit path dependence that makes the application of PDE methods impractical. In other cases, as we shall describe in greater depth later in the book, models that admit an efficient PDE representation are often too inflexible for

216   5 Fixed Income Instruments

application to anything but the simplest of CLEs. Hence, CLEs more often than not must be valued using Monte Carlo methods. We have seen a preview of how optimal exercise can be handled in Monte Carlo in Section 3.5; many more details are provided in Chapter 18.

5.14.3 Types of Callable Libor Exotics

Any exotic swap can be used as an underlying for a callable Libor exotic. For our purposes, the taxonomy of callable Libor exotics can follow closely that of exotic swaps, see Section 5.13. We can thereby distinguish various types of CLEs, e.g. Libor-based, CMS-based, multi-rate, callable range accruals, callable snowballs, and so on. Many variations on the basic CLE design exist, most of which are driven by a desire to increase the value of the option to cancel that the investor sells to the trading desk, in order for a higher coupon to be paid. It is difficult to classify all the features that have been invented; we content ourselves with merely listing some of the more popular ones.

5.14.4 Callable Snowballs

A callable snowball is a CLE with a snowball (or snowrange, etc.) underlying. From a modeling prospective, they are notable for being one of the first widely popular instruments that combine both strong path-dependence and optimal exercise. It is possible to incorporate snowball-type path-dependence into a PDE framework by introducing auxiliary variables, following the principles of Section 2.7.5; Section 18.4.5 discusses details specific to snowballs. Alternatively — and often preferably — optimal exercise can be incorporated into the Monte Carlo method, as discussed in Section 3.5 and later on in Chapter 18.

5.14.5 CLEs Accreting at Coupon Rate

Typically the notional of the underlying swap of a CLE is fixed throughout the life of the deal, but it does not have to be. For instance, it is not uncommon for the notional to vary deterministically, e.g. increase or decrease by non-random additive or multiplicative amounts each coupon period. Such deterministic accretion/amortization rarely adds extra complications from a modeling prospective. Occasionally, however, a contract specifies that the notional of the swap accretes at the structured coupon rate, in which case the accretion rate will be random. For such CLEs, the exercise value in (5.24) must be amended. Specifically, if q_i is the notional in place for the period $[T_i, T_{i+1}]$, then q_i is obtained from the notional over the previous period q_{i-1} by multiplication with the structured coupon over the previous period. Formally,

$$U_n(t) = \beta(t) \sum_{i=n}^{N-1} \tau_i \mathrm{E}_t \left(\beta\left(T_{i+1}\right)^{-1} \times q_i \times (C_i - L_i(T_i)) \right),$$

$$q_i = q_{i-1} \times (1 + \tau_{i-1} C_{i-1}),$$

where the initial notional q_0 is contractually specified.

Interestingly, the random accretion feature above can be incorporated into a PDE-based scheme without any extra cost, see Section 18.4.5.

5.14.6 Multi-Tranches

The more optionality an investor can sell to the issuer, the better coupon she can receive. As described earlier, the option to call the note is already present in a callable structured note. Another option that is sometimes embedded is a right for the issuer to increase the size of the note, i.e. to put more of the same note to the investor, whether she wants it or not. The name of this feature, a "multi-tranche" callable structured note, originates with the fact that these possible notional increases are formalized as tranches[17] of the same note that the issuer has the right to put to the investor. The times when the issuer has the right to increase the notional of the note typically come before the times when the note can be canceled altogether. Callability usually applies jointly to all tranches of the note.

By itself, the multi-tranche feature rarely presents modeling issues, although one must be mindful of it and plan for a pricing infrastructure that is flexible enough to handle it.

5.15 TARNs and Other Trade-Level Features

While sometimes the precise split is a little arbitrary, it is often helpful to think of a Libor exotic as being defined by

- A definition of its coupon, i.e. a formula that converts rates observed during a coupon period (and, sometimes, previous coupons) into the amount of money paid to the investor.
- A collection of trade-level features, i.e. features that cannot conveniently be expressed as coupon definitions, but instead "act" on the whole trade.

We have already seen examples of both features. For instance, a callable snowball CLE has a coupon definition given by the formula (5.22) on top of which callability has been added as a trade-level feature. In the next few sections we review some other trade-level features.

---

[17] "Slices" in French; here meaning "similar securities offered as part of the same transaction".

5.15.1 Knock-out Swaps

A knock-out swap is just an exotic swap that disappears on the first fixing date on which some reference rate is above (or below) a given barrier. If the knock-out rate for the period n is denoted by $X_n(t)$, the coupon by C_n, the Libor rate by L_n, and the knock-out barrier by R, the value of a down-and-out[18] knock-out swap is given by

$$V_{\text{KO}}(t) = \beta(t)\text{E}_t \left(\sum_{n=0}^{N-1} \tau_n \beta\left(T_{n+1}\right)^{-1} \times (C_n - L_n(T_n)) \times \prod_{i=0}^{n} 1_{\{X_i(T_i) > R\}} \right).$$

5.15.2 TARNs

Callable structured notes have proved to be popular with investors, but suffer from the drawback that investors rarely know when the issuer will call the note — indeed, the decision to exercise a Bermudan-style option is driven by a model rarely accessible to the average investor. A relatively recent innovation, the *Targeted Redemption Note* (TARN), presents one possible solution to the problem. In a TARN (see Piterbarg [2004c]), the total investor return, defined as the sum of all structured coupons paid to date, is recorded over time. When the total return exceeds a pre-specified target level— hence the name of the structure — the note is terminated and the principal is returned to the investor.

As with callable notes, issuers do not keep TARN structures on their books, but swap them out with a trading desk. Since the principal payment from investors is invested at the Libor rate, to a trading desk a TARN looks like an exotic swap that knocks out on the total sum of structured coupons. Formally, let the structured coupon[19] for the period $[T_n, T_{n+1}]$ be C_n. The coupon over the period $[T_n, T_{n+1}]$ is only paid if the sum of structured coupons up to (and not including) time T_n is below a total return R. Thus, the value of the TARN at time 0 from the investor's viewpoint is given by

$$V_{\text{tarn}}(t) = \beta(t)\text{E}_t \left(\sum_{n=1}^{N-1} \tau_n \beta\left(T_{n+1}\right)^{-1} \times (C_n - L_n(T_n)) \times 1_{\{Q_n < R\}} \right), \tag{5.25}$$

$$Q_n = \sum_{i=1}^{n-1} \tau_i C_i, \quad Q_1 = 0.$$

---

[18] I.e. disappearing upon some variable breaching a barrier from above; compare to up-and-out options discussed in Chapter 2.

[19] In the original TARN product, an inverse floating coupon (5.17) was used, but any structured coupon can be employed.

We note that a TARN typically pays some fixed coupons to an investor before the knock-out feature starts, mirroring the non-call structure of CLEs, see Section 5.14. We omit these from the contract description as their present value can be computed statically off an interest rate curve separately, as the payments are known in advance.

Various features can be added to the description of the TARN we just described. For example, the last coupon, i.e. the coupon that pushes the total return over the target R, can be paid only partially to make the total return exactly R and not more. This feature is knowns as a *cap at trigger* or a *lifetime cap*. Also, if the total return over the life of the TARN never reaches the target R, a TARN equipped with a so-called *lifetime floor* will make a *make whole* payment at the TARN maturity to ensure that the total return exactly equals R, and not less. These features do not alter the general modeling framework for TARNs that we develop in later chapters, and we shall generally ignore them.

While it could be argued that a TARN is really just a swap with a different coupon definition (namely, $C_n \times 1_{\{Q_n < R\}}$)[20] we prefer to classify a TARN feature as trade-level, reflecting its historic importance and its relationship with the callability feature.

5.15.3 Global Cap

As discussed, a TARN can have a feature that restricts total return to an investor to be exactly the trigger level R. This feature can be decoupled from the TARN definition and used by itself, often called a *global cap*. Specifically, an exotic swap with a global cap R pays a structured coupon C_n to an investor until the sum of the coupons has reached the level R. Note that a swap typically does not terminate at this point, i.e. the trading desk will continue to receive Libor until the maturity of the trade or some other agreed termination event.

5.15.4 Global Floor

A *global floor* guarantees the investor a minimum cumulative return of the note. Specifically, if the sum of structured coupons paid to the investor does not reach a global floor value of F by the maturity of the deal, the issuer will pay a make whole amount to the client, equal to F minus the sum of actual coupons paid. The payment is made at the maturity of the swap or some other termination event (such as when the swap is canceled, in case of a callable global floor note).

---

[20] The same could be said about knock-out swaps previously defined.

5.15.5 Pricing and Trade Representation Challenges

Trade-level features are often combined with each other, and with various coupon formulas. For example, one can be asked to price a callable, TARN'ed snowball with a global cap. As we shall argue in later chapters, the only modeling solution that is sufficiently scalable to accommodate such arbitrary combinations of various trade features involves a generic, flexible model that is calibrated to a broad collection of market volatility information. In such a framework, adding extra features to a trade is ultimately not much of a modeling problem, but could be a significant trade representation challenge. While outside the scope of this book, such a challenge should be addressed by a software framework that is flexible enough to represent any, current or future, trade-level features, and incorporate them into a pricing engine without significant extra effort. A successful implementation of such a trade representation framework requires careful planning and considerable investment. One fairly of common route is to use a domain-specific programming language for trade representation, see for example Jones et al. [2000] for a commercially available version or Frankau et al. [2009] for an example of an in-house one.

5.16 Volatility Derivatives

In a nutshell, a *volatility derivative* is a contingent claim whose underlying is the volatility of a financial observable, rather than a financial observable itself. The simplest example of such a derivative is the *variance swap* (see Carr and Lee [2009b]), a structure that first emerged in equity and foreign exchange trading. In the last few years, similar ideas and structures have entered the fixed income derivative arena.

The demand for volatility derivatives in interest rates is driven by the same factors as in other asset classes; a common motivation is the desire of some market participants — hedge funds in particular — to have direct exposure to interest rate volatility, but not to the outright level of rates, say. In other cases, the product development follows the usual path of creating structured notes with appealing payoff profiles.

Different interest rates constitute different financial observables for defining a volatility, hence one needs to be rather specific when defining a volatility-linked payoff.

5.16.1 Volatility Swaps

A *volatility swap* in interest rates is a contract that measures realized volatility (or a related quantity) of a given rate, although it is structured somewhat differently from volatility swaps in equity or FX markets. Let

5.16 Volatility Derivatives

$X_n(t)$ be the rate used for period n; then the most common coupon of a volatility swap is given by

$$C_n = |X_{n+1}(T_{n+1}) - X_n(T_n)|,$$

or a capped version

$$C_n = \min\left(|X_{n+1}(T_{n+1}) - X_n(T_n)|, c\right).$$

The value of the (structured) leg of a volatility swap measures realized variation of the rate $X_n(\cdot)$,

$$V_{\text{volswap}}(t) = \beta(t) E_t \left(\sum_{n=0}^{N-1} \tau_n \beta(T_{n+1})^{-1} \times |X_{n+1}(T_{n+1}) - X_n(T_n)| \right)$$
$$- V_{\text{floatleg}}(t), \qquad (5.26)$$

where

$$V_{\text{floatleg}}(t) \triangleq \sum_{n=0}^{N-1} \tau_n P(t, T_{n+1}) L_n(t) = P(t, T_0) - P(t, T_N).$$

There are two common choices for the rate X_n. One choice, a *fixed-tenor* volatility swap, involves a swap rate of the same tenor on each of the fixing dates. Technically speaking, *different* swap rates are therefore used for different periods,

$$X_n(t) = S_{n,m}(t)$$

with a fixed value of m, the number of periods in the swap rate (see the definition (5.14)). For example, a rolling 10 year CMS rate could be used. The other choice, a *fixed-expiry* volatility swap, specifies the swap rate to have a fixed expiry and tenor, i.e.

$$X_n(t) = S_{K,m}(t).$$

With this definition, the volatility swap pays the absolute variation of a rate with the fixing date T_K and spanning m periods of the tenor structure $\{T_n\}_{n=0}^{N+m}$. Often $K = N$, so that the variability of the rate $S_{N,m}$ is measured over the whole of its life.

Recently, volatility swaps on CMS spread rates have appeared. As the name implies, they measure the variation of the spread of two rates, e.g. $X_n = S_{n,m_1} - S_{n,m_2}$.

5.16.2 Volatility Swaps with a Shout

Sometimes, the investor in a volatility swap is given an option to *shout*, that is to choose when the fixing of the rate occurs for the purposes of calculating the coupon payoff. In particular, the payoff of the n-th coupon is then

$$C_n = |X_{n+1}(\eta_n) - X_n(T_n)|,$$

where the random stopping time $\eta_n \in [T_n, T_{n+1}]$ is chosen by the investor coupon-by-coupon. For the uncapped version of this payoff, it is intuitively clear that it is always optimal to postpone the shout until the end of the period, i.e. $\eta_n = T_{n+1}$. So, the option given to the investor is actually worthless, while designed to appear to have some value. Interestingly, for the capped version, it is optimal[21] to shout at the lesser of T_{n+1} and the first time that $|X_{n+1}(\eta_n) - X_n(T_n)| \geq c$, where c is the cap level. As a consequence, a capped volatility swap with a shout option is equivalent to a volatility swap with a barrier on each coupon:

$$C_n = c \times 1_{\left\{\max_{t \in [T_n, T_{n+1}]} |X_{n+1}(t) - X_n(T_n)| \geq c\right\}} \\ + |X_{n+1}(T_{n+1}) - X_n(T_n)| \times 1_{\left\{\max_{t \in [T_n, T_{n+1}]} |X_{n+1}(t) - X_n(T_n)| < c\right\}}.$$

This decomposition follows from results in Broadie and Detemple [1995] and is discussed in more detail in Chapter 20. The fact that one can replace an optimal exercise feature with a known static barrier is quite convenient and allows for easy Monte Carlo valuation.

5.16.3 Min-Max Volatility Swaps

The structured coupon for a min-max volatility swap is given by

$$C_n = M_n - m_n,$$

where

$$M_n = \max_{t \in [T_n, T_{n+1}]} X_n(t),$$
$$m_n = \min_{t \in [T_n, T_{n+1}]} X_n(t).$$

The coupon represents the spread between the maximum and the minimum values that a given rate achieves during a coupon period.

While the two products appear quite different at a first glance, there is an interesting connection between min-max and standard volatility swaps. We shall explore this further in Chapter 20.

5.16.4 Forward Starting Options and Other Forward Volatility Contracts

The value of a standard European swaption and a standard fixed-expiry volatility swap are both linked to the volatility of the swap rate over its

---

[21] Ignoring some small convexity effects and the difference between discrete vs. continuous shout option rights.

entire life, i.e. from the valuation date to the fixing date of the swap rate. Some clients, however, prefer securities that are linked to the volatility of a swap rate as measured over only a sub-period of this time; in effect, the clients want exposure to what is often known as *forward* volatility. Precise definitions and the importance of forward volatility are left for future chapters; for now we content ourselves by listing a few relevant varieties of forward volatility derivatives.

Midcurve swaptions are swaptions whose expiry T^e is strictly before the fixing date T_0 of the underlying swap rate. Their value depend on the volatility of the swap rate over the period $[t, T^e] \subset [t, T_0]$.

Given a swap rate $S(\cdot)$ with the fixing date T_0, and a date $T^s < T_0$, a *forward-starting swaption straddle*[22] is given by the payoff

$$A(T_0) \times |S(T_0) - S(T^s)|$$

paid at T_0. Essentially, this contract is a combination of a receiver and a payer swaption, both of which will have their strikes fixed at time T^s to the then-prevailing level of the underlying swap rate. That is, the contract pays the value of the at-the-money straddle on the rate $S(\cdot)$ at time T^s with expiry T_0. The value of a forward-starting swaption straddle is driven by the volatility of the swap rate over the period $[T^s, T_0]$.

Recall that European swaptions are typically quoted in terms of their implied volatilities. Due to this convention, some clients find the forward starting straddle too indirect and instead want to receive implied volatility itself. Particularly popular are contracts that pay implied Normal[23] volatility as defined on p. 204. Fortunately, the at-the-money Normal volatility has a direct relationship to the swaption price, and the payoff of an *implied Normal volatility contract* is

$$\sqrt{\frac{\pi}{2(T_0 - T^s)}} \times |S(T_0) - S(T^s)|$$

paid at T_0. Apart from the factor $A(T_0)$ and a non-consequential scaling factor, the payoffs of a forward-starting straddle and the implied Normal volatility contract are identical. The differences in their prices are just a matter of a minor convexity correction, a topic we return to in Chapter 20.

---

[22]The term *straddle* is used to denote the sum of a put and a call option with identical strikes.

[23]Contracts paying implied log-normal, or Black, volatility are possible but less common, due to the common perception that interest rates are more Gaussian than log-normal, i.e. the implied Gaussian volatility is less sensitive to the changes in the level of interest rates than the implied Black volatility.

5.A Appendix: Day Counting Rules and Other Trivia

In this appendix, we very briefly cover some of the finer details of how schedules are constructed and how interest rate payments accrue under market conventions. We generally ignore these details in the main body of the book, and our treatment here only scratches the surface. For a full account, see Mayle [1993] or Stigum and Robinson [1996].

5.A.1 Libor Rate Definitions

Consider the 6 month Libor rate L fixing at time T. According to (4.2), we would compute this rate as simply

$$L(T) = L(T, T, T + 1/2) = \left(P(T, T + 1/2)^{-1} - 1\right)/\tau, \quad \tau = 1/2. \quad (5.27)$$

In reality, this computation ignores a number of quoting conventions. First, a 6 month USD Libor rate that fixes at time T, does not truly cover an *accrual period* of $[T, T + 1/2]$. Instead, the start date T^s of the accrual is set to be $T^s = T + \delta^s$, where δ^s is a delay of two business days[24]. In other words, the quoted spot Libor rate is in actuality based on a *forward starting* CD that is entered into with time lag of δ^s after the quotation date T. As for the end date T^e of the accrual period, it is normally determined by counting 6 months ahead starting from T^s, adjusting the resulting date to ensure that it is a valid business day. The precise mechanism used to make such a business day adjustments of T^s is determined from a *date rolling convention*. For USD Libor, one always uses the so-called "Modified-Following" convention where weekend or holiday dates are rolled forward to the next business day, unless doing so would cause T^e to lie in the next calendar month, in which case the payment date is rolled to the previous business day. Other rolling conventions are discussed in Mayle [1993] and Stigum and Robinson [1996].

Once the correct accrual period $[T^s, T^e]$ has been determined, to compute rate accrual it remains to compute the proper *year fraction* (or *accrual factor* or *day count fraction*) τ representing how many whole years are spanned by $[T^s, T^e]$. For the purposes of our book, we normally write simply $\tau = T^e - T^s$, but a little thought shows that expressions like this are ambiguous when T^e and T^s are thought of as actual (discrete) calendar dates, rather than as arbitrary numbers on the real line. For instance, given the existence of leap years, how many days are there in a standard year? For quant purposes, it is common to assume that a calendar year has 365.25 days, such that $T^e - T^s$ is obtained by simply counting the number of days between T^s and T^e and then dividing this number by 365.25. This "convention" is sometimes known as Actual/365.25 (or sometimes just A365.25), and is rarely, if ever, used for actual market quoting purposes. Instead, for quotation of Libor

---

[24] Libor rates in other currencies may have different delays. For instance, GBP Libor has zero business day delay.

5.A Appendix: Day Counting Rules and Other Trivia

rates the standard is to use an Actual/360 (A360) convention, where the number of days between T^s and T^e are converted to a year fraction by dividing by 360. As a consequence, the true value of τ used for 6 month Libor quotation purposes is typically slightly larger than $1/2$. For additional year-count conventions (of which there are many), see Mayle [1993] and Stigum and Robinson [1996].

Due to the quoting standards used in real Libor markets, the relationship between discount bonds and quoted Libor rates is more complicated than (5.27). Specifically, if $L_{\mathrm{mkt}}(T)$ represents the true quoted 6 month Libor rate at time T, we instead have

$$L_{\mathrm{mkt}}(T) = \left((P(T,T^e)/P(T,T^s))^{-1} - 1\right) \frac{360}{D(T^s,T^e)}, \qquad (5.28)$$

where by $D(T^s, T^e)$ we denoted the number of days between T^s and T^e according to the convention used. Notice in particular how the formula now involves a forward starting zero-coupon bond $P(T, T^s, T^e) = P(T, T^e)/P(T, T^s)$, as a reflection of the settlement delay δ^s. Using existing "idealized" Libor rate notation (see (4.2)), we may write this expression as

$$L_{\mathrm{mkt}}(T) = L(T, T^s, T^e) \times \frac{360}{365.25} \approx L(T, T^s, T^e) \times 0.986.$$

The difference between $L_{\mathrm{mkt}}(T)$ and $L(T)$ is small enough for us to ignore it in most of this book, but any real system implementation obviously should use precise day counting rules when computing Libor fixings.

5.A.2 Swap Payments

The payments on swaps (and other instruments, such as CDs and FRAs) are subject to similar conventions as the Libor rate. Consider for instance a standard fixed-for-floating interest rate swap issued at time t (today). First, a[25] schedule $\{T_i\}_{i=0}^N$ for interest rate accrual must be constructed, starting from a given base frequency of the swap (e.g.: semi-annual). As was the case above, the schedule normally starts one or two business days after time t, i.e. $T_0 = t + \delta_0$ where δ_0 is some contractually specified delay. Date T_0 is known as the *effective date* of the swap. Given T_0, the remaining T_i, $i = 1, \ldots, N$, are computed by first laying out "unadjusted" dates according to the swap base frequency, and then applying a date rolling convention (typically Modified-Following) to each of the dates. As part of the swap contract, associated with each accrual period $[T_i, T_{i+1}]$ are then:

- A fixing date T_i^f: the date on which the floating leg index (Libor, most often) is observed. Typically T_i^f is two business days before time T_i.

---

[25] We assume that the fixed and floating legs pay interest on the same schedule, but in reality, this may not be the case. For instance, in USD, the standard frequency for the fixed leg is six months, and three months for the floating leg.

226   5 Fixed Income Instruments

- A payment date T_i^p: the date on which the swap payments are made. Normally $T_i^p = T_{i+1}$, but it is not uncommon to have payment delays of 1 or 2 business days after T_{i+1}.
- A fixed leg year fraction τ_i^{fix}: the year fraction used to determine the payment at time T_i^p on the fixed leg. In the US, the most common convention for the fixed leg is[26] 30/360.
- A floating leg year fraction τ_i^{flt}: the year fraction used to determine the payment at time T_i^p on the floating leg[27]. In the US, the most common convention for the floating leg is Actual/360.

At time t, the value of a payer swap paying a fixed coupon c against Libor is therefore

$$V_{\text{swap}}(t) = \sum_{i=1}^{N} P(t, T_i^p) \mathrm{E}^{T_i^p} \left(\tau_i^{flt} L_{\text{mkt}}(T_i^f) - \tau_i^{fix} c \right),$$

where we have used the T_i^p-forward measure to state the valuation, and where L_{mkt} is defined in (5.28). In the book, we normally simplify this to

$$V_{\text{swap}}(t) = \sum_{i=1}^{N} P(t, T_i) \tau_i \mathrm{E}^{T_i} \left(L(T_{i-1}) - c \right), \quad \tau_i = T_i - T_{i-1}.$$

---

[26] When counting days in the 30/360 convention, each month is assumed to have 30 days. The number of days used to determine interest rate accrual can therefore differ from the *actual* number of days (which distinguishes 30/360 from Actual/360).

[27] Another, often ignored, complication with the floating leg is that the periods that define the (payment) year fractions τ_i^{flt} are sometimes slightly different from those that define forward Libor rates, due to certain conventions surrounding date adjustments. Again, see details in Mayle [1993] or Stigum and Robinson [1996].

Part II

Vanilla Models

6
Yield Curve Construction and Risk Management

In a nutshell, the job of an interest rate model is to describe the random movement of a curve of discount bond prices through time, starting from a known initial condition. In reality, however, only a few short-dated discount bonds are directly quoted in the market at any given time, a long stretch from the assumption of many models that an initial curve of discount bond prices is observable for a continuum of maturities out to 20–30 years or more. Fortunately, a number of liquid securities depend in relatively straightforward fashion on discount bonds, opening up the possibility of uncovering discount bond prices from prices of such securities. Still, as only a finite set of securities are quoted in the market, constructing a continuous curve of discount bond prices will inevitably require us to complement market observations with an interpolation rule, based perhaps on direct assumptions about functional form or perhaps on a regularity norm to be optimized on. A somewhat specialized area of research, discount curve construction relies on techniques from a number of fields, including statistics and computer graphics. While we cannot possibly do the subject full justice, discount curve construction is a fundamental step in the modeling exercise, and no book on fixed income models is complete without a discussion of basic techniques.

As mentioned in the Preface to this book, the crisis of 2007–2009 have lead to changes in the foundations of interest rate modeling, not least in the area of yield curve construction and risk management. Pre-crisis, it was often sufficient to construct only a single (Libor) discount curve, but nowadays the task is more complicated as a whole collection of inter-related curves is required. Nevertheless, the traditional techniques used for single-curve construction are by no means obsolete, and their mastery is required before more ambitious curve algorithms can be attempted. Accordingly, we have split this chapter into three parts. In the first, and most significant, part, we introduce notations and cover a number of curve construction techniques, moving from simply bootstrapped C^0 curves through "local spline" C^1 curves to full C^2 smoothing splines with and without tension.

Perturbation locality is discussed, as are methods to control behavior under perturbations. In the second part we discuss the management of interest rate curve risk, covering both basic approaches as well as more advanced methods based on Jacobian techniques. In the last part, we discuss a number of specialized issues and contemporaneous extensions, most notably turn-of-year adjustments and techniques to construct separate discount and forward curves. The need for such a separation has long been recognized (albeit neglected in the literature) as a requirement to avoid arbitrages in markets for foreign exchange forwards and for floating-floating cross-currency swaps. More recently, similar issues have appeared in purely domestic markets where the Libor rate is no longer considered a good proxy for the risk-free discount rate, and where a significant *tenor basis* has developed in floating-floating single-currency swaps. Accordingly, we conclude the chapter with a description of techniques for building a *multi-index curve group*, a self-consistent arbitrage-free collection of discount and forward curves suitable for valuation of different types of swaps and other interest rate derivatives.

6.1 Notations and Problem Definition

6.1.1 Discount Curves

Throughout this chapter, we use the abbreviated notation $P(T) = P(0,T)$ where $P : [0, \mathcal{T}] \to (0,1]$ is a continuous, monotonically decreasing *discount curve*. $\mathcal{T}$ denotes the maximum maturity considered, typically given as the longest maturity in the set of securities the curve is built from. Let there be N such securities — the *benchmark set* — with observable prices $V_1, \ldots, V_N$. We assume that the time 0 price $V_i = V_i(0)$ of security i can be written as a linear combination of discount bond prices at different maturities,

$$V_i = \sum_{j=1}^{M} c_{i,j} P(t_j), \quad i = 1, \ldots, N, \tag{6.1}$$

where $0 < t_1 < t_2 < \ldots < t_M \leq \mathcal{T}$ is a given finite set of dates, in practice obtained by merging together the cash flow dates of each of the N benchmark securities. Let $T_1, T_2, \ldots, T_N$ denote the final maturities of the N benchmark securities, in which case we necessarily must have

$$c_{i,j} = 0, \quad t_j > T_i.$$

Securities that can be represented by pricing expressions of the form (6.1) obviously include coupon and discount bonds, but also FRAs and fixed-floating interest rate swaps. For instance, consider a newly issued unit-notional fixed-floating swap, paying a coupon of $c\tau$ at times $\tau, 2\tau, 3\tau, \ldots, n\tau$. If no spread is paid on the floating rate, the time 0 total swap value to the fixed payer is

6.1 Notations and Problem Definition

$$V_{\text{swap}} = 1 - P(n\tau) - \sum_{j=1}^{n} c\tau P(j\tau),$$

as already discussed in Chapter 5. We can rewrite this as

$$1 - V_{\text{swap}} = P(n\tau) + \sum_{j=1}^{n} c\tau P(j\tau), \tag{6.2}$$

which is in the form[1] (6.1) once we interpret $V_i = 1 - V_{\text{swap}}$. In practice, swaps used for discount curve construction are nearly always newly issued and par-valued, in the sense that the coupon c is set to make $V_{\text{swap}} = 0$. To give another example, consider an FRA on the $[T, T+\tau]$ Libor rate, for which formula (5.2) in Chapter 5 gives, at $t = 0$,

$$V_{\text{FRA}} = \tau \left(L(0, T, T+\tau) - k \right) P(T+\tau), \tag{6.3}$$

where k is the quoted FRA rate. From the definition of $L(0, T, T+\tau)$ this is just

$$V_{\text{FRA}} = P(T) - P(T+\tau) - k\tau P(T+\tau) = P(T) - (1 + k\tau) P(T+\tau),$$

which is in the form (6.1). As for swaps, FRAs used for curve construction are newly issued and typically have k set such that $V_{\text{FRA}} = 0$.

The choice of the securities to be included in the benchmark set depends on the market under consideration. For instance, to construct a Treasury bond curve, it is natural to choose a set of Treasury bonds and T-Bills. On the other hand, if we are interested in constructing a discount curve applicable for bonds issued by a particular firm, we would naturally use bonds and loans issued by the firm in question. For our purposes, the most important discount curve is the *Libor curve*, constructed out of market quotes for Libor deposits, swaps and Eurodollar futures. In the construction of this curve, most firms would use a few certificates of deposit for the first 3 months of the curve, followed by a strip of Eurodollar futures[2] (with maturities staggered 3 months apart) out to 3 or 4 years. Par swaps are then used for the rest of the curve, with typical maturities being 5, 7, 10, 12, 15, 20, 25, and 30 years.

---

[1] For swaps where payment schedules do not coincide perfectly with the accrual periods of the Libor rates, the expression (6.2) is only an approximation, albeit a very good one. In practice we can construct the yield curve assuming that (6.2) holds, and then perform a small post-processing clean-up iteration, along the lines of the algorithm in Section 6.5.2.4.

[2] We note that Eurodollar futures contracts do not allow for a pricing expression of the form (6.1), so a pre-processing step is normally employed to convert the futures rate quote to a forward rate (FRA) quote. See Proposition 4.5.3 or Chapter 16 for more on this.

6.1.2 Matrix Formulation

Define the M-dimensional discount bond vector[3]

$$\mathbf{P} = (P(t_1), \ldots, P(t_M))^\top,$$

and let $\mathbf{V} = (V_1, \ldots, V_N)^\top$ be the vector of observable security prices. Also let $\mathbf{c} = \{c_{i,j}\}$ be an $(N \times M)$-dimensional matrix containing all the cash flows produced by the chosen set of securities. The matrix $\mathbf{c}$ would typically be quite sparse, with many rows containing only a few non-zero entries. A typical, albeit simplified, form of the matrix $\mathbf{c}$ might be ($\times$ marks a non-zero element)

$$\mathbf{c} = \begin{pmatrix} \times & & & & & & & & & & & & & & \\ & \times & & & & & & & & & & & & & \\ & & \times & \times & & & & & & & & & & & \\ & & & \times & \times & & & & & & & & & & \\ & & & & \times & \times & & & & & & & & & \\ & & & & & \times & \times & & & & & & & & \\ \times & \times & \times & \times & \times & \times & \times & \times & \times & & & & & & \\ \times & \times & \times & \times & \times & \times & \times & \times & \times & \times & \times & & & & \\ \times & \times & \times & \times & \times & \times & \times & \times & \times & \times & \times & \times & \times & \times & \times \end{pmatrix},$$

corresponding to two certificates of deposit (first two rows); four FRAs or Eurodollar futures (next four rows); and three swaps (last three rows).

In a consistent, friction-free market without arbitrage opportunities, the fundamental relation

$$\mathbf{V} = \mathbf{cP} \tag{6.4}$$

must be satisfied, giving us a starting point to determine $\mathbf{P}$.

6.1.3 Construction Principles and Yield Curves

In practice, we normally have more cash flow dates than benchmark security prices, i.e. $M > N$, in which case (6.4) is insufficient to uniquely determine $\mathbf{P}$. The problem of curve construction essentially boils down to supplementing (6.4) with enough additional assumptions to allow us to extract $\mathbf{P}$ and to determine $P(T)$ for values of T not in the cash flow timing set $\{t_j\}_{j=1}^M$.

As it is normally easier to devise an interpolation scheme on a curve that is reasonably flat (rather than exponentially decaying), it is common to perform the curve fitting exercise on *discount yields*, rather than directly on bond prices[4]. Specifically, we introduce a continuous yield function $y : [0, \mathcal{T}] \to \mathbb{R}_+$ given by

---

[3] For extra clarity, throughout this chapter we use boldface type for vectors and matrices.

[4] See e.g. Shea [1984] for a discussion of the pitfalls associated with curve interpolators that work directly on the discount function $P(T)$.

$$e^{-y(T)T} = P(T) \quad \Rightarrow \quad y(T) = -T^{-1} \ln P(T), \tag{6.5}$$

such that in (6.4)

$$\mathbf{P} = \left(e^{-y(t_1)t_1}, \ldots, e^{-y(t_M)t_M} \right)^\top.$$

The mapping $T \mapsto y(T)$ is known as the *yield curve*; it is related to the discount curve by the simple transformation (6.5). Of related interest is also the *instantaneous forward curve* $f(T)$, given by

$$P(T) = e^{-\int_0^T f(u)du}. \tag{6.6}$$

Notice that

$$f(T) = y(T) + \frac{dy(T)}{dT}T. \tag{6.7}$$

For alternative transformations, and a discussion of their relative merits, see Andersen [2005]. Unless explicitly stated, in the remainder of this chapter we shall work with yields, i.e. we treat $y(T)$ as the fundamental curve to be estimated.

Whatever space we elect to work in, we have at least three options for solving (6.4).

1. We can introduce new and unspanned securities such that $N = M$ and (6.4) allows for exactly one solution.
2. We can use a parameterization of the yield curve with precisely N parameters, using the N equations in (6.4) to recover these parameters.
3. We can search the space of all solutions to (6.4) and choose the one that is "optimal" according to a given criterion.

Let us provide some comments to these three ideas. First, in option 1 above, introduction of new securities might not truly be possible — such securities may simply not exist — but sometimes interpolation rules applied to the given benchmark set may allow us to provide reasonable values for an additional set of "fictitious" securities. Although it can occasionally be useful in pre-processing to pad an overly sparse benchmark set, this idea will often require some quite ad-hoc decisions about the specifics of the fictitious securities, and excessive use may ultimately lead to odd-looking curves and suboptimal hedge reports. When an interpolation rule is to be used, it is typically better to apply it directly on more fundamental quantities such as zero-coupon yields or forward rates, thereby maintaining a higher degree of control over the resulting discount curve.

In option 2 above, parametric functional forms (e.g. Nelson and Siegel [1987]) are sometimes used, but it is far more common to work with a spline representation with N user-selected knots (typically at the maturity dates of the benchmark securities), with the level of the yield curve at these knots constituting the N unknowns to be solved for. We discuss the details of this approach in Section 6.2, using a number of different spline types. Some required elements of basic spline theory can be found in Appendix 6.A of this chapter.

Option 3 constitutes the most sophisticated approach and can often be stated in completely non-parametric terms, with the yield curve emerging naturally as the solution to an optimization problem. If carefully stated, this approach can easily be modified to handle the situation where the system of equations (6.4) is (near-) singular, in the sense that either no solutions exist or all solutions are irregular and non-smooth[5]. Technically, we handle this by working with *smoothing splines*, in the process replacing (6.4) with a penalized least-squares optimization problem. We discuss elements of this idea in Section 6.3 below.

6.2 Yield Curve Fitting with N-Knot Splines

In this section we discuss a number of well-known yield curve algorithms based on polynomial and exponential (tension) splines of various degrees of differentiability. Throughout, we assume that we can select and arrange our benchmark set of securities to guarantee that the maturities of the benchmark securities satisfy

$$T_i > T_{i-1}, \quad i = 2, 3, \ldots, N, \tag{6.8}$$

where the inequality is strict. Equation (6.8) constitutes a "spanning" condition and allows us to select the N maturities as distinct knots in our yield curve splines.

6.2.1 C^0 Yield Curves: Bootstrapping

If continuity of the yield curve is all that we require, we can work with a common iterative procedure known as *bootstrapping*. The basic idea is encapsulated in the following iteration:

1. Let $P(t_j)$ be known for $t_j \leq T_{i-1}$, such that prices for benchmark securities $1, \ldots, i-1$ are matched.
2. Make a guess for $P(T_i)$.
3. Use an interpolation rule to fill in $P(t_j)$, $T_{i-1} < t_j < T_i$.
4. Compute V_i from the now-known values of $P(t_j)$, $t_j \leq T_i$.
5. If V_i equals the value observed in the market, stop. Otherwise return to Step 2.
6. If $i < N$, set $i = i+1$ and repeat.

---

[5] Intuitively, this situation can arise if, say, two or more securities in the benchmark set have near-identical cash flows, yet have significantly different present values.

6.2 Yield Curve Fitting with N-Knot Splines

The updating of guesses when iterating over Steps 2 through 5 can be handled by any standard one-dimensional root-search algorithm (e.g., the Newton-Raphson or secant methods).

There are strong limitations on what kind of interpolation rule can be applied in Step 3. For instance, one might consider using a representation in terms of instantaneous forwards $f(T)$ (see (6.6)), with the assumption that $f(T)$ is a continuous piecewise linear function on the maturity grid $\{T_i\}_{i=1}^N$. While based on seemingly natural assumptions, this interpolation rule can, however, be shown to be numerically unstable and prone to oscillations. Some stable, and standard, choices for interpolation rules are covered in the next two sections; common for both is that the resulting yield curve is continuous, but non-differentiable. This, in turn, implies that the instantaneous forward curve is discontinuous (see (6.7)).

6.2.1.1 Piecewise Linear Yields

The most common discount curve bootstrap algorithm assumes that the continuously compounded yield $y(T)$ in (6.5) is a continuous piecewise linear function on $\{T_i\}_{i=1}^N$. Formally, the interpolation rule in Step 3 of the algorithm in Section 6.2.1 writes $P(T) = e^{-y(T)T}$, where

$$y(T) = y(T_i)\frac{T_{i+1} - T}{T_{i+1} - T_i} + y(T_{i+1})\frac{T - T_i}{T_{i+1} - T_i}, \quad T \in [T_i, T_{i+1}]. \quad (6.9)$$

To initiate the iterative bootstrap algorithm, we note that the interpolation rule (6.9) may require us to provide an equation for $y(t)$, $t < T_1$. There are a number of ways to do this; one common choice is to simply set $y(t) = y(T_1)$, $t < T_1$.

To give a feel for the types of yield curves produced by linear yield bootstrapping, let us consider a simple example with a benchmark set of $N = 10$ swaps, with maturities and quoted par swap rates as given in Table 6.1[6].

The swaps are assumed to pay on a semi-annual basis,

$$t_j = j \cdot 0.5, \quad j = 1, 2, \ldots, 50.$$

Setting $y(t) = y(1)$, $t < 1$, and then running the bootstrap procedure on the swap price expression (6.2) results in the yield shown in Figure 6.1. The same figure also shows the continuously compounded forward curve, as computed by equation (6.7). The discontinuous "saw-tooth" shape of the forward curve is characteristic for bootstrapped yield curves with piecewise linear yield.

---

[6] In actual markets, swap yields are most often increasing functions of the swap maturity, rather than humped as in Table 6.1. The data in Table 6.1 was picked to stress the curve construction algorithms, in order to emphasize their strengths and weaknesses.

236   6 Yield Curve Construction and Risk Management

Maturity (Years)	Swap Par Rate
1	4.20%
2	4.30%
3	4.70%
5	5.40%
7	5.70%
10	6.00%
12	6.10%
15	5.90%
20	5.60%
25	5.55%

Table 6.1. Swap Benchmark Set for Numerical Tests

Fig. 6.1. Yield and Forward Curve

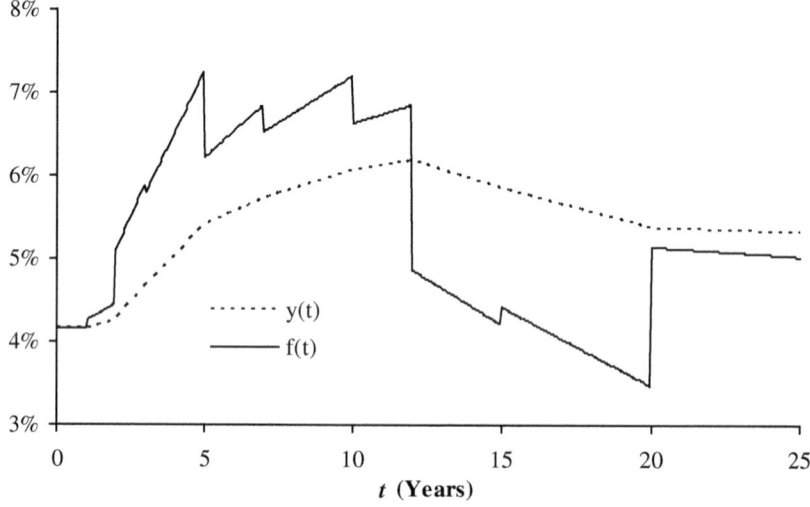

Notes: Yield curve is constructed by bootstrapping, assuming piecewise linear yields. Swap data is in Table 6.1.

6.2.1.2 Piecewise Flat Forward Rates

Assume now that the instantaneous forward curve is piecewise flat, switching to a new level at each point in $\{T_i\}$, i.e.

$$f(T) = f(T_i), \quad T \in [T_i, T_{i+1}), \tag{6.10}$$

with $T_0 \triangleq 0$. This corresponds to an interpolation rule where $\ln P(T)$ is linear in T, or

6.2 Yield Curve Fitting with N-Knot Splines

$$P(T) = P(T_i)e^{-f(T_i)(T-T_i)}, \quad T \in [T_i, T_{i+1}),$$

where a bootstrap algorithm can be used to establish the values of the N unknown constants $f(T_0), f(T_1), \ldots, f(T_{N-1})$. From the equation

$$y(T)T = \int_0^T f(u)\,du,$$

we see that the assumption of piecewise flat forwards gives, for $T \in [T_i, T_{i+1})$,

$$y(T) = \frac{y(T_i)T_i + f(T_i)(T-T_i)}{T} = f(T_i) + \frac{(y(T_i) - f(T_i))T_i}{T},$$

or

$$y(T) = \frac{1}{T}\left(T_i y(T_i)\frac{T_{i+1}-T}{T_{i+1}-T_i} + T_{i+1}y(T_{i+1})\frac{T-T_i}{T_{i+1}-T_i}\right).$$

The yield curve will remain continuous.

Figure 6.2 below shows the results of applying (6.10) to the swap data in Table 6.1. Notice the non-linear behavior of yields between maturity dates and the staircase shape of the forward curve.

Fig. 6.2. Yield and Forward Curve

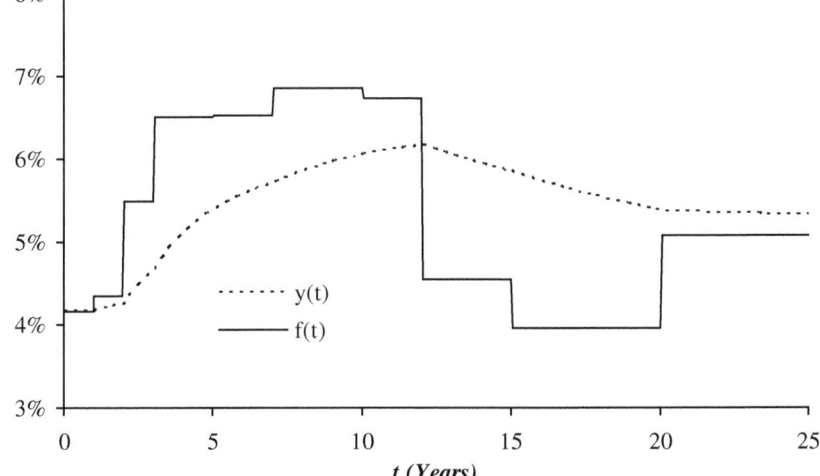

Notes: Yield curve is constructed by bootstrapping, assuming piecewise flat forward rates. Swap data is in Table 6.1.

6.2.2 C^1 Yield Curves: Hermite Splines

As we have seen, simply bootstrapped curves generally result in a discontinuous forward curve. From an empirical/economic perspective, such discontinuities are often unrealistic, and may also result in distortions of derivative prices[7] and technical difficulties in dynamic yield curve models. In this section, we consider a simple scheme to extend the bootstrapping technique to produce a once-differentiable yield curve and a continuous forward curve. Our scheme relies on *Hermite cubic splines*, where we write

$$y(T) = a_{3,i}(T-T_i)^3 + a_{2,i}(T-T_i)^2 + a_{1,i}(T-T_i) + a_{0,i}, \quad T \in [T_i, T_{i+1}], \tag{6.11}$$

for a series of constants $a_{3,i}$, $a_{2,i}$, $a_{1,i}$, $a_{0,i}$ to be determined from given values of $y(T_i)$, $y(T_{i+1})$, $y'(T_i)$, and $y'(T_{i+1})$. Appendix 6.A.1 contains a review of Hermite spline theory.

A particularly popular choice among Hermite splines is the *Catmull-Rom spline*, where derivatives $y'(T_i)$, $i = 1, \ldots, N$, are constructed by finite differences, relieving the user from directly specifying them. As shown in Appendix 6.A.1, for the Catmull-Rom spline we can organize (6.11) in a vector/matrix form as

$$y(T) = \mathbf{D}_i(T)^\top \mathbf{A}_i \begin{pmatrix} y_{i-1} \\ y_i \\ y_{i+1} \\ y_{i+2} \end{pmatrix}, \quad T \in [T_i, T_{i+1}], \quad i = 1, \ldots, N-1, \tag{6.12}$$

where, adapting as necessary the notation from the previous section,

$$\mathbf{D}_i(T) = \begin{pmatrix} d_i^3 \\ d_i^2 \\ d_i \\ 1 \end{pmatrix}, \quad d_i = \frac{T - T_i}{h_i}, \quad y_i = y(T_i), \quad h_i = T_{i+1} - T_i,$$

and the matrix $\mathbf{A}_i$ is as given in (6.54)–(6.56) in Appendix 6.A.1. While nominally (6.12) involves the values y_{N+1} and y_0, the matrices $\mathbf{A}_{N-1}$ and $\mathbf{A}_1$ are such that these values are irrelevant.

The Catmull-Rom spline prescription (6.12) completely specifies the yield curve on the interval $[T_1, T_N]$, given the N constants $y_1, \ldots, y_N$. To extend the yield curve to cover the interval $[0, T_1)$, we need to supply additional extrapolation assumptions. As in bootstrapping, possible choices for this additional equation is $y_0 = y(0) = y_1$, or perhaps the slope condition

$$\frac{y_1 - y_0}{h_0} = \frac{y_2 - y_1}{h_1}. \tag{6.13}$$

---

[7] For instance, as deal maturity crosses a point of discontinuity on the forward curve, the price of an FRA or a caplet on a short-tenor rate will jump.

6.2 Yield Curve Fitting with N-Knot Splines

Away from the boundaries, we notice that the price of security i depends only on $y_1, \ldots, y_{i+1}$, as the pricing equations take the diagonal form

$$V_1 = F_1(y_1, y_2, y_3),$$
$$V_2 = F_2(y_1, y_2, y_3),$$
$$V_3 = F_3(y_1, y_2, y_3, y_4)$$
$$\vdots$$
$$V_{N-1} = F_{N-1}(y_1, \ldots, y_N),$$
$$V_N = F_N(y_1, \ldots, y_N),$$

for non-linear functions F_i. Here F_i is typically only mildly sensitive to y_{i+1}, so the system of equations is nearly, but not quite, in bootstrap form. This makes solving for the y_i's an easy fare for a standard non-linear root-search algorithm (see Press et al. [1992] for several algorithms). We can also consider an iteration on a series of bootstrap procedures. To describe this idea, let $y_i^{(k)}$ be the value for y_i found in the k-th iteration, and consider then the following algorithm:

1. Let $y_j^{(k)}$, $j = 1, \ldots, i-1$, and $y_{i+1}^{(k-1)}$ all be known.
2. Make a guess for $y_i^{(k)}$.
3. Compute $V_i = F_i(y_1^{(k)}, \ldots, y_i^{(k)}, y_{i+1}^{(k-1)})$.
4. If V_i equals the market value stop. Otherwise return to Step 2.
5. If $i < N$, set $i = i+1$ and repeat.

We emphasize that the iteration over Steps 2–4 is still only one-dimensional, as in the bootstrapping algorithm of Section 6.2.1. Upon completion, the algorithm above yields $y_1^{(k)}, \ldots, y_N^{(k)}$. Iterating over k, we repeat the algorithm until the differences between the yields found at the k-th and $(k+1)$-th iteration are sufficiently small, say when

$$N^{-1} \sum_{i=1}^{N} \left(y_i^{(k+1)} - y_i^{(k)} \right)^2 < \varepsilon^2,$$

where ε is a given tolerance. To initialize the iteration over k, we need a starting guess $y_1^{(0)}, \ldots, y_N^{(0)}$; a good choice is the yield curve constructed by regular bootstrapping.

In Figure 6.3, we show the results of applying the algorithm above (using the boundary choice (6.13)) to the numerical example of Sections 6.2.1.1 and 6.2.1.2. We see that, as desired, the yield curve is smooth and the instantaneous forward curve is continuous. As the yield curve by construction is only once differentiable, equation (6.7) shows that the forward curve is not differentiable; this is obvious from the figure.

We can easily extend the procedure above beyond Catmull-Rom splines to more complicated C^1 cubic splines in the Hermite class, using results

240   6 Yield Curve Construction and Risk Management

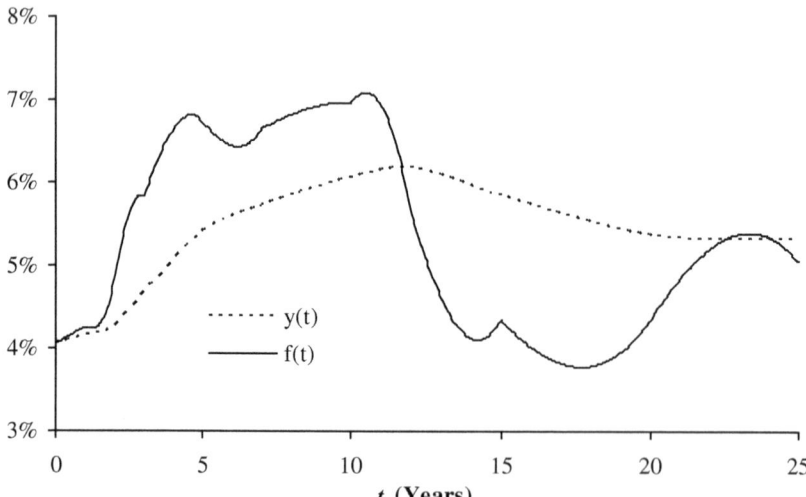

Fig. 6.3. Yield and Forward Curve

Notes: Yield curve is assumed to be a Catmull-Rom cubic spline. Swap data is in Table 6.1.

from Appendix 6.A. For instance, it is relatively straightforward to add *tension* to the Catmull-Rom spline. We cover twice-differentiable tension splines later in this chapter.

6.2.3 C^2 Yield Curves: Twice Differentiable Cubic Splines

While the spline method introduced in the previous section often produces acceptable yield curves, the method is heuristic in nature and ultimately does not produce a smooth forward curve. To improve on the latter, one alternative is to remain in the realm of cubic splines, but now insist that the curve is twice differentiable everywhere on $[T_1, T_N]$. We then write (see Appendix 6.A.2)

$$y(T) = \frac{(T_{i+1} - T)^3}{6h_i} y_i'' + \frac{(T - T_i)^3}{6h_i} y_{i+1}'' + (T_{i+1} - T)\left(\frac{y_i}{h_i} - \frac{h_i}{6} y_i''\right)$$
$$+ (T - T_i)\left(\frac{y_{i+1}}{h_i} - \frac{h_i}{6} y_{i+1}''\right), \quad T \in [T_i, T_{i+1}], \quad (6.14)$$

where $y_i'' = d^2 y(T_i)/dT^2$, $y_i = y(T_i)$, and $h_i = T_{i+1} - T_i$. The appendix demonstrates that continuity of the second derivative across the $\{T_i\}$ knots requires that the y_i'' and y_i are connected through a tri-diagonal linear system of equations, see equation (6.62). To state the expressions explicitly in matrix format, let $\mathbf{y}'' = (y_2'', y_3'', \ldots, y_{N-2}'', y_{N-1}'')^\top$ and

$\mathbf{y} = (y_2, y_3, \ldots, y_{N-2}, y_{N-1})^\top$ such that

$$\mathbf{B}\mathbf{y}'' = \mathbf{C}\mathbf{y} + \mathbf{M}\left(y_1, y_N, y_1'', y_N''\right), \qquad (6.15)$$

where the matrices $\mathbf{B}$ and $\mathbf{C}$ are both $(N-2) \times (N-2)$ tri-diagonal, with elements given by

$$B_{i,i} = \frac{h_i + h_{i+1}}{3}, \quad B_{i,i+1} = \frac{h_{i+1}}{6}, \quad B_{i,i-1} = \frac{h_i}{6},$$

and

$$C_{i,i} = -\left(\frac{1}{h_i} + \frac{1}{h_{i+1}}\right), \quad C_{i,i+1} = \frac{1}{h_{i+1}}, \quad C_{i,i-1} = \frac{1}{h_i}.$$

The $(N-2)$-dimensional vector $\mathbf{M}(y_1, y_N, y_1'', y_N'')$ captures boundary terms at T_1 and T_N. The most important — and, as discussed later, in a sense *best* — boundary specification is that of the *natural spline*, where we set $y_1'' = y_N'' = 0$. In this case, we have

$$\mathbf{M}(y_1, y_N, y_1'', y_N'') = \mathbf{M}(y_1, y_N) = \left(\frac{y_1}{h_1}, 0, 0, \ldots, 0, 0, \frac{y_N}{h_{N-1}}\right)^\top.$$

Notice that application of a natural boundary condition at time T_1 allows us to recover yields inside the time period $[0, T_1]$ by linear interpolation, using the gradient $y'(T_1)$ at time T_1 (which can easily be found by differentiating (6.14)).

We notice that (6.14) combined with (6.15) allows us to turn any guess of $y_1, y_2, \ldots, y_N$ into a guess for the vector $\mathbf{P}$ in (6.4). Specifically, we perform the following steps:

1. Compute the right-hand side of (6.15).
2. Use a standard tri-diagonal LU solver (see Press et al. [1992]) to invert (6.15) and recover $\mathbf{y}''$.
3. Apply (6.14) to determine[8] all values of $y(t_j)$, $j = 1, \ldots, M$, extrapolating as necessary when $t_j < T_1$.
4. Use (6.5) to establish $\mathbf{P}$.

The computational effort of Steps 1 through 4 are $O(N)$, $O(N-2)$, $O(M)$, and $O(M)$, respectively.

To solve for the correct values of $y_1, y_2, \ldots, y_N$, we iterate on Steps 1–4 using a non-linear root-search algorithm, terminating when (6.4) is satisfied to within acceptable tolerances. The fitting problem is typically good-natured, and virtually all standard root-search packages (see Press et al. [1992]) can tackle it successfully. Tanggaard [1997], for instance, uses

---

[8] For computational reasons, the terms multiplying the various y and y'' in (6.14) should be pre-cached, to avoid wasting effort when we ultimately perform an iteration.

242   6 Yield Curve Construction and Risk Management

a simple Gauss-Newton scheme with good results. Whatever root-search algorithm is selected, a good first guess can always be found by simple bootstrapping.

In Figure 6.4, we show the results of applying the algorithm above to a natural cubic spline representation of the yield curve example used in earlier sections. The yield curve is smooth and, unlike the Hermite spline case in Figure 6.3, the instantaneous forward curve is now differentiable, as desired.

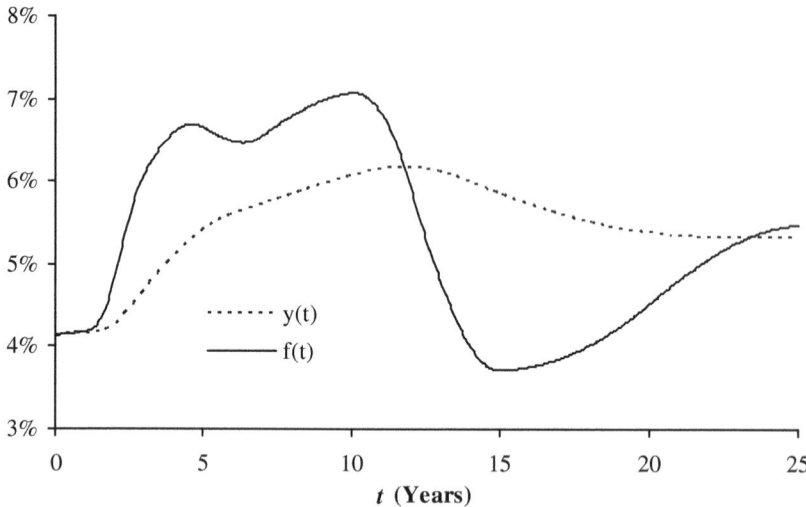

Fig. 6.4. Yield and Forward Curve

Notes: Yield curve is assumed to be a C^2 natural cubic spline. Swap data is in Table 6.1.

While the C^2 cubic spline discussed here has attractive smoothness, it is not necessarily an ideal representation of the yield curve. As discussed in Andersen [2005] and Hagan and West [2004], among others, twice differentiable cubic spline yield curves are often subject to oscillatory behavior, spurious inflection points, poor extrapolatory behavior, and non-local behavior when prices in the benchmark set are perturbed. We shall return to the concept of non-local perturbation effects in Section 6.4 below, but for now just note that perturbation of a single benchmark price can cause a slow-decaying "ringing" effect on the C^2 cubic yield curve, with the effect of the perturbation of the benchmark instrument price spilling into the entire yield curve. This behavior is not surprising, given that the spline is constructed through a full $(N-2) \times (N-2)$ matrix system, where interpolation behavior on the interval $[T_i, T_{i+1}]$ depends on *all* values y_j, $j = 1, \ldots, N$. In contrast, the simple linear-yield bootstrapping method in Section 6.2.1 interpolation on the interval $[T_i, T_{i+1}]$ involves only the two points y_i and y_{i+1}, and the Hermite spline approach involves only the four points $y_{i-1}, y_i, y_{i+1}, y_{i+2}$.

6.2.4 C^2 Yield Curves: Twice Differentiable Tension Splines

C^1 Hermite cubic splines are less prone to non-local behavior than C^2 cubic splines, but accomplish this in a somewhat ad-hoc fashion by giving up one degree of differentiability. Rather than taking such a draconian step, one wonders whether there may be a way to retain the C^2 feature of the cubic spline in Section 6.2.3, yet still allow control of curve locality and "stiffness". As it turns out, an attractive remedy to the shortcomings of the pure C^2 cubic spline is to insert some tension in the spline, that is, to apply a tensile force to the end-points of the spline. Appendix 6.A.3 lists the necessary details of this approach, using the classical *exponential tension spline* construction[9] in Schweikert [1966]. When applied to the yield-curve setting, the construction involves a modification of the cubic equation (6.14) for $y(T)$, $T \in [T_i, T_{i+1}]$, to

$$y(T) = \left(\frac{\sinh(\sigma(T_{i+1} - T))}{\sinh(\sigma h_i)} - \frac{T_{i+1} - T}{h_i} \right) \frac{y_i''}{\sigma^2}$$
$$+ \left(\frac{\sinh(\sigma(T - T_i))}{\sinh(\sigma h_i)} - \frac{T - T_i}{h_i} \right) \frac{y_{i+1}''}{\sigma^2}$$
$$+ y_i \frac{T_{i+1} - T}{h_i} + y_{i+1} \frac{T - T_i}{h_i}, \quad (6.16)$$

where $\sigma \geq 0$ is the *tension factor*, and where we recall the definition $h_i = T_{i+1} - T_i$.

Appendix 6.A.3 discusses a number of properties of tension splines, the most important perhaps being the fact that setting $\sigma = 0$ will recover the ordinary C^2 cubic spline, whereas letting $\sigma \to \infty$ will make the tension spline uniformly approach a linear spline (i.e. the spline we used in Section 6.2.1.1). Loosely, we can thus think of a tension spline as a twice differentiable hybrid between a cubic spline and a linear spline. Equally loosely: as we increase σ, spurious inflections and ringing in the cubic spline are gradually "stretched" out of the curve, accompanied by rising (absolute values of) second derivatives at the knot points. More details on tension splines can be found in Andersen [2005], who also discusses application of computationally efficient local spline bases and the usage of a T-dependent tension factor to gain further control of the curve.

We observe that (6.16) is structurally similar to (6.14), and allows for a matrix representation of the same form as (6.62), albeit with suitably

---

[9]The exponential tension spline is not the only class of twice differentiable tension splines, but is probably the most common. Other classes are discussed in Kvasov [2000] and Andersen [2005].

modified definitions of the vector **M** and the matrices **B** and **C**; we leave these modifications as an exercise to the reader. Suffice to say that once a value of σ has been decided upon, the numerical search for the unknown levels y_i, $i = 1, \ldots, N$, can proceed along the same principles as in Section 6.2.3 above. Figure 6.5 below shows an example; notice how increasing the tension parameter gradually moves us from cubic spline behavior to bootstrap behavior.

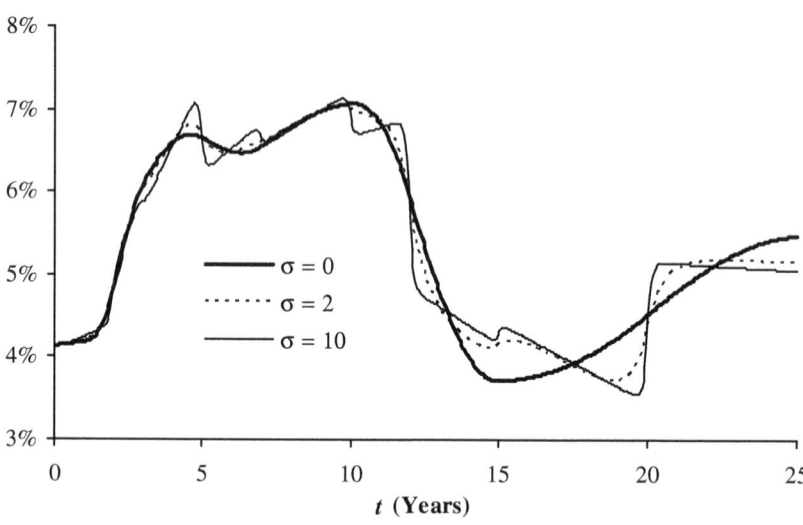

Fig. 6.5. Forward Curve

Notes: The yield curve is constructed as a C^2 natural tension spline, with tension parameters as given in the graph (only the forward curve $f(t)$ is shown). Swap data is in Table 6.1.

Remark 6.2.1. If the tension spline is applied not to yields, but to the logarithm of discount factors $\ln P(t)$, the limit of $\sigma \to \infty$ will produce a piecewise flat forward curve, as in Figure 6.2.

The reader may at this point wonder whether there are any firm rules as to what σ should be. We have no definitive answers to this question, and we do not try to determine σ automatically (although such routines do exist, see Renka [1987]). Instead, we normally treat σ as an "extra knob" that allow users to balance curve smoothness, shape preservation, and perturbation locality to their particular tastes. Inevitably some element of experimentation is required here.

6.3 Non-Parametric Optimal Yield Curve Fitting

The techniques we have outlined so far generally suffice for constructing a discount curve from a "clean" set of non-duplicate benchmark securities, including the carefully selected set of liquid staggered-maturity deposits, futures, and swaps most banks assemble for the purpose of constructing a Libor yield curve. In some settings, however, the benchmark set may be significantly less well-structured, involving illiquid securities with little order in their cash flow timing and considerable noise in their prices. This situation may, say, arise when one attempts to construct a yield curve from corporate bonds. While construction of a Libor curve is the most important task for the purposes of this book, we nevertheless wish to say a few words about techniques suitable for less cooperative benchmark security sets. These techniques can also be applied to Libor curve construction, of course, and are particularly relevant for applications where we are willing to sacrifice some precision in the fit to benchmark prices in return for a smoother yield curve.

6.3.1 Norm Specification and Optimization

When the input benchmark set is noisy, a direct solution of (6.4) may be erratic or may not exist. To overcome this, and to reflect that noise in the input data may make us content to solve (6.4) only to within certain error bounds, we now proceed to replace this equation with a problem of minimization of a penalized least-squares norm. Specifically, define the space $\mathcal{A} = C^2([t_1, t_M])$ of all functions $[t_1, t_M] \to \mathbb{R}$ that are twice differentiable with continuous second derivative, and introduce the M-dimensional discount vector

$$\mathbf{P}(y) = \left(e^{-y(t_1)t_1}, \ldots, e^{-y(t_M)t_M}\right)^\top.$$

Also, let $\mathbf{W}$ be a diagonal $N \times N$ weighting matrix. Then, as our best estimate $\widehat{y}$ of the yield curve we will here use

$$\widehat{y} = \operatorname*{argmin}_{y \in \mathcal{A}} \mathcal{I}(y), \tag{6.17}$$

with

$$\mathcal{I}(y) \triangleq \frac{1}{N}\left(\mathbf{V} - \mathbf{cP}(y)\right)^\top \mathbf{W}^2 \left(\mathbf{V} - \mathbf{cP}(y)\right)$$
$$+ \lambda \left(\int_{t_1}^{t_M} \left[y''(t)^2 + \sigma^2 y'(t)^2\right] dt\right), \tag{6.18}$$

where λ and σ^2 are positive constants. The norm $\mathcal{I}(y)$ consists of three separate terms:

6 Yield Curve Construction and Risk Management

- A least-squares penalty term

$$\frac{1}{N} \left(\mathbf{V} - \mathbf{cP}(y)\right)^\top \mathbf{W}^2 \left(\mathbf{V} - \mathbf{cP}(y)\right)$$

$$= \frac{1}{N} \sum_{i=1}^{N} W_i^2 \left(V_i - \sum_{j=1}^{M} c_{i,j} e^{-y(t_j) t_j}\right)^2,$$

where W_i is the i-th diagonal element of $\mathbf{W}$. This term is an outright precision-of-fit norm and measures the degree to which the constructed discount curve can replicate input security prices. The weights W_i can be used to assign different importance to the various securities in the benchmark set, and/or to translate the precision of the fit from raw dollar amounts into more intuitive quantities, such as security-specific quoted yields[10]. Clearly, if (6.4) can be satisfied, then the least-squares penalty term will attain its minimum (of zero) for all yield curves that satisfy (6.4).
- A weighted smoothness term $\lambda \int_{t_1}^{t_M} y''(t)^2 \, dt$, penalizing high second-order derivatives of y to avoid kinks and discontinuities.
- A weighted curve-length term $\lambda \sigma^2 \int_{t_1}^{t_M} y'(t)^2 \, dt$, penalizing oscillations and excess convexity/concavity.

Our choice of calibration norm is, we believe, an attractive one, but other choices obviously are available as well. For instance, in Adams and van Deventer [1994] the norm contains no curve-length term and the smoothing norm is expressed on the forward curve, rather than on the yield curve. Due to the lack of the curve-length penalty term, the resulting curve will tend to behave like the C^2 cubic spline in Section 6.2.3; see Hagan and West [2004] for some numerical tests.

The following result is shown by variational methods in Andersen [2005]:

Proposition 6.3.1. *The curve $\widehat{y}$ that satisfies (6.17) is a natural exponential tension spline with tension factor σ and knots at all cash flow dates $t_1, t_2, \ldots, t_M$.*

Proposition 6.3.1 establishes that the curve we are looking for is a tension spline with tension factor σ, but does not in itself allow us to identify the optimal spline directly, beyond the fact that i) it is a natural spline with boundary conditions $y''(t_1) = y''(t_M) = 0$; and ii) it has knots at all t_j, $j = 1, \ldots, M$. Identification of the correct spline involves solving for unknown

---

[10] Most fixed-income securities are quoted through some type of yield, e.g. $V_i = g_i(r_i)$ where r_i is the quoted yield and g_i is a function that encapsulates the quoting convention. The quantity $D_i = -(dg_i/dr_i)/g_i$ is known as the *duration* of V_i. Setting $W_i = 1/D_i$ in the least-squares norm will turn price deviations into yield deviations.

levels[11] $y(t_1), y(t_2), \ldots, y(t_M)$ to optimize directly (6.18). In this exercise, the following lemma is useful.

Lemma 6.3.2. *For a natural tension spline interpolating the values* $y(t_1)$, $y(t_2)$, ..., $y(t_M)$, *we have*

$$y'(t_j) = \left(-\sigma \frac{\cosh\left(\sigma\left(t_{j+1} - t_j\right)\right)}{\sinh\left(\sigma\left(t_{j+1} - t_j\right)\right)} + \frac{1}{t_{j+1} - t_j}\right) \frac{y''(t_j)}{\sigma^2}$$
$$+ \left(\frac{\sigma}{\sinh\left(\sigma\left(t_{j+1} - t_j\right)\right)} - \frac{1}{t_{j+1} - t_j}\right) \frac{y''(t_{j+1})}{\sigma^2} + \frac{y(t_{j+1})}{t_{j+1} - t_j} - \frac{y(t_j)}{t_{j+1} - t_j},$$

and

$$\lambda \left(\int_{t_1}^{t_M} [y''(t)^2 + \sigma^2 y'(t)^2]\, dt\right) = -\lambda \sum_{j=1}^{M-1} d_j \left(y(t_{j+1}) - y(t_j)\right), \quad (6.19)$$

where $y''(t_1) = y''(t_M) = 0$, *and*

$$d_j \triangleq \frac{y''(t_{j+1}) - \sigma^2 y(t_{j+1})}{t_{j+1} - t_j} - \frac{y''(t_j) - \sigma^2 y(t_j)}{t_{j+1} - t_j}. \quad (6.20)$$

Proof. The result for $y'(t_j)$ follows from direct differentiation of the basic equations for a tension spline (see (6.16) above, applied to the knot grid $\{t_j\}$). To show (6.19), consider the interval $[t_j, t_{j+1}]$ and the integral

$$\int_{t_j}^{t_{j+1}} \left(y''(t)^2 + \sigma^2 y'(t)^2\right) dt = \int_{t_j}^{t_{j+1}} \left(y''(t) \cdot y''(t) + \sigma^2 y'(t) \cdot y'(t)\right) dt.$$

Integration by parts yields

$$\int_{t_j}^{t_{j+1}} \left(y''(t)^2 + \sigma^2 y'(t)^2\right) dt$$
$$= [y''(t)y'(t)]_{t_j}^{t_{j+1}} - \int_{t_j}^{t_{j+1}} \left(y^{(3)}(t) - \sigma^2 y'(t)\right) y'(t)\, dt$$
$$= y''(t_{j+1})y'(t_{j+1}) - y''(t_j)y'(t_j) - d_j \left(y(t_{j+1}) - y(t_j)\right), \quad (6.21)$$

where d_j is given in (6.20). Here, we have used that, by definition, hyperbolic tension splines have $y^{(3)}(t) - \sigma^2 y'(t)$ piecewise constant and equal to d_j on each interval $[t_j, t_{j+1}]$ (see equation (6.63) in Appendix 6.A). The result (6.20) follows by addition of the terms (6.21) and using the condition $y''(t_1) = y''(t_M) = 0$. □

---

[11] In Andersen [2005], the search for yield levels has been replaced by the more contemporary idea of searching for weights in a local basis representation of the spline.

Lemma 6.3.2 shows us that we can compute the value of the integral penalty term in (6.18) directly from knowledge of yield levels $y(t_1), \ldots, y(t_M)$ and second derivatives $y''(t_2), \ldots, y''(t_{M-1})$. For each guess for the M unknown levels $y(t_1), y(t_2), \ldots, y(t_M)$ we can proceed as follows.

1. Compute the least-squares penalty term $\frac{1}{N}(\mathbf{V} - \mathbf{cP}(y))^\top \mathbf{W}^2 (\mathbf{V} - \mathbf{cP}(y))$ directly from the definition of $\mathbf{P}(y)$.
2. Use the results in Section 6.2.4 to solve for $y''(t_2), \ldots, y''(t_{M-1})$ by solving a tri-diagonal set of equations.
3. Use Lemma 6.3.2 to compute $\lambda(\int_{t_1}^{t_M} [y''(t)^2 + \sigma^2 y'(t)^2] \, dt)$, thereby completing the computation of the norm $\mathcal{I}(y)$.

Embedding Steps 1–3 above in a multi-variate numerical optimizer ultimately allows us to determine the optimal solution $\widehat{y}$. A good generic routine for this optimization step would be the Levenberg-Marquardt algorithm; see Press et al. [1992]. The optimization problem at hand is generally good-natured, and one can also use a simpler Gauss-Newton method, as discussed in Andersen [2005]. If possible, it is often useful to use a simpler method (e.g. bootstrapping) to establish a good guess for the yield curve levels $y(t_1), y(t_2), \ldots, y(t_M)$. A proper implementation of the algorithm should typically construct a yield curve in less than one-tenth of a second on a standard PC.

Remark 6.3.3. If we let $\sigma = 0$, the solution to the optimization problem becomes a *cubic smoothing spline*; see Tanggaard [1997] for more details on this case.

Remark 6.3.4. If we let $\lambda \downarrow 0$, the resulting spline will often end up hitting all benchmark prices exactly, i.e. will satisfy (6.4) in the limit. The resulting spline is then the optimal *interpolating* curve, in the sense that of all twice differentiable yield curves that match the benchmark prices, the spline is the minimizer of the regularity term $\int_{t_1}^{t_M} [y''(t)^2 + \sigma^2 y'(t)^2] \, dt$. If, for $\lambda \downarrow 0$, we do not satisfy (6.4), then the resulting spline can be considered a *least-squares regression spline* solution.

6.3.2 Choice of λ

So far, we have assumed that the parameter λ has been specified exogenously by the user. In practice, however, a good magnitude of λ may sometimes be hard to ascertain by inspection, and a procedure to estimate λ directly from the data is often useful. One possibility is to use a cross-validation approach, either outright or through the more efficient Generalized Cross-Validation (GCV) criterion by Craven and Wahba [1979]. Some results along these lines can be found in Tanggaard [1997] and Andersen [2005], but are outside the scope of our treatment here. A more pragmatic approach is to replace the optimization problem (6.17) with the constrained optimization problem

6.3 Non-Parametric Optimal Yield Curve Fitting

$$\widehat{y} = \operatorname*{argmin}_{y \in \mathcal{A}} \int_{t_1}^{t_M} [y''(t)^2 + \sigma^2 y'(t)^2]\, dt, \qquad (6.22)$$

$$\text{subject to } \frac{1}{N} (\mathbf{V} - \mathbf{cP}(y))^\top \mathbf{W}^2 (\mathbf{V} - \mathbf{cP}(y)) = \gamma^2, \qquad (6.23)$$

where γ is an exogenously specified constant. Note that γ is just the allowed weighted root-mean-square (RMS) error in the fit to benchmark securities, an intuitive quantity that most users should have no problem specifying directly based on, say, observed bid-offer spreads. The Lagrangian for the above problem becomes

$$\widehat{y} = \operatorname*{argmin}_{y \in \mathcal{A}} \left(\int_{t_1}^{t_M} [y''(t)^2 + \sigma^2 y'(t)^2]\, dt \right.$$
$$\left. + \rho \left[\frac{1}{N} (\mathbf{V} - \mathbf{cP}(y))^\top \mathbf{W}^2 (\mathbf{V} - \mathbf{cP}(y)) - \gamma^2 \right] \right), \quad (6.24)$$

where the Lagrange multiplier ρ must be determined such that the constraint (6.23) is satisfied at the optimum of (6.24). Apart from a constant scale, (6.24) is identical to (6.17), so we solve the constrained optimization problem (6.22)–(6.23) through the following iteration over λ:

1. Given a guess for λ, find the optimum value of $y(t_1), y(t_2), \ldots, y(t_M)$, as a solution of (6.17).
2. Compute $\mathcal{S} = \frac{1}{N}(\mathbf{V} - \mathbf{cP}(y))^\top \mathbf{W}^2(\mathbf{V} - \mathbf{cP}(y))$.
3. If $\mathcal{S} = \gamma^2$, stop; otherwise update λ and go to Step 1.

In Step 1, we can proceed as discussed in Section 6.3.1 above. In general, the precision norm $\mathcal{S} = \mathcal{S}(\lambda)$ will be a declining function in λ and, provided that a root to $\mathcal{S}(\lambda) = \gamma^2$ exists[12], the updating in Step 3 can be done by any standard root search algorithm.

6.3.3 Example

To illustrate the effect of λ, we now apply the algorithm in Section 6.3.2 to the test data in Table 6.1 above. In doing so, we use the matrix $\mathbf{W}$ to normalize (see footnote 10) all price errors to yield-to-maturity errors, allowing us to consider γ in (6.23) as the root-mean-square (RMS) yield error. Setting $\sigma = 0$, the forward curves for various choices of γ are shown in Figure 6.6. As one would expect, the higher we allow γ to be, the smoother the forward (and yield) curves become.

For our test case, the zero-RMS optimal (M-knot) forward curve in Figure 6.6 is virtually identical to the N-knot cubic spline solution in Figure 6.4. In general, the N-knot interpolating curve can be interpreted as a

---

[12] There may be instances where $\mathcal{S}(0) > \gamma^2$. If the desired precision is unattainable, we can either increase γ^2 or perhaps prune the benchmark security set.

Fig. 6.6. Forward Curve

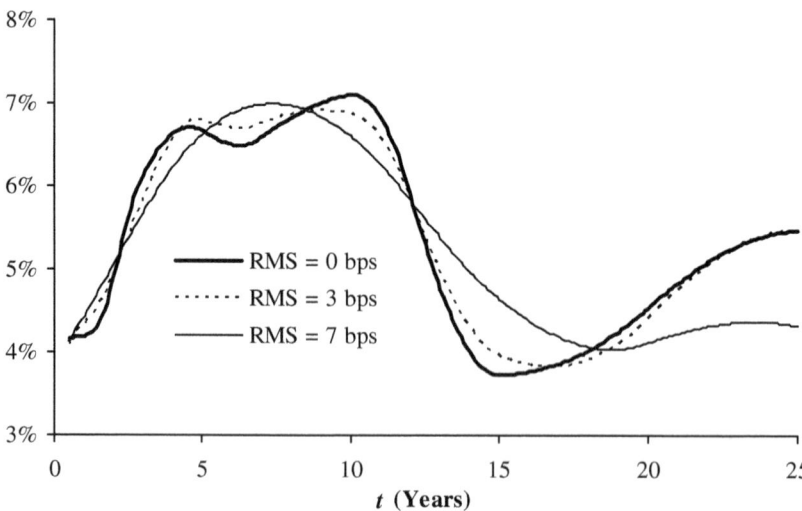

Notes: The yield curve is constructed as an optimal C^2 natural tension spline, with an RMS yield error constraint as listed in the graph (only the forward curve $f(t)$ is shown). The tension parameter is set to $\sigma = 0$ for all curves. Swap data is in Table 6.1.

constrained solution to (6.17) with $\lambda = 0$, with the constraint requiring that knots be placed only at benchmark maturities $\{T_i\}_{i=1}^N$, rather than at all cash flow dates $\{t_j\}_{j=1}^M$. The effect of enforcing this additional constraint is often rather small, at least for the purposes of constructing a Libor curve.

6.4 Managing Yield Curve Risk

Consider a portfolio of securities with value V_0, where V_0 is a function of the yield curve $y(t)$. The securities in V_0 would typically not be in the benchmark set and could contain, say, interest rate options, seasoned swaps, and so forth. As the yield curve is a function of the benchmark set values $\mathbf{V} = (V_1, \ldots, V_N)^\top$, we may write

$$V_0 = V_0\left(V_1, \ldots, V_N; \theta\right),$$

where the vector θ contains model parameters (e.g. volatilities) and where the function $V_0(\cdot)$ is determined both from the valuation model of the security in question, and from the curve construction algorithm employed. Clearly, then

$$dV_0 = \sum_{i=1}^N \frac{\partial V_0}{\partial V_i} dV_i + \sum_i \frac{\partial V_0}{\partial \theta_i} d\theta_i,$$

or, for non-infinitesimal moves,

$$\Delta V_0 \approx \sum_{i=1}^{N} \frac{\partial V_0}{\partial V_i} \Delta V_i + \sum_i \frac{\partial V_0}{\partial \theta_i} \Delta \theta_i. \tag{6.25}$$

For the purpose of managing first-order risk exposure to moves in the yield curve, (6.25) suggests that the collection of derivatives $\partial V_0/\partial V_i$, $i = 1, \ldots, N$ — often called *(bucketed) interest rate deltas* — forms a natural metric for portfolio risk. In particular, if all these derivatives are zero, our portfolio would, to first order, be immunized against any move in the yield curve that is consistent with the chosen curve construction algorithm. On the other hand, if some or all of the derivatives are non-zero, we could manage our risk by setting up a hedge portfolio of benchmark securities, with notional $-\partial V_0/\partial V_i$ on the i-th security. We emphasize that the resulting hedge would typically *not* be model-consistent: most interest models assume that yield curve risk originates from only a few stochastic yield curve factors that tend to move the curve smoothly[13], in a predominantly parallel fashion. Theoretically, a bucket-by-bucket immunization against all terms ΔV_i may then be considered an overkill — we typically hedge against far too many risk factors (N) — but is nevertheless standard industry practice and has proven to be robust. Notice that bucket hedging along these lines would, for instance, correctly reject the notion that we could perfectly hedge a 20 year swap with a 1 month FRA, something that a one-factor interest rate model (see Chapter 4 and Chapter 10) would happily accept. We pick up this subject again in Chapter 22.

6.4.1 Par-Point Approach

The simplest approach to computation of the delta $\partial V_0/\partial V_i$ involves a manual bump[14] to V_i, followed by a reconstruction of the yield curve, and a subsequent repricing of the portfolio V_0. This procedure is sometimes known as the *par-point approach*, and resulting derivatives *par-point deltas*. For the approach to work properly, it is important that the yield curve construction algorithm is fast and produces clean, local perturbations of the yield curve when benchmark prices are shifted. For instance, perturbing a short-dated FRA price should not cause noticeable movements in long-term yields, lest we reach the erroneous conclusion (again) that we can perfectly hedge a 20 year swap with a 1 month FRA. As we have discussed earlier, Hermite splines and bootstrapped yield curves both exhibit good perturbation locality, but cubic C^2 splines often do not. To illustrate this, Figure 6.7 considers the

---

[13] See the principal components analysis in Chapter 14 for more on this.

[14] In practice, rather than bumping the price V_i outright, one may instead bump the yield of the i-th benchmark security (typically by 1 basis point). See also footnote 10.

252   6 Yield Curve Construction and Risk Management

effect on the forward curves in Figures 6.1, 6.3, and 6.4 from a 1 basis point up-move in the par yield of the 2 year swap in Table 6.1. As we can see, the move causes a noisy, ringing perturbation in the C^2 cubic spline solution, spreading into short- and long-dated parts of the forward curve.

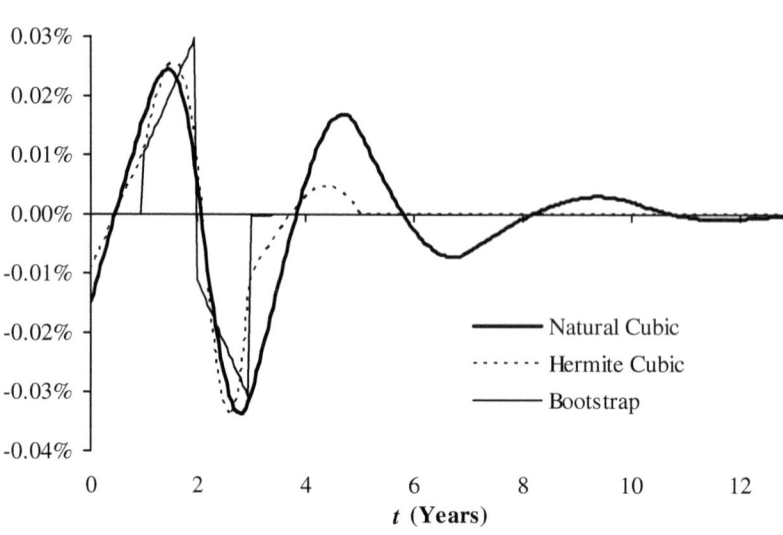

Fig. 6.7. Forward Curve Move

Notes: Change in instantaneous forward curve, from a 1 basis point shift in the 2 year swap yield in Table 6.1. The curve construction methods tested are: bootstrapping with piecewise linear yields ("Bootstrap"), Hermite C^1 cubic spline ("Hermite"), and C^2 natural cubic spline ("Natural Cubic"). Swap data is in Table 6.1.

In Figure 6.8, we have followed the recommendations of Section 6.2.4 and added tension to the C^2 spline, causing a dampening of the perturbation noise. Clearly, the usage of a tension factor can have a beneficial impact on risk reports produced by the par-point approach.

6.4.2 Forward Rate Approach

As an alternative to direct perturbation of benchmark security prices, we can consider applying perturbations directly to the discount curve, thereby mostly avoiding the introduction of artifacts specific to the curve construction algorithm. In practice, this technique typically focuses on the forward curve[15] $f(t)$, to which we apply certain functional shifts $\mu_k(t)$, $k = 1, \ldots, K$. Writing

---

[15] Perturbations may also be performed on discretely, rather than continuously, compounded forward rates.

Fig. 6.8. Forward Curve Move

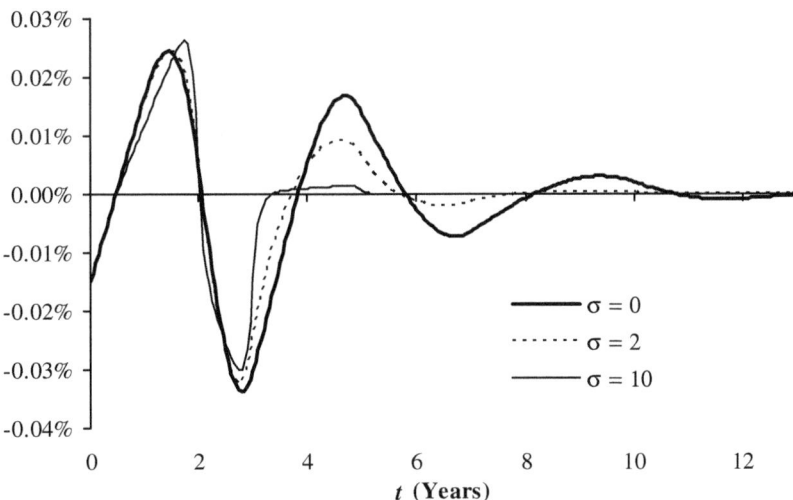

Notes: Change in instantaneous forward curve, from a 1 basis point shift in the 2 year swap yield in Table 6.1. The yield curve was constructed as a tension spline, with tension factors as given in the graph. Swap data is in Table 6.1.

(loosely) $V_0 = V_0(f)$ to highlight the dependence of V_0 on the forward curve, we then compute functional (Gâteaux) derivatives[16] for V_0:

$$\partial_k V_0 = \left. \frac{dV_0\left(f(t) + \varepsilon \mu_k(t)\right)}{d\varepsilon} \right|_{\varepsilon=0}, \quad k = 1, \ldots, K. \tag{6.26}$$

Standard choices for $\mu_k(t)$ are

$$\text{Piecewise Triangular:} \quad \mu_k(t) = \frac{t - t_{k-1}}{t_k - t_{k-1}} 1_{\{t \in [t_{k-1}, t_k)\}}$$

$$+ \frac{t_{k+1} - t}{t_{k+1} - t_k} 1_{\{t \in [t_k, t_{k+1})\}}, \tag{6.27}$$

$$\text{Piecewise Flat:} \quad \mu_k(t) = 1_{\{t \in [t_k, t_{k+1})\}}, \tag{6.28}$$

where $\{t_k\}$ is a user-specified discretization grid. The resulting sensitivities are often called *forward rate deltas*.

It is common practice to use $\{t_k\}$ grids spaced three months apart, with dates on Eurodollar futures maturities. The number of deltas K is thus typically a rather large number, and the K derivatives $\partial_k V_0$ give a detailed picture of where the portfolio risk is concentrated on the forward curve. As forward rate contracts and Eurodollar futures cease to be liquid beyond 4

---

[16] For a proper definition of the Gâteaux derivative, see Gâteaux [1913].

or 5 year maturities, the forward rate deltas do not directly suggest hedging instruments for the medium and long end of the yield curve exposure; however it is not difficult to translate forward rate deltas into a hedging portfolio (see the next section). The choice of par point versus forward rate deltas is largely a matter of personal preference, and it is not uncommon for traders to use both at the same time.

6.4.3 From Risks to Hedging: The Jacobian Approach

A collection of forward rate shifts $\mu_k(t)$, $k = 1, \ldots, K$, defines a certain view on the (first-order) risk on the portfolio $V_0(f)$ via the functional derivatives (6.26). In order to be useful, this risk view ultimately needs to be translated into a portfolio of hedging instruments that offsets the risks of V_0. While fixed income traders normally are quite adept at mentally translating forward rate risk into actual hedge transactions, some linear algebra can help out in this exercise, as we now show.

Suppose a set of L hedging instruments is available, with values $\mathbf{H} = (H_1, \ldots, H_L)^\top$. This set may or may not coincide with the benchmark set used for curve construction; for example, one may want to exclude some benchmark securities from the hedging set due to poor liquidity, or one may want to add instruments to the benchmark set to fine-tune hedging. Using (6.26), we denote the sensitivities of hedging instruments to the shifts $\mu_k(t)$ by $\partial_k H_l$, $l = 1, \ldots, L$, $k = 1, \ldots, K$. If the l-th hedging instrument is included in the hedging portfolio with notional weight p_l, and $\mathbf{p} = (p_1, \ldots, p_L)^\top$, then the sensitivity of the hedge portfolio value

$$H_0(\mathbf{p}) = \mathbf{p}^\top \mathbf{H}$$

to the k-th perturbation is given by

$$\partial_k H_0(\mathbf{p}) = \mathbf{p}^\top \partial_k \mathbf{H},$$

where we have denoted

$$\partial_k \mathbf{H} = (\partial_k H_1, \ldots, \partial_k H_L)^\top.$$

In most cases[17] we would like to choose the weights $\mathbf{p}$ in such a way that $\partial_k H_0(\mathbf{p})$ offsets as much of $\partial_k V_0$ as possible, for all $k = 1, \ldots, K$. Let W_k be the relative importance of offsetting the k-th derivative, and U_l a relative "reluctance" to using the l-th hedging instrument (a function of the bid-ask spread, for example). Then, the optimal hedging weights $\widehat{\mathbf{p}}$ can be defined by the condition

---

[17] Sometimes traders deliberately wish to keep some risk on their books, as a way to speculate on interest rate movements. A non-zero target risk profile is easily accommodated by a change of the optimization target in (6.29).

6.4 Managing Yield Curve Risk

$$\widehat{\mathbf{p}} = \underset{\mathbf{p}}{\operatorname{argmin}} \left(\sum_{k=1}^{K} W_k^2 \left(\partial_k H_0(\mathbf{p}) - \partial_k V_0 \right)^2 + \sum_{l=1}^{L} U_l^2 p_l^2 \right). \tag{6.29}$$

Define the matrix $\partial \mathbf{H}$ to have columns $\partial_1 \mathbf{H}, \ldots, \partial_K \mathbf{H}$, the vector $\partial \mathbf{V}_0$ by

$$\partial \mathbf{V}_0 = (\partial_1 V_0, \ldots, \partial_K V_0)^\top,$$

the matrix $\mathbf{W}$ to be diagonal with W_k's on the diagonal, and the matrix $\mathbf{U}$ to be diagonal with U_l's on the diagonal. With this notation (6.29) can be recast as a least-squares problem,

$$\left(\partial \mathbf{H}^\top \mathbf{p} - \partial \mathbf{V}_0 \right)^\top \mathbf{W}^2 \left(\partial \mathbf{H}^\top \mathbf{p} - \partial \mathbf{V}_0 \right) + \mathbf{p}^\top \mathbf{U}^2 \mathbf{p} \to \min. \tag{6.30}$$

The problem (6.30) can be solved by standard methods; a formal solution is given by the linear system

$$\left(\partial \mathbf{H} \, \mathbf{W}^2 \, \partial \mathbf{H}^\top + \mathbf{U}^2 \right) \widehat{\mathbf{p}} = \partial \mathbf{H} \, \mathbf{W}^2 \, \partial \mathbf{V}_0. \tag{6.31}$$

We note that the addition of the $\mathbf{U}$ term to the optimization problem (6.31) is sometimes called *Tikhonov regularization*, a technique that we shall return to in Chapter 18.

When solving (6.30), one should carefully consider the relative dimensions of the matrices involved. First, if there are fewer hedging instruments than shifts to be immunized ($L < K$), then, in general, not all risks can be offset. In this case, the weights $\mathbf{W}$ gain in importance as they allow the user to focus hedging on risk buckets deemed more important, at the expense of other, less critical ones. Also, when $L < K$ the weights $\mathbf{U}$ are less important, and in most cases can safely be set to zero. Second, if there are more hedging instruments than risk buckets to immunize against ($L > K$), then there are typically multiple hedging portfolios that perfectly offset all risks. In this case, the weights $\mathbf{W}$ can normally be ignored (all set to 1), but the weight matrix $\mathbf{U}$ becomes more critical as it allows one to choose which of the possible hedging portfolios one "likes" best (e.g., the least costly). Finally, if $L = K$, then normally there exists exactly one portfolio that hedges all risks. Both $\mathbf{W}$ and $\mathbf{U}$ are then often of little consequence, although one might still want to specify non-zero weights $\mathbf{U}$ to avoid oscillatory or unstable solutions if the linear equations are ill-posed. We note that in the simple case of $L = K$, $\mathbf{W} = \mathbf{1}$, $\mathbf{U} = \mathbf{0}$ and $\partial \mathbf{H}$ invertible, the solution to the optimization problem is given by

$$\mathbf{p} = \left(\partial \mathbf{H}^\top \right)^{-1} \partial \mathbf{V}_0. \tag{6.32}$$

The method of constructing a hedge portfolio from derivatives to arbitrary shocks of the forward curve via the optimization problem (6.30) is known as the *Jacobian method* for interest rate deltas; the name originates from the fact that the matrix $\partial \mathbf{H}$ is the Jacobian matrix of a hedge set with respect to the forward curve shocks. Combined with the forward rate deltas,

256   6 Yield Curve Construction and Risk Management

the Jacobian method helps aggregate fine-grained risks into various sets of hedges. The approach has considerable generality as the risk basis functions μ_k and the hedge portfolio can be chosen freely by the user — note, for instance, that even the par-point approach can be seen as a special type of the Jacobian method where we effectively choose the hedging set to coincide perfectly with the benchmark set and where the μ_k's are set to be the shifts of the forward curve that correspond to the bumps of benchmark securities. In this special case, the Jacobian $\partial \mathbf{H}$ is then a unit matrix and (from (6.32)) the original par-point deltas are recovered.

The Jacobian method serves to decouple risk calculations from curve construction. This, potentially, allows for combining smooth curves with localized risk, a feat that is difficult to achieve by other methods. The Jacobian is also useful in applications where curves need to be rebuild over and over, to address the fact that Libor and Treasury benchmark security prices (or yields) change very quickly, often quicker than a sophisticated curve construction algorithm can rebuild the curves. With the aid of a Jacobian, changes in benchmark prices can be quickly translated into changes of the forward curve via a matrix multiplication. A full curve rebuild needs only be triggered when the benchmark prices have moved sufficiently far from their initial values.

6.4.4 Cumulative Shifts and other Common Tricks

As evident in Figure 6.7 (the bootstrap case), a shift to a single swap rate (while keeping other swap rates fixed) typically results in a strong "see-saw" impact on the forward rate curve. Let us attempt to gain some intuition about the magnitude of the forward rate shock. For a back-of-the-envelope calculation, we can assume that a swap rate is a linear combination of forward Libor rates (see (4.11)),

$$S_n \approx \sum_{i=1}^{n} w_{i,n} L_i,$$

where S_n denotes a swap rate for a swap covering n periods (for simplicity assume that each period is 1 year), L_i denotes a forward Libor rate for i-th period (from year $i-1$ to year i), and $w_{i,n} \approx 1/n$. Inverting this relationship yields

$$L_n \approx n \left(S_n - \frac{n-1}{n} S_{n-1} \right). \qquad (6.33)$$

As part of a par-point report, assume now that S_n is shifted by the amount δ, but S_{n-1} and S_{n+1} remain unchanged. According to (6.33) L_n will then shift by approximately $n\delta$, and L_{n+1} by $-n\delta$. For instance, if a 30 year swap yield is shifted by 1 basis point, while 29 year and 31 year are kept unchanged, then evidently the forward Libor rate L_{30} will move by 30 basis points, and the rate L_{31} will move by -30 basis points. If the portfolio whose deltas we are computing happens to contain, say, a spread option on the difference between L_{30} and L_{31}, the underlying rate of this option would be shifted by 60 basis points (!). And clearly, a shift of 60 basis points (or 30 basis points, for that matter) is not small, and may be inappropriate for calculating a first-order derivative. We emphasize that what appears to be a benign 1 basis point rate shift translates into a much larger forward curve move that can potentially affect underlying instruments in unexpected ways.

The example above highlights the importance of applying shifts to the forward curve that are consistent with real moves of interest rates. Obviously, it is highly unlikely that a forward curve would move in such a way that a 30 year swap rate has changed but the 29 and 31 year rates have not.

One tweak to the standard par-point approach that goes some way towards the goal of realistic curve shifts is the so-called *cumulative par-point approach* (also known as a *waterfall par-point approach*). The idea is simple: the shift to the i-th benchmark security is retained while calculating the derivative to the $(i+1)$-th (and subsequent) securities. In other words, the two curves for the $(i+1)$-th derivative are constructed from the prices

$$(V_1 + \Delta V_1, \ldots, V_i + \Delta V_i, V_{i+1}, V_{i+2}, \ldots, V_N)$$

(base) and

$$(V_1 + \Delta V_1, \ldots, V_i + \Delta V_i, V_{i+1} + \Delta V_{i+1}, V_{i+2,}, \ldots, V_N)$$

(perturbed). The standard deltas are then computed as differences of two consecutive (cumulative) derivatives. While the resulting deltas should coincide with the standard par-point deltas in the limit of $\Delta V \to 0$, they differ for non-vanishing perturbations.

The forward curve shifts implied by the cumulative par-point method are less extreme than those of the ordinary par-point method, making the cumulative par-point method quite attractive in practice. Another practical advantage of the method is the fact that the sum of deltas computed by the method is always (by definition) exactly equal to the *parallel delta*, i.e. the delta that is obtained by shifting all benchmark yields by the same amount at the same time. Because of the second-order effects, the same is only true for the standard par-point method in the limit of vanishing shifts, not for the non-infinitesimal perturbations used in practice.

The cumulative par-point approach is easy to mimic (and even improve) in the Jacobian framework of Section 6.4.3. Clearly, from (6.33), the i-th cumulative shift roughly corresponds to a piecewise flat move of the forward curve between the maturities of $(i-1)$-th and i-th benchmark. Hence, we can define

$$\mu_i(t) = 1_{\{t \in [T_{i-1}, T_i)\}}, \quad i = 1, \ldots, N, \tag{6.34}$$

with $T_0 \triangleq 0$. Note that this specification involves benchmark maturities $\{T_i\}$, in contrast to (6.28) which is typically set on a 3 month grid; in particular, (6.34) involves as many shocks as there are benchmark securities. Application of the Jacobian method to (6.34) yields an attractive variation of the cumulative par-point method where all forward curve shocks are similarly scaled, in contrast to the basic cumulative par-point where the size of forward curve shocks grows linearly with maturity, as implied by (6.33).

We should note that to improve accuracy, one may compute deltas as the average of deltas computed using first positive shocks, then negative shocks. This idea applies to par-point, forward-rate, Jacobian, cumulative-par-point, or any other delta calculation method. For the simple par-point method, this boils down to using two-sided finite difference approximations versus one-sided for approximating derivatives, a standard trick. For other methods the relationship is not as straightforward but the end result is the same: improved accuracy and stability of deltas. Using averaged deltas is typically particularly useful for security prices that depend on yields in a strongly non-linear fashion.

Finally, let us mention another popular trick. We have spent a good part of Section 6.2.4 describing ways to build smooth yield curves that exhibit good locality under perturbations. A more simplistic approach to tackle the same problem is based on building two different curves. One — smooth — is then used for pricing and the other — bootstrapped and with good locality — used for risk computations. While certainly helpful in a pinch, this approach tends to suffer from poor P&L predict, in the sense that changes in valuations of a portfolio between two dates are not well explained by first-order sensitivities (because values and sensitivities are calculated using different curves). We spend more time on P&L predict in Chapter 22.

6.5 Various Topics in Discount Curve Construction

6.5.1 Curve Overlays and Turn-of-Year Effects

Many of the curve construction algorithms so far have been designed around the implicit idea that the forward curve should ideally be *smooth*. While this is, indeed, generally a sound principle, exceptions do exist. For instance, it may be reasonable to expect instantaneous forwards to jump on or around meetings of monetary authorities, such as the Federal Reserve in the US. In addition, other "special" situations may exist that might warrant introduction of discontinuities into the forward curve. A well-known example is the turn-of-year (TOY) effect where short-dated loan premiums spike for loans between the last business day of the year and the first business day of the following calendar year.

One common way of incorporating TOY-type effects is to exogenously specify an *overlay curve* $\varepsilon_f(t)$ on the instantaneous forward curve. Specifically, the forward curve $f(t) = f(0,t)$ is written as

$$f(t) = \varepsilon_f(t) + f^*(t), \qquad (6.35)$$

6.5 Various Topics in Discount Curve Construction

where $\varepsilon_f(t)$ is user-specified — and most likely contains discontinuities around special event dates — and $f^*(t)$ is unknown. The yield curve algorithm is then subsequently applied to the construction of $f^*(t)$. That is, rather than solving $\mathbf{cP} = \mathbf{V}$ (see equation (6.4)), we instead write

$$P(T) = e^{-\int_0^T \varepsilon_f(t)dt} e^{-\int_0^T f^*(t)dt} \triangleq P_\varepsilon(T)P^*(T) \tag{6.36}$$

and solve

$$\mathbf{c}_\varepsilon \mathbf{P}^* = \mathbf{V}, \tag{6.37}$$

where $\mathbf{P}^* = (P^*(t_1), \ldots, P^*(t_M))^\top$, and $\mathbf{c}_\varepsilon$ is a modified $N \times M$ coupon matrix, with elements

$$(\mathbf{c}_\varepsilon)_{i,j} = c_{i,j} P_\varepsilon(t_j).$$

Construction of $\mathbf{c}_\varepsilon$ can be done as a pre-processing step, after which any of the algorithms discussed earlier in this chapter can be applied to attack (6.37). Once the curve $P^*(t)$ (or, equivalently, the yield curve $y^*(t) = -t^{-1} \ln P^*(t)$) has been constructed, any subsequent use of the curve for cash flow discounting requires, according to (6.36), a multiplicative adjustment of time t discount factors by the quantity $P_\varepsilon(t)$.

6.5.2 Cross-Currency Curve Construction

In this section we consider the issues involved in constructing yield curves simultaneously in multiple currencies. As it turns out, the market for foreign exchange (FX) forwards and cross-currency basis swaps imposes certain arbitrage constraints that must be considered in the curve construction exercise.

6.5.2.1 Basic Problem

To provide some motivation, consider a US dollar (USD) based firm receiving $1 for certain at some future time T. Assuming that we have available a risk-free discount curve $P(\cdot)$ for USD-denominated cash flows, we compute the value of this security simply as $P_\$(T)$. Suppose now that the firm enters into a (costless) FX forward where it commits to pay $1 at time T against receipt of a Japanese yen (JPY) amount of ¥$Y(T)$; $Y(T)$ thereby represents the time 0 forward USD/JPY exchange rate for delivery at time T. By transacting the FX forward, the firm has effectively turned the receipt of $1 into receipt of ¥$Y(T)$, the USD PV at time 0 of which is

$$X(0) P_¥(T) Y(T),$$

where $P_¥(T)$ is a JPY discount factor and $X(0) = 1/Y(0)$ is the time 0 foreign exchange rate in $/¥ terms. To avoid an arbitrage, we evidently need

$$P_\$(T) = X(0)P_¥(T)Y(T) \quad \Rightarrow \quad P_¥(T) = \frac{P_\$(T)}{Y(T)X(0)}. \qquad (6.38)$$

Suppose, say, that we have blindly estimated discount curves $P(\cdot)$ and $P_¥(\cdot)$ from the market for USD- and JPY-denominated interest rate swaps, respectively, without paying any attention to FX markets. The discount curves $P_\$(\cdot)$ and $P_¥(\cdot)$ estimated in this fashion will very likely *not* satisfy (6.38), implying the existence of cross-currency arbitrages. The degree to which (6.38) is typically violated is often small, but any such violation can be highly problematic for a firm engaging in trading of significant amounts of both USD- and JPY-denominated assets.

6.5.2.2 Separation of Discount and Forward Rate Curves

It may appear that there is no way out of this conundrum: after all, our curve construction algorithms imply a unique discount curve out of given swap prices and have few, if any, means of incorporating additional requirements such as (6.38). However, built into our assumptions about how to price a swap (see (6.2)) was an implicit assumption that Libor itself is the proper discount rate for flows based on Libor fixings. As Libor rates represent lending rates between banks, they contain a certain amount of credit risk[18] and it is ex-ante unclear that they are suitable proxies for a "risk-free" rate (or, at least, are suitable for discounting of swap cash flows). More details about this can be found in Collin-Dufresne and Goldstein [2001] and Duffie and Huang [1996]. For our purposes, it suffices to introduce the notion that when computing a swap value we may need two curves: i) the Libor "pseudo-discount" curve $P^{(L)}(t) = P^{(L)}(0,t)$, used to project the Libor-based floating cash flows on the floating leg of the swap; and ii) a real discount curve $P(t) = P(0,t)$, used to discount all cash flows. For, say, a regular JPY-based fixed-floating swap paying a coupon c on a schedule $\{t_i\}_{i=1}^n$, the swap valuation equation thus becomes

$$V_¥(0) = \underbrace{\sum_{i=0}^{n-1} c\tau_i P_¥(0, t_{i+1})}_{\text{Fixed Leg}} - \underbrace{\sum_{i=0}^{n-1} L_¥(0, t_i, t_{i+1})\tau_i P_¥(0, t_{i+1})}_{\text{Floating Leg}}, \qquad (6.39)$$

where $\tau_i = t_{i+1} - t_i$, $t_0 = 0$, and where we compute $L_¥(t, t_i, t_{+1})$ as (compare to (4.2))

---

[18] Reflecting the average bank credit rating, it is common to think of the Libor curve as a proxy for a AA-rated funding curve. In reality, however, this is not quite accurate, as banks with deteriorating credit are eliminated from the consortium of banks polled when determining the Libor rates. As such, the medium- and long-term forwards of the Libor curve contain *less* credit risk adjustment than similar forwards for a curve used to discount obligations to a single AA-rated firm. For more on this, see Collin-Dufresne and Goldstein [2001].

$$L_{¥}(t, t_i, t_{i+1}) = \frac{1}{\tau_i} \left(P_{¥}^{(L)}(t, t_i) / P_{¥}^{(L)}(t, t_{i+1}) - 1 \right).$$

A similar construction can be done for any USD swap, by means of introducing curves $P_{\$}^{(L)}(t)$ and $P_{\$}(t)$. Technically speaking, the Libor forwards $L_{¥}(t, t_i, t_{i+1})$ in (6.39) represent expectations in the t_{i+1}-forward measure — i.e. the martingale measure associated with the numeraire price $P_{¥}(t, t_{i+1})$ (not $P_{¥}^{(L)}(t, t_{i+1})$) — of quoted *spot* Libor rates,

$$L_{¥}(t, t_i, t_{i+1}) \triangleq \mathrm{E}_t^{t_{i+1}, ¥} \left(L_{¥}(t_i, t_i, t_{i+1}) \right). \tag{6.40}$$

In this view, the quoted Libor rate is effectively reduced to an observable index that may have little, if any, relationship to a true discount rate. For this reason, the time 0 pseudo-discount curves $P_{\$}^{(L)}(t)$ and $P_{¥}^{(L)}(t)$ are often referred to as *index* curves.

It should be clear that the introduction of the pseudo-discount curves $P_{\$}^{(L)}(t)$ and $P_{¥}^{(L)}(t)$ equips us with enough degrees of freedom to fit both USD-denominated swaps, JPY-denominated swaps, and the market for FX forward contracts. In fact, we have *too many* degrees of freedom: four curves, but only three separate markets to calibrate to. One way of handling this issue is to impose additional assumptions about the relationship between the curves $P(t)$ and $P^{(L)}(t)$ in one chosen currency. Before the 2007–2009 crisis, the following assumption was common.

Assumption 6.5.1. *In USD, the Libor pseudo-discount curve coincides with the real discount curve, i.e.* $P_{\$}^{(L)}(t, T) = P_{\$}(t, T)$ *for all t and T*, $T \geq t$.

Assumption 6.5.1 amounts to a convention where the liquidity and credit basis of non-USD Libor rates should all be measured relative to a neutral "bed-rock" established by USD Libor rates. Embedded into Assumption 6.5.1 also is the notion that most firms world-wide can fund themselves by borrowing in USD at levels close to USD-Libor; in the past this was often not a bad assumption. As we discuss in Section 6.5.3 below, post-crisis a non-trivial basis between index and discounting curves has emerged in the US. For simplicity of exposition we proceed in this section with Assumption 6.5.1, but the index-discounting basis in the US could be easily incorporated into the algorithm. The problem of accounting for this basis in *single-currency* curve construction is postponed until Section 6.5.3.

It is common to measure the difference between $P^{(L)}(t)$ and $P(t)$ in yield space, writing

$$P^{(L)}(t) = P(t) e^{-s(t) t},$$

where $s(t)$ is a yield spread often known as the *cross-currency (CRX) yield spread*. By Assumption 6.5.1, $s(t)$ is zero for USD, but will rarely be zero for any other currency. As indicated earlier, $s(t)$ will generally be quite small, often in the magnitude of a few basis points or less. Occasionally, however, the CRX yield spread may blow out, particularly if banks in a particular country are perceived as having below-average credit quality. For instance, in the late 1990's, the CRX yield spread reached somewhere around -40 basis points in JPY as Japanese banks were perceived as being in economic trouble. During that period of time, foreign banks could generally fund themselves in USD at USD Libor, but in JPY at rates significantly below JPY Libor (due to their superior credit relative to Japanese banks). Had FX forward rates traded without any large CRX basis, foreign banks could have borrowed in JPY and used the FX forward markets to turn their obligations into USD-denominated ones at a borrowing cost below USD Libor, which would have indicated the existence of an inconsistency and an arbitrage. Conversely, in early 2008 the CRX basis spread became significantly positive (up to $+60$ basis points) as the hedging demands of long-dated FX books increased rapidly on the back of significant strengthening of the Yen versus the US Dollar. During the 2007–2009 crisis, many other currencies (including EUR) have experienced similar dramatic moves in the CRX basis spreads against USD.

6.5.2.3 Cross-Currency Basis Swaps

The discussion so far has assumed the existence of a liquid market for FX forwards, as means to observe and tie down the CRX basis between rates in two separate currencies. In reality, the interbank FX forward market is rarely liquid beyond maturities of one year, a far cry from the 30+ year horizons to which we often want to build yield curves. Rather than relying on FX forwards, instead we can turn to the market for *floating-floating cross-currency (CRX) basis swaps*. Briefly speaking, CRX basis swaps are contracts where floating Libor payments in one currency are exchanged for floating Libor payments in another currency, plus or minus a spread. The swaps involve an exchange of notionals at trade inception and at maturity; the ratio between the two notionals is normally set to equal the spot FX exchange rate prevailing at trade inception. CRX basis swaps are closely related to FX forward contracts — indeed a one-period CRX basis swap is identical to an FX forward contract.

As was the case with FX forward contracts, failing to fit to the market for CRX basis swaps can lead to arbitrageable inconsistencies. For instance, consider the pricing of a stream of fixed USD cash flows. One way to determine the JPY price of these cash flows would be through simple discounting by the USD discount curve, followed by a conversion to JPY at the spot exchange rate. Alternatively, the following zero-cost scheme could be implemented to turn the stream of USD cash flows into fixed JPY cash flows:

1. Swap the fixed cash flows to streams of USD Libor plus some spread x, in a regular USD interest rate swap.

6.5 Various Topics in Discount Curve Construction

2. Swap USD Libor + x against JPY Libor + $e + x$ in a CRX basis swap, e being a market-quoted CRX basis swap spread.
3. Swap JPY Libor + $e + x$ against a stream a fixed JPY coupons, in a regular JPY interest rate swap.

The USD value of the cash flows in Step 3 can be determined through discounting with the JPY discount curve, and subsequent conversion to USD at the spot USD/JPY exchange rate. If the JPY discount curve is inconsistent with the basis-swap market, the value computed this way may not equal the value computed by discounting the original USD cash flows at the USD discount curve. Since the swap transactions 1–3 above are costless, this discrepancy will indicate an arbitrage.

We can use the pricing formalism developed in Section 6.5.2.2 to provide an explicit expression for the value of a CRX basis swap. For concreteness, we again turn to the USD/JPY market and consider a CRX basis swap, where a USD-based corporation receives USD Libor flat in exchange for payments of JPY Libor plus a fixed spread, $e_¥$. With payment dates $\{t_i\}_{i=1}^n$, the USD price $V_{\text{basisswap},\$}$ of the basis swap is (assuming a \$1 notional)

$$V_{\text{basisswap},\$}(0) \qquad (6.41)$$

$$= \sum_{i=0}^{n-1} L_\$(0, t_i, t_{i+1}) \tau_i P_\$(0, t_{i+1}) + P_\$(0, t_n)$$

$$- X(0) \left(\sum_{i=0}^{n-1} (L_¥(0, t_i, t_{i+1}) + e_¥) \tau_i P_¥(0, t_{i+1}) + P_¥(0, t_n) \right)$$

$$= 1 - X(0)$$

$$\times \left(\sum_{i=0}^{n-1} \left(\frac{P_¥^{(L)}(0, t_i)}{P_¥^{(L)}(0, t_{i+1})} - 1 + e_¥ \tau_i \right) P_¥(0, t_{i+1}) + P_¥(0, t_n) \right), \qquad (6.42)$$

where we have used the fact that $P_\$$ and $P_\$^{(L)}$ are identical (by Assumption 6.5.1), in order to reduce the time 0 price of the USD floating leg to \$1. The market quotes par values $e_¥^{par}$ — the values of $e_¥$ that will make $V_{\text{basisswap},\$}(0) = 0$ — in a wide range of maturities extending out to 30 years or more.

6.5.2.4 Modified Curve Construction Algorithm

By Assumption 6.5.1, construction of the USD discount and Libor curves can be accomplished through direct application of the routines in Sections 6.2 or 6.3 on benchmark securities consisting of deposits, FRAs, and swaps. For non-USD currencies, however, matters are more complicated as we must now simultaneously estimate both curves $P(t)$ and $P^{(L)}(t)$, $t > 0$, in a manner ensuring that i) Libor benchmark securities are correctly priced; and ii) par-valued cross-currency swaps against USD are correctly priced. In performing this exercise, we apply (6.42) and adjust valuation expressions for the benchmark securities according to the principles of the swap-pricing[19] equation (6.39). We can make the curve construction problem quite complicated if we insist on $P(t)$ and $P^{(L)}(t)$ both being smooth functions of t; instead, here we will show a simpler idea that applies earlier algorithms in this chapter in iterative fashion.

Working as before with JPY as the foreign currency, we start by assuming that we have somehow managed to construct the correct Libor curve $P_¥^{(L)}(t)$. Were we — erroneously — to pretend that $P_¥^{(L)}(t)$ were a proper discount curve, we would get, for our N benchmark securities, a vector of values $\mathbf{V}^{(L)}$ that would generally *not* equal the correct JPY market prices $\mathbf{V}$:

$$\mathbf{c}\mathbf{P}_¥^{(L)} = \mathbf{V}^{(L)}, \quad \mathbf{V}^{(L)} \neq \mathbf{V}. \tag{6.43}$$

As the $P_¥^{(L)}(t)$ discount curve will be used to project forward rates, the yields and forward rates implied by $P_¥^{(L)}(t)$ should ideally be smooth. The smoothness requirements of the discount curve $P_¥(t)$, however, are significantly lower, as we shall never need to (in effect) differentiate this curve to produce forward Libor rates. Assuming that we have CRX basis swaps maturing on the benchmark set maturity dates $\{T_i\}_{i=1}^N$, it is thus, for instance, not unreasonable to write

$$P_¥(t) = P_¥(T_i)\frac{P_¥^{(L)}(t)e^{-\varepsilon_i \cdot (t-T_i)}}{P_¥^{(L)}(T_i)}, \quad t \in [T_i, T_{i+1}), \tag{6.44}$$

which assumes that the instantaneous forward rates generated by $P_¥(t)$ are given by those computed from $P_¥^{(L)}(t)$ plus a piecewise flat function:

$$f_¥(t) = f_¥^{(L)}(t) + \varepsilon(t), \quad \varepsilon(t) = \sum_{i=0}^{N-1} \varepsilon_i 1_{\{t \in [T_i, T_{i+1})\}}, \tag{6.45}$$

where $T_0 = 0$ as before.

In a cross-currency setting, the expression (6.4) no longer holds and must be replaced with something like

$$\mathbf{f}\left(\mathbf{P}_¥^{(L)}, \mathbf{P}_¥\right) = \mathbf{V} \tag{6.46}$$

for a non-linear vector-valued function $\mathbf{f}$. Indeed, according to (6.39), many of the coupon payments ($\mathbf{c}$) become a non-linear function of points on $\mathbf{P}_¥^{(L)}$ and cannot be considered constants. To salvage the methodologies discussed

---

[19] Pricing of short-term deposits only involves the discount curve P, whereas FRAs can be treated as one-period swaps. We leave details to the reader.

in Sections 6.2 or 6.3, we avoid working with (6.46) directly, and instead use an iteration based on equations (6.42), (6.43), and (6.45). The iteration attempts to estimate the unknown quantity $\mathbf{V}^{(L)}$ in (6.43) and works as follows:

1. Let $\mathbf{V}^{(L)}(j)$ be the j-th iteration for $\mathbf{V}^{(L)}$. Use $\mathbf{V}^{(L)}(j)$ along with (6.43) to estimate the curve $P_{¥}^{(L)}(t)$, using any of the curve construction methods discussed in earlier sections of this chapter.
2. Given knowledge of $P_{¥}^{(L)}(t)$, use (6.44)–(6.45) combined with (6.42) to imply the N constants $\varepsilon_0, \varepsilon_1, \ldots \varepsilon_{N-1}$, by calibration to the N par-valued CRX basis swaps maturing at time $T_1, \ldots, T_N$. This calibration exercise can be done by simple bootstrapping, and establishes the current guess for $P_{¥}(t)$.
3. Given estimates for both $P_{¥}^{(L)}(t)$ and $P_{¥}(t)$, compute guessed prices $\mathbf{V}(j)$ of all benchmark securities, i.e. evaluate the left-hand side of (6.46). If $\mathbf{V}(j)$ and $\mathbf{V}$ are within a given tolerance, we are done.
4. Update the guess for $\mathbf{V}^{(L)}$ according to $\mathbf{V}^{(L)}(j+1) = \mathbf{V}^{(L)}(j) - (\mathbf{V}(j) - \mathbf{V})$, and proceed to Step 1.

The iteration is initiated at $j = 0$ with the estimate $\mathbf{V}^{(L)}(0) = \mathbf{V}$ and runs until the termination criterion in Step 3 is satisfied. As the approximation $\mathbf{V}^{(L)} \approx \mathbf{V}$ is normally very accurate, only a few iterations are needed to reach acceptable precision.

In this book we shall mostly ignore the existence of a non-zero CRX basis spread. In construction of a model for the evolution of the Libor curve, the reader should, however, keep in mind that it may be necessary to adjust the curve slightly before using it to discount any cash flows. In a dynamic setting, it is quite common to perform this adjustment by simply assuming that $\varepsilon(t)$ in (6.45) is deterministic. A discussion of how to incorporate both stochastic and deterministic spreads in a dynamic model for interest rate evolution can be found in Section 15.5. For now, we note that using deterministic spreads is generally safe, unless pricing securities with strong convexity in $\varepsilon(t)$ — e.g. an option on a CRX basis swap — in which case a separate stochastic model for $\varepsilon(t)$ may be needed.

6.5.3 Tenor Basis and Multi-Index Curve Group Construction

Section 6.5.2 relied extensively on the notion of separating the discount curves used for Libor projection and for outright discounting. This idea is quite powerful and has applications in other settings, including some where only a single currency is involved. For instance, for swaps that pay a non-Libor index — e.g. the Bond Market Association (BMA) index in the US — it is natural to introduce a basis spread that measures the difference between forward rates of the non-Libor index curve and the Libor curve itself.

More recently, a similar technique has become important even for curves used for pricing standard Libor-based contracts. We have already mentioned (Section 5.1) that the Fed funds rate, the overnight rate used for balances of bank deposits with the Federal reserve, is often considered the closest proxy to the risk-free rate in the US (with Eonia and Sonia rates, see Section 5.1, fulfilling the same function for Euro and GBP). One argument for this choice is that most inter-dealer transactions are collateralized under the *International Swaps and Derivatives Association* (ISDA) Master Agreement, with the rate paid on collateral being the Fed funds rate (for USD; Eonia and Sonia for Euro and GBP), see Piterbarg [2010] (and also the discussion in Section 5.1). While the spread between the Fed funds rate and 3 month Libor rate used to be very small — in the order of a few basis points — after September 2007 it went up to as much as 275 basis points, and it is now generally accepted now that the Libor rate is no longer a good proxy for a discounting rate on collateralized trades. Uncollateralized derivative contracts are subject to credit risk, and a fully consistent pricing approach needs to incorporate the cost of hedging this risk (the co-called *credit valuation adjustment* or *CVA*). These computations are outside the scope of this book and can get very complex, in part because collateral rules can be complicated and are normally enforced on entire counterparty portfolios and not just on individual trades. See Gregory [2009] for further details.

As we discussed in Section 6.5.2, if we make an assumption on the index-discounting basis in one currency (say, USD), we can translate it into the index-discounting basis in any other currency through the market quotes for forward FX contracts and cross-currency basis swaps. However, to estimate this basis in USD (say), we need to rely on domestic markets only; doing otherwise will introduce a circularity into our arguments. Fortunately, the market in the appropriate instruments, the OIS (overnight index swap, a swap of payments based on a compounded Fed funds rate versus fixed rate, see (5.7)–(5.8)) and the Fed funds/Libor basis swaps (see Section 5.7) — has developed in the US with a range of maturities actively traded. Hence, using techniques that we already discussed, we can construct a pair of curves — a curve for discounting and a curve for projecting 3 month (say) forward Libor rates — in a self-consistent way from the market quotes on deposits, FRAs, swaps with 3 month frequency, and overnight index swaps.

Currently there are no countries where both the OIS market and the cross-currency basis swap (vs. USD) market are very liquid, and we can always use one or the other to find the index-discounting basis. As the markets evolve, there may come a time when there will be two liquid sources of discounting curve information. It turns out that potential conflict between the two can be resolved by carefully analyzing the collateral mechanisms used in the two markets and the implications for yield curve construction. This discussion is outside of scope of the current edition of our book, but the interested reader could consult Fujii et al. [2010] for details.

6.5 Various Topics in Discount Curve Construction

The challenges of curve construction do not end with building separate discount and forecasting curves to take into account the index-discounting basis. We also need to account for the *tenor basis* between vanilla single-currency swaps trading at different frequencies, e.g. 1 month, 3 months, and 6 months. Before proceeding, let us explain in more detail what we mean by tenor basis.

Suppose we construct Libor and discount curves based on, say, vanilla swaps (and for non-USD currencies also CRX basis swaps) paying 3 month Libor on a quarterly schedule. If the resulting index and discount curves are subsequently used to price a vanilla swap paying 1 month Libor on a monthly schedule, the resulting price is typically different from actual market quotes. In other words, there is a basis between the 3 month and 1 month Libor index curves, a basis arising partly from credit considerations and partly liquidity considerations (banks have a natural desire to have longer-term deposits to better match their loan commitments). Historically this basis has also been low; for example, the difference between 1 month and 3 month Libor rates was in the order of one basis point up until September 2007, but since then has been as wide as 50 basis points.

When various basis levels were small, the small discrepancies between different Libor-tenor swaps were often accounted for by building a unique discount curve for the subset of swaps referencing the Libor rate of a particular tenor; this curve would, in addition to generating the floating leg forward rates, then be used to discount floating and fixed cash flows of swaps of that frequency. In a swap pricing framework, this can create an arbitrage since it implies that fixed flows (from the fixed leg) will be discounted at different discount curves, depending on which Libor tenor the fixed flows happen to be paid against. Moreover, it is not clear how to deal with swaps that involve multiple Libor tenors[20], or how to aggregate risks coming from unrelated, individually constructed curves. Again, when the differences were small, these issues were largely ignored.

More recently, the naive approach above has evolved into the idea of using a *multi-index curve group*, a collection consisting of a single discount curve and multiple index curves, one for each Libor tenor covered by the multi-index curve group. The index curves are used in a tenor-specific manner to project Libor forward rates, and the universal discount curve serves to discount all floating and fixed cash flows, irrespective of tenor. The index curves are built sequentially as spreads off previously-built curves, which provides linkage between index curves and also a convenient risk parameterization. This relatively recent idea is discussed in Traven [2008] in considerable details, from where most of the material of this section is derived. Another good reference here is Fujii et al. [2010].

---

[20] The most common example of this is a swap with a short front stub, i.e. a swap where at inception the first payment period is shorter than subsequent ones.

268   6 Yield Curve Construction and Risk Management

To discuss multi-index curve groups in detail, let us introduce a superscript k, $k = 1, \ldots, K$, to distinguish quantities related to different tenors. In particular, let t_i^k be the i-th date in the tenor structure for tenor k; $\tau_i^k = t_{i+1}^k - t_i^k$ the corresponding tenor offset; $L^k(t_i^k, t_i^k, t_{i+1}^k)$ the spot Libor rate of tenor k for the i-th period; and so on. If we denote the expected value of $L^k(t_i^k, t_i^k, t_{i+1}^k)$ under the t_{i+1}^k-forward measure by

$$L^k\left(t, t_i^k, t_{i+1}^k\right) \triangleq \mathrm{E}_t^{t_{i+1}^k}\left(L^k\left(t_i^k, t_i^k, t_{i+1}^k\right)\right)$$

(compare to (6.40)), then the value of a fixed-floating k-Libor-tenor swap with n periods and rate c is given by

$$V^k(0) = \underbrace{\sum_{i=0}^{n-1} c\tau_i^k P(t_{i+1}^k)}_{\text{Fixed Leg}} - \underbrace{\sum_{i=0}^{n-1} L^k(0, t_i^k, t_{i+1}^k)\tau_i^k P(t_{i+1}^k)}_{\text{Floating Leg}}. \tag{6.47}$$

Here $P(t)$ is the universal discounting curve. The time 0 index curve for tenor k, $P^k(t)$, is defined by the condition that forward Libor rates (of tenor k) be defined by the familiar formula

$$L^k\left(0, t_i^k, t_{i+1}^k\right) = \frac{1}{\tau_i^k}\left(P^k\left(t_i^k\right)/P^k\left(t_{i+1}^k\right) - 1\right). \tag{6.48}$$

A multi-index curve group is defined as a collection $\{P(\cdot), P^1(\cdot), \ldots, P^K(\cdot)\}$ of the universal discounting curve and one index curve per tenor, with swaps priced via (6.47) (and equivalent formulas for other linear instruments) and (6.48).

Let us outline how to calibrate a multi-index curve group to market instruments referencing rates of different tenors. For each market, fixed-floating swaps referencing a Libor rate of one particular tenor L^k are usually the most liquid. Assume that this curve is the first index curve in the group, i.e. $k = 1$. The method from Section 6.5.2 can be used to construct a discounting curve, and a *base* index curve $P^1(\cdot)$ from

- Funding instruments such as deposits, forward FX contracts, OIS, overnight rate basis swaps (i.e. floating-floating swaps of an overnight rate versus L^1, see Section 5.7 and a discussion of more general floating-floating single-currency basis swaps in the next paragraph), cross-currency basis swaps, and the like.
- Vanilla instruments referencing L^1 such as FRAs on L^1 and fixed-floating swaps on L^1 versus a fixed rate.

To construct $P^2(\cdot)$, we assume that prices of *floating-floating single-currency basis swaps* are available in the market. A floating-floating basis swap is a swap of payments linked to a Libor rate of a particular tenor — such as L^1 — versus payments based on a Libor rate of different tenor —

such as L^2. Each leg pays on its own schedule of a corresponding tenor. A fixed spread is typically added to one of the legs to make the swap value at inception equal to zero. If a floating-floating basis swap is not traded or not liquid, it can be synthesized from two fixed-floating swaps referencing L^1 and L^2.

If a floating-floating basis swap is traded at par in the market, the values of both legs should be the same at time 0:

$$\sum_{i=0}^{n^2(T)-1} L^2(0, t_i^2, t_{i+1}^2) \tau_i^2 P(t_{i+1}^2)$$

$$= \sum_{i=0}^{n^1(T)-1} \left(L^1(0, t_i^1, t_{i+1}^1) + e^{1,2}(T) \right) \tau_i^1 P(t_{i+1}^1). \quad (6.49)$$

Here $n^k(T)$ is the number of periods in the tenor structure to date T for tenor k, and $e^{1,2}(T)$ is the quoted floating-floating basis spread for exchanging L^1 for L^2 to maturity T, quoted on the L^1 leg. It could be positive or negative, depending on perceived desirability of payments linked to L^1 versus L^2.

Similar to (6.44)–(6.45), we represent $P^2(\cdot)$ as a multiplicative spread to $P^1(\cdot)$:

$$P^2(t) = P^1(t) e^{-\int_0^t \eta^{1,2}(s)\, ds}, \quad t \geq 0$$

for a given, usually piecewise-constant, spread function $\eta^{1,2}(\cdot)$. With the discounting curve $P(\cdot)$ already constructed and $L^1(0, t_i^1, t_{i+1}^1)$ known for all i from the already-built index curve $P^1(\cdot)$, it is a simple exercise to obtain the spread function $\eta^{1,2}(\cdot)$ by solving (6.49) for different T's.

Having built $P^2(\cdot)$, the remaining index curves $P^k(\cdot)$, $3 \leq k \leq K$ may be constructed in a similar fashion, always using floating-floating basis swap spreads for L^k versus L^1 basis swaps or, more generally, for whatever L^k versus L^i basis swaps are the most liquid with $i < k$. In particular, each index curve $P^k(\cdot)$ for $k > 1$ is built as a *spread*, or *basis*, curve to one of the previous curves.

In the presence of multiple curves, it is not entirely clear from the outset how to most sensibly define risk sensitivities. Fortunately, with this spread-based method of curve group construction, sensitivities to instruments used in the curve group have clear, and orthogonal, meaning:

- Perturbations to instruments used in building the base index curve, e.g. non-basis swaps and FRAs referencing L^1, define risk sensitivities to the overall levels of interest rates. Clearly, with basis spreads for L^1-versus-L^k floating-floating basis swaps, $k = 2, \ldots, K$, kept constant, shifts to fixed-floating L^1-versus-fixed swap rates will move all index curves together by the same amount in forward rate space. These sensitivities are the direct analog of the standard interest rate deltas in the traditional, single-curve, world, see Section 6.4.

270   6 Yield Curve Construction and Risk Management

- Perturbations to funding instruments define sensitivities to discounting.
- Perturbations to basis swap spreads for L^k-versus-L^1 floating-floating basis swaps define *basis risk*, i.e. the risk that index curves of different tenors do not move in lock step.

The parameterization allows us to naturally aggregate "similar" risks such as overall rate level risks, discounting risks, basis risks, while keeping different kinds separate for efficient risk management. Had we constructed all index curves in separation from each other (from multiple sets of vanilla fixed-floating swaps, say) such automatic aggregation would not be possible.

6.A Appendix: Spline Theory

6.A.1 Hermite Spline Theory

Consider a given set of data points (x_i, f_i), $i = 1, \ldots, N$, where $x_1 < x_2 < \ldots < x_N$. We wish to apply an interpolation rule such that a continuous function $f(x)$, $x \in [x_1, x_N]$, is created. We require that f be piecewise cubic, be at least once differentiable (C^1), and be a true interpolating function, i.e. $f(x_i) = f_i$ for all $i = 1, \ldots, N$.

In the Hermite spline description, tangents at points x_i, $i = 1, \ldots, N$, are assumed exogenously specified. Let f_i' denote the tangent df/dx at $x = x_i$, $i = 1, \ldots, N$. We write f as a piecewise cubic polynomial

$$f(x) = a_{3,i}(x - x_i)^3 + a_{2,i}(x - x_i)^2 + a_{1,i}(x - x_i) + a_{0,i}, \quad x \in [x_i, x_{i+1}],$$

with unknown coefficients $a_{j,i}$ specific to each interval $[x_i, x_{i+1}]$. Expressing that both f and f' should be continuous across each point x_i allows us, after a little rearrangement, to write the spline specification as simply

$$f(x) = \mathbf{D}_i(x)^\top \mathbf{M} \begin{pmatrix} f_i \\ f_{i+1} \\ f_i' h_i \\ f_{i+1}' h_i \end{pmatrix}, \quad x \in [x_i, x_{i+1}], \tag{6.50}$$

where $h_i \triangleq x_{i+1} - x_i$,

$$\mathbf{D}_i(x) = \begin{pmatrix} \delta_i^3 \\ \delta_i^2 \\ \delta_i \\ 1 \end{pmatrix}, \quad \delta_i \triangleq \frac{x - x_i}{h_i},$$

and $\mathbf{M}$ is the *Hermite matrix*

$$\mathbf{M} = \begin{pmatrix} 2 & -2 & 1 & 1 \\ -3 & 3 & -2 & -1 \\ 0 & 0 & 1 & 0 \\ 1 & 0 & 0 & 0 \end{pmatrix}.$$

One drawback of the Hermite specification is the need to directly specify tangents df/dx. A number of approaches exist to compute these directly from the given data points or by adding additional control points. For our purposes, we highlight the so-called *Catmull-Rom spline* (Catmull and Rom [1974]), where the derivatives are computed as[21]

$$f'_i = \frac{f_{i+1} - f_{i-1}}{x_{i+1} - x_{i-1}}, \quad i = 2, \ldots, N-1. \tag{6.51}$$

At end points (x_1, f_1) and (x_N, f_N) forward and backward differences are used instead:

$$f'_1 = \frac{f_2 - f_1}{x_2 - x_1}; \quad f'_N = \frac{f_N - f_{N-1}}{x_N - x_{N-1}}. \tag{6.52}$$

Notice that with (6.51), the Hermite representation (6.50) can be rewritten in the derivative-free form

$$f(x) = \mathbf{D}_i(x)^\top \mathbf{A}_i \begin{pmatrix} f_{i-1} \\ f_i \\ f_{i+1} \\ f_{i+2} \end{pmatrix}, \quad x \in [x_i, x_{i+1}], \tag{6.53}$$

where

$$\mathbf{A}_i = \begin{pmatrix} -\alpha_i & 2 - \beta_i & -2 + \alpha_i & \beta_i \\ 2\alpha_i & \beta_i - 3 & 3 - 2\alpha_i & -\beta_i \\ -\alpha_i & 0 & \alpha_i & 0 \\ 0 & 1 & 0 & 0 \end{pmatrix}, \quad i = 2, \ldots, N-2, \tag{6.54}$$

with

$$\alpha_i = \frac{h_i}{h_i + h_{i-1}}, \quad \beta_i = \frac{h_i}{h_{i+1} + h_i}.$$

As indicated, equation (6.53) only holds for $i = 2, \ldots, N-2$, with external boundary conditions needed to establish the curve in the segments $[x_1, x_2]$ and $[x_{N-1}, x_N]$. Applying boundary conditions of the type (6.52), we get

$$\mathbf{A}_1 = \begin{pmatrix} 0 & 1 - \beta_1 & -1 & \beta_1 \\ 0 & -1 + \beta_1 & 1 & -\beta_1 \\ 0 & -1 & 1 & 0 \\ 0 & 1 & 0 & 0 \end{pmatrix}, \tag{6.55}$$

and

---

[21] Variations exist which use more elaborate finite difference style derivatives, taking into account the fact that the grid may be non-equidistant; see Chapter 2. Given the semi-heuristic nature of the Catmull-Rom spline, it is doubtful that much is gained by such extensions.

$$\mathbf{A}_{N-1} = \begin{pmatrix} -\alpha_{N-1} & 1 & -1+\alpha_{N-1} & 0 \\ 2\alpha_{N-1} & -2 & 2-2\alpha_{N-1} & 0 \\ -\alpha_{N-1} & 0 & \alpha_{N-1} & 0 \\ 0 & 1 & 0 & 0 \end{pmatrix}. \qquad (6.56)$$

While Catmull-Rom splines shall suffice for the yield curve applications we have in mind here, let us note that it is possible to introduce further parameters to control local spline behavior. For completeness, let us quickly list one popular extension. First, we allow for the possibility that the curve is locally only C^0 by having incoming and outgoing tangents be different. That is, we define

$$f'_{i,I} = \lim_{\varepsilon \downarrow 0} \frac{f(x_i) - f(x_i - \varepsilon)}{\varepsilon}; \quad f'_{i,O} = \lim_{\varepsilon \downarrow 0} \frac{f(x_i + \varepsilon) - f(x_i)}{\varepsilon},$$

and rewrite the Hermite equation (6.50) as

$$f(x) = \mathbf{D}_i(x)^\top \mathbf{M} \begin{pmatrix} f_i \\ f_{i+1} \\ f'_{i,O} h_i \\ f'_{i+1,I} h_i \end{pmatrix}, \quad x \in [x_i, x_{i+1}]. \qquad (6.57)$$

Only when $f'_{i,I} = f'_{i,O}$ for all i is the curve C^1 everywhere. The *Kochanek-Bartels spline* — also known as the *TCB spline* — is defined through the expressions

$$f'_{i,I} = \frac{(1-\sigma_i)(1+c_i)(1-b_i)}{2} \frac{f_{i+1} - f_i}{x_{i+1} - x_i}$$
$$+ \frac{(1-\sigma_i)(1-c_i)(1+b_i)}{2} \frac{f_i - f_{i-1}}{x_i - x_{i-1}}, \qquad (6.58)$$

$$f'_{i,O} = \frac{(1-\sigma_i)(1-c_i)(1-b_i)}{2} \frac{f_{i+1} - f_i}{x_{i+1} - x_i}$$
$$+ \frac{(1-\sigma_i)(1+c_i)(1+b_i)}{2} \frac{f_i - f_{i-1}}{x_i - x_{i-1}}, \qquad (6.59)$$

for parameters $\sigma_i, c_i, b_i \in [-1, 1]$, $i = 1, \ldots, N$. The parameters σ_i, c_i, and b_i are used to control curve *tension*, *continuity*, and *bias*, respectively; clearly, when $\sigma_i = c_i = b_i = 0$, the Kochanek-Bartels spline reduces to the Catmull-Rom spline. A positive value of σ_i will tend to "tighten" the curve around the point (x_i, f_i), and a negative value will generate "slack". The parameters b_i are measures of over- and undershoot. To see this, set $\sigma_i = c_i = 0$ and note that when $b_i = 0$, left and right one-sided tangents are weighted equally producing a regular Catmull-Rom spline; when b_i is close to -1 (1), however, the outgoing (incoming) tangent dominates the path of the curve through the point (x_i, f_i) producing undershoot (overshoot). The parameters c_i control the degree of differentiability of the resulting spline: if a parameter $c_i \neq 0$, the resulting spline will develop a corner (the direction of which depends on the sign of c_i) at point (x_i, f_i), losing its differentiability. Kochanek-Bartels splines are used extensively in computer graphics applications.

6.A.2 C^2 Cubic Splines

The cubic splines in Section 6.A.1 are generally not twice differentiable, and their second derivatives will jump across each knot. We wish to remedy this, and now consider a twice differentiable C^2 cubic spline $f(x)$ interpolating a set of data points (x_i, f_i), $i = 1, \ldots, N$. By necessity, such a cubic spline interpolant is piecewise linear in its second derivative

$$f''(x) = \frac{x_{i+1} - x}{x_{i+1} - x_i} f_i'' + \frac{x - x_i}{x_{i+1} - x_i} f_{i+1}'', \quad x \in [x_i, x_{i+1}], \tag{6.60}$$

where we use primes to denote differentiation and where $f_i'' \triangleq f''(x_i)$. We emphasize that for a C^2 cubic spline, the second derivative is continuous across knot points: $\lim_{x \downarrow x_i} f''(x) = \lim_{x \uparrow x_i} f''(x) = f''(x_i)$. Integrating (6.60) twice and requiring the curve to pass through data points results in the classical spline equation

$$f(x) = \frac{(x_{i+1} - x)^3}{6h_i} f_i'' + \frac{(x - x_i)^3}{6h_i} f_{i+1}'' + (x_{i+1} - x) \left(\frac{f_i}{h_i} - \frac{h_i}{6} f_i'' \right)$$
$$+ (x - x_i) \left(\frac{f_{i+1}}{h_i} - \frac{h_i}{6} f_{i+1}'' \right), \quad x \in [x_i, x_{i+1}], \tag{6.61}$$

where $h_i \triangleq x_{i+1} - x_i$. The second derivatives f_i'', $i = 1, \ldots, N$, needed to evaluate (6.61) can be obtained by requiring $f'(x)$ to be continuous across data points. The result is

$$\frac{h_{i-1}}{6} f_{i-1}'' + \frac{h_{i-1} + h_i}{3} f_i'' + \frac{h_i}{6} f_{i+1}'' = \frac{f_{i+1} - f_i}{h_i} - \frac{f_i - f_{i-1}}{h_{i-1}}, \quad i = 2, \ldots, N-1. \tag{6.62}$$

The set of equations (6.62) is a tri-diagonal system for f_i'' that can be solved in $O(N-2)$ operations once we have specified boundary conditions[22] for f_1'' and f_N''. A classical boundary condition is $f_1'' = f_N'' = 0$, leading to the so-called *natural cubic spline*.

While C^2 cubic splines have a number of useful features, they have, loosely speaking, a built-in aversion to make tight turns (since this will cause large values of f''). This, in turn, will often produce extraneous inflection points and non-local behavior, in the sense that perturbation of a single f_i will significantly affect the appearance of the curve for x-values far from x_i. Also, monotonicity and convexity properties of the original data-set will typically not be preserved.

---

[22] Such boundary conditions may be indirect and can, among other choices, take the form of specification of a gradient f' at x_1 or x_N. By differentiation of (6.61), a gradient specification can always be turned into a condition on f'' at x_1 or x_N.

6.A.3 C^2 Exponential Tension Splines

An attractive remedy to the shortcomings of the cubic spline is to insert some *tension* in the cubic spline, that is, to apply a tensile force to the end-points of the spline. Formally, this can be accomplished (see Schweikert [1966]) by replacing the equation (6.60) with, $x \in [x_i, x_{i+1}]$,

$$f''(x) - \sigma^2 f(x) = \frac{x_{i+1} - x}{x_{i+1} - x_i}\left(f''_i - \sigma^2 f_i\right) + \frac{x - x_i}{x_{i+1} - x_i}\left(f''_{i+1} - \sigma^2 f_{i+1}\right), \quad (6.63)$$

where $\sigma > 0$ is a measure of the tension applied to the cubic spline[23]. Notice that we have replaced the assumption of a piecewise linear second derivative with the assumption that the quantity $f''(x) - \sigma^2 f(x)$ is linear on each sub-interval $[x_i, x_{i+1}]$.

Integrating (6.63) twice and requiring that the curve pass through the given data points, one obtains (after some rearrangements)

$$\begin{aligned} f(x) = & \left(\frac{\sinh\left(\sigma\left(x_{i+1} - x\right)\right)}{\sinh(\sigma h_i)} - \frac{x_{i+1} - x}{h_i}\right)\frac{f''_i}{\sigma^2} \\ & + \left(\frac{\sinh\left(\sigma\left(x - x_i\right)\right)}{\sinh(\sigma h_i)} - \frac{x - x_i}{h_i}\right)\frac{f''_{i+1}}{\sigma^2} \\ & + f_i\frac{x_{i+1} - x}{h_i} + f_{i+1}\frac{x - x_i}{h_i}, \quad x \in [x_i, x_{i+1}], \quad (6.64) \end{aligned}$$

where $h_i = x_{i+1} - x_i$ as before. Requiring continuity of the first derivative we then get for the f''_i,

$$\begin{aligned} & \left(\frac{1}{h_{i-1}} - \frac{\sigma}{\sinh(\sigma h_{i-1})}\right)\frac{f''_{i-1}}{\sigma^2} \\ & + \left(\frac{\sigma\cosh(\sigma h_{i-1})}{\sinh(\sigma h_{i-1})} - \frac{1}{h_{i-1}} + \frac{\sigma\cosh(\sigma h_i)}{\sinh(\sigma h_i)} - \frac{1}{h_i}\right)\frac{f''_i}{\sigma^2} \\ & + \left(\frac{1}{h_i} - \frac{\sigma}{\sinh(\sigma h_i)}\right)\frac{f''_{i+1}}{\sigma^2} = \frac{f_{i+1} - f_i}{h_i} - \frac{f_i - f_{i-1}}{h_{i-1}}. \end{aligned}$$

Again, this is a tri-diagonal system of equations that can be solved in $O(N - 2)$ operations once we specify f''_1 and f''_N.

From the representation (6.64), it is clear that on all intervals $[x_i, x_{i+1}]$ hyperbolic tension splines can be written as linear combinations of the basis functions 1, x, $e^{-\sigma x}$, $e^{\sigma x}$. The representation (6.64), however, has better behavior for large and small values of σ (see Renka [1987] and Rentrop [1980] for details about proper evaluation of the hyperbolic functions in (6.64) for large and small values of σ).

---

[23] Extension to non-uniform tension parameter is straightforward and involves replacing σ with σ_i in (6.63), with σ_i then being a measure of the tension applied locally to the curve in the interval $[x_i, x_{i+1}]$.

We notice that when the tension parameter $\sigma = 0$, equations (6.63) and (6.60) are identical, i.e. the tension spline degenerates into a regular cubic spline. On the other hand, when $\sigma \gg 1$ (6.63) reduces to piecewise linear interpolation, as

$$\lim_{\sigma \to \infty} f(x) = \frac{x_{i+1} - x}{x_{i+1} - x_i} f_i + \frac{x - x_i}{x_{i+1} - x_i} f_{i+1}, \quad x \in [x_i, x_{i+1}]. \tag{6.65}$$

Evidently, the equation (6.63) defines a twice differentiable curve that is a hybrid between a cubic spline and a piecewise linear spline.

The convergence of the tension spline towards a piecewise linear curve as $\sigma \to \infty$ can be shown to be *uniform*, i.e. (6.65) holds uniformly in $[x_i, x_{i+1}]$ for $i = 1, \ldots, N - 1$. Similarly

$$\lim_{\sigma \to \infty} f'(x) = \frac{f_{i+1} - f_i}{x_{i+1} - x_i} \quad \text{and} \quad \lim_{\sigma \to \infty} f''(x) = 0$$

uniformly in any closed subinterval of $[x_i, x_{i+1}]$. See Pruess [1976] for details and a proof. The uniform convergence is important as it guarantees that the monotonicity and convexity properties of the underlying discrete data set are preserved, simply by choosing a sufficiently high value of the tension factor. Due to this property, hyperbolic tension splines are said to be *shape-preserving*[24]. As the tension factor increases, the resulting spline will also behave in increasingly local fashion towards input perturbations. In the limit $\sigma \to \infty$ each point $f(x)$ on the spline will only be associated with the two nearest-neighbor knots.

---

[24]Generalizing, suppose we introduce constraints on function values, first derivatives, or second derivatives. As long as these constraints are satisfied by piecewise linear interpolation, there will exist some value of the tension parameter σ (possibly $\sigma = 0$) which will make the tension spline satisfy the constraints. This observation is key to algorithms for automatic selection of σ from externally specified function constraints. See, for instance, Lynch [1982] and Renka [1987] for details and efficient algorithms for automatic tension selection.

7

Vanilla Models with Local Volatility

We have shown in Section 5.10 that European swaptions (and caplets/floorlets which we equate with one-period swaptions) can be valued as European options on forward swap rates. As a consequence, a full term structure model that specifies the dynamics of the whole yield curve through time is essentially unnecessary for European swaption valuation. Instead, we only need a model for the evolution — in fact, just a terminal distribution — of a single swap rate in isolation. Models of this type shall be denoted *vanilla models*, to distinguish them from full term structure models. Vanilla models can be extended by copula methods to describe joint terminal distributions of more than one rate, as we discuss in Chapter 17. Ultimately, however, their primary purpose in this book is to serve as a foundation for development of more widely applicable full term structure models, that is, models which provide consistent dynamics for *all* points on the yield curve simultaneously. Term structure models are extensively covered later in the book.

In this chapter, we review one-factor diffusive models where our ability to alter the terminal distribution stems from a single source: a swap rate dependent diffusion function. Models of this type are often known as *deterministic volatility function* (DVF), or sometimes *local volatility function* (LVF), models. We first discuss the most common tractable specifications of such models — the CEV, displaced diffusion, and quadratic models — and then move on to efficient numerical or expansion-based methods for European option pricing within the general DVF model class. The listed techniques and results will be frequently referenced in later chapters, especially in the context of model calibration.

7.1 General Framework

7.1.1 Model Dynamics

Let $S(t)$ denote a forward Libor or swap rate, and let $W(t)$ be a one-dimensional Brownian motion under a measure P in which $S(\cdot)$ is a martingale. We assume that $S(t)$ follows the one-dimensional SDE

$$dS(t) = \lambda \varphi(S(t)) \, dW(t), \qquad (7.1)$$

where λ is a positive constant[1] and $\varphi : \mathbb{R} \to \mathbb{R}$ satisfies regularity conditions, such as those in Theorem 1.6.1. In most applications we would ideally want $S(t)$ to be non-negative, which is easily seen to impose the restriction

$$\varphi(0) = 0. \qquad (7.2)$$

In some cases we may consciously decide to violate (7.2) for the sake of model tractability.

When dealing with vanilla models, we primarily work in the measure P, so we typically abbreviate E^{P} to E when there is no possibility of confusion.

7.1.2 Volatility Smile and Implied Density

The role of the function φ is to match the distribution of S to that observed through puts and calls traded in the market. Specifically, let $c(t, S; T, K)$ denote the (non-deflated) time t value of a T-maturity European call option struck at K with $S(t) = S$, i.e.

$$c(t, S(t); T, K) = \mathrm{E}_t\left((S(T) - K)^+\right). \qquad (7.3)$$

The time t probability density of $S(T)$ can be derived from time t observed values of c, as (proceeding heuristically)

$$\mathrm{P}_t(S(T) \in dK)/dK = \mathrm{E}_t(\delta(S(T) - K)) \qquad (7.4)$$

$$= \mathrm{E}_t\left(\frac{\partial^2 c(T, S(T); T, K)}{\partial K^2}\right) = \frac{\partial^2 c(t, S(t); T, K)}{\partial K^2}, \qquad (7.5)$$

where δ is the Dirac delta function. This classical result is due to Breeden and Litzenberger [1978] and allows us to construct the marginal density of $S(T)$ from prices of T-maturity call options for a continuum of strikes K.

In option markets, it is common to express the strike dependency of call (and put) options in terms of the so-called *implied volatilities*. Specifically,

---

[1] We allow for time dependence later in the chapter, starting in Section 7.6.

for a given option price c at strike K and maturity T, we define the time t implied volatility function $\sigma_B(t, S; T, K)$ as the solution to

$$c(t, S; T, K) = S\Phi(d_+) - K\Phi(d_-), \tag{7.6}$$

$$d_\pm = \frac{\ln(S/K) \pm \frac{1}{2}\sigma_B(t, S; T, K)^2(T-t)}{\sigma_B(t, S; T, K)\sqrt{T-t}}.$$

We recognize the right-hand side of (7.6) as the Black-Scholes-Merton formula for a martingale process, i.e. the Black model (see Remark 1.9.4), with constant volatility $\sigma_B(t, S; T, K)$. The mapping $K \mapsto \sigma_B(t, S; T, K)$ is known as the T-maturity *volatility smile*[2]. In most established fixed income markets, the volatility smile is predominantly downward-sloping[3] in K, although it is not uncommon for σ_B to eventually increase in K for sufficiently large values of K.

7.1.3 Choice of φ

Had we allowed φ to depend on time, results by Dupire [1994] and Andersen and Brotherton-Ratcliffe [1998] demonstrate that any arbitrage-free marginal distribution of $S(T)$ can be realized by suitable choice of $\varphi = \varphi(t, S)$, $t \in [0, T]$. Indeed, non-parametric expressions exist to uniquely imply $\varphi(t, K)$ from observations of $\sigma_B(0, S(0); t, K)$ for the double continuum $(t, K) \in [0, T] \times [0, \infty)$. Unless the resulting φ happens to be monotonically increasing or decreasing in S, however, the resulting model will imply non-stationary volatility smile behavior, which is contrary to typical behavior of actual markets. To expand on this issue, consider setting

$$\varphi(S) = a + (S - S(0))^2, \tag{7.7}$$

where $a > 0$. The function $\varphi(S)$ is thus a U-shaped function with a minimum value of a at $S = S(0)$. Using formulas from Section 7.3 below, it can be verified (and is intuitively obvious) that the time 0 volatility smile σ_B produced by this parameterization is also U-shaped. Moving forward to time $t > 0$, consider the smile generated at t by (7.7) if $S(t) \gg S(0)$. At a large level of $S(t)$, $\varphi(S)$ will appear to be a strongly increasing function of S, causing (7.7) to produce a volatility smile no longer U-shaped, but instead monotonically increasing at all statistically relevant strikes. Conversely, if $S(t)$ diffuses below $S(0)$ such that $S(t) \ll S(0)$, a monotonically *decreasing* smile will arise at time t.

---

[2] In case the smile is monotonically downward or upward sloping, i.e. not U-shaped, it is often called a *volatility skew*. Skew then refers to the slope of the smile.

[3] This is not necessarily true for emerging markets where the volatility smile, when observed, can be significantly upward sloping or convex.

Strong level-dependence of the basic volatility smile shape is often at odds with observable market behavior, and non-monotonic specifications of $\varphi(S)$ — such as (7.7) — should consequently be approached with some care. As a consequence, the basic model (7.1) is most appropriate for markets where the volatility smile is (close to) a monotonic function of K. A classical monotonic choice for φ is the *constant elasticity of variance* (CEV) specification

$$\varphi(S) = S^p, \qquad (7.8)$$

for some constant p. As we proceed to show, this specification is analytically tractable.

7.2 CEV Model

7.2.1 Basic Properties

In this section, we examine the CEV specification (7.8) in detail. We start out with the following proposition:

Proposition 7.2.1. *Consider the stochastic differential equation*

$$dS(t) = \lambda S(t)^p \, dW(t), \qquad (7.9)$$

where $p > 0$ is constant and $W(t)$ is a one-dimensional Brownian motion. The following holds:

1. *All solutions to (7.9) are non-explosive.*
2. *For $p \geq 1/2$, the SDE (7.9) has a unique solution.*
3. *For $0 < p < 1$, $S = 0$ is an attainable boundary for (7.9); for $p \geq 1$, $S = 0$ is an unattainable boundary for (7.9).*
4. *For $0 < p \leq 1$, $S(t)$ in (7.9) is a martingale; for $p > 1$, $S(t)$ is a strict supermartingale.*

Proof. Property 1 follows from a remark on page 332 and equation (5.5.19) in Karatzas and Shreve [1991], and Property 2 follows from Example 5.2.14 in Karatzas and Shreve [1991]. Property 3 can be proven using the classical Feller boundary classification techniques based on speed/scale measure integral, see Section 5.5 of Karatzas and Shreve [1991]; Andersen and Andreasen [2000b] have the details. Property 4 is proven in Sin [1998]. □ More details on boundary characterization for CEV processes can be found in Davydov and Linetsky [2001].

Remark 7.2.2. For $p \geq 1/2$, the solution to (7.9) is unique. Hence, if the solution ever reaches the origin ($S = 0$), it stays there, i.e. is *absorbed*. For $0 < p < 1/2$, however, there are solutions that stay at origin if they reach it, and there are solutions that jump out if it. Hence, to define a unique solution, a boundary condition at $S = 0$ must be specified for (7.9). In practice, we set $S = 0$ to be an *absorbing* barrier: if $S(t)$ hits 0 for the first time at $t = \tau$, $S(u) = 0$ for all $u \geq \tau$. This condition is not only imposed to be consistent with the case of $p \geq 1/2$, but is also the only boundary condition consistent with the absence of arbitrage.

Remark 7.2.3. While it is common to require the parameter p to be positive, the process is well-defined for negative p, $p < 0$, as well, with the same absorbing boundary condition at $S = 0$ as for the case $0 < p < 1/2$ above. This enlargement of the domain of applicability of the process is occasionally useful in the fixed income markets, although much less so than in equity or FX markets where the smiles can generally be much more downward sloping.

For $p < 1$ and $t > 0$, the time 0 probability that $S(t) = 0$ is non-zero. In fact, it can be shown (see, for example, Cox [1996]) that if τ, the first time $S(\cdot)$ hits 0, is greater than t, then

$$P_t\left(\tau < T | \tau > t\right) = G\left(|\vartheta|, \frac{X(t)}{2\lambda^2(T-t)}\right), \quad T > t,$$

where

$$\vartheta = \frac{1}{2(p-1)}, \tag{7.10}$$

$$X(t) = \frac{S(t)^{2(1-p)}}{(1-p)^2}, \tag{7.11}$$

and G is the *complementary Gamma function*

$$G(a, x) \triangleq \frac{\Gamma(a, x)}{\Gamma(a)},$$

with the incomplete Gamma function $\Gamma(a, x)$ given by

$$\Gamma(a, x) = \int_x^\infty u^{a-1} e^{-u}\, du, \quad \Gamma(a) = \Gamma(a, 0). \tag{7.12}$$

If the absorption probability is substantial, one may want to consider regularizing the process to prevent absorption; see Section 7.2.3 for this.

Due to the result in Proposition 7.2.1, Property 4, we normally prefer to avoid using $p > 1$. As $p > 1$ will produce volatility smiles increasing in K (and thereby different from those in fixed income markets), this restriction on p is often of little practical concern.

The transition density of $S(\cdot)$ in (7.9) is known in closed form and is listed below for reference, along with a short proof that highlights the relationship between CEV processes and squared Bessel processes.

Lemma 7.2.4. *Consider the SDE (7.9) for any $p \neq 1$ (including $p < 0$ and $p > 1$), and let ϑ and $X(t)$ be as in (7.10)–(7.11). Let $q(X(T)|X(t))$ be the conditional P-density of $X(T)$ given $X(t) > 0$, $t < T$. If the level $S = 0$ is defined to be an absorbing boundary for (7.9) when $p \leq 1/2$, then*

$$q(X(T)|X(t)) = \frac{1}{2\lambda^2(T-t)} \exp\left(-\frac{X(T) + X(t)}{2\lambda^2(T-t)}\right)$$
$$\times \left(\frac{X(t)}{X(T)}\right)^{-\vartheta/2} I_{|\vartheta|}\left(\frac{\sqrt{X(T)X(t)}}{\lambda^2(T-t)}\right),$$

where $I_a(x)$ is the modified Bessel function of the first kind of order a:

$$I_a(x) = \sum_{j=0}^{\infty} \frac{(x/2)^{a+2j}}{j!\Gamma(a+j+1)}.$$

Proof. According to Ito's lemma, the process $X(t)$ satisfies the SDE

$$dX(t) = \lambda^2 \frac{1-2p}{1-p} dt + 2\lambda \sqrt{X(t)}\, dW(t).$$

Define the process $Y(v)$ by $Y(v) = X(v/\lambda^2)$. Applying a time change, it follows that

$$dY(v) = \frac{1-2p}{1-p} dv + 2\sqrt{Y(v)}\, d\widetilde{W}(v),$$

where $\widetilde{W}(\cdot)$ is a Brownian motion, up to the absorption time $\inf\{v > 0 : Y(v) = 0\}$. The process for Y can be identified as a so-called *squared Bessel process of index ϑ*. Standard results for this process (see e.g. p. 117 of Borodin and Salminen [1996]) give the result in the lemma. $\square$

Remark 7.2.5. By the usual transformation rules for densities, the density for $S(T)$ conditional on $S(t)$ is

$$q(X(T)|X(t)) \cdot 2S(T)^{2(1-p)-1}/|1-p|.$$

7.2.2 Call Option Pricing

Consider now the valuation of European call options in the CEV model, requiring evaluation of the expectation

$$c_{\text{CEV}}(t, S(t); T, K) \triangleq E_t\left((S(T) - K)^+\right)$$

for $S(\cdot)$ that follows (7.9). Using the definition (7.11), we can rewrite this as

$$c_{\text{CEV}}(t, S(t); T, K) = E_t\left(\left([(1-p)^2 X(T)]^{-\vartheta} - K\right)^+\right)$$
$$= \int_0^\infty \left([(1-p)^2 x]^{-\vartheta} - K\right)^+ q(x|X(t))\, dx,$$

where we have assumed $p \neq 1$ and the density $q(x|X(t))$ is given in Lemma 7.2.4. A straightforward, but tedious, integration exercise (see e.g. Schroder [1989] or Andersen and Andreasen [2000b]) yields the following result:

Proposition 7.2.6. *Consider the CEV model (7.9). Let $\chi^2_\nu(\gamma)$ be a non-central chi-square distributed variable with ν degrees of freedom and non-centrality parameter γ, and let $\Upsilon(x,\nu,\gamma) = \mathrm{P}(\chi^2_\nu(\gamma) \leq x)$ be the cumulative distribution function for $\chi^2_\nu(\gamma)$. Also define*

$$a = \frac{K^{2(1-p)}}{(1-p)^2 \lambda^2 (T-t)}, \quad b = |p-1|^{-1}, \quad c = \frac{S^{2(1-p)}}{(1-p)^2 \lambda^2 (T-t)}.$$

Then, for $0 < p < 1$ and an absorbing boundary at $S = 0$ we have, for $K > 0$,

$$c_{\mathrm{CEV}}(t,S;T,K) = S\left(1 - \Upsilon(a, b+2, c)\right) - K\Upsilon(c, b, a). \tag{7.13}$$

Remark 7.2.7. The result above in fact holds for all $p < 1$, including negative p. A complimentary result holds for $p > 1$,

$$c_{\mathrm{CEV}}(t,S;T,K) = S\left(1 - \Upsilon(c, b, a)\right) - K\Upsilon(a, b+2, c). \tag{7.14}$$

Remark 7.2.8. The special case $p = 1$ leads to the Black pricing formula with volatility λ, see (1.43) and Remark 1.9.4, so that

$$c_{\mathrm{B}}(t,S;T,K;\lambda) = S\Phi(d_+) - K\Phi(d_-), \tag{7.15}$$

where

$$d_\pm = \frac{\ln(S/K) \pm \lambda^2(T-t)/2}{\lambda\sqrt{T-t}},$$

and $\Phi(\cdot)$ is the standard Gaussian CDF.

Remark 7.2.9. For the case $p = 0$, if we remove the assumption of an absorbing barrier at the origin, $S(t)$ is a Gaussian process. In this case, it is straightforward to compute that the option pricing formula, sometimes called the *Normal*, *Gaussian* or *Bachelier* pricing formula with (Normal) volatility[4] λ, becomes

$$c_{\mathrm{N}}(t,S;T,K;\lambda) = (S-K)\Phi(d) + \lambda\sqrt{T-t}\phi(d), \quad d = \frac{S-K}{\lambda\sqrt{T-t}}, \tag{7.16}$$

where $\Phi(\cdot)$ and $\phi(\cdot)$ are the standard Gaussian CDF and PDF, respectively.

Further details about the non-central chi-square distribution can be found in Chapter 3. A number of efficient numerical algorithms exist to compute $\Upsilon(x,\nu,\gamma)$; see Johnson et al. [1995] for a survey. A standard algorithm can be found in Ding [1992]. Figure 7.1 on page 298 gives some examples of volatility skews produced by the CEV model.

---

[4] Also known as Gaussian volatility; when applied to interest rates, Gaussian volatilities are often called *basis-point*, or *bp*, volatilities.

7.2.3 Regularization

As discussed earlier, the CEV process implies a positive probability of absorption at $S = 0$ (for $p < 1$). This phenomenon is not necessarily a problem for pricing of simple European call options, but is obviously not desirable from an empirical standpoint[5], and might also create some difficulties in pricing of more exotic structures. To avoid absorption, we can specify a regularized version of the CEV model by letting,

$$\varphi(x) = x \min\left(\varepsilon^{p-1}, x^{p-1}\right), \quad \varepsilon > 0, p < 1. \tag{7.17}$$

Roughly speaking, when $S(t)$ crosses the level ε, the resulting process becomes (locally) a geometric Brownian motion with finite volatility ε^{p-1}. With $\varphi(x)$ now Lipschitz continuous, it is straightforward to verify that the process for $S(t)$ can no longer reach the origin. On the other hand, the specification (7.17) will not allow for closed-form call option pricing but will, in principle at least, require the usage of numerical methods such as the finite difference method (see Section 7.4). On the other hand, for small to moderate values of ε, we would expect the CEV pricing formulas from Proposition 7.2.6 to hold as a good approximation. Andersen and Andreasen [2000b] verify numerically that this holds quite robustly, for strikes not too far from the spot value of S. More formally, we have the following result:

Proposition 7.2.10. *For $p < 1$ and $\varepsilon > 0$, let*

$$dx(t) = \lambda x(t)^p \, dW(t),$$
$$dy(t) = \lambda y(t) \min\left(\varepsilon^{p-1}, y(t)^{p-1}\right) dW(t),$$

where $x(0) = y(0) > 0$ and $W(t)$ is a one-dimensional Brownian motion in measure P. For $p < 1/2$, 0 is assumed to be an absorbing boundary for x. For some $T > t$ and some constant K, we then have

$$\lim_{\varepsilon \downarrow 0} |\mathrm{P}\left(x(T) < h\right) - \mathrm{P}\left(y(T) < h\right)| = 0,$$

$$\lim_{\varepsilon \downarrow 0} \left|\mathrm{E}\left((x(T) - K)^+\right) - \mathrm{E}\left((y(T) - K)^+\right)\right| = 0.$$

The result is intuitive, but the proof is somewhat technical, and we skip it. Details can be found in Andersen and Andreasen [2000b].

---

[5] As the measure P is equivalent to the real-life (statistical) measure, a non-zero probability of absorption under P implies a non-zero probability of absorption under the real-life measure.

7.2.4 Displaced Diffusion Models

An easy extension of the CEV model that is sometimes useful involves adding a displacement constant to the CEV specification. Specifically, we write

$$\varphi(x) = (\alpha + x)^p \tag{7.18}$$

for some constant α. In the process (7.1), (7.18), let us set $Z(t) = \alpha + S(t)$. By Ito's lemma, $Z(t)$ then satisfies the CEV SDE

$$dZ(t) = \lambda Z(t)^p \, dW(t).$$

With $Z(t)$ having an absorbing boundary at 0, $S(t)$ then must have an absorbing boundary at $-\alpha$. Call option pricing with (7.18) is straightforward:

Proposition 7.2.11. *Let*

$$c_{\mathrm{DCEV}}(t, S(t); T, K, \alpha) = \mathrm{E}_t\left((S(T) - K)^+\right)$$

be the call option price associated with the displaced CEV process (7.1), (7.18). Then

$$c_{\mathrm{DCEV}}(t, S; T, K, \alpha) = c_{\mathrm{CEV}}(t, S + \alpha; T, K + \alpha), \quad S, K > -\alpha, \tag{7.19}$$

where the right-hand side is given by Proposition 7.2.6.

Proof. The result follows directly from the observation that

$$\mathrm{E}_t\left((S(T) - K)^+\right) = \mathrm{E}_t\left((Z(T) - (K + \alpha))^+\right),$$

where $Z(t) = \alpha + S(t)$ follows a regular CEV process. □

Introduction of the displacement constant α allows for a (somewhat) richer family of volatility smiles than those of the pure CEV specification. In practice, however, the main use of displacement constants is for the special case of the *displaced log-normal*, or *shifted log-normal*, process where $p = 1$. The call option price formula for this case is listed below, for later reference.

Proposition 7.2.12. *Consider the displaced log-normal process*

$$dS(t) = \lambda\left(\beta + \zeta S(t)\right) dW(t), \tag{7.20}$$

where $W(t)$ is a one-dimensional Brownian motion in measure P, and $\zeta, \lambda \neq 0$. Assuming $S(t), K > -\beta/\zeta$, we have

$$c_{\mathrm{DLN}}(t, S(t); T, K) \triangleq \mathrm{E}_t\left((S(T) - K)^+\right)$$

$$= \left(S(t) + \frac{\beta}{\zeta}\right)\Phi(d_+) - \left(K + \frac{\beta}{\zeta}\right)\Phi(d_-),$$

$$d_\pm = \frac{\ln\left(\frac{S(t) + \beta/\zeta}{K + \beta/\zeta}\right) \pm \frac{1}{2}\zeta^2\lambda^2(T - t)}{|\zeta\lambda|\sqrt{T - t}}.$$

Proof. The result follows directly from the Black-Scholes equation (see Section 1.9) and (7.19), after setting $\alpha = \beta/\zeta$ and writing $\lambda(\beta + \zeta S(t)) = \lambda\zeta(\alpha + S(t))$. □

Remark 7.2.13. It is often convenient to rewrite the displaced log-normal process in a slightly different form

$$dS(t) = \sigma\left(bS(t) + (1-b)L\right)dW(t). \tag{7.21}$$

The parameter L is often set to near, or exactly at, the initial value $S(0)$. In this parameterization, σ is expressed in the units of relative volatility, just like in the Black model, because $bS(0) + (1-b)L \approx S(0)$. In particular, σ always has the same scale for all values of b. Moreover, the effects of σ and b are almost "orthogonal", in the sense that the parameter σ changes the overall level of the implied volatility smile but not its slope, whereas b only changes the slope (skew) of the implied volatility smile but not its overall level (i.e. not the at-the-money implied volatility). We use the parameterization (7.21) extensively in later chapters.

Remark 7.2.14. Consider the general local volatility model (7.1). Expanding the local volatility function $\varphi(\cdot)$ around at-the-money to the first order, we obtain

$$dS(t) \approx \lambda\left(\varphi\left(S(0)\right) + \varphi'\left(S(0)\right)\left(S(t) - S(0)\right)\right)dW(t),$$

which we identify as being of the form (7.21) with

$$\sigma = \lambda\frac{\varphi\left(S(0)\right)}{S(0)}, \quad b = \varphi'\left(S(0)\right)\frac{S(0)}{\varphi\left(S(0)\right)}, \quad L = S(0).$$

Hence, a first-order approximation to any local volatility process is of displaced log-normal type. In view of this, displaced log-normal processes are extensively used in various types of approximations and asymptotic expansions.

The previous remark can be applied to the CEV process:

$$\sigma = \lambda S(0)^{p-1}, \quad b = p, \quad L = S(0).$$

The approximation of the CEV process with (7.21) turns out to be particularly close, and we later use it to increase the tractability of certain stochastic volatility models. We also use it as a justification to freely switch from one type of process to the other. It is worth noting, however, that (7.20) has certain drawbacks relative to a pure CEV process. First, the process for $S(\cdot)$ can become negative if $\beta\zeta$ (as is usual) is positive. Second, in stochastic volatility applications the asymptotic linear growth of $\varphi(x)$ in x can sometimes lead to technical problems and unbounded second moments of $S(\cdot)$. We shall return to this issue shortly, in Chapter 8.

7.3 Quadratic Volatility Model

In practice, volatility smiles in fixed income markets are not always perfectly monotonic in strike; indeed, as mentioned earlier, for sufficiently high strikes it is not uncommon for the smile to reverse direction and start increasing in strike. This type of behavior is inconsistent with a pure CEV model, but can, to some extent, be captured by the displaced CEV specification $\varphi(x) = (\alpha + x)^p$. Often, however, this model is hard to fit to actual data. A more powerful approach involves overlaying the CEV process with stochastic volatility, something that we turn to in Chapter 8. If we here wish to stay within the realm of DVF processes, one way to generate arbitrarily convex smiles is to use a *quadratic volatility model*, where

$$\varphi(x) = \alpha + \beta x + \gamma x^2, \tag{7.22}$$

for constants α, β, γ. We develop some aspects of this model here, but remind the reader of the caveats discussed in Section 7.1.3; in particular, for the model to be realistic, γ should probably be small.

Before commencing with derivations, let us note that the behavior of a DVF model (7.1) equipped with volatility function (7.22) will depend strongly on the root configuration in the quadratic polynomial $\alpha + \beta x + \gamma x^2$. For instance, if φ has two real roots l, u, $l < u$, straddling the initial value $S(0)$, it is clear that $S(t)$ will itself be bound to this range, i.e. $S(t) \in [l, u]$. Specifically, whenever $S(t)$ gets close to either l or u, $\varphi(x)$ will approach zero and the diffusion for $S(t)$ will gradually slow down. As such range-bound dynamics are rather unrealistic for interest rate applications[6], we do not consider it in the following.

7.3.1 Case 1: Two Real Roots to the Left of $S(0)$

We first consider the case where $\alpha + \beta x + \gamma x^2$ has two real roots l and u, $l < u$, both lying to the left of $S(0)$. Without loss of generality, we may then consider the normalized process

$$dS(t) = \frac{(S(t) - u)(S(t) - l)}{u - l} dW(t), \quad S(0) > u > l. \tag{7.23}$$

We start by listing a few lemmas.

Lemma 7.3.1. *The range for $S(t)$ in (7.23) is $S(t) \in (u, \infty)$. In particular, the process for $S(t)$ does not explode in measure* P.

---

[6]For an application of the range-bound quadratic model to FX markets (where currency controls may potentially create upper and lower bounds), see Ingersoll [1997].

Proof. That $S(t)$ cannot go below u is obvious; further, Feller's boundary criteria (e.g. Karlin and Taylor [1981], Chapter 15.6) establishes that u is not accessible when $S(0) > u$. As $S(t)$ is described by a time-homogeneous one-dimensional SDE, it cannot explode (Karatzas and Shreve [1991], p. 332]). □

While the process for $S(t)$ is non-explosive, the super-linear growth[7] of $\varphi(x)$ causes some interesting technical problems, In particular, we have the following result, proved in Andersen [2010].

Lemma 7.3.2. *The process (7.23) is a strict supermartingale in measure* P.

As the process for S is not a martingale, the usual pricing results require some modifications. For the purpose of pricing puts and calls, we need use the following.

Lemma 7.3.3. *Suppose that $S(t)$ satisfies (7.23) in some measure* P *and assume that put-call parity holds. Then the prices at time 0 for the put (p) and call (c) are*

$$p(0, S(0); T, K) = \mathrm{E}\left((K - S(T))^+\right),$$

$$c(0, S(0); T, K) = p(0, S(0); T, K) + S(0) - K > \mathrm{E}\left((S(T) - K)^+\right).$$

Proof. (Sketch only). In the absence of arbitrage, the put price is a local martingale in measure P. As a bounded local martingale is a martingale and the put payout is bounded between 0 and $K - u$, it follows then that the put price in fact must be a true P-martingale. The expression for $p(0, S(0); T, K)$ follows. Applying put-call parity (see Chapter 1) yields the result for $c(0, S(0); T, K)$, where the inequality follows from Lemma 7.3.2. □

We emphasize the non-standard result $c(0, S(0); T, K) > \mathrm{E}(c(T, S(T); T, K))$ which is a consequence of the supermartingale property of $S(t)$. The inequality holds for arbitrarily large strikes; indeed, rather counter-intuitively, $\lim_{K \to \infty} c(0, S(0); T, K) = S(0) - \mathrm{E}(S(T)) > 0$. We should also note that our assumption of put-call parity being valid is critical here, as it allows us to produce unique prices of both puts and calls. As described in Heston et al. [2007] and Andersen [2010], it is, however, possible to work with other assumptions without violating no-arbitrage.

With 7.3.3 we are now ready to tackle the derivation of an option pricing formula. We will be using the shorthand

$$p(t) \triangleq p(t, S(t); T, K),$$

and so forth. First, notice the useful relationship

---
[7] A similar issue is present in CEV processes with $p > 1$, as noted earlier.

7.3 Quadratic Volatility Model

$$S - K = \frac{(S-u)(K-l) - (K-u)(S-l)}{u-l}, \quad (7.24)$$

which allows us to write

$$\begin{aligned}
p(T) &= \frac{1}{u-l}\left((K-u)(S(T)-l) - (S(T)-u)(K-l)\right)^+ \\
&= \frac{(K-u)(S(T)-l)}{u-l} 1_{\{(K-u)(S(T)-l)-(S(T)-u)(K-l)>0\}} \\
&\quad - \frac{(S(T)-u)(K-l)}{u-l} 1_{\{(K-u)(S(T)-l)-(S(T)-u)(K-l)>0\}} \\
&\triangleq p_1(T) - p_2(T). \quad (7.25)
\end{aligned}$$

The payouts p_1 and p_2 have identical structure, so it suffices to focus our attention on pricing one of them, e.g. p_1.

From Lemma 7.3.3, we have $p_1(0) = E(p_1(T))$, which we rewrite as

$$p_1(0) = \frac{K-u}{u-l} E\left((S(T)-l)1_{\{(S(T)-u)/(S(T)-l)<(K-u)/(K-l)\}}\right). \quad (7.26)$$

At this point our first instinct would be to perform a measure shift that eliminates that factor $S(T) - l$ in the expectation, i.e. we would like to introduce a new measure $\widetilde{P}$ such that

$$\widetilde{P}(B) = \frac{1}{S(0)-l} E((S(T)-l)1_{\{B\}}),$$

for any $\mathcal{F}_T$-measurable event B. We recall, however, that $S(t)$ (and therefore $S(t)-l$) is not a martingale in P, so such a measure shift cannot be performed outright. Let us nevertheless try. Proceeding mechanically as if $S(t)$ were a martingale, we would get, for the process $Y(t) \triangleq (S(t)-u)/(S(t)-l)$,

$$dY(t) \stackrel{?}{=} Y(t)\,d\widetilde{W}(t), \quad Y(0) = \frac{S(0)-u}{S(0)-l} < 1, \quad (7.27)$$

where $\widetilde{W}$ is a Brownian motion in $\widetilde{P}$. Clearly, however, there are technical problems here: the range for $Y(t)$ in (7.27) is $[0, \infty)$, whereas we know that in measure P we have $Y(t) \in (0,1)$ (since $S(t) \in (u, \infty)$); the two measures therefore cannot be equivalent. For option pricing purposes, it turns out that the correct way to handle the technical conflict involves inserting an *absorbing boundary* at $Y = 1$ in (7.27).

Proposition 7.3.4. *Let*

$$dY(t) = Y(t)\,d\widetilde{W}(t), \quad Y(0) = \frac{S(0)-u}{S(0)-l} < 1,$$

be geometric Brownian motion in $\widetilde{P}$. Define $\tau = \inf\{t > 0 : Y(t) = 1\}$, and let $K > u$. Then $p_1(0)$ in (7.26) is given by

$$p_1(0) = \frac{(K-u)(S(0)-l)}{u-l} E^{\tilde{P}} \left(1_{\{Y(T)<(K-u)/(K-l)\}} 1_{\{\tau>T\}} \right). \quad (7.28)$$

Stated explicitly,

$$p_1(0) = K_1 \Phi \left(\frac{-\ln(X_1/K_1) + T/2}{\sqrt{T}} \right) - X_2 \Phi \left(\frac{\ln(X_2/K_2) + T/2}{\sqrt{T}} \right), \quad (7.29)$$

with Φ being the Gaussian cumulative distribution function, and

$$K_1 = \frac{(K-u)(S(0)-l)}{u-l}, \quad X_1 = \frac{(S(0)-u)(K-l)}{u-l},$$
$$K_2 = \frac{(K-l)(S(0)-l)}{u-l}, \quad X_2 = \frac{(S(0)-u)(K-u)}{u-l}.$$

Proof. The result (7.28) is proven in Andersen [2010]. The result (7.29) follows by direct calculations, similar to those leading to the Black-Scholes-Merton formula. □

Following similar steps leads to an expression for $p_2(0)$, which in turn leads to the following result for $p(0) = p_1(0) - p_2(0)$.

Proposition 7.3.5. *Let K_i, X_i, $i = 1, 2$, be given as in Proposition 7.3.4. Assuming $K > u$, the put price $p(0)$ for the model (7.23) has the explicit representation*

$$p(0, S(0); T, K) = K_1 \Phi \left(-d_-^{(1)} \right) - X_2 \Phi \left(d_+^{(2)} \right) - X_1 \Phi \left(-d_+^{(1)} \right) + K_2 \Phi \left(d_-^{(2)} \right),$$

$$d_\pm^{(i)} = \frac{\ln(X_i/K_i) \pm T/2}{\sqrt{T}}, \quad i = 1, 2.$$

An application of put-call parity then immediately gives the call price:

Corollary 7.3.6. *Assuming put-call parity holds, the call price for the model (7.23) is*

$$c(0, S(0); T, K) = S(0) - K + p(0, S(0); T, K),$$

with $p(0, S(0); T, K)$ given in Proposition 7.3.5.

We recall that Proposition 7.3.5 applies to (7.23), rather than our original process which, at the root configuration in question, is

$$dS(t) = \lambda\gamma\left(S(t)-u\right)\left(S(t)-l\right) dW(t) = q \frac{(S(t)-u)(S(t)-l)}{u-l} dW(t), \quad (7.30)$$

where $q = \lambda\gamma(u-l)$. The constant in front of the quadratic polynomial is easily handled by time-scaling: to price options in (7.30) we simply set the put price equal to $p(0, S(0); q^2 T, K)$, where $p(0, S(0); \cdot, K)$ is given by the formula in Proposition 7.3.5.

7.3.2 Case 2: One Real Root to the Left of $S(0)$

If we let the single root to $\alpha + \beta x + \gamma x^2$ be denoted u, $u < S(0)$, it suffices to consider the normalized process

$$dS(t) = (S(t) - u)^2 \, dW(t). \tag{7.31}$$

But this process is a special case, with power equal to 2, of the displaced CEV model in Section 7.2.4, and the option pricing formulas from that section then apply directly. As these formulas are rather complicated in their dependence on the non-central chi-square distribution, it is worthwhile noticing that simple expressions exist for the special case of power equal to 2. The result is listed below.

Proposition 7.3.7. *For the process (7.31), the put option price is*

$$p(0, S(0); T, K) = (S(0) - u)(K - u)\sqrt{T} \\ \times \{d_+ \Phi(d_+) + \phi(d_+) - d_- \Phi(d_-) - \phi(d_-)\},$$

where $\phi(x)$ is the Gaussian density, and

$$d_\pm = \frac{\pm \frac{1}{S(0)-u} - \frac{1}{K-u}}{\sqrt{T}}.$$

Proof. We observe that for the process

$$dS(t) = (S(t) - u)(S(t) - l) \, dW(t), \quad l < u < S(0),$$

the put price can be computed from the result in Proposition 7.3.5, after a time-change, from T to $T(u-l)^2$; see the comments at the end of Section 7.3.1. Taking the limit of the put price as $l \uparrow u$ then establishes the result. □

The call option price can, as before, be found by put-call parity. To establish put and call option prices for the original diffusion

$$dS(t) = \lambda \left(\alpha + \beta S(t) + \gamma S(t)^2 \right) dW(t) = \lambda \gamma \left(S(t) - u \right)^2 dW(t),$$

we simply change T to $\lambda^2 \gamma^2 T$ in Proposition 7.3.7.

7.3.3 Extensions and Other Root Configurations

The results listed in Sections 7.3.1 and 7.3.2 have given a flavor of how to deal with a quadratic volatility process, and shall suffice for the purposes of this book. Other root configurations are treated in detail in Andersen [2010], including the case where $\varphi(x)$ has no roots (in which case the put and call option price formulas are infinite sine-series). Andersen [2010] also discusses the case where an absorbing barrier has been inserted at the origin to prevent $S(t)$ from going negative.

7.4 Finite Difference Solutions for General φ

For general specifications of φ, closed-form solutions for European options will not exist. In such cases, we may instead rely on the finite difference methods discussed in Chapter 2. Consider again the evaluation of

$$c(t, S(t); T, K) = \mathrm{E}_t \left((S(T) - K)^+ \right),$$

with $S(t)$ following (7.1). With suitable regularity conditions on φ, the Feynman-Kac theorem of Section 1.8 shows that $c(t, S) = c(t, S; T, K)$ (with T, K fixed) satisfies the PDE

$$\frac{\partial c(t, S)}{\partial t} + \frac{1}{2} \lambda^2 \varphi(S)^2 \frac{\partial^2 c(t, S)}{\partial S^2} = 0, \qquad (7.32)$$

subject to the terminal condition

$$c(T, S) = (S - K)^+. \qquad (7.33)$$

This PDE can be solved numerically using, say, the Crank-Nicholson finite difference grid method in Chapter 2. A direct discretization of (7.32) is normally sufficient, but we note that it may occasionally be possible to take advantage of special forms of φ and introduce transformations of S to improve the properties of the finite difference scheme. For example, as we have already seen in Chapter 2, when $\varphi(S) = S$, it is customary (and appropriate) to introduce $y(S) = \ln S$ and discretize in y. More generally, for sufficiently regular φ, the transformation

$$y(S) = \int \frac{dS}{\varphi(S)} \qquad (7.34)$$

(see (2.81)–(2.82)) might offer numerical advantages over a direct discretization provided, of course, that the inverse in (7.34) exists. The following semi-heuristic argument explains the rationale. With the transform (7.34), the SDE for $y(t) = y(S(t))$ is (ignoring the drift)

$$dy(t) = O(dt) + \lambda \, dW(t).$$

The diffusion coefficient in the process for y is independent of the state of S, suggesting that a differential operator expressed in terms of y may have better numerical properties than the one expressed in terms of S. Even if y is not used for discretization, the transformation (7.34) suggests the discretization grid in the S-domain. In particular, $\{S_n\}_{n=0}^{m+1}$ can be defined by the condition that $y_n = y(S_n)$, $n = 0, \ldots, m+1$, are equidistant over $[y(S_0), y(S_{m+1})]$. For $n = 0, \ldots, m+1$ this gives ($y^{-1}(\cdot)$ is the inverse transform of (7.34))

$$y_n = y(S_0) + \frac{n}{m+1} \left(y(S_{m+1}) - y(S_0) \right),$$

$$S_n = y^{-1}(y_n) = y^{-1} \left(y(S_0) + \frac{n}{m+1} \left(y(S_{m+1}) - y(S_0) \right) \right).$$

7.4.1 Multiple λ and T

In applications, we often need to compute the values of $c(t, S; T, K)$ for many different values of T and/or λ. This need arises, for instance, in a standard model calibration exercise where we use a root-search algorithm to determine the value of λ that will make the computed call prices at different maturities T equal to values observed in the market. In such cases, we note that one should *not* simply solve (7.32) over and over (at great computational expense), but instead rely on the following observation:

Proposition 7.4.1. *Let $g(\tau, x)$ solve the following PDE*

$$-\frac{\partial g(\tau, x)}{\partial \tau} + \frac{1}{2}\varphi(x)^2 \frac{\partial^2 g(\tau, x)}{\partial x^2} = 0, \qquad (7.35)$$

with initial condition

$$g(0, x) = (x - K)^+. \qquad (7.36)$$

Let $c(t, S)$ solve the backward PDE (7.32)–(7.33) for a given value of λ. Then

$$c(t, S; T, K) = g\left(\lambda^2(T - t), S\right). \qquad (7.37)$$

Proof. Follows directly from a variable transformation $\tau(t) = \lambda^2(T - t)$ in (7.32)–(7.33), taking advantage of the time-homogeneity of φ. □

Using finite difference techniques to solve the PDE (7.35), we can construct the function g on a (τ, S)-grid; once this grid is stored in memory, (7.37) is used to recover $c(t, S; T, K)$ for arbitrary choices of S, λ and T by simple lookup or interpolation. We emphasize that this approach involves the numerical solution of only a *single* PDE. Also note that PDE is solved forward in time from a known *initial* condition, rather than backwards from a *terminal* condition.

7.4.2 Forward Equation for Call Options

While the function g from (7.35) is conveniently independent of T and λ, it does depend on K through the initial condition (7.36). In some applications, we may wish to use different strikes for different values of T, in which case the approach in Section 7.4.1 requires us to numerically solve as many finite difference grids as there are different values of K. We can improve on this by replacing the backward equation (7.32) with the forward equation of Dupire [1994]. In this approach, calendar time t and the initial value of S are considered fixed, whereas maturity T and strike K are variable. In view of this, we define $c(T, K) = c(t, S; T, K)$ for fixed t, S. We need the following proposition:

Proposition 7.4.2. *Define the function $c(T, K) \triangleq c(t, S; T, K)$ where t, S are fixed and $c(t, S; T, K)$ is defined by (7.3) for the model (7.1). Then $c(T, K)$ satisfies the forward PDE*

$$-\frac{\partial c(T, K)}{\partial T} + \frac{1}{2}\lambda^2 \varphi(K)^2 \frac{\partial^2 c(T, K)}{\partial K^2} = 0, \qquad (7.38)$$

for $T > t$, subject to the time t initial condition

$$c(t, K) = (S - K)^+.$$

Proof. In Dupire [1994], the result is proven by combining the Fokker-Planck equation (see Section 1.8) with the result (7.5), followed by a series of integrations. A more intuitive line of attack proceeds as follows. Consider the function $H(t) = (S(t) - K)^+$. While $H(t)$ clearly does not satisfy the smoothness requirements of Ito's lemma, the Tanaka extension nevertheless justifies the following result, obtained by formally applying Ito's lemma to H:

$$dH(t) = 1_{\{S(t) > K\}} \lambda \varphi(S(t)) \, dW(t) + \frac{1}{2}\delta(S(t) - K) \lambda^2 \varphi(S(t))^2 \, dt. \quad (7.39)$$

That is,

$$H(T) = H(t) + \int_t^T 1_{\{S(u) > K\}} \lambda \varphi(S(u)) \, dW(u)$$
$$+ \frac{1}{2}\int_t^T \delta(S(u) - K) \lambda^2 \varphi(S(u))^2 \, du$$
$$= H(t) + M(T) + \frac{1}{2}\int_t^T \delta(S(u) - K) \lambda^2 \varphi(K)^2 \, du,$$

where δ is the Dirac delta function and $M(t)$ is a continuous martingale with $M(t) = 0$. From (7.3), we have that

$$c(t, S(t); T, K) = E_t(H(T))$$
$$= H(t) + \frac{1}{2}\int_t^T E_t(\delta(S(u) - K)) \lambda^2 \varphi(K)^2 \, du$$
$$= H(t) + \frac{1}{2}\lambda^2 \varphi(K)^2 \int_t^T \frac{\partial^2 c(t, S(t); u, K)}{\partial K^2} \, du,$$

where we have used the martingale property of M as well as the result (7.5). Differentiating this equation with respect to T gives the result in Proposition 7.4.2. □

As mentioned in Chapter 1, the term $\int_t^T \delta(S(u) - K) \, du$ in the expression for $H(T)$ is known as the *local time* of $S(\cdot)$ at the level K. Local time and the Tanaka extension are deep subjects (see Karatzas and Shreve [1991] for a formal discussion) and have many interesting applications in finance, see for instance Andersen et al. [2002], Andersen and Andreasen [2000a], Andersen and Buffum [2003], Henderson and Hobson [2000], Carr and Jarrow [1990], Carr and Wu [2003], among many others.

We emphasize that while the backward equation (7.32) holds for European derivative securities on S in general, the forward equation (7.38) is unique to calls and puts, as only put and call payouts allow for the basic result (7.5).

Equipped with Proposition 7.4.2, the following result immediately follows from the proof of Proposition 7.4.1. Notice the difference in the initial conditions (7.36) and (7.40).

Proposition 7.4.3. *Let $h(\tau, x)$ solve the following PDE*

$$-\frac{\partial h(\tau, x)}{\partial \tau} + \frac{1}{2}\varphi(x)^2 \frac{\partial^2 h(\tau, x)}{\partial x^2} = 0,$$

with initial condition

$$h(0, x) = (S - x)^+. \tag{7.40}$$

Then

$$c(t, S; T, K) = h\left(\lambda^2(T-t), K\right).$$

As long as the initial value of $S(t)$ is kept constant, the result in Proposition 7.4.3 allows us use a single finite difference grid to price call options with multiple maturities, strikes, and λ's. We note, however, that in many applications $S(t)$ may in fact be T-dependent, as S will often represent, say, T-maturity Libor forward rates. In such cases, the question of whether Proposition 7.4.3 leads to a more efficient numerical scheme than Proposition 7.4.1 is settled by comparing the number of strikes and the number of spot levels involved.

7.5 Asymptotic Expansions for General φ

As we have shown, there are a number of "tricks" that can be employed to make the application of finite difference methods a computationally viable approach to pricing a large number of European call options. Nevertheless, there is significant convenience and computer code simplification associated with closed-form pricing formulas, so we now turn to the development of asymptotic approximations for the solution to the generic backward PDE (7.32). There are a number of approaches that can be taken, including the "most likely path" method in Gatheral [2001] (see also Gatheral [2006]) and Section 22.1.7, and the singular perturbation techniques in Hagan and Woodward [1999b], Henry-Labordère [2008], Gatheral et al. [2009], to name a few. Our presentation here is based on a fairly straightforward, yet often highly accurate, asymptotic expansion in time to maturity.

7.5.1 Expansion around Displaced Log-Normal Process

As in Proposition 7.4.1, we start by writing $c(t, S; T, K) = g(\tau, S)$, where $\tau = \lambda^2(T-t)$ and g satisfies (7.35). Inspired by the known solution of (7.35) in Proposition 7.2.12 for the case $\varphi(x) = \beta + \zeta x$, $\zeta \neq 0$, let us guess at a solution of (7.35) of the form

$$g(\tau, S) = \left(S + \frac{\beta}{\zeta}\right)\Phi(z_+) - \left(K + \frac{\beta}{\zeta}\right)\Phi(z_-), \tag{7.41}$$

$$z_\pm = \frac{\ln\left(\frac{S+\beta/\zeta}{K+\beta/\zeta}\right) \pm \frac{1}{2}\Omega(\tau,S)^2}{\Omega(\tau, S)},$$

where the function $\Omega(\tau, S)$ is to be determined. In (7.41), note that we obviously must assume that $S, K > -\beta/\zeta$. Substituting (7.41) into (7.35) gives the following PDE for $\Omega(\tau, S)$:

$$\left(S + \frac{\beta}{\zeta}\right)^2 \Omega \frac{\partial \Omega}{\partial \tau}$$
$$= \frac{1}{2}\varphi(S)^2 \left[\left(S + \frac{\beta}{\zeta}\right)^2 \Omega \frac{\partial^2 \Omega}{\partial S^2} + (1 - h_{-3}) - h_1(1 - h_{-1})\right], \tag{7.42}$$

where

$$h_i \triangleq \left(S + \frac{\beta}{\zeta}\right)\frac{\partial \Omega}{\partial S}\left(\Omega^{-1}\ln\left(\frac{S + \beta/\zeta}{K + \beta/\zeta}\right) + \frac{1}{2}i\Omega\right), \quad i = -3, -1, 1.$$

The PDE (7.42) does not generally allow for an explicit solution, so we resort to an asymptotic expansion in τ.

Proposition 7.5.1. *An asymptotic expansion for the solution of (7.35) is given by (7.41), with*

$$\Omega(\tau, S) = \Omega_0(S)\tau^{1/2} + \Omega_1(S)\tau^{3/2} + O\left(\tau^{5/2}\right), \tag{7.43}$$

$$\Omega_0(S) = \ln\left(\frac{S + \beta/\zeta}{K + \beta/\zeta}\right)\left(\int_K^S \varphi(u)^{-1}\,du\right)^{-1}, \tag{7.44}$$

$$\Omega_1(S) = -\frac{\Omega_0(S)}{\left(\int_K^S \varphi(u)^{-1}\,du\right)^2} \ln\left(\Omega_0(S)\frac{(S + \beta/\zeta)(K + \beta/\zeta)}{\varphi(S)\varphi(K)}\right)^{1/2},$$

$$\tag{7.45}$$

where the parameters β and ζ can be chosen arbitrarily, subject to the constraints $S, K > -\beta/\zeta$ and $\zeta \neq 0$.

Proof. In (7.41) we clearly require $\Omega(\tau, S) \sim \tau^{1/2}$ as $\tau \to 0$, so we seek a small-time solution of the form

$$\Omega(\tau, S) = \sum_{i \geq 0} \tau^{i+1/2} \Omega_i(S). \tag{7.46}$$

Notice that (7.46) omits all integer powers of τ — it turns out that their weights are all identically 0. Substituting (7.46) into (7.42) and matching terms of order $O(1)$ gives

$$(S + \beta/\zeta)^2 \Omega_0^2 = \varphi(S)^2 \left(1 - \frac{\Omega_0'}{\Omega_0} (S + \beta/\zeta) \ln\left(\frac{S + \beta/\zeta}{K + \beta/\zeta}\right)\right)^2, \tag{7.47}$$

where the prime denotes differentiation with respect to S. Taking the square root of the above equation and rearranging leads to two first-order ordinary differential equations of the Bernoulli type. Solving (7.47) subject to the boundary condition that the limit of Ω_0 must be finite for $S \to K$ (and discarding the negative solution) leads to (7.44).

Progressing now to the $O(\tau)$ term in (7.42), we get

$$2(S + \beta/\zeta) \Omega_1 = \frac{1}{2}\varphi(S)^2 \left((S + \beta/\zeta) \Omega_0'' + \Omega_0'\right)$$
$$- \varphi(S)(S + \beta/\zeta) \ln\left(\frac{S + \beta/\zeta}{K + \beta/\zeta}\right) \left(\frac{\Omega_1'}{\Omega_0} - \frac{\Omega_1 \Omega_0'}{\Omega_0^2}\right).$$

Inserting the result for Ω_0 and rearranging again leads to a Bernoulli-type ODE, the explicit solution of which is (7.45). As before, we have ensured that the limit $S \to K$ is finite. □

Remark 7.5.2. We notice that $\Omega_0(K)$ and $\Omega_1(K)$ in Proposition 7.5.1 exist by construction. Taking the limit $S \to K$ explicitly, we get

$$\Omega_0(K) = \frac{\varphi(K)}{K + \beta/\zeta},$$
$$\Omega_1(K) = \frac{1}{24} \Omega_0(K)^3 \left[1 + (K + \beta/\zeta)^2 \varphi(K)^{-2} \left(2\varphi(K)\varphi''(K) - \varphi'(K)^2\right)\right].$$

While Proposition 7.5.1 only includes two terms in the expansion for Ω, it is possible to compute further terms if necessary. Such terms become increasingly cumbersome however, and typically do not add much further accuracy.

The best choice of the parameters β and ζ is not always obvious. One choice is to use Remark 7.2.14. Alternatively, we could think of a more global approach and, roughly speaking, set them in such a way that the straight line $\beta + \zeta x$ would provide as good a fit to $\varphi(x)$ as possible, over the statistically relevant range of x. Sometimes, we can use a Taylor expansion around $x = (S+K)/2$, say, and set

$$\zeta = \varphi'\left((S+K)/2\right), \quad \beta = \varphi\left((S+K)/2\right) - \zeta\left(S+K\right)/2.$$

We note that when $\beta = 0$, $\Omega(\lambda^2(T-t), S(t))/\sqrt{T-t}$ in Proposition 7.5.1 conveniently becomes the time t implied Black volatility σ_B discussed earlier. For a few selected φ, Figure 7.1 below compares σ_B computed from the expansion in Proposition 7.5.1 (with $\beta = 0$) against exact results. Despite the long option maturity used in the figure, precision of the expansion is excellent, especially for the CEV case.

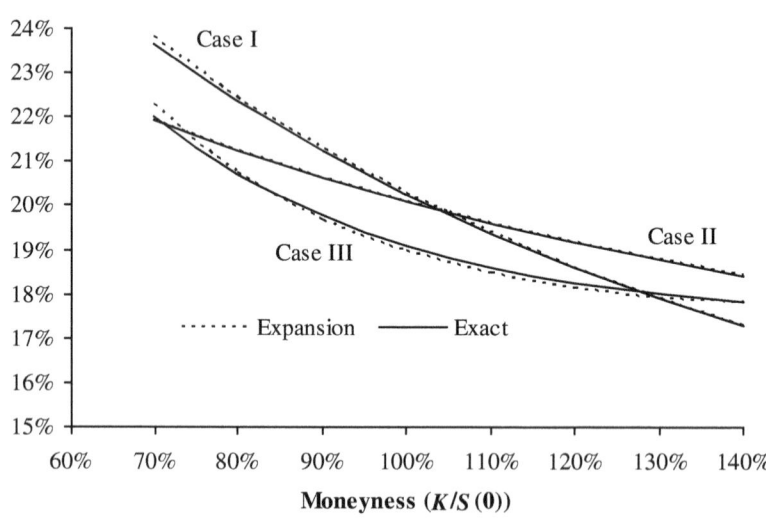

Fig. 7.1. Implied Volatility

Notes: The graph shows the implied volatility for a 10 year option, as a function of option moneyness $K/S(0)$. The initial value of the underlying is $S(0) = 6\%$. Three DVF models are considered in the figure. Case I: $\varphi(x) = x^{0.1}$, $\lambda = 1.59\%$. Case II: $\varphi(x) = x^{0.5}$, $\lambda = 4.90\%$. Case III: $\varphi(x) = x(1+30e^{-10x})$, $\lambda = 16.75\%$. The "Expansion" numbers in the graph were computed from the result in Proposition 7.5.1 with $\beta = 0$. For Case I and Case II, the "Exact" numbers were computed by the known CEV pricing formula in Proposition 7.2.6; for Case III the "Exact" numbers were computed in a Crank-Nicholson finite difference grid with 150 time steps and 250 spatial steps.

7.5.2 Expansion around Gaussian Process

For cases where φ is close to a constant, one might like to base the asymptotic expansion on $\varphi(x) = \beta$, for some constant β. In this case (which violates one of the restrictions in Proposition 7.5.1), we use the Gaussian formula (7.16), and write

$$g(\tau, S) = (S - K)\Phi(w) - \Psi(\tau, S)\phi(w), \quad w = \frac{S - K}{\Psi(\tau, S)}. \quad (7.48)$$

For completeness, an asymptotic expansion of $\Psi(\tau, S)$ is given below.

Proposition 7.5.3. *An asymptotic expansion for the solution of (7.35) is given by (7.48), with*

$$\Psi(\tau, S) = \Psi_0(S)\tau^{1/2} + \Psi_1(S)\tau^{3/2} + O\left(\tau^{5/2}\right),$$

$$\Psi_0(S) = (S - K)\left(\int_K^S \varphi(u)^{-1}\,du\right)^{-1},$$

$$\Psi_1(S) = -\frac{\Psi_0(S)^3}{(S-K)^2} \ln\left(\Psi_0(S)\left(\varphi(S)\varphi(K)\right)^{-1/2}\right).$$

In Proposition 7.5.3, the limit $S \to K$ leads to the following expressions

$$\Psi_0(K) = \varphi(K),$$

$$\Psi_1(K) = \frac{1}{24}\varphi(K)\left(2\varphi(K)\varphi''(K) - \varphi'(K)^2\right).$$

The proof of Proposition 7.5.3 is similar to that of Proposition 7.5.1 and is omitted. Note that $\Psi(\lambda^2(T-t), S(t))/\sqrt{T-t}$ can be interpreted as an *implied Normal volatility*.

7.6 Extensions to Time-Dependent φ

So far, we have limited our discussion to the case where the function φ is independent of calendar time t. While there is some danger in making φ a function of t — the model inevitably becomes less time-stationary — there are a number of applications where such an extension is necessary to improve the fit to market data. Unlike the non-parametric approaches in Dupire [1994], Derman and Kani [1994], and Andersen and Brotherton-Ratcliffe [1998] (and many others) where $\varphi(t, S)$ is calibrated to fit a double continuum of call option prices, the applications we have in mind are normally parametric, and are inspired by typical requirements of calibrating term structure models to swaptions and caplets.

By itself, swaption and caplet pricing does not require time-dependent parameters, as only the terminal distribution is relevant. From that point of view, vanilla models with time-dependent local volatility functions may appear to have limited use in fixed income modeling. However, they often arise as describing approximate dynamics of swap or Libor rates in term structure models. Many examples of such approximations are given in later chapters (see Chapters 13 and 14, for instance), and handling time-dependent parameters in local volatility models is important for term structure model calibration.

7.6.1 Separable Case

Recall the basic SDE (7.1). Its simplest time-dependent extension specifies a time-dependent scaling volatility λ, $\lambda = \lambda(t)$:

$$dS(t) = \lambda(t)\varphi\left(S(t)\right) dW(t). \tag{7.49}$$

This is the so-called *separable case*, as the local volatility function is represented as a product of two functions: $\lambda(\cdot)$, a function of the time variable only, and $\varphi(\cdot)$, a function of the state variable only. The separable form allows for application of the following simple time change argument:

Proposition 7.6.1. *Define*

$$\tau(t) = \int_0^t \lambda(u)^2\, du,$$

and define $s(\cdot)$ by $S(t) = s(\tau(t))$, with $S(t)$ following (7.49). Then

$$ds(\tau) = \varphi\left(s(\tau)\right) d\widetilde{W}(\tau), \quad s(0) = S(0), \tag{7.50}$$

where $\widetilde{W}$ is a Brownian motion.

Proof. The result follows directly from standard results for time-changed Brownian motion, see e.g. Karatzas and Shreve [1991]. □

Consider now the valuation of

$$c\left(t, S(t); T, K\right) = \mathrm{E}_t\left((S(T) - K)^+\right),$$

which in the notation of Proposition 7.6.1 can be written as

$$c\left(t, s(\tau(t)); T, K\right) = \mathrm{E}\left((s\left(\tau(T)\right) - K)^+ \Big| \widetilde{\mathcal{F}}_{\tau(t)}\right),$$

where $\widetilde{\mathcal{F}}$ is the filtration generated by $\widetilde{W}$. As the process for $s(\cdot)$ in (7.50) is of the type (7.1) (with $\lambda = 1$), all results from previous sections hold unchanged after the simple substitutions $\lambda \mapsto 1$ and $(T-t) \mapsto (\tau(T) - \tau(t))$. Equivalently, whenever the European option price results for constant λ involve terms of the form $\lambda^2(T-t)$, they should be replaced with $\int_t^T \lambda(u)^2\, du$ to accommodate a time-varying $\lambda(\cdot)$.

7.6.2 Skew Averaging

While the separable case can be handled quite easily, it is often too restrictive to be truly useful. Consider therefore the general case

$$dS(t) = \varphi(t, S(t))\, dW(t), \tag{7.51}$$

for $\varphi(t,x)$ satisfying the standard regularity conditions. European options could be valued in this model by PDE methods without much difficulty. However, with calibration applications in mind, this may be too slow or insufficiently accurate.

In this section, we develop European option approximations based on the idea of *time averaging*. Given the SDE (7.51), we look for a model with a *time-independent* local volatility function that yields European option prices approximately matching prices from the time-dependent model. The time-independent local volatility function can then be interpreted as a time average of the time-dependent function. This reduces the problem to one we know how to solve.

We have already seen a flavor of the averaging results that we are looking for. As demonstrated in Section 7.6.1, the values of T-expiry European options in the model

$$dS(t) = \lambda(t)\varphi(S(t))\, dW(t) \tag{7.52}$$

are the same as in the model

$$dS(t) = \overline{\lambda}\varphi(S(t))\, dW(t),$$

where $\overline{\lambda}$ is given by

$$\overline{\lambda}^2 = \frac{1}{T} \int_0^T \lambda(u)^2 \, du.$$

Thus, $\overline{\lambda}$ is an *effective volatility* for expiry T for the model (7.52).

Given the comments on U-shaped local volatility functions in Section 7.1.3, our initial focus shall be on functions that are monotonic in the state variable (see Section 7.6.3 for extensions). Such functions are typically well-described by two parameters, with the first parameter governing the overall level of volatility and the second the slope of the volatility smile (or skew). In the general case, both parameters are time-dependent. Let us concentrate on finding the averaging result for the time-dependent skew or, equivalently, on finding the *effective skew* formula.

We apply asymptotic expansion techniques with the *slope of the local volatility function* being the small parameter. Let us denote

$$X_0 = S(0), \quad \lambda(t) = \varphi(t, X_0), \quad g(t,x) = \frac{\varphi(t,x)}{\varphi(t,X_0)}.$$

Then (7.51) can be rewritten as

$$dS(t) = \lambda(t)g(t, S(t))\, dW(t), \tag{7.53}$$

where

$$g(t, X_0) = 1. \tag{7.54}$$

Let us fix a time horizon $T > 0$ and attempt to derive conditions that a time-independent function $\bar{g}(x)$ needs to satisfy so that the SDE (7.53) can be replaced with

$$dS(t) = \lambda(t)\bar{g}(S(t))\, dW(t) \tag{7.55}$$

for the purposes of valuing T-expiry European options of all strikes. Without loss of generality, the function $\bar{g}(x)$ is assumed to satisfy

$$\bar{g}(X_0) = 1.$$

Choose $\epsilon \geq 0$, the small slope parameter, and define

$$g^\epsilon(t, x) = g(t, X_0 + (x - X_0)\epsilon), \quad \bar{g}^\epsilon(x) = \bar{g}(X_0 + (x - X_0)\epsilon).$$

Next, define two sets of processes

$$dX^\epsilon(t) = \lambda(t)g^\epsilon(t, X^\epsilon(t))\, dW(t), \quad X^\epsilon(0) = X_0,$$
$$dY^\epsilon(t) = \lambda(t)\bar{g}^\epsilon(Y^\epsilon(t))\, dW(t), \quad Y^\epsilon(0) = X_0.$$

The requirement that the prices of European options on $X^\epsilon(T)$ and $Y^\epsilon(T)$ across all strikes be close can be reformulated as the requirement that the distributions of $X^\epsilon(T)$ and $Y^\epsilon(T)$ be close. This can be formalized as finding $\bar{g}(\cdot)$ such that

$$q(\epsilon) \to \min,$$

where

$$q(\epsilon) \triangleq \mathrm{E}\left((X^\epsilon(T) - Y^\epsilon(T))^2\right). \tag{7.56}$$

Considering the small slope limit $\epsilon \to 0$, we expand $q(\epsilon)$ in powers of ϵ to obtain

$$q(\epsilon) = q(0) + q'(0)\epsilon + \frac{1}{2}q''(0)\epsilon^2 + O(\epsilon^3).$$

As part of the proof below we will show that $q(0) = q'(0) = 0$. Hence, the minimization problem simplifies to

$$q''(0) \to \min.$$

The (necessary) minimum condition is given in the following result.

Proposition 7.6.2. *Any function $\bar{g}$ that minimizes $q''(0)$ must satisfy the condition*

$$\frac{\partial \bar{g}(X_0)}{\partial x} = \int_0^T \frac{\partial g(t, X_0)}{\partial x} w_T(t)\, dt, \tag{7.57}$$

where

$$w_T(t) = \frac{v(t)^2 \lambda(t)^2}{\int_0^T v(t)^2 \lambda(t)^2\, dt}, \quad v(t)^2 \triangleq \mathrm{E}\left((X^0(t) - X_0)^2\right). \tag{7.58}$$

7.6 Extensions to Time-Dependent φ

Proof. By Theorem 1.1.3, $q(\epsilon)$ as defined by (7.56) must equal

$$q(\epsilon) = \mathrm{E}\left(\int_0^T (g^\epsilon(t, X^\epsilon(t)) - \bar{g}^\epsilon(Y^\epsilon(t)))^2 \lambda(t)^2 \, dt\right).$$

Differentiating with respect to ϵ, we get (omitting arguments on g^ϵ and $\bar{g}^\epsilon$ for brevity)

$$q'(\epsilon) = 2\mathrm{E}\left(\int_0^T (g^\epsilon - \bar{g}^\epsilon) \times \left(\frac{\partial}{\partial \epsilon} g^\epsilon - \frac{\partial}{\partial \epsilon} \bar{g}^\epsilon\right) \lambda(t)^2 \, dt\right), \qquad (7.59)$$

$$q''(\epsilon) = 2\mathrm{E}\left(\int_0^T \left(\frac{\partial}{\partial \epsilon} g^\epsilon - \frac{\partial}{\partial \epsilon} \bar{g}^\epsilon\right)^2 \lambda(t)^2 \, dt\right)$$

$$+ 2\mathrm{E}\left(\int_0^T (g^\epsilon - \bar{g}^\epsilon) \left(\frac{\partial^2}{\partial \epsilon^2} g^\epsilon - \frac{\partial^2}{\partial \epsilon^2} \bar{g}^\epsilon\right) \lambda(t)^2 \, dt\right).$$

Since $g^0(t, x) = \bar{g}^0(x) = 1$, it follows that $q(0)$, $q'(0)$ and the second term in the expression for $q''(0)$ are zero. Hence,

$$q''(0) = 2\mathrm{E}\left(\int_0^T \left(\frac{\partial}{\partial \epsilon} g^\epsilon(t, X^\epsilon(t))\bigg|_{\epsilon=0} - \frac{\partial}{\partial \epsilon} \bar{g}^\epsilon(Y^\epsilon(t))\bigg|_{\epsilon=0}\right)^2 \lambda(t)^2 \, dt\right).$$

Note that

$$\frac{\partial}{\partial \epsilon} g^\epsilon(t, X^\epsilon(t)) = \left[\epsilon \left(\frac{\partial X^\epsilon}{\partial \epsilon}\right)(t) + (X^\epsilon(t) - X_0)\right]$$
$$\times \frac{\partial g}{\partial x}(t, X_0 + \epsilon(X^\epsilon(t) - X_0)),$$

$$\frac{\partial}{\partial \epsilon} \bar{g}^\epsilon(Y^\epsilon(t)) = \left[\epsilon \left(\frac{\partial Y^\epsilon}{\partial \epsilon}\right)(t) + (Y^\epsilon(t) - X_0)\right]$$
$$\times \frac{\partial \bar{g}}{\partial x}(X_0 + \epsilon(Y^\epsilon(t) - X_0)).$$

In particular, as $Y^0(t) = X^0(t)$,

$$\frac{\partial}{\partial \epsilon} g^\epsilon(t, X^\epsilon(t))\bigg|_{\epsilon=0} = (X^0(t) - X_0) \frac{\partial g}{\partial x}(t, X_0),$$

$$\frac{\partial}{\partial \epsilon} \bar{g}^\epsilon(Y^\epsilon(t))\bigg|_{\epsilon=0} = (X^0(t) - X_0) \frac{\partial \bar{g}}{\partial x}(X_0).$$

Thus,

$$q''(0) = 2\int_0^T \mathrm{E}\left((X^0(t) - X_0)^2 \lambda(t)^2\right)\left(\frac{\partial g}{\partial x}(t, X_0) - \frac{\partial \bar{g}}{\partial x}(X_0)\right)^2 dt$$

$$= 2\int_0^T v(t)^2 \lambda(t)^2 \left(\frac{\partial g}{\partial x}(t, X_0) - \frac{\partial \bar{g}}{\partial x}(X_0)\right)^2 dt,$$

with $v(t)^2$ defined in (7.58). Differentiating with respect to the slope $\partial \bar{g}(X_0)/\partial x$ and setting the resulting derivative to zero, we obtain a condition for the minimum of $q''(0)$. This gives (7.57). □

It follows from the proposition that for the purposes of (approximately) pricing T-expiry European options, (7.53) can be replaced with (7.55), where $\bar{g}(\cdot)$ is a function whose slope (skew) at-the-money, $\partial \bar{g}(S(0))/\partial x$, is a weighted average of the time-dependent at-the-money slopes (skews) of the original function $\partial g(t, S(0))/\partial x$, $t \in [0, T]$. The weights $w_T(t)$ to be used in forming the slope-average are the weights $w_T(t)$ in (7.58). Once the SDE of the form (7.53) has been approximated with (7.55), various tools developed in the first part of the chapter become available, and European option prices can be computed efficiently.

7.6.2.1 Examples

The time-dependent local volatility function $g(t, x)$ is often defined to be a time-indexed collection of functions from the same family. Examples include the time-dependent displaced log-normal function

$$g(t, x) = b(t)\frac{x}{S(0)} + (1 - b(t)), \quad t \in [0, T], \tag{7.60}$$

or the time-dependent CEV function

$$g(t, x) = \left(\frac{x}{S(0)}\right)^{p(t)}, \quad t \in [0, T]. \tag{7.61}$$

Note that the functions in the formulas have been scaled to satisfy (7.54). The condition (7.57) does not define the function $\bar{g}$ uniquely. To improve the accuracy of the approximation, it is often beneficial to choose $\bar{g}$ from the same family as the functions they approximate. In particular, for g of the type (7.60), the function $\bar{g}$ is best chosen to be of the same displaced log-normal type

$$\bar{g}(x) = \bar{b}\frac{x}{S(0)} + (1 - \bar{b}). \tag{7.62}$$

In the same vein, for the CEV case (7.61), a natural choice for $\bar{g}$ is

$$\bar{g}(x) = \left(\frac{x}{S(0)}\right)^{\bar{p}}. \tag{7.63}$$

Both the displaced log-normal parameter b and the CEV parameter p are used as a measure of the skew in the implied volatility smile. The next corollary expressed the averaging result directly in terms of these parameters, and also explicitly derives the averaging weights.

Corollary 7.6.3. *Over the time-horizon $[0, T]$, the effective skew $\bar{b}$ in (7.62) for the model defined by the time-dependent local volatility function (7.60) is given by*

$$\bar{b} = \int_0^T b(t) w_T(t) \, dt, \qquad (7.64)$$

where

$$w_T(t) = \frac{v(t)^2 \lambda(t)^2}{\int_0^T v(t)^2 \lambda(t)^2 \, dt}, \quad v(t)^2 = \int_0^t \lambda(s)^2 \, ds. \qquad (7.65)$$

Proof. For $g(t, x)$ and $\bar{g}(x)$ given by (7.60) and (7.62), we have

$$\frac{\partial g}{\partial x}(t, S(0)) = \frac{b(t)}{S(0)}, \quad \frac{\partial \bar{g}}{\partial x}(S(0)) = \frac{\bar{b}}{S(0)}.$$

Thus, (7.64) follows from (7.57). The formula (7.65), and in particular the expression for $v(t)^2$, follows from the definition

$$v(t)^2 = \mathrm{E}\left(\left(X^0(t) - X_0 \right)^2 \right)$$

and the fact that $X^0(t)$ satisfies

$$dX^0(t) = \lambda(t) g^0\left(t, X^0(t)\right) dW(t)$$

with

$$g^0\left(t, X^0(t)\right) \equiv 1.$$

□

Remark 7.6.4. An identical result holds for the effective CEV parameter $\bar{p}$,

$$\bar{p} = \int_0^T p(t) w_T(t) \, dt,$$

where $p(\cdot)$ and $\bar{p}$ are the parameters in (7.61) and (7.63), and $w_T(t)$ is as given in (7.65).

Example 7.6.5. Assuming constant volatility $\lambda(t) \equiv \lambda$, we obtain particularly simple formulas for the effective skew,

$$v(t)^2 = \lambda^2 t, \quad w_T(t) = \frac{t}{\int_0^T t \, dt} = \frac{t}{T^2/2},$$

so that

$$\bar{b} = \frac{1}{T^2/2} \int_0^T t b(t) \, dt.$$

This demonstrates that instantaneous skews $b(t)$ for larger t contribute more to $\bar{b}$ than those for lower t. Intuitively, the process needs to build up its variance before the changes in the instantaneous slopes start having an effect on the effective slope of the local volatility.

7.6.2.2 A Caveat About the Process Domain

Even though the skew averaging result is obtained in the small slope limit, practical experience validates its broad applicability in option pricing problems. Some typical results can be found in Piterbarg [2005c] and Piterbarg [2006]. Still, the equivalence between the original time-dependent model and the time-averaged one should not be taken too far, as we now proceed to demonstrate. For this, we focus on the simple displaced diffusion model from the previous section, i.e. we consider the time-dependent SDE

$$dS(t) = \lambda \left(b(t)S(t) + (1 - b(t))S(0)\right) dW(t), \qquad (7.66)$$

and approximate it with

$$dS(t) = \lambda \left(\bar{b}S(t) + (1 - \bar{b})S(0)\right) dW(t), \qquad (7.67)$$

where $\bar{b}$ is set as in Corollary 7.6.3. While the two SDEs (7.66) and (7.67) may have similar properties in the neighborhood of $S(0)$, they generally do not even have the same range for $S(t)$. For the constant parameter case (7.67) with $\bar{b} > 0$, the process $S(t)$ has a lower bound, the root of the local volatility function: $S(t) \in (S(0)(\bar{b} - 1)/\bar{b}, \infty)$. The same is not necessarily true for the time-dependent SDE (7.66), as should be reasonably clear from the following heuristic argument. If at a given time t, $S(t)$ is close to the root of the local volatility function but still above it, i.e.

$$S(t) \gtrsim S(0)(b(t) - 1)/b(t),$$

it may so happen that at $t + dt$, $S(t + dt)$ is actually *below* the root of the local volatility function,

$$S(t + dt) < S(0)(b(t + dt) - 1)/b(t + dt)$$

due to the change in the function $b(\cdot)$. The range

$$(-\infty, S(0)(b(t + dt) - 1)/b(t + dt))$$

will then be reachable by $S(\cdot)$. The following proposition provides formal justification.

Proposition 7.6.6. *Consider the SDE*

$$dX(t) = (a(t) + b(t)X(t)) dW(t) \qquad (7.68)$$

with $X(0) \geq -a(0)/b(0)$. If $(a(u)/b(u))' \geq 0$ for all $u \in [0, t]$, then $X(t) > -a(t)/b(t)$ a.s. If there exists u, $0 < u < t$, such that $(a(u)/b(u))' < 0$, then $P(X(t) < l) > 0$ for any $l \in \mathbb{R}$.

Proof. Define
$$\zeta(t) = \int_0^t b(u)\, dW(u) - \frac{1}{2}\int_0^t b^2(u)\, du, \quad Z(t) = \exp\left(\zeta(t)\right).$$
Then the solution to the SDE (7.68) is given by
$$X(t) = Z(t)\left[X(0) - \int_0^t (a(u)/b(u))\, d(1/Z(u))\right],$$
as can either be checked directly or obtained from Section 5.6.C of Karatzas and Shreve [1991]. Integrating by parts yields
$$X(t) = Z(t)\left(X(0) + \frac{a(0)}{b(0)}\right) - \frac{a(t)}{b(t)} + Z(t)\int_0^t \frac{(a(u)/b(u))'}{Z(u)}\, du.$$
With $X(0) \geq -a(0)/b(0)$,
$$Z(t)\left(X(0) + \frac{a(0)}{b(0)}\right) - \frac{a(t)}{b(t)}$$
is bounded from below by $-a(t)/b(t)$. If $(a(u)/b(u))' \geq 0$ for all $u \in [0,t]$ then the remaining term
$$Z(t) \int_0^t \frac{(a(u)/b(u))'}{Z(u)}\, du$$
is non-negative and $X(t)$ is bounded from below by $-a(t)/b(t)$. If, however, there exists u such that $(a(u)/b(u))' < 0$, this term can be arbitrarily negative with positive probability. □

In practice, the likelihood of actually breaching the lower boundary is typically small and we can often safely ignore this possibility. If needed, one can always "regularize" the time-dependent process to limit its range, along the same lines as done in Section 7.2.3.

7.6.3 Skew and Convexity Averaging by Small-Noise Expansion

The technique used in the previous section to derive Proposition 7.6.2 is not the only route to go. An alternative approach relies on *small-noise expansion*, a concept closely related to the Ito-Taylor expansion in Chapter 3. To illustrate the versatility of this method, we shall use it to derive not only the skew averaging result in Corollary 7.6.3, but also to demonstrate how to compute *average convexity* in a time-dependent quadratic model.

As our starting point, we define, for some constant X_0, the quadratic form
$$\varphi(t, X(t)) = \varphi(b(t), c(t), X(t))$$
$$= (1 - b(t))X_0 + b(t)X(t) + \frac{1}{2}c(t)(X(t) - X_0)^2,$$

and then introduce the following two processes:

$$dX(t) = \lambda(t)\varphi\left(b(t), c(t), X(t)\right) dW(t), \quad X(0) = X_0, \tag{7.69}$$
$$dY(t) = \lambda(t)\varphi\left(\bar{b}, \bar{c}, Y(t)\right) dW(t), \quad Y(0) = X_0, \tag{7.70}$$

where $W(t)$ is a Brownian motion in some probability measure. We can characterize the process for $X(t)$ as having quadratic local volatility with time-dependent slope $b(t)$ and time-dependent convexity $c(t)$; for a fixed value of T, we are interested in establishing how to set the constants $\bar{b}$ and $\bar{c}$ in the process for Y such that $Y(T)$ is a good approximation to $X(T)$.

We will answer the question posed above in the small-noise limit. For that, set

$$dX^\epsilon(t) = \epsilon\lambda(t)\varphi\left(b(t), c(t), X^\epsilon(t)\right) dW(t), \tag{7.71}$$
$$dY^\epsilon(t) = \epsilon\lambda(t)\varphi\left(\bar{b}, \bar{c}, Y^\epsilon(t)\right) dW(t), \tag{7.72}$$

with $Y^\epsilon(0) = X^\epsilon(0) = X_0$. Notice that $X^1(t) = X(t)$ and $Y^1(t) = Y(t)$, and that $X^0(t) = Y^0(t) = X_0$.

Lemma 7.6.7. *For the SDE (7.71), we have the formal expansion*

$$X^\epsilon(T) = X_0 + \epsilon A_X(T) + \frac{1}{2}\epsilon^2 B_X(T) + \frac{1}{6}\epsilon^3 C_X(T) + O\left(\epsilon^4\right),$$

where

$$A_X(T) = X_0 \int_0^T \lambda(t)\, dW(t),$$

$$B_X(T) = 2\int_0^T \lambda(t)b(t)A_X(t)\, dW(t),$$

$$C_X(t) = 3\int_0^T \lambda(t)c(t)A_X(t)^2\, dW(t) + 3\int_0^T \lambda(t)b(t)B_X(t)\, dW(t).$$

Proof. We rely on standard asymptotic expansion techniques (e.g. Yoshida [1992]) to construct a Taylor series of $X^\epsilon(T)$ around $\epsilon = 0$. Dropping the arguments of $\varphi(t) = \varphi(b(t), c(t), X^\epsilon(t))$ for brevity, we get

$$A_X(T) = \left.\frac{\partial X^\epsilon(T)}{\partial \epsilon}\right|_{\epsilon=0}$$
$$= \left.\left(\int_0^T \lambda(t)\varphi(t)dW(t) + \epsilon\int_0^T \lambda(t)\frac{\partial\varphi(t)}{\partial X^\epsilon(t)}\frac{\partial X^\epsilon(t)}{\partial \epsilon}dW(t)\right)\right|_{\epsilon=0}$$
$$= \int_0^T \lambda(t)\varphi\left(b(t), c(t), X_0\right) dW(t) = X_0\int_0^T \lambda(t)dW(t).$$

Similarly,

$$B_X(T) = \frac{\partial^2 X^\epsilon(T)}{\partial \epsilon^2}\bigg|_{\epsilon=0}$$

$$= \left(\int_0^T \lambda(t) \frac{\partial \varphi(t)}{\partial X^\epsilon(t)} \frac{\partial X^\epsilon(t)}{\partial \epsilon} dW(t) + \int_0^T \lambda(t) \frac{\partial \varphi(t)}{\partial X^\epsilon(t)} \frac{\partial X^\epsilon(t)}{\partial \epsilon} dW(t) \right) \bigg|_{\epsilon=0}$$

$$+ \epsilon \int_0^T \lambda(t) \frac{\partial^2 \varphi(t)}{\partial X^\epsilon(t)^2} \left(\frac{\partial X^\epsilon(t)}{\partial \epsilon} \right)^2 dW(t) \bigg|_{\epsilon=0}$$

$$+ \epsilon \int_0^T \lambda(t) \frac{\partial \varphi(t)}{\partial X^\epsilon(t)} \frac{\partial^2 X^\epsilon(t)}{\partial \epsilon^2} dW(t) \bigg|_{\epsilon=0}$$

$$= 2 \int_0^T \lambda(t) b(t) A_X(t) \, dW(t),$$

where we have used the fact that $\partial \varphi(t)/\partial X^\epsilon(t) = b(t)$ when $X^\epsilon(t) = X_0$. The result for $C_X(T)$ follows in the same fashion. □

For the variable Y^ϵ in (7.72), we get

$$Y^\epsilon(T) = X_0 + \epsilon A_X(T) + \frac{1}{2}\epsilon^2 B_Y(T) + \frac{1}{6}\epsilon^3 C_Y(T) + O\left(\epsilon^4\right),$$

where B_Y and C_Y are found by substituting $\bar{b}$ for $b(t)$ and $\bar{c}$ for $c(t)$ in the expressions for B_X and C_X in Lemma 7.6.7. We therefore immediately have the following result.

Lemma 7.6.8. *Consider the ϵ-indexed processes (7.71)–(7.72). Then, for $T > 0$,*

$$X^\epsilon(T) - Y^\epsilon(T) = \epsilon^2 I_1(\bar{b}; T) + \epsilon^3 I_2(\bar{b}, \bar{c}; T) + O(\epsilon^4),$$

where we have defined zero-mean random variables

$$I_1(\bar{b}; T) = \int_0^T \lambda(t) \left(b(t) - \bar{b} \right) A_X(t) \, dW(t),$$

$$I_2(\bar{b}, \bar{c}; T) = \frac{1}{2} \int_0^T \lambda(t) \left(c(t) - \bar{c} \right) A_X(t)^2 \, dW(t)$$

$$+ \frac{1}{2} \int_0^T \lambda(t) b(t) B_X(t) \, dW(t) - \frac{1}{2} \bar{b} \int_0^T \lambda(t) B_Y(t) \, dW(t).$$

There are numerous ways in which we can use the results of the previous section to determine the values of $\bar{b}$ and $\bar{c}$ that will make $Y^\epsilon(T)$ best approximate $X^\epsilon(T)$. Starting with $\bar{b}$, we here elect to set it such that the variance of the $O(\epsilon^3)$ term (the "skew term") in Lemma 7.6.8 is minimized. That is, our optimal choice $\bar{b}^*$ for $\bar{b}$ is characterized by

$$\bar{b}^* = \operatorname*{argmin}_{\bar{b}} \mathrm{E}\left(I_1\left(\bar{b}; T\right)^2 \right). \tag{7.73}$$

Proposition 7.6.9. *The solution to (7.73) is*

$$\bar{b}^* = \int_0^T b(t) w_T(t) dt, \quad w_T(t) = \frac{\lambda(t)^2 v(t)^2}{\int_0^T \lambda(t)^2 v(t)^2 dt},$$

where $v(\cdot)^2$ is defined in (7.65).

Proof. First, we need to establish the expectation of the random variable $I_1(\bar{b}; T)^2$. From elementary properties of the Ito integral (see Theorem 1.1.3), we know that

$$E\left(I_1\left(\bar{b}; T\right)^2\right) = E\left(\left(\int_0^T \lambda(t) \left(b(t) - \bar{b}\right) A_X(t) dW(t)\right)^2\right)$$

$$= \int_0^T \lambda(t)^2 \left(b(t) - \bar{b}\right)^2 E\left(A_X(t)^2\right) dt.$$

Since $A_X(t)$ is a Gaussian random variable with mean 0 and variance $X_0^2 v(t)^2$, it follows that

$$E\left(I_1\left(\bar{b}; T\right)^2\right) = X_0^2 \int_0^T \lambda(t)^2 \left(b(t) - \bar{b}\right)^2 v(t)^2 dt.$$

The (necessary) condition for a minimum is

$$\frac{1}{X_0^2} \frac{\partial E\left(I_1\left(\bar{b}; T\right)^2\right)}{\partial \bar{b}} = 2\bar{b} \int_0^T \lambda(t)^2 v(t)^2 dt - 2 \int_0^T \lambda(t)^2 b(t) v(t)^2 dt = 0,$$

from which the result in Proposition 7.6.9 follows. □

As advertised, the result of Proposition 7.6.9 is identical to that of Corollary 7.6.3.

It remains to find $\bar{c}$. We fundamentally wish to fix it such that the variance of the $O(\epsilon^4)$ term (the "convexity term") in Lemma 7.6.8 is minimized, given $\bar{b} = \bar{b}^*$. When $\bar{b} = \bar{b}^*$, however, we can observe that

$$I_2(\bar{b}^*, \bar{c}; T) \approx \frac{1}{2} \int_0^T \lambda(t) \left(c(t) - \bar{c}\right) A_X(t)^2 dW(t),$$

which suggests the simplified condition[8]

$$\bar{c}^* = \operatorname*{argmin}_{\bar{c}} E\left(\left(\frac{1}{2} \int_0^T \lambda(t) \left(c(t) - \bar{c}\right) A_X(t)^2 dW(t)\right)^2\right). \tag{7.74}$$

---

[8]More rigorous results can be found in Andersen and Hutchings [2010], but the accuracy of (7.74) is typically sufficient for applications.

Proposition 7.6.10. *The value $\bar{c}^*$ that satisfies (7.74) is*

$$\bar{c}^* = \int_0^T c(t) q_T(t)\, dt, \quad q_T(t) \triangleq \frac{\lambda(t)^2 v(t)^4}{\int_0^T \lambda(t)^2 v(t)^4 dt},$$

where $v(\cdot)^2$ is defined in (7.65).

Proof. We note that

$$E\left(\left(\frac{1}{2}\int_0^T \lambda(t)\left(c(t)-\bar{c}\right) A_X(t)^2\, dW(t)\right)^2\right)$$

$$= \frac{1}{4}\int_0^T \lambda(t)^2 \left(c(t)-\bar{c}\right)^2 E\left(A_X(t)^4\right)\, dt.$$

From a standard property of Gaussian random variables, we have

$$E\left(A_X(t)^4\right) = 3 X_0^4 v(t)^4.$$

Applying this result, the (necessary) condition for a minimum is

$$\frac{1}{2}\bar{c}\int_0^T \lambda(t)^2 v(t)^4\, dt - \frac{1}{2}\int_0^T \lambda(t)^2 c(t) v(t)^4\, dt = 0.$$

The Proposition 7.6.10 follows. □

Remark 7.6.11. For the special case where λ is constant, we have $v(t)^2 = \lambda^2 t$ and therefore

$$\bar{b}^* = \frac{2\int_0^T b(t)t\, dt}{T^2}, \quad \bar{c}^* = \frac{3\int_0^T c(t)t^2\, dt}{T^3}.$$

Note that the contribution of the instantaneous convexity $c(t)$ to the effective local volatility convexity grows with t at a faster rate $(O(t^2))$ than the contribution of $b(t)$ to the effective local volatility skew $(O(t))$.

7.6.4 Numerical Example

A brief numerical example is now in order. To provide a simple setup in which we can test our averaging results, we consider a two-period case where

$$\lambda(t) = \begin{cases} \lambda_0, & t \in [0, T'], \\ \lambda', & t \in (T', T], \end{cases} \quad b(t) = \begin{cases} b_0, & t \in [0, T'], \\ b', & t \in (T', T], \end{cases} \quad c(t) = \begin{cases} 0, & t \in [0, T'], \\ c', & t \in (T', T]. \end{cases} \tag{7.75}$$

The advantage of this setup is that it allows for high precision call option pricing without the need for finite difference grids or Monte Carlo methods. In particular, by having $c(t) = 0$ for $t \in [0, T']$, it follows that

$$dX(t) = (b_0 X(t) + (1-b_0)X_0)\lambda_0\,dW(t), \quad t \in [0, T']$$

such that, from the fact that these dynamics are those of a simple displaced log-normal process,

$$X(T') = \begin{cases} b_0^{-1}\left(X_0 \exp\left(-\tfrac{1}{2}b_0^2\lambda_0^2 T' + b_0\lambda_0 W(T')\right) - (1-b_0)X_0\right), & b_0 \neq 0 \\ X_0 + X_0\lambda_0 W(T'), & b_0 = 0. \end{cases} \quad (7.76)$$

Let[9] $C(t, x; K, T)$ be the time t price of a K-strike, T-maturity call option when $X(t) = x$. Clearly (assuming zero interest rates)

$$C(0, X_0; K, T) = \mathrm{E}\left(C(T', X(T'); K, T)\right).$$

At time T', process parameters switch to constant values λ', c', b' so for any value of $X(T')$ computation of $C(T', X(T'); K, T)$ can be done using the formulas for call options in the quadratic model (see Section 7.3 and Andersen [2010]). From (7.76), computation of $C(0, X_0; K, T)$ can then easily be performed by numerical integration. Figure 7.2 below shows a sample fit for a high-convexity case ($X_0 = 1, c' = 4$).

The constant-parameter approximation here does an excellent job of matching the volatility smile of the true model. For even higher precision — especially for the (rare) case where convexity is very large and rapidly changing in time — additional correction terms may be required; see Andersen and Hutchings [2010] for the details and more numerical tests.

---

[9] We temporarily use notation C (rather than the usual c) for a call option, to distinguish it from the convexity function $c(t)$.

Fig. 7.2. Implied Volatility Smile

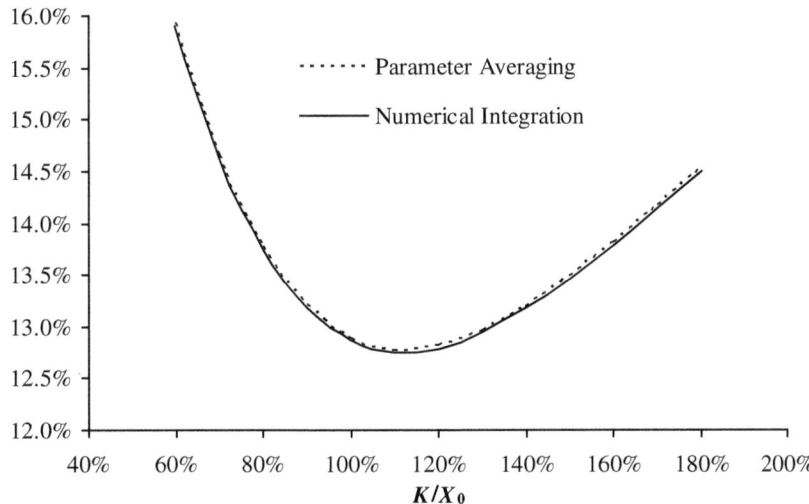

Notes: Parameters are as in (7.75), with $T' = 1$, $T = 2$, $\lambda_0 = 10\%$, $\lambda' = 15\%$, $b_0 = 0$, $b'=0.75$, $c' = 4$. The x-axis denotes relative strike K/X_0, with $X_0 = 1$. The "Numerical Integration" graph is the 2 year implied volatility smile for the time-dependent model, computed as outlined in the text (100 integration nodes). The "Parameter Averaging" graph computes the 2 year volatility smile from a constant-parameter quadratic model with parameters set as in Propositions 7.6.9 and 7.6.10.

8

Vanilla Models with Stochastic Volatility I

In Chapter 7 we introduced and studied diffusive single-factor vanilla models where the volatility is a deterministic function of the underlying rate. While such level-dependence of volatility is observable in interest rate markets — implied Black volatility is normally higher when rates are low — there is strong empirical evidence for additional sources of randomness in interest rate volatilities. To make our model setup more realistic, and to improve our ability to fit models to market-implied volatility smiles, we continue our investigation of vanilla models by enlarging the DVF models from the previous chapter to allow the volatility to be driven by a separate Brownian motion. The resulting models are said to have *stochastic volatility*.

Beyond raising the dimension of our models dynamics from one to two, the introduction of stochastic volatility brings with it a number of technical complications and, for many important models, the need to work with Fourier transforms when pricing options. We discuss these issues in detail in this chapter, paying particular attention to the *displaced log-normal Heston model* which has good analytical tractability and often provides an excellent fit to market observations.

Stochastic volatility constitutes a large and important topic in contemporary fixed income modeling, and we shall need two chapters of this book to lay the proper foundation for later work. In this chapter, our focus is on basic material and on the development of Fourier integration methods in a time-homogeneous setting. More advanced topics — including many numerical methods and the extension to time-dependent parameters — are postponed to Chapter 9.

8.1 Model Definition

As in Chapter 7, let $S(t)$, the "underlying" as we shall often call it, denote a forward Libor or swap rate. Also, let $Z(t)$, $W(t)$ be two different one-dimensional Brownian motions under a measure P in which $S(t)$ is a martingale; we assume that $Z(t)$ and $W(t)$ are correlated with constant correlation ρ. As before, we use E instead of E^{P} for the expected value operator under measure P if there is no possibility of confusion. A fairly general family of stochastic volatility models[1] is obtained by specifying

$$dS(t) = \lambda \varphi\left(S(t)\right) \sqrt{z(t)}\, dW(t), \tag{8.1}$$
$$dz(t) = \theta \left(m(t) - z(t)\right) dt + \eta \psi\left(z(t)\right) dZ(t), \quad z(0) = z_0, \tag{8.2}$$

where λ, θ, η are positive constants, $m(\cdot)$ a positive deterministic function of time, and $\varphi(\cdot)$ and $\psi(\cdot)$ two smooth deterministic functions. In these SDEs, $z(\cdot)$ is a *stochastic variance* process, the square root of which scales a DVF diffusion term similar to that discussed in Chapter 7.

We notice that the drift term of $z(\cdot)$ is such that $z(t)$ gets pulled towards the level $m(t)$ at an exponential rate of θ, known as the *mean reversion speed* (or sometimes just the *mean reversion*). The parameter η is the *volatility of variance*, and $\psi(z)$ is a skew function for the stochastic variance. We shall later discuss in more detail the roles and effects of the individual parameters in the dynamics for $z(t)$, but before doing so let us try to indicate what constitutes a reasonable model specification. First, since the effect of $z(\cdot)$ on the volatility of $S(\cdot)$ is multiplicative, the initial value z_0 and the value $m(t)$ to which it mean-reverts can be scaled to arbitrary level; for convenience[2] we typically set $m(t) \equiv z_0 = 1$. As for the functions $\psi(\cdot)$ and $\varphi(\cdot)$, there are many empirically reasonable choices, but convenience and efficiency of available valuation algorithms for European options need to be considered. Typically, the function $\psi(\cdot)$ is chosen to be the square root function, making the process for $z(t)$ *affine* and improving analytical tractability. That said, other power functions, nevertheless, can be used and sometimes may be preferred, for reasons explained later (see, e.g., the end of Section 8.3). Analytical tractability also suggests using a linear function for $\varphi(\cdot)$, such that the underlying DVF model is a standard displaced log-normal model, see Section 7.2.4.

It only remains to comment on the correlation parameter ρ. In interest rate applications, the correlation ρ between the Brownian motions driving the stochastic variance and the underlying is often set to 0 due to undesirable effects of common measure changes on the stochastic variance process when correlation is non-zero, see Proposition 8.3.9. This is rarely a limitation,

---

[1] For non-linear functions $\varphi(x)$ or $\varphi(t,x)$ such models are sometimes called *local stochastic volatility*, or LSV, models. Occasionally the name is also used for models with linear φ.

[2] Note that setting $m(\cdot)$ to a constant different from z_0 defines a model with constant coefficients that has a somewhat richer term volatility structure than with $m(\cdot) \equiv z_0$. The utility of this is limited as we are ultimately interested in time-dependent model extensions anyway.

as the effect of correlation on option prices and their implied volatilities[3] can typically be captured in parameters of the function $\varphi(\cdot)$. From the perspective of matching the implied volatility smile, non-zero correlation is thus largely superfluous. Provided that we define our hedge sensitivities in a certain, natural way, this observation also holds for hedging, a point we shall return to in Section 8.9.2. To keep our discussion general, we nevertheless keep correlation non-zero for much of the discussion that follows.

With the parameter specializations described above, the simplified model we shall concentrate most of our efforts on is defined as

$$dS(t) = \lambda \left(bS(t) + (1-b) L \right) \sqrt{z(t)} \, dW(t), \tag{8.3}$$
$$dz(t) = \theta \left(z_0 - z(t) \right) dt + \eta \sqrt{z(t)} \, dZ(t), \quad z(0) = z_0 = 1, \tag{8.4}$$

with $\langle dZ(t), dW(t) \rangle = \rho \, dt$. Going forward, this model will be referred to as simply the *SV model*. For the case where $b = 1$, the model becomes identical to the so-called *Heston model*; see Heston [1993]. To avoid degenerate situations, we make the following assumption:

Assumption 8.1.1. *All parameters b, θ, η, λ are strictly positive, and $|\rho| < 1$.*

8.2 Model Parameters

We proceed to a more detailed discussion of the parameters in the model (8.3)–(8.4). First, recall that in the local volatility model of the displaced log-normal type (7.21), the parameter λ is responsible for the overall level of the implied volatility smile, while the parameter b is responsible for its slope. This interpretation of the parameters carries over to the stochastic volatility case (8.3)–(8.4), and we often refer to λ and b as the *SV volatility* and the *skew*, respectively.

The volatility of variance parameter η controls the curvature of the volatility smile, see Section 8.7. The effect of η on the volatility smile is similar to that of the second-order, or convexity, term in a quadratic DVF model of Section 7.3, although the dynamics of the volatility smile are quite different in the two models, a point we shall return to later, in Section 8.8.

The mean reversion of variance, θ, controls the speed at which deviations of $z(\cdot)$ away from z_0 are pulled back towards this level. Increasing θ decreases the long-term variance of $z(\cdot)$ and limits the effect of the stochastic variance process on the volatility smile for medium- and long-dated maturities. In essence, θ controls the speed of decay of the volatility smile convexity.

---

[3] If the correlation is negative — i.e. if $z(t)$ tends to be high when $S(t)$ is low — the model will imply a downward-sloping volatility smile, as should be intuitively clear.

The local volatility function $\varphi(x) = bx + (1-b)L$ involves a quantity L, the *level* parameter. As discussed in the previous chapter, we normally set this to a number equal or close to[4] $S(0)$, to ensure that λ will have the dimension of implied Black volatility, irrespective of the setting of b. This decoupling of parameters is particularly convenient in a calibration context.

As in the (local volatility) displaced log-normal model, λ is expressed in the units of relative volatility, while the skew b is typically confined to a range between 0% to 100%, although the "super-Normal" ($b < 0$) and "super-log-normal" ($b > 1$) settings may occasionally be useful. For $b < 0$ or $b > 1$, our earlier discussion in Section 7.6.2.2 shows that if $L > 0$, the state space for $S(\cdot)$ is bounded (above or below depending on the sign of b) by the value $-L(1-b)/b$. The existence of such a bound is somewhat unrealistic; however, the advantages of being able to use values of b outside of $[0,1]$ usually outweigh this concern.

The parameter η is expressed in the units of annualized relative volatility of *variance*. Sometimes it is more natural to think in terms of the volatility of *volatility*, i.e. the volatility of the process for $\sqrt{z(t)}$. By Ito's lemma,

$$d\sqrt{z(t)} = O\left(dt\right) + \frac{\eta}{2} dZ(t).$$

When $z(t)$ has unit magnitude, $\eta/2$ can loosely be thought of as the volatility of volatility. For example, a value of 100% for η associates the implied Black volatility of the model with an instantaneous relative annualized volatility of about 50%. The related parameter θ, the speed of mean reversion, is expressed in percentage points per year. The inverse quantity θ^{-1} is measured in years and is related to the time over which a volatility shock dissipates. Specifically, the half-life of a volatility shock is

$$t_{1/2} = \frac{\ln 2}{\theta}.$$

All major interest rate markets exhibit high volatility of variance/low mean reversion of variance parameters, with $\eta = 150\%$ and $\theta = 10\%$ being typical parameter settings. While a half-life of a volatility shock of $10 \ln 2 \approx 7$ years may appear quite unrealistic, one should not forget that the pricing measure P will rarely represent real-world probabilities whereby the drift in the process for $z(\cdot)$ will likely contain strong market price of risk adjustment. The impact of measure changes on the speed of mean reversion for the variance is highlighted in Proposition 8.3.9.

---

[4] The rationale for not letting $L = S(0)$ always is that computation of delta sensitivities $\partial/\partial S(0)$ would then perturb the constant in the linear form $\varphi(x)$ which may or may not be desirable. See Sections 16.1.1 and 16.1.2 for more details.

8.3 Basic Properties

In this section we collect several important facts about the distribution and other relevant characteristics of $z(\cdot)$ and $S(\cdot)$ in the model (8.3)–(8.4). First, we look at the regularity properties of the process for the stochastic variance $z(\cdot)$; the results below should be compared to Proposition 7.2.1.

Proposition 8.3.1. *The SDE (8.4) has a unique solution. If $2z_0\theta \geq \eta^2$, i.e. the so-called* Feller condition *holds, $z = 0$ is unattainable. If the Feller condition is violated, $2z_0\theta < \eta^2$, then $z = 0$ is an attainable boundary but is strongly reflecting.*

Proof. See Revuz and Yor [1999] or Andersen and Piterbarg [2007]. □

The transition distribution for the variance process $z(\cdot)$ given by (8.4) was derived in Cox et al. [1985] is listed below.

Proposition 8.3.2. *Let $\Upsilon(z;\nu,\gamma)$ be the cumulative distribution function for the non-central chi-square distribution with ν degrees of freedom and non-centrality parameter γ:*

$$\Upsilon(z;\nu,\gamma) = e^{-\gamma/2} \sum_{j=0}^{\infty} \frac{(\gamma/2)^j}{j! 2^{\nu/2+j} \Gamma(\nu/2+j)} \int_0^z y^{\nu/2+j-1} e^{-y/2}\, dy. \tag{8.5}$$

For the process (8.4) define

$$d = 4\theta z_0/\eta^2, \quad n(t,T) = \frac{4\theta e^{-\theta(T-t)}}{\eta^2 \left(1 - e^{-\theta(T-t)}\right)}, \quad T > t. \tag{8.6}$$

Let $T > t$. Conditional on $z(t)$, $z(T)$ is distributed as $e^{-\theta(T-t)}/n(t,T)$ times a non-central chi-square distributed random variable with d degrees of freedom and non-centrality parameter $z(t)n(t,T)$,

$$\mathrm{P}\left(z(T) < x \,|\, z(t)\right) = \Upsilon\left(\frac{x \cdot n(t,T)}{e^{-\theta(T-t)}}; d, z(t)n(t,T)\right).$$

Of particular importance, especially in Monte Carlo methods discussed later in Section 9.5, are the conditional moments of $z(\cdot)$. From the known properties of the non-central chi-square distribution, the following corollary easily follows[5]:

Corollary 8.3.3. *For $T > t$, $z(T)$ has the following first two conditional moments:*

$$\mathrm{E}\left(z(T)|z(t)\right) = z_0 + (z(t) - z_0) e^{-\theta(T-t)}, \tag{8.7}$$

$$\mathrm{Var}\left(z(T)|z(t)\right) = \frac{z(t)\eta^2 e^{-\theta(T-t)}}{\theta}\left(1 - e^{-\theta(T-t)}\right) + \frac{z_0 \eta^2}{2\theta}\left(1 - e^{-\theta(T-t)}\right)^2. \tag{8.8}$$

---

[5]In Appendix A.A, p.1152, we also derive an expression for, and a numerical approximation to, $\mathrm{E}(\sqrt{z(t)})$.

The transition distribution is useful for setting numerical bounds for PDE and Monte Carlo methods. Because it is somewhat complicated, we often find it convenient to use the stationary distribution of $z(t)$ (that is, the distribution of $z(\infty)$) instead, as an approximation.

Proposition 8.3.4. *The stationary distribution of $z(\cdot)$ in (8.4) is a Gamma distribution, see (3.9), and the stationary density $\pi(z)$ is given by*

$$\pi(z) = \frac{z^{\alpha-1} e^{-\beta z}}{\Gamma(\alpha) \beta^{-\alpha}},$$

where

$$\alpha = 2\theta z_0 / \eta^2, \quad \beta = 2\theta / \eta^2.$$

In particular, the mean of the stationary distribution is given by

$$\int_0^\infty z \pi(z)\, dz = \frac{\alpha}{\beta} = z_0,$$

and the variance by

$$\int_0^\infty (z - z_0)^2 \pi(z)\, dz = \frac{\alpha}{\beta^2} = \frac{z_0 \eta^2}{2\theta}.$$

Proof. Follows directly from Proposition 8.3.2 and Corollary 8.3.3, by taking the limit $T - t \to \infty$.

Now let us look at the properties of the process $S(\cdot)$ for the underlying. The martingale property for $S(\cdot)$ should not be taken for granted in stochastic volatility models, but fortunately holds in our case:

Proposition 8.3.5. *The process $S(\cdot)$ given by (8.3)–(8.4) is a proper martingale.*

Proof. See Andersen and Piterbarg [2007]. □

The SDE (8.3) for $S(\cdot)$ can be integrated explicitly:

Proposition 8.3.6. *In the model (8.3)–(8.4), we have*

$$S(t) = \frac{1}{b}\left[(bS(0) + (1-b)L) X(t) - (1-b)L\right],$$

where

$$dX(t)/X(t) = \lambda b \sqrt{z(t)}\, dW(t), \quad X(0) = 1,$$

i.e.,

$$\ln X(t) = \lambda b \int_0^t \sqrt{z(s)}\, dW(s) - \frac{1}{2}\lambda^2 b^2 \int_0^t z(s)\, ds. \tag{8.9}$$

Proof. Follows by applying Ito's lemma to $\ln(bS(t) + (1-b)L)$. □

The moment-generating function of $\ln X(t)$ in (8.9) is of fundamental importance for European option pricing in the model (8.3)–(8.4), and is linked to the moment-generating function of the integrated variance process, as the following proposition demonstrates.

Proposition 8.3.7. *Define*

$$\Psi_X(u;t) \triangleq \mathrm{E}\left(e^{u \ln X(t)}\right) = \mathrm{E}\left(X(t)^u\right). \tag{8.10}$$

In the model (8.3)–(8.4), for any $u \in \mathbb{C}$ for which the right-hand side exists, we have

$$\Psi_X(u;t) = \Psi_{\bar{z}}\left(\frac{1}{2}(\lambda b)^2 u(u-1), u; t\right),$$

where we have denoted

$$\Psi_{\bar{z}}(v,u;t) \triangleq \mathrm{E}^{\widetilde{\mathrm{P}}}\left(e^{v\bar{z}(t)}\right), \quad \bar{z}(t) \triangleq \int_0^t z(s)\, ds. \tag{8.11}$$

Under the new probability measure $\widetilde{\mathrm{P}}$ the process for $z(\cdot)$ is

$$dz(t) = (\theta(z_0 - z(t)) + \rho\eta\lambda buz(t))\, dt + \eta\sqrt{z(t)}\, d\widetilde{Z}(t), \quad z(0) = z_0, \tag{8.12}$$

with $\widetilde{Z}$ a $\widetilde{\mathrm{P}}$-Brownian motion. If $\rho = 0$, then $\widetilde{\mathrm{P}} = \mathrm{P}$ and $z(\cdot)$ in (8.11) follows (8.4) rather than (8.12).

Proof. From (8.9) we get

$$\mathrm{E}\left(e^{u \ln X(t)}\right) = \mathrm{E}\left(\varsigma(t) \exp\left(\frac{1}{2}u(u-1)\lambda^2 b^2 \int_0^t z(s)\, ds\right)\right),$$

where $\varsigma(t)$ is the exponential martingale

$$\varsigma(t) = \mathcal{E}\left(u\lambda b \int_0^t \sqrt{z(s)}\, dW(s)\right)$$

$$= \exp\left(u\lambda b \int_0^t \sqrt{z(s)}\, dW(s) - \frac{1}{2}u^2\lambda^2 b^2 \int_0^t z(s)\, ds\right).$$

Letting $\varsigma(t)$ be the density process for a measure change, Proposition 8.3.7 follows from Girsanov's theorem, see Theorem 1.5.1. □

A version of the proposition above also holds for a more general process (8.2) for $z(\cdot)$, see Andersen and Piterbarg [2007]. What makes the specification (8.4) particularly useful is the availability of a closed-form expression for $\Psi_{\bar{z}}(v,u;t)$.

Proposition 8.3.8. *For $\Psi_{\bar{z}}(v, u; t)$ defined by (8.11) we have that*

$$\ln \Psi_{\bar{z}}(v, u; T) = A(v, u) + B(v, u) z_0,$$

$$A(v, u) = \frac{\theta z_0}{\eta^2} \left[2 \ln \left(\frac{2\gamma}{\theta' + \gamma - e^{-\gamma T}(\theta' - \gamma)} \right) + (\theta' - \gamma) T \right],$$

$$B(v, u) = \frac{2v(1 - e^{-\gamma T})}{(\theta' + \gamma)(1 - e^{-\gamma T}) + 2\gamma e^{-\gamma T}},$$

$$\gamma = \gamma(v, u) = \sqrt{(\theta')^2 - 2\eta^2 v},$$

$$\theta' = \theta'(u) = \theta - \rho \eta \lambda b u.$$

Proof. The process (8.12) is of the form

$$dz(t) = (\theta z_0 - \theta' z(t)) dt + \eta \sqrt{z(t)} d\widetilde{Z}(t), \quad \theta' = \theta - \rho \eta \lambda b u,$$

which is of the same form as the short rate process in Cox et al. [1985], see Section 10.2. As demonstrated by, e.g., Dufresne [2001], the discount bond pricing result from Cox et al. [1985] (derived via PDE methods) immediately establishes the moment-generating function of the time integral of $z(\cdot)$. □

Beyond being useful in the proof of Proposition 8.3.7, measure changes are of primary importance in interest rate modeling, where a stochastic volatility model would typically be "embedded" in a full term structure model. To get a feel for issues that arise in this context, let us consider the impact of measure changes on the stochastic variance process. For this, let $V(t, X(t))$ be the numeraire-deflated price process for some asset in the model (8.3)–(8.4), where $V(t, x)$ is a deterministic function. Implicit in the notation is the assumption that the price does not depend on the stochastic variance process $z(\cdot)$, an assumption that holds true in the cases of interest to us. Assuming the price process is positive, it can be used as a numeraire, defining a new measure $\widetilde{P}$, see Section 1.3. Since we have

$$dV(t, X(t))/V(t, X(t)) = \lambda b X(t) \frac{\partial \ln(V(t, X(t)))}{\partial x} \sqrt{z(t)} \, dW(t),$$

the process

$$\left(d\widetilde{W}(t), d\widetilde{Z}(t) \right)^\top = (dW(t), dZ(t))^\top$$

$$- \left(\lambda b X(t) \frac{\partial \ln(V(t, X(t)))}{\partial x} \sqrt{z(t)}, \rho \lambda b X(t) \frac{\partial \ln(V(t, X(t)))}{\partial x} \sqrt{z(t)} \right)^\top dt$$

is a two-dimensional Brownian motion under the measure $\widetilde{P}$, see Theorem 1.5.1, and we obtain the following result.

Proposition 8.3.9. *In the model (8.3)–(8.4), the dynamics of the stochastic variance process $z(\cdot)$ under a measure $\widetilde{P}$ defined by a numeraire $V(t, X(t))$ are given by*

$$dz(t) = \widetilde{\theta}(t, X(t)) \left(\widetilde{f}(t, X(t)) - z(t)\right) dt + \eta \sqrt{z(t)} \, d\widetilde{Z}(t),$$

where

$$\widetilde{\theta}(t, x) = \theta - \eta \rho \lambda b x \frac{\partial \ln(V(t, x))}{\partial x}, \quad \widetilde{f}(t, x) = \frac{\theta z_0}{\widetilde{\theta}(t, x)},$$

and $\widetilde{Z}(\cdot)$ is a $\widetilde{\mathbf{P}}$-Brownian motion.

We note that if $\rho \neq 0$, not only do the speed of mean reversion and the mean reversion level get altered by the measure change, they become dependent on the process for the underlying $S(\cdot)$ itself. As mentioned before, this makes it difficult, if not impossible, to relate statistically-observed stochastic variance parameters to the risk-neutral ones. Additionally, non-zero value of the correlation ρ introduces technical complications in interest rate modeling due to the heavy use of measure change machinery, complications that we normally avoid by setting ρ to 0.

Returning to the examination of the properties of the S-process, we note that while $S(\cdot)$ in (8.3)–(8.4) is always a martingale (see Proposition 8.3.5), some of its higher-order moments may become infinite with time. This has important implications in interest rate modeling where values of some common types of contracts require finite second-order moments, see Chapter 16 on convexity derivatives. The following proposition gives sharp conditions on moment existence.

Proposition 8.3.10. *Consider the model (8.3)–(8.4). For a given $u > 1$, set $v = (\lambda b)^2 u(u-1)/2 \geq 0$ and define*

$$\beta = 2v/\eta^2 > 0, \quad \alpha = 2(\rho \eta \lambda b u - \theta)/\eta^2, \quad D = \alpha^2 - 4\beta.$$

The moment $\mathrm{E}(S(T)^u)$ will be finite for $T < T^*$ and infinite for $T \geq T^*$, where T^* is given by

1. $D \geq 0, \alpha < 0$:

$$T^* = \infty;$$

2. $D \geq 0, \alpha > 0$:

$$T^* = \gamma_+^{-1} \eta^{-2} \ln\left(\frac{\alpha/2 + \gamma_+}{\alpha/2 - \gamma_+}\right), \quad \gamma_+ \triangleq \frac{1}{2}\sqrt{D};$$

3. $D < 0$:

$$T^* = 2\gamma_-^{-1}\eta^{-2} \times \left(\pi 1_{\{\alpha < 0\}} + \arctan(2\gamma_-/\alpha)\right), \quad \gamma_- \triangleq \frac{1}{2}\sqrt{-D}.$$

Proof. See Andersen and Piterbarg [2007]. □

The problem of moment explosions in the SV model (8.3)–(8.4) can be resolved by replacing (8.4) with a slightly more general specification (8.2) with $\psi(z) = z^p$ for $p < 1/2$, at a cost of losing some analytical tractability.

There are a number of subtle but important issues related to stochastic volatility processes with $\psi(z) = z^p$; the reader is referred to Andersen and Piterbarg [2007] for a comprehensive discussion. While somewhat outside the main focus of our exposition, we list some relevant results in Appendix 8.A.

8.4 Fourier Integration

Having covered the basics, we now turn to the problem of establishing of accurate pricing methods for the SV model. The method we present here is based on the application of Fourier integration methods, and is largely taken from Lewis [2000], with some modifications. Carr and Madan [1999], Lipton [2002], and Lee [2004], among many others, can be consulted for additional details.

8.4.1 General Theory

The following general result shows how to calculate call option prices when a moment-generating function is available for the logarithm of the underlying.

Theorem 8.4.1. *Let ξ be a random variable, and define its moment-generating function by $\chi(u)$,*

$$\chi(u) = \mathrm{E}\left(e^{u\xi}\right).$$

Then for $k \in \mathbb{R}$,

$$\mathrm{E}\left(\left(e^{\xi} - e^{k}\right)^{+}\right) = \chi(1) - \frac{e^{k}}{2\pi} \int_{-\infty}^{\infty} \frac{e^{-k(\alpha+i\omega)}\chi(\alpha + i\omega)}{(\alpha + i\omega)(1 - \alpha - i\omega)} d\omega \quad (8.13)$$

for any $0 < \alpha < 1$ for which the right-hand side exists.

Proof. Let

$$c(k) = \mathrm{E}\left(\left(e^{\xi} - e^{k}\right)^{+}\right).$$

To improve regularity of our eventual numerical scheme, we split out a bounded component $\min(e^{\xi-k}, 1)$ from the unbounded function $(e^{\xi} - e^{k})^{+}$, writing

$$\begin{aligned} c(k) &= \mathrm{E}\left(\max\left(e^{\xi} - e^{k}, 0\right)\right) \\ &= \mathrm{E}\left(e^{\xi} - e^{k}\min\left(e^{\xi-k}, 1\right)\right) \\ &= \chi(1) - e^{k}\mathrm{E}\left(\min\left(e^{\xi-k}, 1\right)\right). \end{aligned}$$

Our intention is now to apply Fourier transforms in the computation of $\mathrm{E}(\min(e^{\xi-k}, 1))$. While the function $\min(e^{x-k}, 1)$ is bounded by design, it

8.4 Fourier Integration

is not integrable — it equals 1 for all $x \geq k$. To work around this, we can follow Carr and Madan [1999] and write, with $p(x)$ being the density of ξ,

$$E\left(\min\left(e^{\xi-k}, 1\right)\right) = e^{-\alpha k} \int_{-\infty}^{\infty} \left[\min\left(e^{-(k-x)}, 1\right) e^{\alpha(k-x)}\right] \left[e^{\alpha x} p(x)\right] dx,$$

where $\alpha > 0$ is a classical *dampening constant*. Note that this integral is a convolution

$$(f_1 * f_2)(k) \triangleq \int_{-\infty}^{\infty} f_1(k-x) f_2(x) \, dx$$

of two functions,

$$f_1(x) = \min\left(e^{-x}, 1\right) e^{\alpha x}$$

and

$$f_2(x) = e^{\alpha x} p(x),$$

evaluated at k. Let $\mathcal{F}$ be Fourier transform and $\mathcal{F}^{-1}$ its inverse, i.e.,

$$(\mathcal{F} f)(\omega) \triangleq \int_{-\infty}^{\infty} e^{i\omega x} f(x) \, dx, \tag{8.14}$$

$$(\mathcal{F}^{-1} g)(x) \triangleq \frac{1}{2\pi} \int_{-\infty}^{\infty} e^{-i\omega x} g(\omega) \, d\omega. \tag{8.15}$$

As is well known, the Fourier transform of a convolution is a product of Fourier transforms, so

$$\int_{-\infty}^{\infty} \left[\min\left(e^{-(k-x)}, 1\right) e^{\alpha(k-x)}\right] \left[e^{\alpha x} p(x)\right] dx$$
$$= (f_1 * f_2)(k) = \left(\mathcal{F}^{-1}\left(\mathcal{F}(f_1 * f_2)\right)\right)(k) = \left(\mathcal{F}^{-1}\left(g_1(\omega) g_2(\omega)\right)\right)(k),$$

where

$$g_1(\omega) = \int_{-\infty}^{\infty} e^{i\omega x} \min\left(e^{-x}, 1\right) e^{\alpha x} \, dx,$$

$$g_2(\omega) = \int_{-\infty}^{\infty} e^{i\omega x} e^{\alpha x} p(x) \, dx.$$

Simple calculations lead to

$$g_1(\omega) = \int_{-\infty}^{0} e^{x(\alpha+i\omega)} dx + \int_{0}^{\infty} e^{x(-1+\alpha+i\omega)} dx$$
$$= \frac{1}{\alpha + i\omega} - \frac{1}{\alpha - 1 + i\omega}$$
$$= \frac{1}{(\alpha + i\omega)(1 - \alpha - i\omega)},$$
$$g_2(\omega) = \chi(\alpha + i\omega),$$

where the convergence of integrals follows from the fact that $0 < \alpha < 1$. Therefore,

$$E\left(\min\left(e^{\xi-k}, 1\right)\right) = \frac{1}{2\pi} \int_{-\infty}^{\infty} \frac{e^{-k(\alpha+i\omega)} \chi(\alpha+i\omega)}{(\alpha+i\omega)(1-\alpha-i\omega)} d\omega$$

and the theorem follows. □

Remark 8.4.2. The formula (8.13) from Theorem 8.4.1 can be re-written as

$$E\left(\left(e^\xi - e^k\right)^+\right) = \chi(1) - \frac{e^k}{\pi} \int_0^\infty \mathrm{Re}\left(\frac{e^{-k(\alpha+i\omega)} \chi(\alpha+i\omega)}{(\alpha+i\omega)(1-\alpha-i\omega)}\right) d\omega,$$

a form that is used in, say, Attari [2004] and may yield computational benefits.

Proof. Let $\bar{x}$ be the complex conjugate of x, $x \in \mathbb{C}$. If $H(\omega)$ is such that

$$H(-\omega) = \overline{H(\omega)}, \tag{8.16}$$

then

$$\int_{-\infty}^{\infty} H(\omega)\, d\omega = \int_{-\infty}^{0} H(\omega)\, d\omega + \int_{0}^{\infty} H(\omega)\, d\omega$$

$$= \int_{0}^{\infty} H(-\omega)\, d\omega + \int_{0}^{\infty} H(\omega)\, d\omega$$

$$= \overline{\int_{0}^{\infty} H(\omega)\, d\omega} + \int_{0}^{\infty} H(\omega)\, d\omega$$

$$= 2\mathrm{Re}\left(\int_{0}^{\infty} H(\omega)\, d\omega\right).$$

Since

$$\overline{\chi(\alpha+i\omega)} = \mathrm{E}\left(\overline{e^{(\alpha+i\omega)\xi}}\right) = \mathrm{E}\left(e^{(\alpha-i\omega)\xi}\right) = \chi(\alpha-i\omega),$$

the integrand in (8.13) satisfies (8.16) and the result follows. □

A result complementary to Theorem 8.4.1 holds for a call option on ξ rather than e^ξ.

Theorem 8.4.3. *In the notations of Theorem 8.4.1,*

$$E\left((\xi - k)^+\right) = \left.\frac{d\chi(k)}{dk}\right|_{k=0} - k + \frac{1}{2\pi} \int_{-\infty}^{\infty} \frac{e^{-k(-\alpha+i\omega)} \chi(-\alpha+i\omega)}{(-\alpha+i\omega)^2} d\omega$$

for any $\alpha > 0$ for which the right-hand side exists.

Proof. As in the proof of Theorem 8.4.1, denote
$$c(k) = \mathrm{E}\left((\xi - k)^+\right).$$

While not strictly necessary, to keep the presentation consistent with the proof of Theorem 8.4.1, we manipulate this expression to obtain a bounded payoff inside the expected value,
$$\begin{aligned} c(k) &= \mathrm{E}\left(\max\left(\xi - k, 0\right)\right) \\ &= \mathrm{E}\left(\xi - \min\left(\xi, k\right)\right) \\ &= \chi'(0) - k - \mathrm{E}\left(\min\left(\xi - k, 0\right)\right), \end{aligned}$$

where $\chi'(\cdot)$ is the first-order derivative of the moment-generating function. Choosing $\alpha > 0$ and dampening the integrand with an exponential function, we obtain
$$\mathrm{E}\left((\xi - k)^+\right) = \chi'(0) - k$$
$$- e^{\alpha k} \int_{-\infty}^{\infty} \left[\min\left(-(k - x), 0\right) e^{-\alpha(k-x)}\right] \left[e^{-\alpha x} p(x)\right] dx,$$

where $p(x)$ is the density of ξ. By the same arguments as in the proof of Theorem 8.4.1,
$$\mathrm{E}\left((\xi - k)^+\right) = \chi'(0) - k - e^{\alpha k}\left(\mathcal{F}^{-1}\left(g_1(\omega) g_2(\omega)\right)\right)(k),$$

where
$$g_1(\omega) = \int_{-\infty}^{\infty} e^{i\omega x} \min(-x, 0) e^{-\alpha x} dx,$$
$$g_2(\omega) = \int_{-\infty}^{\infty} e^{i\omega x} e^{-\alpha x} p(x) dx.$$

Simple calculations lead us to
$$g_1(\omega) = -\int_0^{\infty} x e^{x(-\alpha + i\omega)} dx = -\frac{1}{(-\alpha + i\omega)^2},$$
$$g_2(\omega) = \chi(-\alpha + i\omega),$$

and the theorem follows. □

8.4.2 Applications to SV Model

Combining Theorem 8.4.1 with the closed-form expression for the moment-generating function in the SV model (Propositions 8.3.6, 8.3.7, and 8.3.8), we obtain an efficient formula for pricing European call and put options in the model (8.3)–(8.4). As suggested in Andersen and Andreasen [2002], its numerical properties can be enhanced by a type of control variate method where we add the Black formula and subtract its Fourier representation, reducing the discretization errors in the process. We present the call price result in this form.

Theorem 8.4.4. *The price of a call option $c_{SV}(0, S; T, K)$ in the SV model (8.3)–(8.4) is given by*

$$c_{SV}(0, S; T, K) = \frac{1}{b} c_B(0, S'; T, K', \lambda b)$$
$$- \frac{K'}{2\pi b} \int_{-\infty}^{\infty} \frac{e^{(1/2+i\omega)\ln(S'/K')} q(1/2 + i\omega)}{\omega^2 + 1/4} d\omega, \quad (8.17)$$

where $c_B(0, S'; T, K', \sigma)$ is the Black formula for spot S', strike K', expiry T and volatility σ, with

$$S' = bS + (1-b)L, \quad K' = bK + (1-b)L.$$

Also, we have defined

$$q(u) = \Psi_{\bar{z}}\left(\frac{1}{2}(\lambda b)^2 u(u-1), u; T\right) - e^{\frac{1}{2}\lambda^2 b^2 z_0 T u(u-1)}, \quad (8.18)$$

with $\Psi_{\bar{z}}$ given in Proposition 8.3.8.

Remark 8.4.5. In (8.17) we use volatility λb in the Black model. As a further refinement, one can use the ATM volatility implied by the SV model instead. The ATM volatility can, for instance, be approximated by an expansion approach, as explained in Sections 8.7 and 9.2.

Proof. From Proposition 8.3.6,

$$c_{SV}(0, S; T, K) = E(S(T) - K)^+$$
$$= \frac{1}{b} E\left(S' e^{\ln X(T)} - K'\right)^+$$
$$= \frac{S'}{b} E\left(e^{\ln X(T)} - e^{\ln(K'/S')}\right)^+.$$

By Theorem 8.4.1 and the definition (8.10) of $\Psi_X(u; t)$,

$$c_{SV}(0, S; T, K) = \frac{1}{b}\left(S' - \frac{K'}{2\pi} \int_{-\infty}^{\infty} \frac{e^{-(\alpha+i\omega)\ln(K'/S')} \Psi_X(\alpha + i\omega; T)}{(\alpha + i\omega)(1 - \alpha - i\omega)} d\omega\right), \quad (8.19)$$

where we have used the fact that $\Psi_X(1; T) = 1$. Applying this result to the SV model with $\eta = 0$, we find that the value of the option in the displaced log-normal model $c_{DLN}(0, S; T, K)$ is given by

$$c_{\mathrm{DLN}}(0,S;T,K) = \frac{1}{b}\left(S' - \frac{K'}{2\pi}\int_{-\infty}^{\infty}\frac{e^{-(\alpha+i\omega)\ln(K'/S')}\Psi_X^0(\alpha+i\omega;T)}{(\alpha+i\omega)(1-\alpha-i\omega)}d\omega\right),$$
(8.20)

where

$$\Psi_X^0(u;T) \triangleq \mathrm{E}\left(e^{u\left(\lambda b\sqrt{z_0}W(T)-\frac{1}{2}\lambda^2 b^2 z_0 T\right)}\right) = e^{\frac{1}{2}\lambda^2 b^2 z_0 T(u^2-u)}. \quad (8.21)$$

On the other hand,

$$c_{\mathrm{DLN}}(0,S;T,K) = \frac{1}{b}c_{\mathrm{B}}(0,S';T,K',\lambda b),$$

so that

$$\frac{1}{b}c_{\mathrm{B}}(0,S';T,K',\lambda b)$$
$$-\frac{1}{b}\left(S' - \frac{K'}{2\pi}\int_{-\infty}^{\infty}\frac{e^{-(\alpha+i\omega)\ln(K'/S')}\Psi_X^0(\alpha+i\omega;T)}{(\alpha+i\omega)(1-\alpha-i\omega)}d\omega\right) = 0.$$

Adding the left-hand side of this identity, which is zero, to the right-hand side of (8.19), we obtain

$$c_{\mathrm{SV}}(0,S;T,K) = \frac{1}{b}c_{\mathrm{B}}(0,S';T,K',\lambda b)$$
$$- \frac{K'}{2\pi b}\int_{-\infty}^{\infty}\frac{e^{-(\alpha+i\omega)\ln(K'/S')}q(\alpha+i\omega)}{(\alpha+i\omega)(1-\alpha-i\omega)}d\omega,$$

where

$$q(u) = \Psi_X(u;T) - \Psi_X^0(u;T).$$

Using Propositions 8.3.7 and 8.3.8 for $\Psi_X(u;T)$ and (8.21) for $\Psi_X^0(u;T)$, and setting $\alpha = 1/2$, the result follows. □

Remark 8.4.6. The choice of $\alpha = 1/2$ in Theorem 8.4.4 is common in practice (see Lipton [2002]) and appears to give robust and stable results in most situations. As first pointed out by Lewis [2001], the value of α can be seen to define an integration contour in the complex plane, and values of α other than $1/2$ can be used as long as $\alpha + i\omega$ for all $\omega \in \mathbb{R}$ lie in the so-called *strip of convergence*[6]. One can attempt to optimize α to improve the numerical properties of the integral, see, e.g., Lee [2004] or Lord and Kahl [2007] for details. Moreover, integration contours are not restricted to straight lines. Lucic [2007] shows that all singularities of the function $q(u)$ are real (for our definition of q), paving the way for finding better — curvilinear — contours.

---

[6]The region of $u \in \mathbb{C}$ for which the moment-generating function $\chi(u)$ exists. Heston [1993] and Lewis [2000] establish the strip of convergence for the Heston model. The strip is directly related to moment existence, for the latter see Proposition 8.3.10.

Remark 8.4.7. Integrating complex values functions, such as $q(\alpha + i\omega)$, in a complex domain typically requires some care. Particularly troublesome are multi-valued functions such as the complex logarithm, as present in the expression for $\Psi_{\bar{z}}$ in Proposition 8.3.8. Should an integration contour cross a branching cut of such a function, the value will jump to a different branch, typically leading to wrong results. Fortunately the moment-generating function as presented in Proposition 8.3.8 is free of such problems. This is not the case for other, mathematically equivalent, expressions, such as, say, the one given in the original paper Heston [1993] — the reader is referred to Albrecher et al. [2007] for proofs and a detailed discussion of related issues.

Remark 8.4.8. By Assumption 8.1.1, Theorem 8.4.4 does not cover the case $b = 0$. If needed, this case can be handled by utilizing Theorem 8.4.3 instead of Theorem 8.4.1.

8.4.3 Numerical Implementation

The Fourier integral in (8.17) can be evaluated directly by any numerical integration scheme, in what is sometimes called the *direct integration approach*, see Kilin [2007]. With suitable restrictions on the integration technique and the integration grid spacing, one can formulate the pricing formula as a *discrete Fourier transform* (DFT), allowing for the usage of the *Fast Fourier Transform* (FFT) method, see Press et al. [1992]. The FFT method is developed in Section 8.4.5 below for applications requiring calculations of option prices for multiple strikes — such as volatility smile calibration or evaluation of European payoffs beyond simple puts and calls. The FFT method is certainly not competitive for calculating a *single* call option price, so here we focus on the direct integration method.

A direct numerical integration of (8.17) involves a scheme to discretize the integral and to handle the infinite integration domain. Many algorithms of varying degrees of sophistication have been proposed, some of which involve adaptive error control, optimal choice of dampening parameter α, and the mapping of the infinite integration domain on to a finite one. Lee [2004], Kilin [2007], Kahl and Jäckel [2005], Lord and Kahl [2007] contain sample algorithms, none of which employ the Black control variate inherent in our formulation (Theorem 8.4.4). As the control variate produces powerful error cancellations, we find that its inclusion allows for excellent results even when much simpler integration schemes are employed. We outline one such approach here.

Turning first to the integration bounds, we focus on the behavior of the integrand in (8.17) for large $|\omega|$; in fact, by Remark 8.4.2, only the limit $\omega \to +\infty$ needs to be explored. It turns out that the function $q(1/2 + i\omega)$ decays exponentially for large ω. In particular, as we can write

$$|q(1/2 + i\omega)| = e^{\text{Re}(\ln(q(1/2+i\omega)))},$$

we have the following result for $\ln(q(1/2 + i\omega))$.

Proposition 8.4.9. *Under our standard assumption that $|\rho| < 1$, for $q(\cdot)$ defined as in Theorem 8.4.4 we have*

$$\lim_{\omega \to +\infty} \frac{1}{\omega} \ln\left(q\left(1/2 + i\omega\right)\right) = -q_\infty,$$

where we have defined

$$q_\infty \triangleq \frac{\lambda b z_0}{\eta}\left(\sqrt{1-\rho^2} + i\rho\right)(1+\theta T). \tag{8.22}$$

Proof. The proof is obtained by applying simple calculus to formulas from Proposition 8.3.8; here we merely sketch it following the ideas of Kahl and Jäckel [2005]. We consider the limit of large positive ω. Let us denote

$$u(\omega) = 1/2 + i\omega, \quad v(\omega) = \frac{1}{2}(\lambda b)^2 u(\omega)(u(\omega) - 1) = -\frac{1}{2}(\lambda b)^2(\omega^2 + 1/4).$$

Using the notations of Proposition 8.3.8, we have (we use "$\sim$" to denote equivalence in the limit $\omega \to +\infty$),

$$\theta'(u(\omega)) \sim -i\rho\eta\lambda b\omega, \quad \gamma(v(\omega), u(\omega)) \sim \rho^c \eta \lambda b\omega,$$

where

$$\rho^c \triangleq (1-\rho^2)^{1/2}. \tag{8.23}$$

From the asymptotic behavior of $\gamma(\cdot, \cdot)$ it follows that the term $e^{-\gamma T}$ in the expressions for $A(\cdot, \cdot)$, $B(\cdot, \cdot)$ in Proposition 8.3.8 tends to zero as $\omega \to +\infty$. Therefore,

$$B(v(\omega), u(\omega)) \sim -\frac{\lambda b}{\eta}(\rho^c + i\rho)\omega,$$

and the logarithm in the definition of $A(\cdot, \cdot)$ tends to a constant,

$$\lim_{\omega \to +\infty} \ln\left(\frac{2\gamma}{\theta' + \gamma - e^{-\gamma T}(\theta' - \gamma)}\right) = \ln\left(\frac{2\rho^c}{\rho^c - i\rho}\right).$$

Therefore, only the term $(\theta' - \gamma)T$ in the expression for $A(\cdot, \cdot)$ grows with ω, and thus

$$A(v(\omega), u(\omega)) \sim -\frac{\lambda b z_0}{\eta}\theta(i\rho + \rho^c)T\omega.$$

Hence,

$$-\frac{1}{\omega}\ln(\Psi_X(1/2 + i\omega; T)) = -\frac{1}{\omega}\ln(\Psi_{\bar{z}}(v(\omega), u(\omega); T))$$

$$= -\frac{1}{\omega}(A(v(\omega), u(\omega)) + z_0 B(v(\omega), u(\omega)))$$

$$\to \frac{\lambda b z_0}{\eta}(\rho^c + i\rho)(1 + \theta T)$$

as $\omega \to +\infty$. Clearly, $\Psi_X^0(1/2+i\omega;T)$ decays faster than that, as $e^{-\text{const}\times\omega^2}$, so $q(\cdot)$ inherits its tail behavior from $\Psi_X(\cdot;T)$, and the result follows. □

The indefinite integral in Theorem 8.4.4 needs to be truncated before it can be evaluated numerically. Let $\omega_{\max} > 0$ be the upper truncation limit. We have the following simple tail estimate,

$$\left|\int_{\omega_{\max}}^{\infty} \frac{e^{(1/2+i\omega)\ln(S'/K')}q(1/2+i\omega)}{\omega^2+1/4}d\omega\right|$$

$$\leq \int_{\omega_{\max}}^{\infty} \left|e^{(1/2+i\omega)\ln(S'/K')}\right| \frac{|q(1/2+i\omega)|}{\omega^2}d\omega$$

$$\leq \sqrt{\frac{S'}{K'}}e^{-\operatorname{Re}(q_\infty)\omega_{\max}}\int_{\omega_{\max}}^{\infty}\frac{d\omega}{\omega^2}$$

$$= \sqrt{\frac{S'}{K'}}\frac{e^{-\operatorname{Re}(q_\infty)\omega_{\max}}}{\omega_{\max}}.$$

If $\varepsilon_\omega > 0$ is the absolute tolerance for computing the option price via (8.17) (a value of $\varepsilon_\omega = 10^{-3}$ to 10^{-6} is a reasonable choice), then we set the upper truncation limit $\omega_{\max}$ by the condition

$$\frac{e^{-\operatorname{Re}(q_\infty)\omega_{\max}}}{b\omega_{\max}} = \varepsilon_\omega, \tag{8.24}$$

where q_∞ is as given in Proposition 8.4.9. With Remark 8.4.2 in mind and a computational budget of N_ω points (N_ω is usually of the order of 100), we proceed to discretize uniformly over $[0,\omega_{\max}]$ and apply the rectangular rule

$$\operatorname{Re}\left(\int_0^\infty \frac{e^{(1/2+i\omega)\ln(S'/K')}q(1/2+i\omega)}{\omega^2+1/4}d\omega\right)$$

$$\approx \frac{\omega_{\max}}{N_\omega}\sum_{n=0}^{N_\omega-1}\frac{e^{\ln(S'/K')/2}}{\omega_n^2+1/4}\operatorname{Re}\left(e^{i\omega_n\ln(S'/K')}q(1/2+i\omega_n)\right), \tag{8.25}$$

where

$$\omega_n = \omega_{\max}n/N_\omega, \quad n=0,\ldots,N_\omega-1.$$

Other quadrature rules (e.g. the trapezoidal rule) can, of course, be used instead of the rectangular one.

8.4.4 Refinements of Numerical Implementation

While the method of Section 8.4.3 is simple and robust, numerical experiments show that the integration interval $[0,\omega_{\max}]$, with $\omega_{\max}$ obtained in (8.24), is often too wide, in the sense that a large proportion of the N_ω integration points are located in the region of integration where the integrand is so small that contributions to the integral are immaterial. To rectify this, we can contemplate using an adaptive integration scheme, which by design would focus the computational work in regions where the integrand is material. Alternatively, we can refine our analysis of the integrand to provide guidance for where an ordinary integration scheme should spend its time. The latter is more involved but also more illuminating, so we pursue this approach here. Much of the material is based on Bang [2009], which can be consulted for additional details. As noted earlier, the ultimate benefit of sophisticated integration schemes (including the one proposed here) tends to be rather limited in practice, as long as the Black-Scholes control variate is properly employed.

We start by stating the following refinement of Proposition 8.4.9.

Proposition 8.4.10. *Let $q(\cdot)$ be defined as in Theorem 8.4.4 and assume, as always, that $|\rho| < 1$. Then for any $\epsilon > 0$ there exists $\Omega_\epsilon > 0$ such that, for any ω that satisfies*

$$\omega \geq \max\left(\Omega_\epsilon, \frac{5}{\eta\lambda b \rho^c T}\right),$$

we have

$$\frac{1-\epsilon}{\omega^2} \leq \left|\ln\left(q\left(1/2 + i\omega\right)\right) - \left(-q_\infty \omega + q_0 - \frac{q_{-1}}{\omega}\right)\right| \leq \frac{1+\epsilon}{\omega^2}, \quad (8.26)$$

where (compare to (8.22))

$$q_\infty = \frac{\lambda b z_0}{\eta}(\rho^c + i\rho)(1 + \theta T),$$

$$q_0 = \frac{z_0}{\rho^c \eta^2}(\rho^c + i\rho)\widehat{\theta}(1 + \theta T) + \frac{2\theta z_0}{\eta^2}\left(\ln(2\rho^c) + i\arctan\left(\frac{\rho}{\rho^c}\right)\right),$$

$$q_{-1} = \frac{\theta z_0}{\eta^2}\left(T\mu\eta\lambda b + 2\widehat{\theta}\frac{\rho^c + i\rho}{(\rho^c)^2 \eta\lambda b}\right) + \mu\frac{\lambda b z_0}{\eta},$$

$$\mu = \frac{\widehat{\theta}^2}{2\eta^2(\lambda b)^2 (\rho^c)^3} + \frac{1}{8\rho^c}.$$

Here $\rho^c = (1-\rho^2)^{1/2}$ is given by (8.23) and $\widehat{\theta} = \theta'(1/2)$, where $\theta'(u) = \theta - \rho\eta\lambda bu$ is defined in Proposition 8.3.8.

Proof. The proof is by expanding $\ln(q(1/2 + i\omega))$ into a series in $1/\omega$ for small values of $1/\omega$, along the lines of the proof of Proposition 8.4.9. Full details are available in Bang [2009]. □

Let us denote

$$r(\omega) = \ln\left(q\left(1/2 + i\omega\right)\right)$$

and by $r_\infty(\omega)$ its expansion to the zeroth order for large ω (see (8.26)),

$$r_\infty(\omega) = -q_\infty \omega + q_0.$$

Consider the integral on the left-hand side of (8.25), and let us split out a part that covers the region of (approximate) validity for the asymptotic approximation $\ln(q(1/2+i\omega)) \approx r_\infty(\omega)$. To define this region, let us choose $\varepsilon'_\omega > 0$ reasonably small (of the order 10^{-2}) and pick $\omega'_{max} > 0$ such that the following two conditions are simultaneously met:

$$\omega'_{max} > \max\left(\Omega_\varepsilon, \frac{5}{\eta\lambda b \rho^c T}\right) \tag{8.27}$$

and, for any $\omega > \omega'_{max}$,

$$\frac{|q-1|}{\omega} \leq |r_\infty(\omega)|\,\varepsilon'_\omega. \tag{8.28}$$

Then, from Proposition 8.4.10,

$$\frac{|\ln(q(1/2+i\omega)) - r_\infty(\omega)|}{|r_\infty(\omega)|} \approx \frac{|q-1|}{\omega\,|r_\infty(\omega)|} \leq \varepsilon'_\omega$$

and, thus, for $\omega > \omega'_{max}$, the function $\ln(q(1/2+i\omega))$ is indeed well-approximated by $r_\infty(\omega)$. Accordingly, we write

$$\int_0^\infty \frac{e^{(1/2+i\omega)\ln(S'/K')} q(1/2+i\omega)}{\omega^2 + 1/4}\, d\omega = I_1 + I_2 + I_3, \tag{8.29}$$

where

$$I_1 = \int_0^{\omega'_{max}} \frac{e^{(1/2+i\omega)\ln(S'/K')} q(1/2+i\omega)}{\omega^2 + 1/4}\, d\omega, \tag{8.30}$$

$$I_2 = \int_{\omega'_{max}}^\infty \frac{e^{(1/2+i\omega)\ln(S'/K')}}{\omega^2 + 1/4} \left(q(1/2+i\omega) - e^{r_\infty(\omega)}\right) d\omega, \tag{8.31}$$

$$I_3 = \int_{\omega'_{max}}^\infty \frac{e^{(1/2+i\omega)\ln(S'/K')}}{\omega^2 + 1/4} e^{r_\infty(\omega)}\, d\omega. \tag{8.32}$$

As it turns out, the integral I_3 in (8.32) can be expressed through special functions. Let $E_1(z)$ be the so-called *exponential integral* (see Abramowitz and Stegun [1965]), i.e. an analytic continuation of the integral

$$E_1(z) = \int_1^{+\infty} \frac{e^{-zk}}{k}\, dk$$

to the complex plane. We then have the following result.

Lemma 8.4.11. *Let a and c be two non-negative real numbers and let z be a complex number such that $\mathrm{Re}(z) > 0$. Then*

$$R(z, a, c) \triangleq \int_c^\infty \frac{e^{-zk}}{k^2 + a^2} \, dk$$
$$= \frac{1}{2ia} \left(e^{-iaz} E_1 \left(z(c - ia) \right) - e^{iaz} E_1 \left(z(c + ia) \right) \right).$$

Proof. Follows by standard contour integration methods of complex analysis. Details are in Bang [2009]. □

Remark 8.4.12. The function $E_1(\cdot)$ can be evaluated numerically using an algorithm from Press et al. [1992]. Bang [2009] also recommends an efficient algorithm available from http://jin.ece.uiuc.edu.

With the help of Lemma 8.4.11, we can rewrite I_3 in (8.32) as

$$\int_{\omega'_{\max}}^\infty \frac{e^{(1/2+i\omega)\ln(S'/K')}}{\omega^2 + 1/4} e^{r_\infty(\omega)} \, d\omega = e^{q_0 + \ln(S'/K')/2}$$
$$\times R(q_\infty - i\ln(S'/K'), 1/2, \omega'_{\max}),$$

and calculate it efficiently using Remark 8.4.12.

Turning next to the integral I_2 in (8.29), we wish to employ a quadrature rule designed to handle the oscillations of the integrand in (8.31). To that end, and following Bang [2009], we introduce a step size

$$\delta_\omega = \frac{\left(\lambda b z_0 \sqrt{T}\right)^{-1}}{2 N_{\text{stdev}}},$$

where N_{stdev} is a user-specified range in standard deviations[7] (typically 5–6), set the number of points to be N''_ω (to be specified shortly), define

$$\omega''_n = \omega'_{\max} + \delta_\omega n, \quad n = 0, \ldots, N''_\omega,$$

and write

$$I_2 \approx e^{q_0} \sqrt{\frac{S'}{K'}} \int_{\omega'_{\max}}^{\omega'_{\max} + \delta_\omega N''_\omega} \frac{e^{\omega(-q_\infty + i\ln(S'/K'))}}{\omega^2 + 1/4} \left(e^{r(\omega) - r_\infty(\omega)} - 1 \right) d\omega$$
$$= e^{q_0} \sqrt{\frac{S'}{K'}} \sum_{n=0}^{N''_\omega - 1} \int_{\omega''_n}^{\omega''_{n+1}} \frac{e^{\omega(-q_\infty + i\ln(S'/K'))}}{\omega^2 + 1/4} \left(e^{r(\omega) - r_\infty(\omega)} - 1 \right) d\omega,$$

so that

---

[7] This step size in Fourier space is inspired by a Fourier transform of a Gaussian distribution. If the "width" of a Gaussian PDF is given by its standard deviation σ, then the "width" of its characteristic function is given by $1/\sigma$.

336   8 Vanilla Models with Stochastic Volatility I

$$I_2 \approx e^{q_0} \sqrt{\frac{S'}{K'}} \sum_{n=0}^{N''_\omega - 1} \frac{e^{r(\omega''_n) - r_\infty(\omega''_n)} - 1}{(\omega''_n)^2 + 1/4} \int_{\omega''_n}^{\omega''_{n+1}} e^{\omega(-q_\infty + i \ln(S'/K'))} d\omega$$

$$= e^{q_0} \sqrt{\frac{S'}{K'}} \sum_{n=0}^{N''_\omega - 1} \frac{e^{r(\omega''_n) - r_\infty(\omega''_n)} - 1}{(\omega''_n)^2 + 1/4}$$

$$\times \frac{e^{\omega''_{n+1}(-q_\infty + i \ln(S'/K'))} - e^{\omega''_n(-q_\infty + i \ln(S'/K'))}}{-q_\infty + i \ln(S'/K')}. \tag{8.33}$$

Note how we integrated analytically the oscillatory part of the integrand on the last step. With this scheme in place, we calculate I_2 using the quadrature rule (8.33) with N''_ω terms of the sum where we choose N''_ω adaptively by stopping when incremental changes from new terms in the sum are small enough.

Finally, let us discuss the term I_1 in (8.29), defined by (8.30). Here nothing special[8] is needed and we can just use a quadratic or trapezoidal rule with a given budget of N'_ω points (say, around 50 or so) along the same lines as we did in (8.25).

In conclusion, let us summarize the complete algorithm for calculating the integral in (8.29). First we choose a small $\varepsilon'_\omega > 0$ (of the order 10^{-2}) and find the cutoff point ω'_{max} that satisfies (8.27)–(8.28). Then we decompose the integral in (8.29) into three parts. The first integral I_1 is calculated by the standard quadratic or trapezoidal rule, similarly to (8.25). The second integral I_2 is calculated by the quadrature rule (8.33) with the number of points determined by the convergence criteria (relative or absolute). Finally the term I_3 is calculated per Remark 8.4.12. We note that while this scheme is more complex than what we described in Section 8.4.3, it does result in a faster and more accurate algorithm with a better utilization of the computational budget.

8.4.5 Fourier Integration for Arbitrary European Payoffs

Consider the problem of computing prices of European-style options with arbitrary payoffs. In particular, let $f(x)$ be a payoff function, and consider the problem of computing the following expected value,

$$\mathrm{E}\left(f\left(S(T)\right)\right).$$

Clearly,

---

[8]Of the two terms in the definition of $q(1/2 + i\omega)$ in (8.18), the (second) one related to the Gaussian distribution decays much faster than the (first) one related to the SV model, as we already noted. Hence, we can stop sampling the second term for smaller values of ω, to save a bit on calculation time. This is described in Bang [2009].

$$E(f(S(T))) = \int f(K) P(S(T) \in dK)$$

and, by (7.5),

$$E(f(S(T))) = \int f(K) \frac{\partial^2 c(0, S(0); T, K)}{\partial K^2} dK, \tag{8.34}$$

where $c(0, S; T, K)$ is the European call option value for the process $S(\cdot)$. Integrating by parts, we obtain a useful representation of a general European payoff in terms of European calls and puts.

Proposition 8.4.13. *For any twice-continuously differentiable[9] $f(x)$, the value of a European option with payoff $f(\cdot)$ and expiry T is equal to the weighted integral of call and put options with weights equal to the second derivative of $f(\cdot)$,*

$$E(f(S(T))) = f(K^*) + f'(K^*)(S(0) - K^*)$$
$$+ \int_{-\infty}^{K^*} p(0, S(0); T, K) f''(K) dK + \int_{K^*}^{\infty} c(0, S(0); T, K) f''(K) dK, \tag{8.35}$$

for any K^*.

Proof. Follows by integration by parts of (8.34). □

A combination of the suitably-discretized integral representation from Proposition 8.4.13 and Theorem 8.4.4 gives us an algorithm for computing values of European-style options with arbitrary payoffs. With the need to simultaneously compute call option prices of different strikes, the FFT method may deserve a closer look. In order to apply FFT, the discretization scheme of the integrals in (8.35) should be chosen carefully. From Theorem 8.4.4, the integrals to evaluate are

$$I(K') = \int_{-\infty}^{\infty} \frac{e^{(1/2+i\omega)\ln(S'/K')}}{\omega^2 + 1/4} q(1/2 + i\omega) d\omega \tag{8.36}$$

for various K'. We set $K^* = S(0)$ in (8.35) and discretize K in such a way that $\ln(S'/K')$ in (8.36) are equidistant. In particular, we choose $\delta > 0$, the discretization step, and define

$$x_n = \delta n, \quad K'_n = S' e^{x_n}, \quad n = -N, \ldots, N.$$

This leads to
$$bK_n + (1-b)L = (bS + (1-b)L) e^{x_n},$$

or

---

[9] But see Section 16.6.1 for extensions.

$$K_n = \left(S + \frac{1-b}{b}L\right)e^{x_n} - \frac{1-b}{b}L.$$

Then

$$I_n \triangleq I(K'_n) = \int_{-\infty}^{\infty} \frac{e^{-(1/2+i\omega)\delta n}}{\omega^2 + 1/4} q(1/2 + i\omega)\, d\omega = e^{-0.5\delta n} J_n,$$

$$J_n \triangleq \int_{-\infty}^{\infty} e^{-i\omega\delta n} \frac{q(1/2+i\omega)}{\omega^2 + 1/4}\, d\omega.$$

At a computational effort of $O(N \ln N)$, all J_n's can now be evaluated by applying (inverse) FFT to the function

$$\frac{q(1/2+i\omega)}{\omega^2 + 1/4}.$$

Once the J_n are computed, all

$$p_{\text{SV}}(0, S; T, K_n), \quad c_{\text{SV}}(0, S; T, K_n), \quad n = -N, \ldots, N,$$

can be calculated easily. The value of the option with any payoff $f(\cdot)$ is then obtained by discretizing the integrals in (8.35). We state the result as a proposition.

Proposition 8.4.14. *Fix $\delta > 0$. Let K_n, K'_n, $n = -N, \ldots, N$, be defined by*

$$K_n = \left(S + \frac{1-b}{b}L\right)e^{\delta n} - \frac{1-b}{b}L, \quad K'_n = S'e^{\delta n}.$$

Then the value of a call option with payoff $f(\cdot)$ at time T in the SV model (8.3)–(8.4), is approximately given by

$$E(f(S(T))) \approx f(S(0)) + \sum_{n=-N}^{-1} p_{\text{SV}}(0, S; T, K_n) f''(K_n)(K_{n+1} - K_n)$$

$$+ \sum_{n=0}^{N-1} c_{\text{SV}}(0, S; T, K_n) f''(K_n)(K_{n+1} - K_n),$$

where

$$c_{\text{SV}}(0, S; T, K_n) = \frac{1}{b} c_{\text{B}}(0, S'; T, K'_n, \lambda b) - \frac{K'_n}{2\pi b} e^{-0.5\delta n} J_n,$$

$$p_{\text{SV}}(0, S; T, K_n) = \frac{1}{b} p_{\text{B}}(0, S'; T, K'_n, \lambda b) - \frac{K'_n}{2\pi b} e^{-0.5\delta n} J_n,$$

with $\{J_n\}_{n=-N}^{N}$ evaluated by an inverse FFT transform of the function

$$\frac{q(1/2+i\omega)}{\omega^2 + 1/4},$$

and $q(u)$ given in Theorem 8.4.4.

Using FFT to compute the $2N + 1$ J_n-integrals improves numerical effort of a direct integration scheme, from $O(N^2)$ to $O(N \ln N)$. On the other hand, FFT has several potential drawbacks, including the fact that it imposes quite onerous requirements on the discretization of the strike domain, requiring that N be a power of 2 and that the grid be equidistant in $\ln(S'/K')$. Also, by the nature of FFT, an equidistant grid of the same size is then used to discretize the frequency domain. Both choices are often suboptimal — for example, we may want to choose a strike grid to take into account particular features of the payoff $f(\cdot)$, and we may want to discretize the frequency domain with a different number of grid points and/or non-equidistant spacing. In fact, Kilin [2007] observes that the integration effort is dominated by the calculation of the values of $q(1/2 + i\omega)$ for different ω and that they, critically, do not depend on strike. Kilin [2007] convincingly demonstrates that a careful implementation of the direct integration method of (8.17), even for multiple strikes, is often more efficient than FFT, provided that i) the values of $q(\cdot)$ are cached and reused when valuing different options, and ii) better discretization schemes are employed in the strike/frequency domains than those required by the FFT method.

8.5 Integration in Variance Domain

Under the assumption $\rho = 0$, a well-known "mixing" result (see e.g. Hull and White [1987]) represents the value of a European call option in the SV model (8.3)–(8.4) as an integral of the values of call options under the displaced log-normal model against the distribution of integrated variance. Specifically, the following lemma holds.

Lemma 8.5.1. *In the SV model (8.3)–(8.4) with $\rho = 0$, the value of a call option is given by*

$$c_{\text{SV}}(0, S; T, K) = \frac{1}{b} \text{E}\left(c_{\text{B}}\left(0, S'; T, K', \lambda b \sqrt{\bar{z}(T)/T}\right)\right), \quad (8.37)$$

where S' and K' are given in Theorem 8.4.4,

$$\bar{z}(T) = \int_0^T z(t)\, dt$$

(see (8.11)), and $c_{\text{B}}(\cdot, \cdot; \cdot, \cdot, \sigma)$ is the value of a call option in the Black model with volatility σ.

Proof. Follows by conditioning on the trajectory of $z(\cdot)$ and using the independence of the Brownian motion $W(\cdot)$ of $z(\cdot)$. □

Remark 8.5.2. An extension of this result to non-zero correlation ρ exists, see Proposition A.3.7 and in particular equation (A.39). Unfortunately it cannot be used for our purposes here, as the more general formula involves not only the time integral of $z(\cdot)$ but also other random variables.

It is natural to treat the function under the expected value operator in (8.37) as a function of $\bar{z}(T)$,

$$c_{SV}(0, S; T, K) = \mathrm{E}\left(C\left(\bar{z}(T)\right)\right), \quad C(U) = \frac{1}{b}c_B\left(0, S; T, K, \lambda b\sqrt{U/T}\right). \tag{8.38}$$

As the moment-generating function $\Psi_{\bar{z}}(u, 0; T)$ of $\bar{z}(T)$ is known from Proposition 8.3.8, the expected value in (8.38) can be computed by Fourier integration. In particular, denoting by $p_{\bar{z}}(U)$ the probability density function of $\bar{z}(T)$, consider using (8.38) to argue that

$$\begin{aligned}
c_{SV}(0, S; T, K) &= \int_0^\infty C(U) p_{\bar{z}}(U) \, dU \\
&= \frac{1}{2\pi} \int_0^\infty C(U) \int_{-\infty}^\infty e^{-i\omega U} \Psi_{\bar{z}}(i\omega, 0; T) \, d\omega \, dU \\
&= \frac{1}{2\pi} \int_{-\infty}^\infty \Psi_{\bar{z}}(i\omega, 0; T) \left(\int_0^\infty C(U) e^{-i\omega U} \, dU \right) d\omega \\
&= \frac{1}{2\pi} \int_{-\infty}^\infty \Psi_{\bar{z}}(i\omega, 0; T) \left(\mathcal{F}C\right)(-\omega) \, d\omega,
\end{aligned}$$

where

$$\left(\mathcal{F}C\right)(\omega) \triangleq \int_0^\infty e^{i\omega U} C(U) \, dU \tag{8.39}$$

is the Fourier transform of $C(U)$ and we have used in the second equality the fact that $\Psi_{\bar{z}}$ is the Fourier transform of $p_{\bar{z}}$.

This argument demonstrates the main idea behind Fourier integration in the variance domain, but suffers from the fundamental problem that the function $C(\cdot)$ is not integrable, whereby the Fourier transform (8.39) is not well-defined. Fortunately we can solve the problem by the standard remedy of introducing a dampening function $e^{-\alpha U}$ in the integrand, as the following proposition demonstrates.

Proposition 8.5.3. *For $\alpha > 0$ such that $\Psi_{\bar{z}}(\alpha, 0; T)$ exists, the following holds,*

$$c_{SV}(0, S; T, K) = \frac{1}{2\pi} \int_{-\infty}^\infty \Psi_{\bar{z}}(\alpha + i\omega, 0; T) \left(\mathcal{F}\widehat{C}\right)(-\omega) \, d\omega,$$

where

$$\widehat{C}(U) = C(U) e^{-\alpha U}, \tag{8.40}$$

and $\Psi_{\bar{z}}(u, 0; T)$ is given in Proposition 8.3.8.

Proof. We have

8.5 Integration in Variance Domain

$$c_{\text{SV}}(0, S; T, K) = \int_0^\infty C(U)e^{-\alpha U}\left(e^{\alpha U}p_{\bar{z}}(U)\right) dU$$

$$= \frac{1}{2\pi}\int_0^\infty C(U)e^{-\alpha U}\left(\int_{-\infty}^\infty e^{-i\omega U}\Psi_{\bar{z}}(\alpha + i\omega, 0; T)\, d\omega\right) dU$$

$$= \frac{1}{2\pi}\int_{-\infty}^\infty \Psi_{\bar{z}}(\alpha + i\omega, 0; T)\left(\int_0^\infty C(U)e^{-\alpha U}e^{-i\omega U}\, dU\right) d\omega$$

$$= \frac{1}{2\pi}\int_{-\infty}^\infty \Psi_{\bar{z}}(\alpha + i\omega, 0; T)\left(\mathcal{F}\hat{C}\right)(-\omega)\, d\omega.$$

□

It is probably the case that the numerical method based on the result of Proposition 8.5.3 is not as speedy as the direct integration method in Section 8.4, but it allows for interesting generalizations to arbitrary payoff functions and arbitrary skew functions, a setup where it compares favorably to Monte Carlo or PDE methods. With this generalization in mind, consider the general model specification (8.1)–(8.2), where we have the following counterpart to Lemma 8.5.1.

Lemma 8.5.4. *For a positive constant v, let $g(t, S; v)$ satisfy the PDE*

$$\frac{\partial g(t, S; v)}{\partial t} + \frac{1}{2}v\varphi(S)^2\frac{\partial^2 g(t, S; v)}{\partial S^2} = 0, \tag{8.41}$$

subject to the terminal boundary condition $g(T, S; v) = f(S)$. For the general stochastic volatility model dynamics (8.1)–(8.2) with $\rho = 0$ we have

$$\mathrm{E}\left(f\left(S(T)\right)\right) = \mathrm{E}\left(g\left(0, S(0); T^{-1}\lambda^2 \bar{z}(T)\right)\right). \tag{8.42}$$

Consistent with (8.39) and (8.40), we proceed to introduce a Fourier transform of a dampened function g,

$$(\mathcal{F}\hat{g})(\omega) = \int_{-\infty}^\infty e^{i\omega U}e^{-\alpha U}g\left(0, S(0); T^{-1}\lambda^2 U\right) dU, \tag{8.43}$$

where $\alpha > 0$ is as in Proposition 8.5.3. Then we have the following generalization of Proposition 8.5.3.

Proposition 8.5.5. *Consider the system (8.1)–(8.2), with $\psi(z) = \sqrt{z}$. Let $g(t, S; v)$ be as in (8.41) and $(\mathcal{F}\hat{g})$ as in (8.43) for $\alpha > 0$ such that $\Psi_{\bar{z}}(\alpha + i\omega, 0; T)$ is finite for all ω. Then*

$$\mathrm{E}\left(f\left(S(T)\right)\right) = \frac{1}{2\pi}\int_{-\infty}^\infty (\mathcal{F}\hat{g})(-\omega)\Psi_{\bar{z}}(\alpha + i\omega, 0; T)\, d\omega,$$

where $\Psi_{\bar{z}}(u, 0; T)$ is given in Proposition 8.3.8.

The proposition gives us a way to compute values of arbitrary European options in a model with an essentially arbitrary volatility function $\varphi(\cdot)$. In calculating the integral in (8.43), we need a way to efficiently compute the function $g(0, S(0); v)$ from (8.41) for many different values of v. Fortunately, in Chapter 7 we developed many such methods, ranging from analytical expressions, to expansions and finite difference methods[10]. We note that if the function $\varphi(\cdot)$ is complicated enough to require finite difference methods, it is crucial that we use the "trick" of Section 7.4.1 to ensure that only a single finite difference grid is solved.

Remark 8.5.6. It can be verified that the moment-generating function $\Psi_{\bar{z}}(u, 0; T)$ is finite in a neighborhood around $u = 0$. Moments of arbitrary order of $\bar{z}(T)$ consequently exist and can be computed by differentiation

$$\mathrm{E}\left(\bar{z}(T)^n\right) = \left.\frac{d^n}{du^n} \Psi_{\bar{z}}(u, 0; T)\right|_{u=0}, \quad n = 1, 2, \ldots.$$

Among other things, these moments can be used to dimension the U-grid used for the integration algorithm. For instance, for a given confidence multiplier γ (e.g. 5 or 10) we can, somewhat crudely, set

$$U_{\max} = \mathrm{E}\left(\bar{z}(T)\right) + \gamma\sqrt{\mathrm{Var}\left(\bar{z}(T)\right)}, \quad U_{\min} = \left(\mathrm{E}\left(\bar{z}(T)\right) - \gamma\sqrt{\mathrm{Var}\left(\bar{z}(T)\right)}\right)^+.$$

More elaborate schemes are also possible.

We note that Proposition 8.5.5 can also be applied to the case $\psi(z) = \sqrt{z - v}$, where $v > 0$ is a constant and where we enforce the additional constraint that $v < z_0$. To see this, consider the SDE

$$dz(t) = \theta\left(z_0 - z(t)\right)dt + \eta\sqrt{z(t) - v}\,dZ(t),$$

and set $z^*(t) = z(t) - v$. Then

$$dz^*(t) = \theta\left(z_0^* - z^*(t)\right)dt + \eta\sqrt{z^*(t)}\,dZ(t), \quad z_0^* = z_0 - v > 0, \quad (8.44)$$

and

$$\mathrm{E}\left(e^{u\int_0^T z(t)\,dt}\right) = \mathrm{E}\left(e^{u\int_0^T (z^*(t) + v)\,dt}\right)$$

$$= e^{uvT}\mathrm{E}\left(e^{u\int_0^T z^*(t)\,dt}\right) = e^{uvT}\Psi_{\bar{z}}(u, 0; T),$$

where $\Psi_{\bar{z}}(u, 0; T)$ is computed as in Proposition 8.3.8 with the substitution $z_0 \to z_0 - v$. The form $\psi(x) = \sqrt{x - v}$ is useful if we wish to keep the process

---

[10] Many of the methods in Chapter 7 were specific to calls, for which the boundary condition on the PDE is $f(S(T)) = (S(T) - K)^+$. Not only is this case by far the most important in practice, but also helps with pricing of other payouts via the replication approach (Proposition 8.4.13).

$z(t)$ away from $z=0$: it easily follows from (8.44) and $z(t) = z^*(t) + v$ that $z(t)$ will never go below v. According to Proposition 8.A.1, another way to keep $z(\cdot)$ away from the origin is to use $\psi(x) = x^p$, $1/2 < p < 1$. This case, however, has no analytical tractability.

For general $\psi(\cdot)$, let us consider ways to characterize the function $\Psi_{\bar{z}}(u, 0; T)$ that we now define by (8.11) for a general $z(\cdot)$ in (8.2). A useful starting point is the following result, easily proven from the Feynman-Kac formula in Section 1.8.

Lemma 8.5.7. *Let*

$$dz(t) = \theta\left(z_0 - z(t)\right) dt + \eta \psi\left(z(t)\right) dZ(t).$$

Then $\Psi_{\bar{z}}(u, 0; T) = L(0, z_0; u)$, where $L(t, z; u)$ satisfies the PDE

$$\frac{\partial L}{\partial t} + \theta\left(z_0 - z\right) \frac{\partial L}{\partial z} + \frac{1}{2} \eta^2 \psi(z)^2 \frac{\partial^2 L}{\partial z^2} + uzL = 0,$$

subject to the boundary condition $L(T, z; u) = 1$.

Solution of the PDE in Lemma 8.5.7 can, of course, be done by finite difference methods, but at considerable numerical expense. An asymptotic expansion approach with decent precision is possible, however, and shall be demonstrated in Section 9.2 for the more general case of time-dependent λ. As it turns out, for many choices of $\psi(\cdot)$ — most notably for $\psi(z) = z^p$ — naively writing

$$\psi\left(z(t)\right) \approx \sqrt{z(t)} \psi\left(z(0)\right) / \sqrt{z(0)}$$

and then using the expression for $\Psi_{\bar{z}}(u, 0; T)$ from Proposition 8.3.8 often gives good results. Indeed, as shown in Andersen and Brotherton-Ratcliffe [2005], for call options, the dependence of option values on p in the specification $\psi(z) = z^p$ is quite mild across a reasonably wide range of strikes.

For complicated functions $\varphi(\cdot)$ and $\psi(\cdot)$ — and for the case where $\rho \neq 0$ — we always have the option of abandoning Fourier methods altogether and instead opting for more generally applicable numerical techniques, such as Monte Carlo and two-dimensional finite difference methods. We cover the application of these schemes to stochastic volatility models later on, in Sections 9.5 and 9.4, respectively.

8.6 CEV-Type Stochastic Volatility Models and SABR

As discussed earlier, certain choices of $\varphi(\cdot)$ and $\psi(\cdot)$ introduce technical problems, such as exploding higher-order moments of $S(\cdot)$, non-zero probability of generating negative $S(\cdot)$, or non-zero probability of the variance process $z(\cdot)$ being absorbed at zero. In practice, moment explosion is often the thorniest of these issues, as it has the potential to produce severe errors for certain common securities (see Section 16.9). As it turns out, a simple switch from a linear function for $\varphi(\cdot)$ to a CEV-type specification prevents moment explosions that exist (Proposition 8.3.10) in the SV model. This is a useful result, so let us state it formally below. The proof is in Andersen and Piterbarg [2007].

Proposition 8.6.1. *Consider the model (8.1)–(8.2) with $\varphi(x) = x^c$ and $\psi(z) = z^p$, with $0 < c < 1$ and $p > 0$. Then for all $T \geq 0$ and $u \geq 0$,*

$$\mathrm{E}\left(S(T)^u\right) < \infty.$$

A particular CEV-type stochastic volatility model that has gained popularity with many practitioners is the so-called *SABR model*, see Hagan et al. [2002]. In Hagan et al. [2002], the SABR model is defined as

$$dS(t) = S(t)^c u(t)\, dW(t), \tag{8.45}$$
$$du(t) = \nu u(t)\, dZ(t), \tag{8.46}$$

with $\langle dW(t), dZ(t)\rangle = \rho\, dt$ and $0 < c < 1$. Note that the stochastic volatility $u(\cdot)$ is here modeled as simple geometric Brownian motion with zero drift. To translate the SDE (8.45)–(8.46) into more familiar terms, set $u(t) = \lambda\sqrt{z(t)}$, where $\lambda = u(0)/\sqrt{z_0}$. Then, with $\eta = 2\nu$,

$$dS(t) = \lambda S(t)^c \sqrt{z(t)}\, dW(t),$$
$$dz(t) = \frac{1}{4}\eta^2 z(t)\, dt + \eta z(t)\, dZ(t).$$

We recognize this as a special case of our set-up (8.1)–(8.2) with $\psi(z) = z$, $m(t) = 0$, and *negative* mean reversion speed $\theta = -\eta^2/4$. The drift term in the process for $z(\cdot)$ is rather unattractive but allows for some tractability, as we shall see below. While higher-order moments can be very large in the SABR model, it follows from Proposition 8.6.1 that all positive moments of $S(t)$ exist (the fact that the mean reversion is negative can be shown to not influence the result in the proposition). Notice also that in the SABR model $S(\cdot)$ cannot go negative (although absorption at zero is a possibility) and that the variance process is strictly positive.

The main justification for the form of the equations (8.1)–(8.2) is that it allows for fairly accurate asymptotic expansions for European option prices. Hagan et al. [2002] obtained the first such expansion result by combining classical perturbation methods with, in the words of Obloj [2008], "impressive intuition". Still, the result in Hagan et al. [2002] suffers from an internal inconsistency as $c \to 1$ and has later been revised by authors relying on more formal approaches. The result we list below is proven in Obloj [2008], based on earlier theoretical results in Berestycki et al. [2004] and Henry-Labordère [2005]. A similar result has been proven by Osajima [2007], using the small-noise expansion technique that we employed in Section 7.6.3.

Proposition 8.6.2. *For the model (8.45)–(8.46), the implied volatility smile is*
$$\sigma_B(t, S(t); K, T) = I^0 \left(1 + (T-t)I^1\right) + O\left((T-t)^2\right),$$
where
$$I^0 = \frac{-\nu \ln(K/S(t))}{\ln\left(\frac{\sqrt{1-2\rho q + q^2} + q - \rho}{1-\rho}\right)}, \quad q = \frac{\nu}{u(t)} \frac{S(t)^{1-c} - K^{1-c}}{1-c},$$
$$I^1 = \frac{(c-1)^2}{24} \frac{u(t)^2}{(S(t)K)^{1-c}} + \frac{1}{4} \frac{\rho \nu u(t) c}{(S(t)K)^{(1-c)/2}} + \frac{2 - 3\rho^2}{24} \nu^2.$$

Due to its lack of a mean reversion parameter, the SABR model often has difficulty matching smiles at different maturities when only a single set of calibration parameters $(\nu, c, \rho, u(0))$ is used. In practice, many financial institutions therefore maintain T-indexed vectors of these parameters, using the model primarily as a tool to interpolate and extrapolate the volatility smile. Some care must be exercised here, since the expansion listed above is not necessarily arbitrage-free; indeed, it is known that the expansion above may imply negative state price densities for low strikes and large maturities[11]. These issues could potentially be rectified by ad-hoc methods for modifying the density, see Section 16.9 for an example.

8.7 Numerical Examples: Volatility Smile Statics

Having established a valuation formula for European options in the SV model, let us proceed to put it to work on some concrete model parameterizations. In doing so, we pay special attention to the way the various parameters of the SV model effect the implied volatility smile $\sigma_B(0, S(0); K, T)$. The results here provide additional color to the qualitative parameter discussion in Section 8.2. To aid our discussion, we start by listing a small-T expansion for the implied volatility of the SV model. The expansion is not particularly precise for medium and long-dated securities, but it suffices for the largely qualitative analysis in this section. As the expansion relies on techniques that we discuss in detail later (in Section 9.2) we skip the proof and also omit, for now, a precise characterization of the approximation convergence as $T \to 0$.

Lemma 8.7.1. *Define log-moneyness $\chi \triangleq \ln(K/S(0))$ and consider writing the implied Black volatility as*

---

[11] Relative to the original SABR expansion in Hagan et al. [2002], the expansion in Proposition 8.6.2 is more robust in the low-strike tail; see Obloj [2008] for some numerical comparisons.

$$\sigma_B(0, S(0); T, K) = \sigma_{\text{ATM}} + R \cdot \chi + \frac{1}{2} B \cdot \chi^2 + \ldots$$

for certain constants R and B. For small T and small χ, in the SV model (8.3)–(8.4) with $L = S(0)$ we have

$$\sigma_{\text{ATM}} \approx \lambda, \quad R \approx \frac{\lambda}{2}\left(-(1-b) + \frac{\eta\rho}{2\lambda}\right),$$

$$B \approx \lambda \left(\frac{1-b^2}{6} + \frac{\eta^2(2-5\rho^2)}{24\lambda^2}\right).$$

Armed with Lemma 8.7.1, we start out with an example of how the volatility of variance parameter η affects the convexity of the volatility smile. As discussed previously, η serves to generate convexity in the volatility smile, an effect that is obvious from the approximation for B in Lemma 8.7.1 and also clearly visible in Figure 8.1.

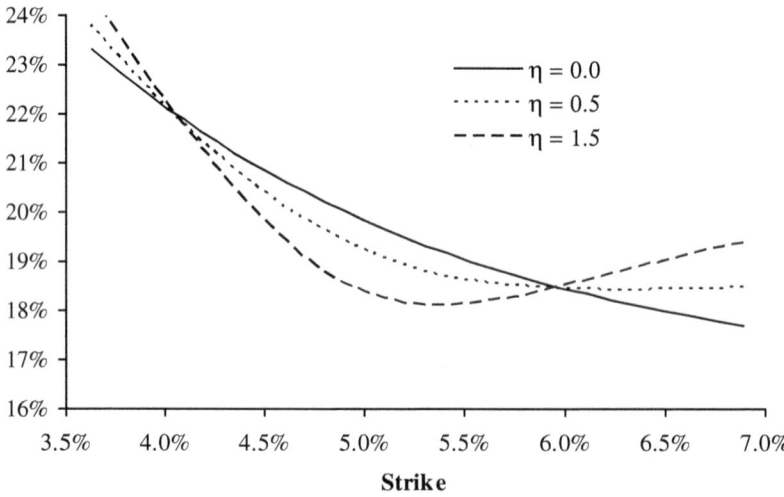

Fig. 8.1. 1 Year Volatility Smile

Notes: Implied volatility smile for SV model with $T = 1$, $S(0) = L = 5\%$, $z_0 = 1$, $b = 0.1$, $\lambda = 20\%$, $\theta = 0.1$, and $\rho = 0$. The volatility of variance parameter η varies as shown in the graph.

In Figure 8.1, the variance process is uncorrelated to the rate process, whereby Lemma 8.7.1 tells us that the slope (or skew) of the volatility smile at the at-the-money strike (5%) is generated solely by the slope parameter $b = 0.1$ in the local volatility function of the SV model. The stochastic volatility process can, of course, contribute to the skew if we use non-zero correlation; see Figure 8.2 for a numerical example. As expected, lowering correlation rotates the smile clockwise, qualitatively similar to the impact of b. Another effect is also evident in Figure 8.2: when ρ moves away from zero, the convexity of the smile around the ATM strike is reduced. This effect is consistent with the expression for B in Lemma 8.7.1 which shows that the convexity (approximately) scales with[12] $2 - 5\rho^2$.

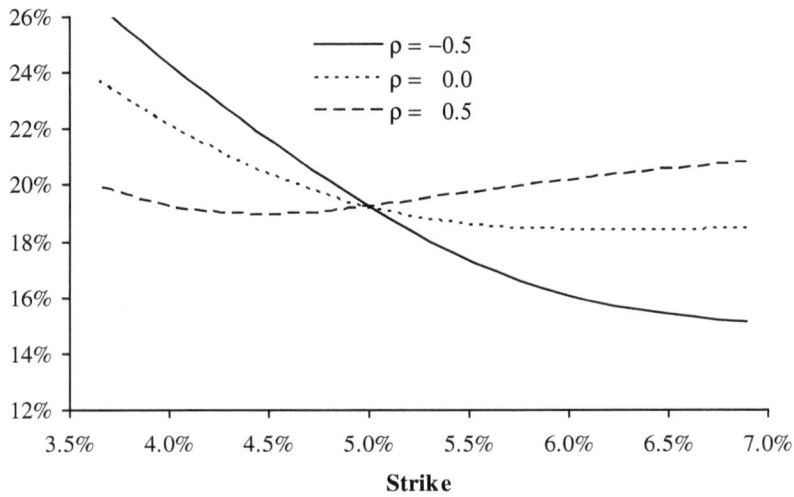

Fig. 8.2. 1 Year Volatility Smile

Notes: Implied volatility smile for SV model with $T = 1$, $S(0) = L = 5\%$, $z_0 = 1$, $b = 0.1$, $\lambda = 20\%$, $\theta = 0.1$, and $\eta = 1$. The correlation parameter ρ varies as shown in the graph.

The examples shown in Figures 8.1 and 8.2 both list the 1 year volatility smile only. To examine how the volatility smile $\sigma_B(0, S(0); K, T)$ in the SV model depends on T, consider first the case where $\rho = 0$; representative data are shown in Figure 8.3. The convexity of the smile, which originates with the stochastic volatility process, here clearly decays away as maturity is increased. As hinted at by Lemma 8.5.1, the convexity of the smile at time T is roughly proportional to the variance of the normalized realized variance $T^{-1} \int_0^T z(t)\, dt$. The convexity decay can therefore be interpreted as a mean reversion effect, since the variance of the normalized realized variance itself decays to a long-term (stationary) level, as can be seen from Corollary 8.3.3.

---

[12] Indeed, according to Lemma 8.7.1 the (short-maturity) smile convexity originating from stochastic volatility can become *negative* is $|\rho| > \sqrt{2/5} \approx 0.632$. This is easily verified numerically.

The speed of the decay is controlled by manipulating mean reversion speed θ; the higher θ is, the quicker the smile convexity decays in the T-direction.

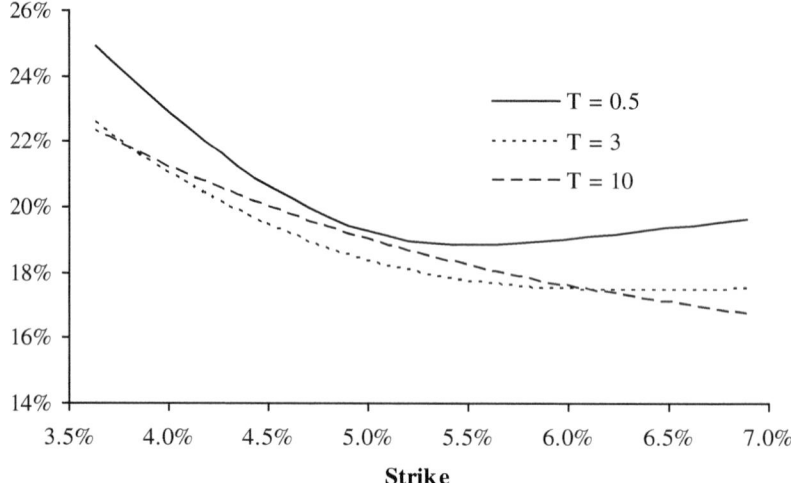

Fig. 8.3. Term Structure of Volatility Smiles

Notes: Implied volatility smile for SV model with $S(0) = L = 5\%$, $z_0 = 1$, $\rho = 0.0$, $\lambda = 20\%$, $\theta = 0.5$, $b = 0.1$, and $\eta = 1.5$. The smile maturity T varies as shown in the graph.

We note in passing that the ATM volatility of a constant parameter SV model is not a monotonic function of option maturity, as a quick glance at Figure 8.3 will confirm. For an analysis of the ATM volatility level and its dependence on maturity, see Lewis [2000].

In Figure 8.3 the slope of the smile around the ATM point is generated only from the parameter b in the local volatility function and consequently shows little decay in T. If, on the other hand, we had used a negative variance-spot correlation to generate the skew, we would expect the volatility smile to flatten out in T, for the same reason that the smile convexity decays. Figure 8.4 confirms this intuition.

8.8 Numerical Examples: Volatility Smile Dynamics

As we mentioned earlier, one rationale for introducing stochastic volatility into an LV model is the desire to generate realistic *smile dynamics*. In Section 7.1.3, we listed some qualitative reasons for the failure of LV models to generate reasonable model dynamics in certain cases; we are now in a

8.8 Numerical Examples: Volatility Smile Dynamics 349

Fig. 8.4. Term Structure of Volatility Smiles

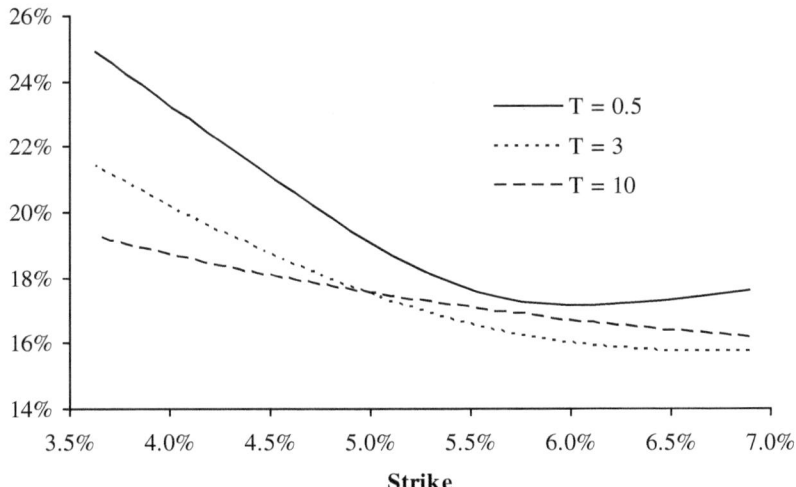

Notes: Implied volatility smile for SV model with $S(0) = 5\%$, $z_0 = 1$, $\rho = -0.5$, $\lambda = 20\%$, $\theta = 0.5$, $b = 1$, and $\eta = 1.5$. The smile maturity T varies as shown in the graph.

position to expand on this discussion and to show some concrete results. Specifically, we here wish to compare how the volatility smile moves with the underlying rate process, for two models: i) an ordinary (log-normal) Heston model obtained by setting $b = 1$ in the SV model (8.3)–(8.4); and ii) a pure LV model with quadratic volatility,

$$dS(t) = \lambda \left((1-b)L + bS(t) + \frac{1}{2}c(S(t) - L)^2 \right) dW(t). \qquad (8.47)$$

For our numerical experiments, we move calendar time forward to some arbitrary value t and examine how the smile looks for several levels of $S(t)$. In performing this analysis for the Heston model, we shall initially assume that $z(t)$ stays equal to its initial value z_0, but we relax this assumption later.

First, we consider the case of a (near) symmetric smile which in the local volatility model (8.47) can be obtained by setting $b = 1$. The effect of a 50 bps downward move in $S(0)$ (i.e. $S(t) = S(0) - 0.5\%$) on a specific LV model is shown in Figure 8.5. Starting from a symmetric smile when the forward rate $S(t) = S(0) = 5\%$, a shift down to 4.5% causes an overall increase in volatility levels, as well as a clock-wise tilt of the previously symmetric smile. This is readily understood, as the quadratic local volatility function will itself increase and loose its symmetry when $S(t)$ is reduced from 5% to 4.5%.

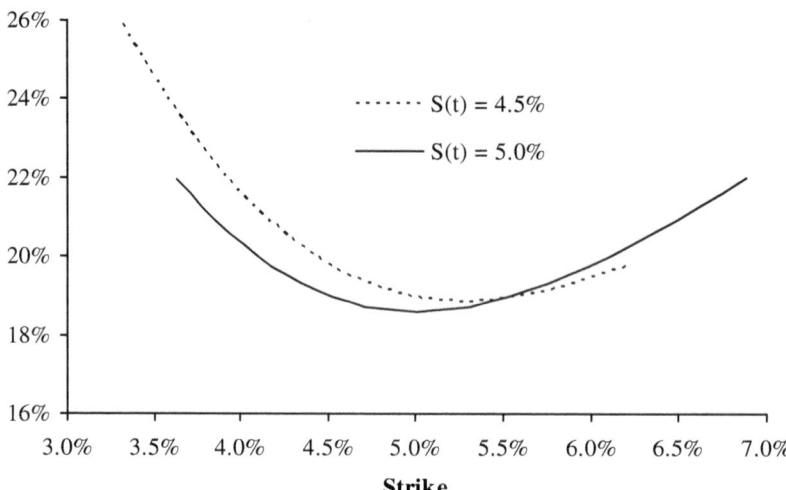

Fig. 8.5. Volatility Smile Dynamics in Quadratic LV Model

Notes: Time t implied volatility smile for quadratic LV model with $T = t + 1$, $S(0) = L = 5\%$, $b = 1$, $\lambda = 18\%$, and $c = 0.6$. Two different values for the forward rate $S(t)$ are used, as indicated in the graph.

Turning now to the Heston model, we first make the observation from Theorem 8.4.4 that European put and call option values normalized by spot S in both the Heston and Black models — and thereby the implied volatility smile of the Heston model — depend on strike K and forward rate $S(t)$ only through the ratio $K/S(t)$. Specifically, we have $\sigma_B(t, S(t); K, T) = g(K/S(t), T - t)$, for some function $g(\cdot, \cdot)$. In trader lingo, this is known as a "sticky delta" volatility smile[13], and implies that the $T = t + \Delta$ volatility smile expressed in moneyness $K/S(t)$, or log-moneyness $\ln(K/S(t))$, is independent of t and $S(t)$, as long as $z(t)$ remains unchanged at its initial value z_0. This fact makes it easy to construct the Heston model dynamics of the volatility smile in strike space; Figure 8.6 shows an example for a case where the correlation ρ has been set to zero to make the smile is symmetric in log-moneyness. Notice that as $S(t)$ drops from 5% to 4.5%, the volatility smile floats to the left, in tandem with the move in $S(t)$ such that the bottom of the smile remains centered at the forward rate.

---

[13] A reflection of the fact that the delta in the Black model, i.e. $\partial c_B / \partial S$, only depends on K/S.

Fig. 8.6. Volatility Smile Dynamics in Heston SV Model

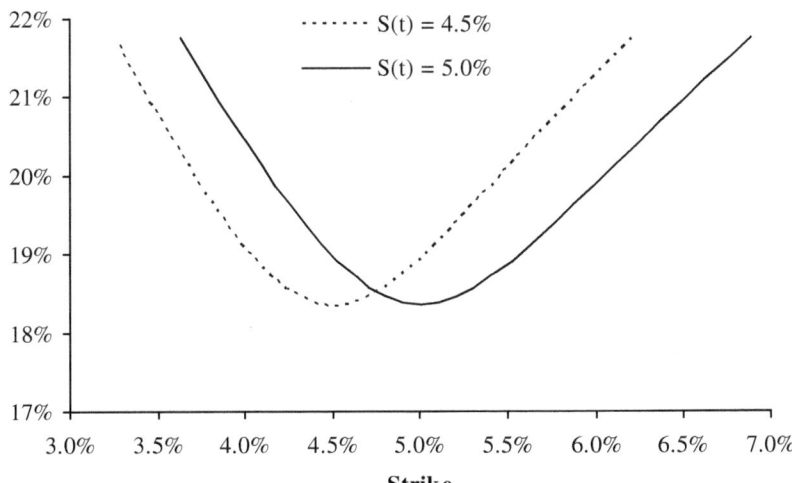

Notes: Time t implied volatility smile for SV model with $T = t+1$, $S(0) = 5\%$, $z(t) = z_0 = 1$, $b = 1$, $\lambda = 20\%$, $\theta = 0.1$, $\rho = 0$, and $\eta = 1.5$. Two different values for the forward rate $S(t)$ are used, as indicated in the graph.

While Figures 8.5 and 8.6 are interesting and highlight some important differences between local and stochastic volatility models, it is more relevant in an interest rate setting to consider the case where the volatility smile has significant skew. First, we consider the local volatility case, see Figure 8.7. A shift down in $S(t)$ will increase the level of the local volatility function and raise the level of the smile; alternatively, we can interpret the move as a slide to the right. As convexity is relatively low in the graph relative to the skew, the move in $S(t)$ has little effect on the slope of the graph.

In Figure 8.8 we examine the smile dynamics of a Heston model with a significant downward skew, induced by a non-zero correlation ρ. The sticky-delta dynamics of the smile are still in effect here, causing a slide to the left when $S(t)$ is lowered, in a manner identical to that of the symmetric case in Figure 8.6.

The dynamics on display in Figures 8.7 and 8.8 appear to be diametrically opposite of each other: the smile shifts to the right in the local volatility model and to the left in the stochastic volatility model. In reality, however, differences in model dynamics are less dramatic than these graphs show. In particular, we recall that when we computed Figure 8.8, we kept $z(t)$ constant at the value z_0. However, as $z(t)$ and $S(t)$ are negatively correlated in the model used in Figure 8.8, keeping one process constant while the other moves will clearly be wrong "on average". A more representative

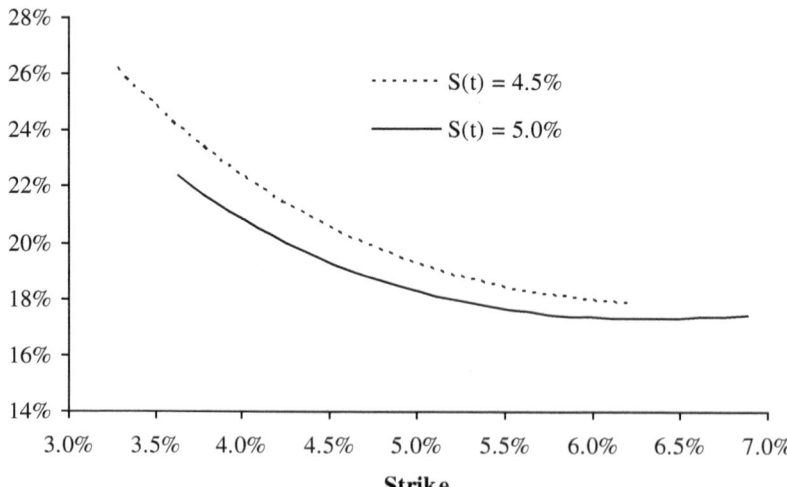

Fig. 8.7. Volatility Smile Dynamics in Quadratic LV Model

Notes: Time t implied volatility smile for quadratic LV model with $T = t+1$, $S(0) = L = 5\%$, $b = 0.1$, $\lambda = 18\%$, and $c = 0.25$. Two different values for the forward rate $S(t)$ are used, as indicated in the graph.

characterization of the smile dynamics of the Heston process would move the variance process to its most likely outcome, given the move in the underlying. That is, we wish to set $z(t)$ equal to

$$E(z(t)|S(t))$$

which we here compute by a simple Gaussian approximation that ignores mean reversion,

$$E(z(t)|S(t)) \approx z_0 + \frac{\eta\rho}{\lambda}\frac{S(t) - S(0)}{S(0)}. \tag{8.48}$$

Performing this modification on the data in Figure 8.8 results in the data in Figure 8.9.

With the rule in (8.48), the volatility smile shift of Figure 8.8 has reversed direction in Figure 8.9 and now looks quite similar to that of the local volatility dynamics of Figure 8.7. In other words, for volatility smiles that are "skew-dominated", i.e. the skew is significant and the convexity is modest, smile dynamics of local and stochastic volatility models are quite similar on average. This observation is emphasized by Dupire [2006] and to some extent goes against common wisdom (see e.g. Hagan et al. [2002]) which tends to emphasize the sticky delta behavior of the stochastic volatility model. Of course, while the behavior in Figure 8.9 may be more

8.9 Hedging in Stochastic Volatility Models

Fig. 8.8. Volatility Smile Dynamics in Heston SV Model

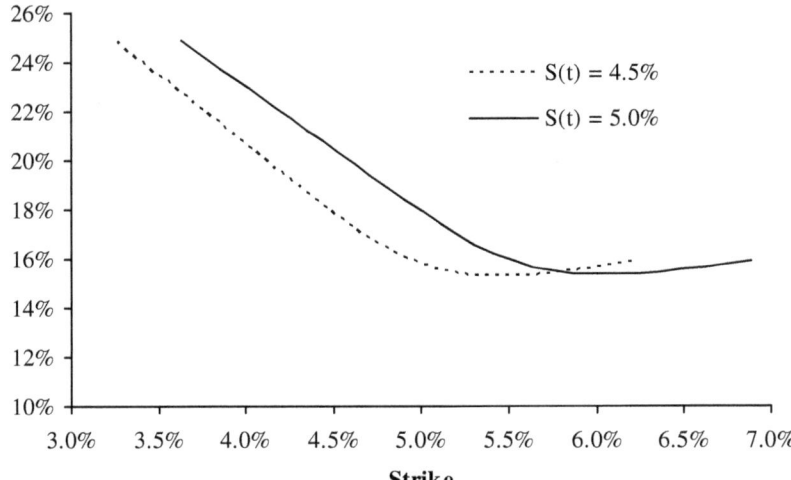

Notes: Time t implied volatility smile for SV model with $T = t+1$, $z(t) = z_0 = 1$, $S(0) = 5\%$, $b = 1$, $\lambda = 20\%$, $\theta = 0.1$, $\rho = -0.6$, and $\eta = 1.5$. Two different values for the forward rate $S(t)$ are used, as indicated in the graph.

likely than that of Figure 8.8, both are feasible in a stochastic variance setting, depending on what value $z(t)$ happens to take. For derivatives that have convexity with respect to volatility smile moves[14], what most reasonably represents "average" smile behavior is obviously less important than the fact that variance is random.

We finish this section by noting that the ideas behind (8.48) are also relevant for hedge construction in presence of stochastic volatility. We return to this topic in Section 8.9.2.

8.9 Hedging in Stochastic Volatility Models

8.9.1 Hedge Construction, Delta and Vega

Having now treated the subject of option pricing with stochastic volatility in quite some detail, let us make a foray into the topic of hedge construction. With their two generally non-collinear sources of randomness W and Z, it

---

[14] An option on implied volatility is an obvious example, although somewhat esoteric in an interest rate setting. A fairly common interest rate product with some volatility convexity is a barrier option. Many examples exist in other asset classes, such as reverse cliquets and Napoleons, see Jeffery [2004].

Fig. 8.9. Volatility Smile Dynamics in Heston SV Model

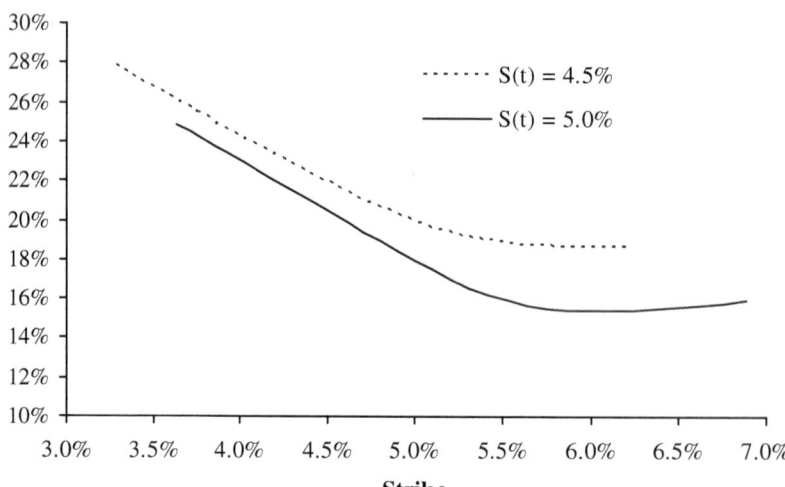

Strike

Notes: Time t implied volatility smile for the SV model in Figure 8.8, but now with $z(t)$ set as computed from formula (8.48).

should be clear that stochastic volatility models of the type (8.1)–(8.2) are not complete (in the sense defined in Section 1.4) if we limit ourselves to simple delta hedging with positions only in $S(t)$ itself. However, if options with volatility sensitivity are available for trading, these can be included into the hedge portfolio to complete the market.

Assuming general dynamics (8.1)–(8.2), we proceed to consider hedging of a contingent claim $V(t)$ that depends on both $S(t)$ and $z(t)$, i.e. we write $V(t) = V(t, S(t), z(t))$. We assume existence of two traded securities $U_1(t) = U_1(t, S(t), z(t))$ and $U_2(t) = U_2(t, S(t), z(t))$. Using the framework of Section 1.7, we associate $U_1(t)$, $U_2(t)$ with the elements of the asset vector $X(t)$ from that section. Forming a hedging portfolio Π consisting of $-\pi_1(t)$ units of $U_1(t)$ and $-\pi_2(t)$ units of $U_2(t)$, we obtain from (1.26) that

$$\pi_i = \frac{\partial V}{\partial U_i}, \quad i = 1, 2.$$

A bit of calculus leads us to expressions for the hedge ratios in terms of sensitivities to the primitives S, z of the model, and the following result follows.

Lemma 8.9.1. *The portfolio* $\Pi(t) = V(t) - \pi_1(t)U_1(t) - \pi_2(t)U_2(t)$ *is locally riskless if*

$$\pi_1 = \left(\frac{\partial V}{\partial S}\frac{\partial U_2}{\partial z} - \frac{\partial U_2}{\partial S}\frac{\partial V}{\partial z}\right)\left(\frac{\partial U_1}{\partial S}\frac{\partial U_2}{\partial z} - \frac{\partial U_2}{\partial S}\frac{\partial U_1}{\partial z}\right)^{-1}, \qquad (8.49)$$

$$\pi_2 = \left(\frac{\partial V}{\partial S}\frac{\partial U_1}{\partial z} - \frac{\partial U_1}{\partial S}\frac{\partial V}{\partial z}\right)\left(\frac{\partial U_2}{\partial S}\frac{\partial U_1}{\partial z} - \frac{\partial U_1}{\partial S}\frac{\partial U_2}{\partial z}\right)^{-1}. \qquad (8.50)$$

Remark 8.9.2. In practice, the first security U_1 would often be chosen to not depend on z — for example the swap from which $S(t)$ is computed could be used as U_1 — in which case the hedge weights simplify. In particular,

$$\pi_1 = \frac{\partial V/\partial S}{\partial U_1/\partial S} - \frac{\partial V/\partial z}{\partial U_2/\partial z}\frac{\partial U_2/\partial S}{\partial U_1/\partial S}, \qquad \pi_2 = \frac{\partial V/\partial z}{\partial U_2/\partial z},$$

as one would expect.

Remark 8.9.3. The sensitivity of a given security to volatility is often called its *vega*. Even for a model with non-stochastic volatility, such as the Black model, a vega can be computed, but will not enter the hedge balance equation (1.28). In a stochastic volatility model, a vega can conveniently[15] be defined to be $\partial/\partial z$ — which *will* enter the hedge balance equation. It follows that the choice (8.49)–(8.50) ensures that the hedged portfolio Π is delta-neutral, in the sense that

$$\frac{\partial \Pi(t)}{\partial S} = 0,$$

as well as *vega-neutral*,

$$\frac{\partial \Pi(t)}{\partial z} = 0.$$

8.9.2 Minimum Variance Delta Hedging

While the theoretical notion of "delta" assumes that the stochastic variance process z is kept fixed under perturbations of S, we saw earlier in Section 8.8 (see, in particular, Figure 8.9 and the discussion around it) that it sometimes might be more natural to let z float along with S, in a manner determined by the correlation between these quantities. Indeed, to the extent that our hedging strategy were to employ a position in S only, and not to separately hedge the exposure to z, the "best" hedging strategy — in the sense of locally minimizing hedging errors — is one based on such a joint move in z and S. We proceed to present this idea, using rather ad-hoc (or "deceptively simple", to paraphrase Ewald et al. [2007]) techniques; for a full account and for a connection to the concept of the *minimal martingale measure*, see Föllmer and Schweizer [1990] and Ewald et al. [2007].

First, let us return to the model (8.1)–(8.2), but now use a Cholesky decomposition to rewrite the process for $z(t)$ as

---

[15] From a theoretical viewpoint. More practical definitions of vega are covered later in the book, see Chapter 26 in particular.

8 Vanilla Models with Stochastic Volatility I

$$dz(t) = O(dt) + \sigma_z(t)\left(\rho\, dW(t) + \sqrt{1-\rho^2}\, dB(t)\right),$$

where B is a Brownian motion that is *independent* of W, and we use $\sigma_z(t) = \eta\psi(z(t))$ and $\sigma_S(t) = \lambda\varphi(S(t))\sqrt{z(t)}$ for notational clarity. Consider now a claim

$$V(t) = V(t, S(t), z(t)),$$

where, by Ito's lemma,

$$dV(t) = O(dt) + \frac{\partial V(t)}{\partial S}\, dS(t) + \frac{\partial V(t)}{\partial z}\sigma_z(t)\left(\rho\, dW(t) + \sqrt{1-\rho^2}\, dB(t)\right).$$

Let us form a portfolio Π of the claim V and a position of $-\pi(t)$ in $S(t)$; that is,

$$d\Pi(t) = -\pi(t)\, dS(t) + dV(t). \tag{8.51}$$

We wish to set $\pi(t)$ such that $\mathrm{Var}_t(d\Pi(t))$ is minimized.

Lemma 8.9.4. *With $d\Pi(t)$ defined in (8.51), the variance $\mathrm{Var}_t(d\Pi(t))$ is minimized by setting $\pi(t) = \pi_{\mathrm{mv}}(t)$, where*

$$\pi_{\mathrm{mv}}(t) = \frac{\partial V(t)}{\partial S} + w(t), \qquad w(t) = \frac{\partial V(t)}{\partial z}\frac{\rho\sigma_z(t)}{\sigma_S(t)},$$

and $\sigma_z(t) = \eta\psi(z(t))$, $\sigma_S(t) = \lambda\varphi(S(t))\sqrt{z(t)}$.

Proof. It is easily seen that

$$\mathrm{Var}_t(d\Pi(t)) = \left(-\pi(t)\sigma_S(t) + \frac{\partial V(t)}{\partial S}\sigma_S(t) + \frac{\partial V(t)}{\partial z}\sigma_z(t)\rho\right)^2 dt$$
$$+ \left(\frac{\partial V(t)}{\partial z}\right)^2 \sigma_z(t)^2 (1-\rho^2)\, dt.$$

The first-order condition for the minimum is therefore

$$0 = -2\sigma_S(t)\left(-\pi(t)\sigma_S(t) + \frac{\partial V(t)}{\partial S}\sigma_S(t) + \frac{\partial V(t)}{\partial z}\sigma_z(t)\rho\right),$$

from which the lemma follows. $\square$

We notice that $w(t)$ in Lemma 8.9.4 can be written informally as

$$w(t) = \frac{\partial V(t)}{\partial z}\frac{\mathrm{E}_t\left(dz(t)|dS(t) = dS\right)}{dS}$$

which shows that the *minimum-variance* (MV) hedge ratio is obtained, in effect, by moving the z-process to its expected value, given an infinitesimal perturbation in the S-process. In other words, the hedge represents our best guess for a position in the underlying that will hedge moves in $V(t)$ caused by changes in *both* $S(t)$ and $z(t)$, as in Figure 8.9.

To further characterize the properties of the MV hedge weight, we insert the result of Lemma 8.9.4 into (8.51), which yields

$$d\Pi(t) = O(dt) + \frac{\partial V(t)}{\partial z}\sigma_z(t)\sqrt{1-\rho^2}\,dB(t).$$

In other words, the MV hedge produces a portfolio that is not exposed to $W(t)$ but only to the orthogonal Brownian motion $B(t)$. If one thinks of $W(t)$ as "market" noise, we can say — in the language of the classical CAPM[16] analysis — that the hedged portfolio has no *beta*. For this reason, the hedge construction in Lemma 8.9.4 is also sometimes known as a *zero-beta hedge*.

8.9.3 Minimum Variance Hedging: an Example

To better understand the practical ramifications of MV hedging, let us do a concrete example based on the SABR model from Section 8.6, which we here parameterize as

$$dS(t) = \lambda\sqrt{z(t)}S(t)^c\,dW(t),$$
$$dz(t) = \frac{1}{4}\eta^2 z(t)dt + \eta z(t)\left(\rho\,dW(t) + \sqrt{1-\rho^2}\,dB(t)\right), \quad z(0) = 1.$$

According to Lemma 8.9.4, the MV hedge ratio in SABR is

$$\pi_{\mathrm{mv}}(t) = \frac{\partial V(t)}{\partial S} + \eta\sqrt{z(t)}\rho\frac{\partial V(t)}{\partial z}\frac{1}{\lambda S(t)^c}.$$

In a typical interest rate application $z(t) \approx 1$, $\lambda S(t)^c \approx 0.01$ and $\eta \approx 1$, such that, as a rule of thumb,

$$\pi_{\mathrm{mv}}(t) \approx \frac{\partial V(t)}{\partial S} + 100 \times \rho\frac{\partial V(t)}{\partial z}.$$

For call and put options, the hedge adjustment to the "pure" delta $\partial V/\partial S$ is here typically negative, as we have $\partial V/\partial z > 0$ and, in normal market conditions, $\rho < 0$. This is consistent with Figure 8.8.

We now perform the following small experiment: we lock the correlation parameter at a pre-fixed value and then least-squares calibrate the SABR model to an actual market Black volatility smile. For a range of correlation parameters, we then compute "pure" deltas ($\partial V/\partial S$) and MV deltas (π_{mv}) for swaptions with different strikes. Using market data roughly consistent with the 5y×5y swaption volatility smile in the summer of 2005, the calibration results are in Table 8.1.

As one would expect, making correlation progressively more negative causes the skew power c to increase, from about 20% at $\rho = 0$ to nearly

---

[16] Capital Asset Pricing Model, see Sharpe [1964].

ρ	0	-0.1	-0.2	-0.3	-0.35
$\lambda S(0)^{1-c}$	0.135	0.136	0.137	0.139	0.140
c	0.223	0.432	0.648	0.877	0.999
η	0.684	0.686	0.696	0.712	0.726

Table 8.1. SABR Calibration Results

90% at $\rho = -0.3$, with other parameters being quite stable across different correlation choices. Figures 8.10 and 8.11 show the pure delta $\partial V/\partial S$ and the minimum variance delta π_{mv} for selected strikes and correlations. Clearly, the MV delta is here virtually independent of the choice of ρ, whereas the pure delta can increase quite substantially as correlation becomes more negative. It is clear from the figures that as long as hedge ratios are computed to be MV deltas, rather than pure deltas, the precise blend of local and stochastic volatility may not be critical, at least not for vanilla-like options in a skew-dominated market. This confirms a point we made earlier, in Section 8.1.

Fig. 8.10. Pure Delta

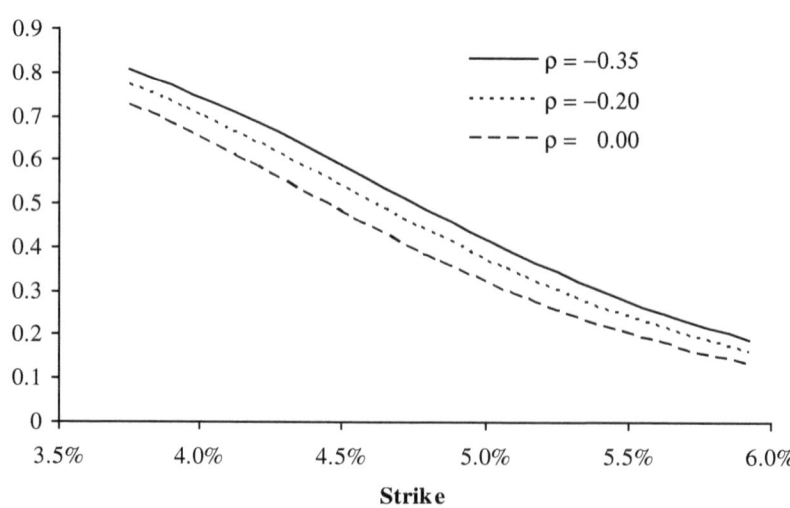

Notes: The figure shows the pure delta for the SABR models in Table 8.1.

Fig. 8.11. Minimum Variance Delta

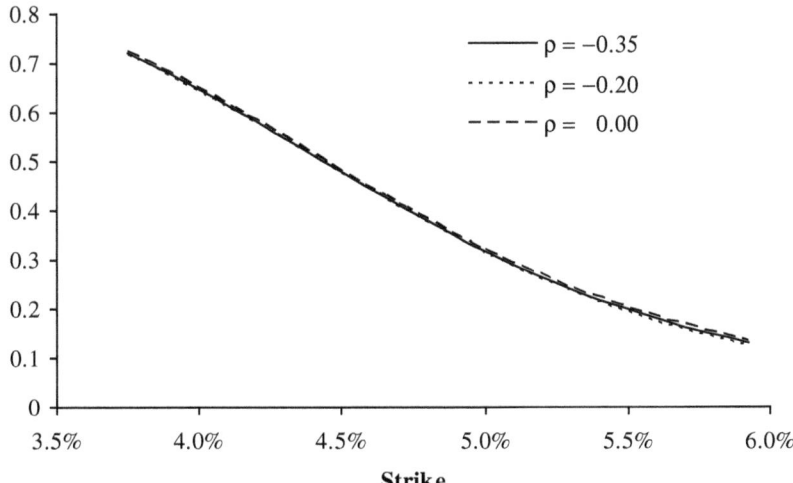

Notes: The figure shows the minimum variance (MV) delta for the SABR models in Table 8.1.

8.A Appendix: Martingale Characterization, Moment Stability, and Other Fundamental Properties for General Variance Processes

As explained in Section 8.3, it is sometimes beneficial to consider a specification of the stochastic volatility model that is more general than (8.3)–(8.4). Let us consider a general power function for $\psi(z)$ in (8.2),

$$dS(t) = \lambda \left(bS(t) + (1-b)L\right) \sqrt{z(t)}\, dW(t), \tag{8.52}$$
$$dz(t) = \theta \left(z_0 - z(t)\right) dt + \eta z(t)^p \, dZ(t), \tag{8.53}$$

with $\langle dZ(t), dW(t) \rangle = \rho\, dt$. We assume $p > 0$. In this section we briefly outline important properties of such models. For more comprehensive treatment the reader is referred to Andersen and Piterbarg [2007]. Our first result spells out the boundary behavior of the stochastic variance process.

Proposition 8.A.1. *For the process (8.53), the following holds:*

1. *0 is always an attainable boundary for $0 < p < 1/2$.*
2. *0 is an attainable boundary for $p = 1/2$, if $2z_0\theta < \eta^2$.*
3. *0 is an unattainable boundary for $p > 1/2$.*
4. *∞ is an unattainable boundary for all values of $p > 0$.*

When $0 < p < 1/2$, the origin is always accessible and we need to impose a boundary condition at $z = 0$ to make the process unique. To ensure that the process for $z(\cdot)$ has a stationary distribution, we make the following natural choice:

Assumption 8.A.2. *For $0 < p < 1/2$, the process (8.53) for $z(\cdot)$ is reflected at the origin.*

The marginal one-dimensional distribution of $z(t)$ can in principle be computed numerically by various methods, such as PDE methods or by Fourier inversion of a characteristic function. It is often convenient, however, to have an easily-computable approximation. For that purpose, a stationary distribution, if one exists, can be useful. A stationary distribution for $z(\cdot)$ does indeed exist and can be easily computed.

Proposition 8.A.3. *Let $\pi(y)$ be the stationary distribution density for $z(\cdot)$ in (8.53). Under the assumptions listed above,*

$$\pi(y) = C(p) y^{-2p} e^{Q(y;p)}, \quad C(p)^{-1} \triangleq \int_0^\infty y^{-2p} e^{Q(y;p)} dy,$$

where the function $Q(y;p)$ is given by

1. $0 < p < 1/2$ or $1/2 < p < 1$ or $p > 1$:

$$Q(y;p) = \frac{2\theta}{\eta^2} \left(\frac{z_0 y^{1-2p}}{1-2p} - \frac{y^{2-2p}}{2-2p} \right).$$

2. $p = 1/2$:

$$Q(y;p) = \frac{2\theta}{\eta^2} (z_0 \ln y - y).$$

3. $p = 1$:

$$Q(y;p) = \frac{2\theta}{\eta^2} (-z_0/y - \ln y).$$

A-priori, $S(\cdot)$ defined by (8.52)–(8.53) is only a local martingale. In fact, under some circumstances, $S(\cdot)$ is a strict local martingale, , usually a significant technical complication. Specifically, we have the following result.

Proposition 8.A.4. *When $p \leq 1/2$ or $p > 3/2$, $S(\cdot)$ is a proper martingale. When $1/2 < p < 3/2$, $S(\cdot)$ is a martingale for $\rho \leq 0$ and a strict supermartingale for $\rho > 0$. For $p = 3/2$, $S(\cdot)$ is a martingale for $\rho \leq \frac{1}{2}\eta(\lambda b)^{-1}$ and a strict supermartingale for $\rho > \frac{1}{2}\eta(\lambda b)^{-1}$.*

What this proposition states is that the set of parameters $1/2 < p < 3/2$, $\rho > 0$, should be avoided in practical modeling. The SV model (8.3)–(8.4), as already noted, has no issues in this regard. If we use $\rho = 0$ — a typical choice in interest rate modeling as explained previously — all values of p

between 0 and 3/2 are acceptable, at least as far as the martingale property is concerned.

In the model with $p = 1/2$, some moments of $S(\cdot)$ can become infinite, as stated in Proposition 8.3.10. With $p < 1/2$, this is no longer an issue:

Proposition 8.A.5. *In the model (8.52)–(8.53), if $p < 1/2$, moments $\mathrm{E}(S(T)^u)$ of all orders $u \geq 1$ for all times T are finite.*

On the other hand, if $p > 1/2$ moments may be unstable. For instance:

Proposition 8.A.6. *In the model (8.52)–(8.53), if $p > 1/2$ and $\rho = 0$, all moments $\mathrm{E}(S(T)^u)$ of all orders $u > 1$ for all times T are infinite.*

The case of non-zero correlation and $p > 1/2$ is more complicated; we refer the reader to Andersen and Piterbarg [2007].

9

Vanilla Models with Stochastic Volatility II

Having covered stochastic volatility models with time-homogeneous dynamics in Chapter 8, we are now ready to proceed with an analysis of the time-dependent case. As we shall see many examples of later in this book, stochastic volatility models with time-dependent parameters emerge naturally when vanilla models are used to approximate interest rate dynamics in a full term structure model.

In this chapter, we start out by modifying the Fourier analysis of Chapter 8 to cover time-dependent model parameters. We then proceed to introduce several approximation techniques that can speed up the calibration of model parameters to observable option prices. In particular, we continue our development of parameter averaging techniques, extending their scope to cover stochastic volatility and outlining in detail their usage in model calibration. Finally, the chapter gives detailed coverage of PDE and MC methods for general derivatives pricing; both of these numerical techniques are, as it turns out, rather tricky to apply to models with stochastic volatility, and an efficient implementation requires careful attention to detail.

9.1 Fourier Integration with Time-Dependent Parameters

As a start, let us consider extending the basic SV model (8.3)–(8.4) to allow for time-dependence of the volatility parameter[1] λ. That is, we now consider the P-measure dynamics

$$dS(t) = \lambda(t) \left(bS(t) + (1-b) L\right) \sqrt{z(t)} \, dW(t), \tag{9.1}$$

$$dz(t) = \theta \left(z_0 - z(t)\right) dt + \eta \sqrt{z(t)} \, dZ(t), \tag{9.2}$$

---

[1] A further extension to time-dependence in η, ρ, and θ is trivial, and is covered in Remark 9.1.3.

364 9 Vanilla Models with Stochastic Volatility II

where $\langle dZ(t), dW(t)\rangle = \rho \, dt$.

The model (9.1)–(9.2) still allows for call option pricing by the Fourier integration method of Section 8.4, provided that we can establish the moment-generating function (mgf) of $\ln X(t)$, with $X(t)$ being the linear function of $S(t)$ defined in Proposition 8.3.6. Let us retain the notation $\Psi_X(u;t)$ for

$$\Psi_X(u;t) = \mathrm{E}\left(e^{u \ln X(t)}\right),$$

where the process for $X(t)$ now is modified from that of Proposition 8.3.6 to include time-dependence in λ:

$$dX(t)/X(t) = b\lambda(t)\sqrt{z(t)}\, dW(t), \quad X(0) = 1.$$

The following counterpart to Proposition 8.3.7 is easily proven.

Proposition 9.1.1. *In the model (9.1)–(9.2), for any $u \in \mathbb{C}$ for which the right-hand side exists, we have*

$$\Psi_X(u;t) = \Psi_{\overline{z\lambda^2}}\left(\frac{1}{2}b^2 u(u-1), u; t\right),$$

where we have defined

$$\Psi_{\overline{z\lambda^2}}(v, u; t) \triangleq \mathrm{E}^{\widetilde{\mathrm{P}}}\left(e^{v\overline{z\lambda^2}(t)}\right), \quad \overline{z\lambda^2}(t) \triangleq \int_0^t z(s)\lambda(s)^2 \, ds, \qquad (9.3)$$

and under the new probability measure $\widetilde{\mathrm{P}}$ the process for $z(t)$ is

$$dz(t) = (\theta(z_0 - z(t)) + \rho \eta \lambda(t) b u z(t))\, dt + \eta \sqrt{z(t)}\, d\widetilde{Z}(t), \quad z(0) = z_0, \tag{9.4}$$

with $\widetilde{Z}(t)$ a $\widetilde{\mathrm{P}}$-Brownian motion. If $\rho = 0$, $\widetilde{\mathrm{P}} = \mathrm{P}$ and $z(t)$ in (9.3) follows (9.2) rather than (9.4).

The following proposition demonstrates how to compute the moment-generating function of $\overline{z\lambda^2}(T)$.

Proposition 9.1.2. *The function $\Psi_{\overline{z\lambda^2}}(v, u; T)$ defined by (9.3) is given by*

$$\Psi_{\overline{z\lambda^2}}(v, u; T) = \exp(A(0, T) + z_0 B(0, T)),$$

where $(A(t,T), B(t,T))$ solve the system of Riccati ODEs

$$\frac{d}{dt} A(t,T) + \theta z_0 B(t,T) = 0, \qquad (9.5)$$

$$\frac{d}{dt} B(t,T) - (\theta - \rho \eta b u \lambda(t)) B(t,T) + \frac{\eta^2}{2} B(t,T)^2 + v \lambda(t)^2 = 0, \qquad (9.6)$$

with the terminal conditions

$$B(T,T) = A(T,T) = 0.$$

9.1 Fourier Integration with Time-Dependent Parameters

Proof. Let us define
$$G(t, z) \triangleq \mathrm{E}^{\widetilde{\mathrm{P}}}\left(e^{v \int_t^T \lambda(s)^2 z(s)\, ds}\,\middle|\, z(t) = z\right).$$

Clearly,
$$\Psi_{\overline{z\lambda^2}}(v, u; T) = G(0, z_0).$$

On the other hand, by the Feynman-Kac formula, $G(t,z)$ satisfies the following PDE,

$$\frac{\partial}{\partial t} G(t, z) + (\theta z_0 - (\theta - \rho \eta b u \lambda(t)) z) \frac{\partial}{\partial z} G(t, z)$$
$$+ \frac{\eta^2}{2} z \frac{\partial^2}{\partial z^2} G(t, z) + v\lambda(t)^2 z G(t, z) = 0, \quad (9.7)$$

with the terminal condition
$$G(T, z) = 1, \quad z \geq 0. \qquad (9.8)$$

The PDE (9.7) is affine in z, i.e. all coefficients are linear functions of z. To solve it, we make the *ansatz* that the solution $G(t, z)$ is of the exponential form
$$G(t, z) = \exp(A(t, T) + zB(t, T)).$$

Substituting this conjectured solution into the PDE (9.7) and dividing by G, we get

$$\frac{d}{dt} A(t, T) + z \frac{d}{dt} B(t, T) + (\theta z_0 - (\theta - \rho \eta b u \lambda(t)) z) B(t, T)$$
$$+ \frac{\eta^2}{2} z B(t, T)^2 + v\lambda(t)^2 z = 0.$$

By collecting the coefficients on different powers of z, the two ODEs (9.5)–(9.6) emerge. Boundary conditions follow from (9.8). □

The system of ODEs (9.5)–(9.6) can be solved numerically using the *Runge-Kutta method*, see e.g. Press et al. [1992]. In practice, it is common for the time-dependent volatility $\lambda(t)$ to be piecewise constant,
$$\lambda(t) = \lambda_i, \quad t \in (t_{i-1}, t_i],$$

for some $0 = t_0 < t_1 < \ldots < t_I = T$. In this case, on each of the intervals $(t_{i-1}, t_i]$, the ODEs (9.5)–(9.6) can be solved in closed form, using the formulas from Proposition 8.3.8. By piecing these solutions together[2], we obtain the exact solution to the ODEs over the whole time interval $[0, T]$. However, for a given tolerance on accuracy, the Runge-Kutta method may still be faster than exact solution of the ODEs, as it avoids expensive evaluations of functions exp, ln, etc.

---
[2] The full procedure is described in Section 10.2.2.2.

Remark 9.1.3. So far, we assumed that η, ρ, and θ were constants. However, it follows easily from the proof of Proposition 9.1.2 that incorporation of time-dependence in η, ρ and θ is merely a matter of changing the ODEs (9.5)–(9.6) to

$$\frac{d}{dt}A(t,T) + \theta(t)z_0 B(t,T) = 0,$$

$$\frac{d}{dt}B(t,T) - (\theta(t) - \rho(t)\eta(t)bu\lambda(t))B(t,T) + \frac{\eta(t)^2}{2}B(t,T)^2 + v\lambda(t)^2 = 0.$$

No matter which scheme is ultimately used to solve (9.5)–(9.6), combining the integration method of Theorem 8.4.4 with the integrand in Proposition 9.1.2 — possibly extended as in Remark 9.1.3 — allows for the pricing of call options by the Fourier methods in Section 8.4.

9.2 Asymptotic Expansion with Time-Dependent Volatility

As demonstrated in previous sections, the Fourier method constitutes a powerful tool for establishing a pricing algorithm for European options, provided that the underlying stochastic volatility process is of a sufficiently simple form. Should, say, the volatility function $\psi(z)$ for $z(t)$ be something other than $\sqrt{z}$, or should the skew function $\varphi(x)$ be more complicated than a linear form, analytic tractability (as in Proposition 9.1.2) is often lost and the Fourier method may not be feasible. However, asymptotic expansion methods can still be used in some situations and may, even for cases where Fourier methods do apply, offer a compelling (and very fast) approach to European option pricing.

To develop the asymptotic expansion approach, we return to the general skew functions $\varphi(x)$ and $\psi(z)$ in (8.1)–(8.2), under the simplifying (yet practically relevant) assumption that $\rho = 0$. As in the previous section, we will assume that the volatility $\lambda(t)$ is time-dependent. To summarize, the SDE system under consideration will be

$$dS(t) = \lambda(t)\varphi(S(t))\sqrt{z(t)}\,dW(t), \tag{9.9}$$
$$dz(t) = \theta(z_0 - z(t))\,dt + \eta\psi(z(t))\,dZ(t), \quad z(0) = z_0, \tag{9.10}$$

where $\langle dZ(t), dW(t)\rangle = 0$. The form of the time-dependence — as introduced here exclusively in $\lambda(t)$ — allows us to use time-change arguments similar to those in Section 7.6.1 to show that Lemma 8.5.4 as well as Proposition 8.5.5 still apply.

Lemma 9.2.1. *For the system (9.9)–(9.10) the results of Lemma 8.5.4 and Proposition 8.5.5 hold unchanged, provided we redefine (8.43) to*

9.2 Asymptotic Expansion with Time-Dependent Volatility

$$(\mathcal{F}\hat{g})(\omega) = \int_{-\infty}^{\infty} e^{i\omega U} e^{-\alpha U} g\left(0, S(0); T^{-1} U\right) dU,$$

and make the substitutions

$$\lambda^2 \overline{z}(T) \to \overline{z\lambda^2}(T), \quad \lambda^2 U \to U, \quad \Psi_{\overline{z}} \to \Psi_{\overline{z\lambda^2}}.$$

For the special case $\psi(z) = \sqrt{z}$, Proposition 9.1.2 derives the expression for $\Psi_{\overline{z\lambda^2}}(u, 0; T)$. For more general choices of $\psi(z)$, we can rely on the PDE from Lemma 8.5.7, appropriately extended to time-dependent $\lambda(t)$. Specifically, $\Psi_{\overline{z\lambda^2}}(u, 0; T) = L(0, z_0; u)$, where $L(t, z; u)$ satisfies the PDE

$$\frac{\partial L}{\partial t} + \theta(z_0 - z)\frac{\partial L}{\partial z} + \frac{\eta^2}{2}\psi(z)^2\frac{\partial^2 L}{\partial z^2} + u\lambda(t)^2 z L = 0, \quad (9.11)$$

subject to the boundary condition $L(T, z; u) = 1$. The equation can be solved numerically, or we can attempt to derive approximations. For the latter, we first introduce a centered transform

$$l(t, z; u) \triangleq L(t, z; u) e^{-u\mu_{\overline{z\lambda^2}}(t, z)}, \quad (9.12)$$

where, under mild regularity conditions on $\psi(z)$,

$$\mu_{\overline{z\lambda^2}}(t, z) \triangleq \mathrm{E}\left(\int_t^T \lambda(s)^2 z(s)\, ds \,\bigg|\, z(t) = z\right)$$

$$= \int_t^T \lambda(s)^2 \mathrm{E}\left(z(s) | z(t) = z\right) ds$$

$$= \int_t^T \lambda(s)^2 \left(z_0 + (z - z_0) e^{-\theta(s-t)}\right) ds.$$

Introduction of $l(t, z; u)$ focuses attention on deviations of $\overline{z\lambda^2}(t)$ away from its mean, which can be expected to be small if η is small — a limit that we shall shortly examine. Insertion of (9.12) into (9.11) reveals that $l(t, z; u)$ satisfies

$$\frac{\partial l}{\partial t} + \theta(z_0 - z)\frac{\partial l}{\partial z} + \frac{\eta^2}{2}\psi(z)^2\left\{\frac{\partial^2 l}{\partial z^2} + lu^2 p(t)^2 + 2up(t)\frac{\partial l}{\partial z}\right\} = 0, \quad (9.13)$$

where

$$p(t) = \int_t^T \lambda(s)^2 e^{-\theta(s-t)} ds \quad (9.14)$$

and $l(T, z; u) = 1$.

Lemma 9.2.2. *Let $p(t)$ be as in (9.14), and define $\tilde{\psi}(z) = \frac{1}{2}\psi(z)^2$ and $h(s, z) = z_0 + (z - z_0)e^{\theta(t-s)}$. An asymptotic expansion for the solution to (9.13) in terms of η^2 is given by*

$$l(t, z; u) = 1 + \eta^2 l_1(t, z; u) + \eta^4 l_2(t, z; u) + O(\eta^6),$$

where

$$l_1(t, z; u) = u^2 l_{1,2}(t, z),$$
$$l_2(t, z; u) = u^2 l_{2,2}(t, z) - u^3 l_{2,3}(t, z) + \frac{1}{2} u^4 \left(l_{1,2}(t, z)\right)^2,$$

and

$$l_{1,2}(t, z) = \int_t^T p(s)^2 \widetilde{\psi}\left(h(s, z)\right) ds,$$
$$l_{2,2}(t, z) = \int_t^T e^{2\theta s} \widetilde{\psi}\left(h(s, z)\right) \int_s^T e^{-2\theta v} p(v)^2 \widetilde{\psi}''\left(h(v, z)\right) dv\, ds,$$
$$l_{2,3}(t, z) = -2 \int_t^T e^{\theta s} p(s) \widetilde{\psi}\left(h(s, z)\right) \int_s^T e^{-\theta v} p(v)^2 \widetilde{\psi}'\left(h(v, z)\right) dv\, ds.$$

Proof. Let

$$l(t, z; u) = 1 + \sum_{i \geq 1} l_i(t, z; u) \eta^{2i}.$$

Notice that odd powers of η are not used in the expansion, as only η^2 figures in the PDE (9.13). Inserting into (9.13) and collecting terms of order η^2 gives

$$\frac{\partial l_1}{\partial t} + \theta (z_0 - z) \frac{\partial l_1}{\partial z} + \frac{1}{2} u^2 p(t)^2 \psi(z)^2 = 0,$$

with terminal condition $l_1(T, z) = 0$. This simple PDE can be solved in closed form, yielding the solution listed in the lemma. The result for l_2 is established by collecting terms of order η^4 and proceeding as for l_1. □

While somewhat complicated in appearance, the expressions for the integrals $l_{1,2}$, $l_{2,2}$, and $l_{2,3}$ are trivial to implement on a computer. Indeed, due to the nested nature of the double integrals $l_{2,2}$ and $l_{2,3}$, all integrals can be computed in a single numerical integration loop, at negligible computational cost. In doing the integrals we start from the back, at time T, allowing us at each integration step to update the outer integral, as well as to resolve the inner integrals. In some cases of practical interest it is also possible to evaluate the integrals analytically.

Apart from potential direct application in the Fourier technique in Proposition 8.5.5, the result of Lemma 9.2.2 allows us to compute central moments as follows:

$$E\left(\left(\overline{z\lambda^2}(T) - \mu_{\overline{z\lambda^2}}(0, z_0)\right)^n\right) = \left.\frac{\partial^n l(0, z_0; u)}{\partial u^n}\right|_{u=0}, \quad n = 1, 2, \ldots. \quad (9.15)$$

There are many ways to turn these moments into an option price expression. For instance, we could rely on a classical Gram-Charlier expansion (see

9.2 Asymptotic Expansion with Time-Dependent Volatility

Ochi [1990]) or perhaps some parametric density family to express the full density of $z\lambda^2(T)$, to be used directly in (time-dependent generalizations of) equations (8.37) or (8.42). Alternatively, we can use Taylor expansions for a closed-form asymptotic result. Specifically, if the function g is defined as in Lemma 8.5.4, we can write

$$\mathrm{E}\left(f\left(S(T)\right)\right) = g\left(0, S(0); \bar{v}\right)$$
$$+ \sum_{n=1}^{\infty} \frac{1}{n! T^n} \left.\frac{\partial^n g}{\partial v^n}\right|_{v=\bar{v}} \mathrm{E}\left(\left(\overline{z\lambda^2}(T) - \mu_{\overline{z\lambda^2}}(0, z_0)\right)^n\right),$$

where the derivatives are to be evaluated at $\bar{v} \triangleq \mu_{\overline{z\lambda^2}}(0, z_0)/T$.

From (9.15) and the expansion formula in Lemma 9.2.2, a few manipulations give the required result.

Lemma 9.2.3. With $g(t, S; v)$ defined as in Lemma 8.5.4, we have to order $O(\eta^4)$

$$\mathrm{E}\left(f\left(S(T)\right)\right) = g\left(0, S(0); \bar{v}\right) + T^{-2} \left(\eta^2 l_{1,2} + \eta^4 l_{2,2}\right) \frac{\partial^2 g}{\partial v^2}$$
$$- \eta^4 T^{-3} l_{2,3} \frac{\partial^3 g}{\partial v^3} + \frac{1}{2}\eta^4 T^{-4} l_{1,2}^2 \frac{\partial^4 g}{\partial v^4},$$

where all derivatives are evaluated at $\bar{v} = \mu_{\overline{z\lambda^2}}(0, z_0)/T$.

To show an application of this lemma, consider the important special case of a call option $f(x) = (x - K)^+$.

Proposition 9.2.4. Define the log-moneyness $k = \ln(K/S(0))$ and set $\tau = \int_0^T \lambda(s)^2 ds$. Also set

$$q_1 = \mu_{\overline{z\lambda^2}}(0, z_0)/T + \alpha_0 \eta^2 + \alpha_1 \eta^2 k^2 + O\left(\eta^4\right), \tag{9.16}$$

$$q_2 = \mu_{\overline{z\lambda^2}}(0, z_0)/T + \left(\alpha_0 \eta^2 + \beta_0 \eta^4\right)$$
$$+ \left(\alpha_1 \eta^2 + \beta_1 \eta^4\right) k^2 + \beta_2 \eta^4 k^4 e^{-\Lambda \eta^2 k^2} + O\left(\eta^6\right), \tag{9.17}$$

where Λ is an arbitrary positive number and the coefficients $\alpha_0, \alpha_1, \beta_0, \beta_1, \beta_2$ are given in Appendix 9.B. Then the value of a European call option in the model (9.9)–(9.10) is given by

$$c(0, S; T, K) \approx S(0)\Phi(d_+) - K\Phi(d_-), \tag{9.18}$$

$$d_\pm = \frac{-k \pm \sigma_{\text{imp}}^2 T/2}{\sigma_{\text{imp}} \sqrt{T}},$$

where, to order η^2,

$$\sigma_{\text{imp}} = \Omega_0 \sqrt{q_1} + \Omega_1 q_1^{3/2} T + O\left(T^2\right),$$

or, to order η^4,

$$\sigma_{imp} = \Omega_0\sqrt{q_2} + \Omega_1 q_2^{3/2}T + O(T^2).$$

Also, we have

$$\Omega_0 = \frac{-k}{\int_K^{S(0)} \varphi(u)^{-1}du},$$

$$\Omega_1 = -\frac{\Omega_0}{\left(\int_K^{S(0)} \varphi(u)^{-1}du\right)^2} \ln\left(\Omega_0 \left(\frac{KS(0)}{\varphi(K)\varphi(S(0))}\right)^{1/2}\right).$$

Proof. (Sketch). For the case of a call option, the function g can be approximated using the small-time expansion result in Proposition 7.5.1; we here choose to expand around a log-normal model, so $\beta = 0$ in the proposition. Using the resulting expression to evaluate the terms in Lemma 9.2.3 yields, after some work, a direct expansion for the call option price. It is often more accurate to convert the price expansion into an expansion in implied "skew variance" v^*, where v^* satisfies

$$\mathrm{E}\left((S(T) - K)^+\right) = g(0, S(0); v^*). \qquad (9.19)$$

We write

$$v^* = \bar{v} + \eta^2 v_1^* + \eta^4 v_2^* + \ldots, \qquad (9.20)$$

insert this expression into (9.19) and Taylor-expand around $\bar{v}$. Matching the resulting expression against the direct expansion for the call option price yields closed-form expressions for v_1^* and v_2^*. These results are such that

$$\bar{v} + \eta^2 v_1^* = q_1, \quad \bar{v} + \eta^2 v_1^* + \eta^4 v_2^* = q_2,$$

where q_1 and q_2 are defined in (9.16) and (9.17), respectively. Another application of Proposition 7.5.1 turns the skew variance into an implied Black volatility,

$$\sigma_{imp}\sqrt{T} = \Omega_0\sqrt{v^*T} + \Omega_1(v^*T)^{3/2} + \ldots.$$

The proposition follows. □

Remark 9.2.5. Full details for the proof of Proposition 9.2.4 and tests of the precision of the expansion can be found in Andersen and Brotherton-Ratcliffe [2005].

9.3 Averaging Methods

The Fourier integration method from Section 9.1 involves numerical integration of a function that itself is calculated numerically by solving a coupled system of ODEs. If both the integral and the ODEs are discretized with N steps, the complexity of the scheme $O(N^2)$, which could be costly. On the other hand, the asymptotic expansion method from Section 9.2 is fast but may not be accurate enough for certain values of model parameters, especially high η. In this section we develop the parameter averaging approach to time-dependent model parameters that is both fast and accurate. We have seen applications of the method to local volatility models already, in Section 7.6.2.

9.3.1 Volatility Averaging

We initially work with the model (9.1)–(9.2) with zero correlation, $\rho = 0$. Our goal is to replace the time-dependent $\lambda(t)$ with a constant $\overline{\lambda}$ in such a way that pricing of vanilla options at a given maturity T is preserved to good approximation. For this, we first notice that a European option price can be represented as an integral of a known function against the distribution of the term stochastic variance, a representation we have already fruitfully used in Sections 8.5 and 9.2. In particular, for an at-the-money option, where $K = S(0)$,

$$\mathrm{E}\left((S(T) - S(0))^+\right) = \mathrm{E}\left(\mathrm{E}\left((S(T) - S(0))^+ \,\big|\, \{z(t),\, t \in [0,T]\}\right)\right). \tag{9.21}$$

Because the Brownian motion that drives $z(t)$ is independent of the Brownian motion that drives $S(t)$, the distribution of $S(T)$ in the model (9.2) is displaced log-normal when conditioned on a particular path of $z(t)$. Hence, the inner conditional expectation in (9.21) can be evaluated easily to yield

$$\mathrm{E}\left((S(T) - S(0))^+\right) = \mathrm{E}\left(h\left(\overline{z\lambda^2}(T)\right)\right), \tag{9.22}$$

where $\overline{z\lambda^2}(T)$ is defined by (9.3) and the function $h(x)$ is the displaced log-normal at-the-money option value as function of variance:

$$h(x) = \frac{bS(0) + (1-b)L}{b}\left(2\Phi\left(b\sqrt{x}/2\right) - 1\right). \tag{9.23}$$

Given the practical importance of correctly pricing at-the-money options, the problem of finding the effective, time-independent model volatility can be cast into the problem of finding such $\overline{\lambda}$ that

$$\mathrm{E}\left(h\left(\int_0^T \lambda(t)^2 z(t)\, dt\right)\right) = \mathrm{E}\left(h\left(\overline{\lambda}^2 \int_0^T z(t)\, dt\right)\right) \tag{9.24}$$

or, in our notations,

$$\mathrm{E}\left(h\left(\overline{z\lambda^2}(T)\right)\right) = \mathrm{E}\left(h\left(\overline{\lambda}^2 \overline{z}(T)\right)\right).$$

Neither of the expected values in (9.24) is available in closed form. However, the moment-generating functions of both $\overline{z\lambda^2}(T)$ and $\overline{z}(T)$ are available in closed form and as a solution to a system of ODEs, respectively (see Propositions 8.3.8 and 9.1.2). This observation suggests approximating $h(x)$ with a function of exponential form

$$h(x) \approx a + be^{cx}. \tag{9.25}$$

We choose the coefficients a, b, c to get the best local second-order fit at the mean of $\overline{z\lambda^2}(T)$,

$$h(\zeta_T) = a + be^{c\zeta_T}, \quad h'(\zeta_T) = bce^{c\zeta_T}, \quad h''(\zeta_T) = bc^2 e^{c\zeta_T}, \tag{9.26}$$

where

$$\zeta_T = \mathrm{E}\left(\overline{z\lambda^2}(T)\right) = \mu_{\overline{z\lambda^2}}(0, z_0) = z_0 \int_0^T \lambda(t)^2 \, dt.$$

Clearly

$$c = \frac{h''(\zeta_T)}{h'(\zeta_T)}, \tag{9.27}$$

and the problem (9.24) can be approximated with

$$a + b\mathrm{E}\left(e^{c\overline{z\lambda^2}(T)}\right) = a + b\mathrm{E}\left(e^{c\overline{\lambda}^2 \overline{z}(T)}\right) \quad \Rightarrow \quad \mathrm{E}\left(e^{c\overline{z\lambda^2}(T)}\right) = \mathrm{E}\left(e^{c\overline{\lambda}^2 \overline{z}(T)}\right), \tag{9.28}$$

which gives us an *effective volatility* approximation result that we formulate as a theorem.

Theorem 9.3.1. *Values of European options with expiry T in the model (9.1)–(9.2) are well approximated by their values in the model (8.3)–(8.4) with λ set to the effective SV volatility $\overline{\lambda}$, which solves the equation*

$$\Psi_{\overline{z}}\left(\frac{h''(\zeta_T)}{h'(\zeta_T)}\overline{\lambda}^2, 0; T\right) = \Psi_{\overline{z\lambda^2}}\left(\frac{h''(\zeta_T)}{h'(\zeta_T)}, 0; T\right), \tag{9.29}$$

where

$$\zeta_T = z_0 \int_0^T \lambda(t)^2 \, dt,$$

the function $h(x)$ is given by (9.23), and the moment-generating functions $\Psi_{\overline{z}}$ and $\Psi_{\overline{z\lambda^2}}$ are given by Propositions 8.3.8 and 9.1.2, respectively.

Proof. Follows after replacing the problem (9.24) with (9.28), using the expression (9.27) for c. □

Remark 9.3.2. The expression on the left-hand side of (9.29) can be computed in closed form; the right-hand side is straightforward to calculate from Proposition 9.1.2 and the accompanying remarks. Equation (9.29) can be solved for $\overline{\lambda}^2$ in just a couple of Newton-Raphson iterations, starting from an initial guess of $T^{-1} \int_0^T \lambda(t)^2 \, dt$.

Remark 9.3.3. The effective volatility $\overline{\lambda}$ as given by Theorem 9.3.1 is second-order accurate in the sense that the approximation (9.25) is second-order accurate with the choice of parameters in (9.26). We note that the method does not readily lend itself to higher-order approximations but this is of little relevance as the quality of the approximation is excellent as is.

9.3.2 Skew Averaging

The slope of the volatility smile in the SV model (8.3)–(8.4) is controlled by the skew parameter b. In this section we make the skew parameter time-dependent, and consider a model driven by the SDEs

$$dS(t) = \lambda(t) \left(b(t) S(t) + (1 - b(t)) L \right) \sqrt{z(t)} \, dW(t), \quad (9.30)$$

$$dz(t) = \theta \left(z_0 - z(t) \right) dt + \eta \sqrt{z(t)} \, dZ(t), \quad (9.31)$$

with $\langle dZ(t), dW(t) \rangle = 0$. In Section 7.6.2 we derived the formula for the effective, or average, skew for local volatility models, see Proposition 7.6.2 and Corollary 7.6.3. The extension of these results to stochastic volatility models is straightforward, leading to a similar expression with somewhat more complicated averaging weights, as the following proposition demonstrates.

Proposition 9.3.4. *The effective skew $\overline{b}$ for the equation*

$$dS(t) = \lambda(t) \left(b(t) S(t) + (1 - b(t)) S(0) \right) \sqrt{z(t)} \, dW(t)$$

over a time horizon $[0, T]$ is given by

$$\overline{b} = \int_0^T b(t) w_T(t) \, dt, \quad (9.32)$$

where the weights $w_T(t)$ are given by

$$w_T(t) = \frac{v(t)^2 \lambda(t)^2}{\int_0^T v(t)^2 \lambda(t)^2 \, dt}, \quad (9.33)$$

$$v(t)^2 = z_0^2 \int_0^t \lambda(s)^2 \, ds + z_0 \eta^2 e^{-\theta t} \int_0^t \lambda(s)^2 \frac{e^{\theta s} - e^{-\theta s}}{2\theta} \, ds.$$

The result in Proposition 9.3.4 can be derived by the same technique that lead to Proposition 7.6.2 and Corollary 7.6.3. Alternatively, it can be found by the small-noise expansion method in Section 7.6.3. We leave the details of these derivations to the reader and, for instructional value, instead list a third proof based on Markovian semi-groups in Appendix 9.A, see also Piterbarg [2005b]. The fact that the same result is obtained as a solution to a number of differently posed problems of skew averaging suggests robustness and general applicability.

It will be useful for the next section to derive an extension of Proposition 9.3.4 to cover the process $z(t)$ with time-dependent volatility of variance. Specifically, let us use the following dynamics for the stochastic variance process

$$dz(t) = \theta\left(z_0 - z(t)\right) dt + \eta(t) \sqrt{z(t)}\, dZ(t). \tag{9.34}$$

Corollary 9.3.5. *The effective skew $\bar{b}$ for the equation*

$$dS(t) = \lambda(t) \left(b(t)S(t) + (1 - b(t)) S(0)\right) \sqrt{z(t)}\, dW(t)$$

with $z(t)$ following (9.34) over a time horizon $[0, T]$ is given by

$$\bar{b} = \int_0^T b(t) w_T(t)\, dt, \tag{9.35}$$

where the weights $w_T(t)$ are given by

$$w_T(t) = \frac{\widehat{v}(t)^2 \lambda(t)^2}{\int_0^T \widehat{v}(t)^2 \lambda(t)^2\, dt}, \tag{9.36}$$

$$\widehat{v}(t)^2 = z_0^2 \int_0^t \lambda(s)^2\, ds + z_0 e^{-\theta t} \int_0^t \lambda(s)^2 e^{-\theta s} \int_0^s \eta(u)^2 e^{2\theta u}\, du\, ds.$$

Proof. The proof or the corollary proceeds as the proof (in Appendix 9.A) of Proposition 9.3.4, but using

$$E\left(z(t)^2\right) = z_0^2 + z_0 \int_0^t \eta(u)^2 e^{-2\theta(t-u)}\, du \tag{9.37}$$

instead of (9.100) in (9.101) for $z(t)$ given by (9.34). □

9.3.3 Volatility of Variance Averaging

Finally, we turn our attention to the problem of averaging the volatility of variance η in (9.1). More precisely, suppose we have a stochastic variance process with time-dependent volatility of variance (9.34). We would like to find a constant parameter $\bar{\eta}$ such that the model (9.30), (9.34) is approximated by the model (9.30), (9.31) with $\eta = \bar{\eta}$.

9.3 Averaging Methods

Before discussing our proposed solution method, we note that usage of time-dependent volatility of variance $\eta(t)$ for model calibration purposes may not be quite as necessary as for other parameters. Fundamentally, a time-dependent η will allow us to control the term structure of volatility smile convexity in the maturity direction. On the other hand, we already have control over the curvatures of volatility smiles at different times T via θ, the mean reversion of variance parameter: higher values of θ make implied volatility smiles flatten faster as option expiries increase, while lower values make them flatten slower, see Sections 8.2 and 8.7. Even though the level of control granted through θ is rather crude, it is often sufficient in practice, all the more so since the volatility smile curvatures are typically not observable to a high degree of precision.

The curvature of the volatility smile is related to the kurtosis of the distribution of $S(T)$ which, in stochastic volatility models, is controlled by the variance of the quantity

$$\overline{z\lambda^2}(T) = \int_0^T \lambda(t)^2 z(t)\, dt,$$

i.e. the integrated stochastic variance to expiry time T. Since the curvature of the smile is the main effect of the volatility of variance parameter η, a representative constant volatility of variance $\overline{\eta}$ should intuitively be chosen as the solution to

$$\mathrm{E}\left(\left(\int_0^T \lambda(t)^2 \widehat{z}(t)\, dt\right)^2\right) = \mathrm{E}\left(\left(\int_0^T \lambda(t)^2 z(t)\, dt\right)^2\right), \quad (9.38)$$

where $z(t)$ follows (9.34) and $\widehat{z}(t)$ follows (9.31).

Theorem 9.3.6. *For (9.34), the effective volatility of variance to maturity T, derived from the condition (9.38), is given by*

$$\overline{\eta}^2 = \frac{\int_0^T \eta(t)^2 \rho_T(t)\, dt}{\int_0^T \rho_T(t)\, dt},$$

where the weight function $\rho_T(t)$ is given by

$$\rho_T(r) = \int_r^T ds \int_s^T dt\, \lambda(t)^2 \lambda(s)^2 e^{-\theta(t-s)} e^{-2\theta(s-r)}.$$

Proof. While the proof is straightforward, we here provide full details in order to demonstrate some generally useful manipulations for the computations of moments in stochastic volatility models. First, we have

$$\mathrm{E}\left(\left(\int_0^T \lambda(t)^2 z(t)\,dt\right)^2\right)$$

$$= 2\int_0^T dt \int_0^t ds\, \lambda(t)^2 \lambda(s)^2 \mathrm{E}(z(t)z(s))$$

$$= 2\int_0^T dt \int_0^t ds\, \lambda(t)^2 \lambda(s)^2 e^{-\theta(t-s)} \mathrm{E}(z(s)^2)$$

$$+ 2\int_0^T dt \int_0^t ds\, \lambda(t)^2 \lambda(s)^2 (1 - e^{-\theta(t-s)}) z_0 \mathrm{E}(z(s)).$$

Using (9.37) for $\mathrm{E}(z(s)^2)$ we get

$$\mathrm{E}\left(\left(\int_0^T \lambda(t)^2 z(t)\,dt\right)^2\right) = 2z_0^2 \int_0^T dt \int_0^t ds\, \lambda(t)^2 \lambda(s)^2 e^{-\theta(t-s)}$$

$$+ 2z_0 \int_0^T dt \int_0^t ds\, \lambda(t)^2 \lambda(s)^2 e^{-\theta(t-s)} \int_0^s \eta(r)^2 e^{-2\theta(s-r)} dr$$

$$+ 2\int_0^T dt \int_0^t ds\, \lambda(t)^2 \lambda(s)^2 \left(1 - e^{-\theta(t-s)}\right) z_0 \mathrm{E}(z(s)).$$

Changing the order of integration for the second term, we obtain

$$\mathrm{E}\left(\left(\int_0^T \lambda(t)^2 z(t)\,dt\right)^2\right) = 2z_0^2 \int_0^T dt \int_0^t ds\, \lambda(t)^2 \lambda(s)^2 e^{-\theta(t-s)}$$

$$+ 2z_0 \int_0^T dr\, \eta(r)^2 \int_r^T ds \int_s^T dt\, \lambda(t)^2 \lambda(s)^2 e^{-\theta(t-s)} e^{-2\theta(s-r)}$$

$$+ 2\int_0^T dt \int_0^t ds\, \lambda(t)^2 \lambda(s)^2 \left(1 - e^{-\theta(t-s)}\right) z_0 \mathrm{E}(z(s)).$$

If we define

$$\rho_T(r) = \int_r^T ds \int_s^T dt\, \lambda(t)^2 \lambda(s)^2 e^{-\theta(t-s)} e^{-2\theta(s-r)},$$

the equation (9.38) can be rewritten in the form

$$\int_0^T \bar{\eta}^2 \rho_T(t)\,dt = \int_0^T \eta(t)^2 \rho_T(t)\,dt.$$

The theorem is proved. □

Remark 9.3.7. While we used zero correlation between the underlying and its stochastic variance both in motivating our results and in deriving them, the same approach can be applied in the non-zero correlation case. Some results, in particular Proposition 9.3.4 and Theorem 9.3.6, remain unchanged. On the other hand, the effective volatility formula in Theorem 9.3.1 is based on the representation (9.22) which, clearly, does not hold with non-zero correlation; despite that, the formula can still be used with good accuracy.

9.3.4 Calibration by Parameter Averaging

The main application of the averaging formulas developed above is in creating efficient model calibration algorithms. In this section, we discuss in some detail how such an algorithm could proceed; the principles that we outline here shall be used repeatedly later in this book. Now, suppose a collection of expiries

$$0 = T_0 < T_1 < T_2 < \ldots < T_N$$

is given, as well as a collection of strikes $K_1, \ldots, K_M$. Let the market values of European call options with expiries T_n and strikes K_m be denoted by

$$\{\widehat{c}_{n,m}, \quad n = 1, \ldots, N, \quad m = 1, \ldots, M\}.$$

Our objective is to find time-dependent model parameters $\lambda(t)$, $b(t)$, and $\eta(t)$ such that the model

$$dS(t) = \lambda(t)\left(b(t)S(t) + (1-b(t))L\right)\sqrt{z(t)}\,dW(t), \qquad (9.39)$$

$$dz(t) = \theta\left(z_0 - z(t)\right)dt + \eta(t)\sqrt{z(t)}\,dZ(t), \qquad (9.40)$$

values European options with expiries T_n, $n = 1, \ldots, N$, and strikes K_m, $m = 1, \ldots, M$, as closely as possible to their market values[3] $\{\widehat{c}_{n,m}\}$.

Let us denote the prices of options in the model (9.39)–(9.40) by

$$c_{n,m} = c_{n,m}\left(\mathcal{X}\right),$$

where by $\mathcal{X}$ we denote the state of the model,

$$\mathcal{X} = \{\lambda(\cdot), b(\cdot), \eta(\cdot)\}.$$

Typically, calibration would be performed by solving the following non-linear optimization problem

$$\{\lambda(\cdot), b(\cdot), \eta(\cdot)\} = \operatorname*{argmin} \sum_{n,m} \left(c_{n,m}\left(\mathcal{X}\right) - \widehat{c}_{n,m}\right)^2, \qquad (9.41)$$

---

[3] In interest rate markets, the underlyings for options of different expiries are often different, in the sense that they represent swap rates of different tenors and fixing dates. We will deal with such complications in due time.

where[4] $c_{n,m}(\mathcal{X})$'s are obtained in some sort of numerical procedure. With the averaging formulas, an appealing alternative is available. To describe it, let us denote triples of SV "market" parameter values by $\{\widehat{\lambda}_n, \widehat{b}_n, \widehat{\eta}_n\}$, $n = 1, \ldots, N$, determined such that the market prices of European options expiring at time T_n, i.e. $\{\widehat{c}_{n,m}, m = 1, \ldots, M\}$, match prices obtained in the model

$$dS(t) = \widehat{\lambda}_n \left(\widehat{b}_n S(t) + \left(1 - \widehat{b}_n\right) L\right) \sqrt{z(t)}\, dW(t), \tag{9.42}$$

$$dz(t) = \theta\left(z_0 - z(t)\right) dt + \widehat{\eta}_n \sqrt{z(t)}\, dZ(t). \tag{9.43}$$

Sets of market parameters are routinely maintained and updated by trading desks, and instead of considering $\{\widehat{c}_{n,m}\}$ to be fundamental market inputs, we can think of $\{\widehat{\lambda}_n, \widehat{b}_n, \widehat{\eta}_n\}$, $n = 1, \ldots, N$, as such. We often refer to them as "term" parameters to highlight the fact that they are constant for the whole "term", or life, of the relevant options.

Critically, the averaging formulas link time-dependent parameters $\{\lambda(t), b(t), \eta(t)\}$ to constant parameters $\{\widehat{\lambda}_n, \widehat{b}_n, \widehat{\eta}_n\}$, $n = 1, \ldots, N$, directly without referencing option values. To take advantage of this, let us denote by

$$\{\overline{\lambda}_n(\mathcal{X}), \overline{b}_n(\mathcal{X}), \overline{\eta}_n(\mathcal{X})\}$$

the averaged parameters (to time T_n) for the model (9.39)–(9.40). Then the optimization problem (9.41) can be replaced by a more convenient one,

$$\{\lambda(\cdot), b(\cdot), \eta(\cdot)\} = \operatorname{argmin}\left(W_\lambda \sum_n \left(\overline{\lambda}_n(\mathcal{X}) - \widehat{\lambda}_n\right)^2\right.$$

$$\left. + W_b \sum_n \left(\overline{b}_n(\mathcal{X}) - \widehat{b}_n\right)^2 + W_\eta \sum_n \left(\overline{\eta}_n(\mathcal{X}) - \widehat{\eta}_n\right)^2\right), \tag{9.44}$$

where W_λ, W_b, and W_η are weights linked to relative importance of matching particular parameters. Compared to (9.41), this norm formulation is both more intuitive to traders — who often tend to think about the state of the market in terms of model parameters, rather than in terms of absolute option prices — and computationally advantageous, insofar as the norm requires no outright computation of option values.

In practice, the calibration (9.44) needs not be performed by brute-force optimization. By carefully choosing the order of calculations, calibration can be split into independent sub-calibrations: one for volatility of variance (η); one for skewness (b); and one for volatility (λ). Skew and volatility of variance calibrations can be performed by matrix manipulations, and the volatility calibration can be split into a sequence of numerically solved one-dimensional equations. To describe this calibration idea in more detail,

---

[4] Often different terms are weighted differently.

9.3 Averaging Methods 379

let us first collect all relevant averaging results for easy reference. For the volatility of variance, we have from Theorem 9.3.6,

$$\bar{\eta}_n(\mathcal{X})^2 = \frac{\int_0^{T_n} \eta(t)^2 \rho_{T_n}(t;\lambda(\cdot))\,dt}{\int_0^{T_n} \rho_{T_n}(t;\lambda(\cdot))\,dt}, \quad n=1,\ldots,N, \qquad (9.45)$$

where we have now explicitly indicated the dependence of weights $\rho_T(t;\lambda(\cdot))$ on the volatility function $\lambda(t)$. For the skews, we have from Corollary 9.3.5,

$$\bar{b}_n(\mathcal{X}) = \int_0^{T_n} b(t) w_{T_n}(t;\lambda(\cdot),\eta(\cdot))\,dt, \quad n=1,\ldots,N, \qquad (9.46)$$

where again the dependence of weights $w_T(t;\lambda(\cdot),\eta(\cdot))$ on model parameters is highlighted. Finally, the equations for volatilities from Theorem 9.3.1 are

$$\bar{\lambda}_n(\mathcal{X}) = F\left(\lambda(\cdot); \bar{b}_n(\mathcal{X}), \bar{\eta}_n(\mathcal{X})\right), \quad n=1,\ldots,N, \qquad (9.47)$$

where, in the notation of Theorem 9.3.1,

$$F(\lambda(\cdot); \bar{b}_n(\mathcal{X}), \bar{\eta}_n(\mathcal{X})) = \sqrt{\frac{h'(\zeta_{T_n})}{h''(\zeta_{T_n})} \times \Psi_{\bar{z}}^{-1}\left(\Psi_{\overline{z\lambda^2}}\left(\frac{h''(\zeta_{T_n})}{h'(\zeta_{T_n})}, 0; T\right), 0; T\right)},$$

$$\zeta_{T_n} = z_0 \int_0^{T_n} \lambda(t)^2\,dt.$$

Note that the function F depends on $\bar{b}_n$ through h, and on $\bar{\eta}_n$ through $\Psi_{\overline{z\lambda^2}}$ and $\Psi_{\bar{z}}$.

Equations (9.45)–(9.47) can be discretized if the model parameters are constant between option expiry dates $\{T_n\}_{n=1}^N$, a common assumption in practice. In this case, we can define λ_i, b_i and η_i by

$$\lambda(t) = \sum_{i=1}^N \lambda_i 1_{\{t \in (T_{i-1}, T_i]\}},$$

$$b(t) = \sum_{i=1}^N b_i 1_{\{t \in (T_{i-1}, T_i]\}},$$

$$\eta(t) = \sum_{i=1}^N \eta_i 1_{\{t \in (T_{i-1}, T_i]\}}.$$

In addition, we discretize the weights and define $\rho_{n,i}(\lambda(\cdot))$ and $w_{n,i}(\lambda(\cdot),\eta(\cdot))$ by

$$\rho_{T_n}(t;\lambda(\cdot)) = \sum_{i=1}^n \rho_{n,i}(\lambda(\cdot)) 1_{\{t \in (T_{i-1}, T_i]\}},$$

$$w_{T_n}(t;\lambda(\cdot),\eta(\cdot)) = \sum_{i=1}^n w_{n,i}(\lambda(\cdot),\eta(\cdot)) 1_{\{t \in (T_{i-1}, T_i]\}}.$$

Denote
$$\bar{p}_{n,i}\left(\lambda(\cdot)\right) = \frac{p_{n,i}\left(\lambda(\cdot)\right)}{\int_0^{T_n} p_{T_n}\left(t;\lambda(\cdot)\right)dt}.$$

Our goal is to solve three systems of equations:

$$\sum_{i=1}^n \bar{p}_{n,i}\left(\lambda(\cdot)\right)(T_i - T_{i-1})\eta_i^2 = (\widehat{\eta}_n)^2, \tag{9.48}$$

$$\sum_{i=1}^n w_{n,i}\left(\lambda(\cdot),\eta(\cdot)\right)(T_i - T_{i-1})b_i = \widehat{b}_n, \tag{9.49}$$

$$F\left(\lambda(\cdot); \bar{b}_n\left(\mathcal{X}\right), \bar{\eta}_n\left(\mathcal{X}\right)\right) = \widehat{\lambda}_n, \tag{9.50}$$

for $n = 1, \ldots, N$. At first glance this does not seem entirely straightforward. For example, the system (9.48) appears to be a linear system of equations in $\eta_1^2, \ldots, \eta_N^2$, but the coefficients $\bar{p}_{n,i}(\lambda(\cdot))$ depend on $\lambda(t)$, another unknown model parameter. However, by iteratively solving these equations in the right order, we can design a very efficient algorithm, which we now proceed to describe in detail.

First, we note that the equations on volatilities (9.50) do not depend on any other model parameters. They do depend on term parameters $\bar{b}_n(\mathcal{X})$, $\bar{\eta}_n(\mathcal{X})$, which we just replace with their market values, thus solving

$$F\left(\lambda(\cdot); \widehat{b}_n, \widehat{\eta}_n\right) = \widehat{\lambda}_n, \quad n = 1, \ldots, N.$$

The n-th equation in this series only involves λ_i's for $i = 1, \ldots, n$, so the n-th equation can be rewritten as

$$F\left(\lambda_1, \ldots, \lambda_n; \widehat{b}_n, \widehat{\eta}_n\right) = \widehat{\lambda}_n.$$

The case $n = 1$ has the trivial solution

$$\lambda_1^* = \widehat{\lambda}_1.$$

Proceeding iteratively in n, the n-th equation is reduced to

$$F\left(\lambda_1^*, \ldots, \lambda_{n-1}^*, \lambda_n; \widehat{b}_n, \widehat{\eta}_n\right) = \widehat{\lambda}_n, \tag{9.51}$$

where the λ_i^*, $i = 1, \ldots, n-1$, are the model parameters already solved for. Thus, the first step of calibration consists of solving the system of equations (9.50) as N decoupled one-dimensional equations (9.51).

On the second step, we solve the linear system (9.48) for η_i^2, $i = 1, \ldots, N$. The coefficients of the system depend on λ_i's which have already been computed, and we solve

$$\sum_{i=1}^n \bar{p}_{n,i}\left(\lambda^*(\cdot)\right)(T_i - T_{i-1})\eta_i^2 = (\widehat{\eta}_n)^2, \quad n = 1, \ldots, N.$$

The solution η_i^*, $i = 1, \ldots, N$, to this system can either be found by matrix methods, or by simple sequential substitution since the n-th equation involves η_i^2 for $i = 1, \ldots, n$ only.

Finally, on the third step, we solve the linear system

$$\sum_{i=1}^{n} w_{n,i}\left(\lambda^*(\cdot), \eta^*(\cdot)\right)(T_i - T_{i-1})b_i = \widehat{b}_n, \quad n = 1, \ldots, N, \quad (9.52)$$

for b_i, $i = 1, \ldots, N$. This system is obtained from (9.49) by substituting $\lambda(\cdot)$, $\eta(\cdot)$ with their solved-for values $\lambda^*(\cdot)$, $\eta^*(\cdot)$. Again, the system can be solved sequentially.

To prevent overfitting, it is often useful to regularize the optimization problem through introduction of smoothing terms in the objective function. This can help to, for example, dampen the noise that could be present in market-observed parameters. Taking (9.52) as an example and fixing a smoothing weight $W > 0$, we can replace (9.52) with the minimization problem

$$\sum_{n=1}^{N} \left(\sum_{i=1}^{n} w_{n,i}\left(\lambda^*(\cdot), \eta^*(\cdot)\right)(T_i - T_{i-1})b_i - \widehat{b}_n \right)^2$$

$$+ W \sum_{i=2}^{N} (b_i - b_{i-1})^2 \to \min. \quad (9.53)$$

This is a simple quadratic minimization problem with no constraints and is easily solved by linear algebra methods, see Golub and van Loan [1989]. The same regularization idea could be applied to the problem of finding $\lambda(t)$ and $b(t)$.

If the regularization weight W in (9.53) is too high then the averaged skew calculated by the model can be significantly different from the market skew, $\bar{b}_n(\mathcal{X}^*) \neq \widehat{b}_n$, $n = 1, \ldots, N$. By itself this may not be such a bad thing as one may prefer a smoother model skew over the exact fit to market skews. However, this poses problems to the *volatility* calibration, as the equation for model volatility (9.51) used the "wrong" skew (and volatility of variance as well, were we to apply regularization to that). The exact fit to market volatilities is often much more important than the exact fit to skews or volatilities of variance. Fortunately, this problem is easy to rectify by solving the system (9.51) again, this time using the true model averaged skews $\bar{b}_n(\mathcal{X}^*)$ (and volatilities of variance) on the left-hand side of (9.51) which are available at this stage of the algorithm.

9.4 PDE Method

In the previous three sections, we discussed the development of methods for efficient model calibration and for the pricing of simple European options. In the remainder of this chapter, we turn our attention to numerical techniques that allow a calibrated model to be used for pricing of general fixed income derivatives. We start out with the application of the PDE methods from Chapter 2.

9.4.1 PDE Formulation

The flexibility of the PDE method makes it applicable to a generalization of the specification (8.1)–(8.2) with a fully general time-dependent volatility function $\varphi(t, S)$. Let us therefore consider the following vector SDE

$$dS(t) = \varphi(t, S(t)) \sqrt{z(t)} \, dW(t), \tag{9.54}$$
$$dz(t) = \theta(z_0 - z(t)) \, dt + \eta(t)\psi(z(t)) \, dZ(t), \tag{9.55}$$

where $\langle dZ(t), dW(t)\rangle = \rho \, dt$ and $z(0) = z_0$. Let $V(T)$ be an $\mathcal{F}_T$-measurable payoff and let $V(t, z, S)$ denote the numeraire-deflated value at time t, given $S(t) = S$ and $z(t) = z$, of a derivative that pays $V(T)$ at time T, $t \leq T$. By the usual arguments, $V(t, z, S)$ satisfies the following partial differential equation

$$\begin{aligned}0 =& \frac{\partial}{\partial t}V(t, z, S) + \theta(z_0 - z)\frac{\partial}{\partial z}V(t, z, S) + \frac{\eta(t)^2}{2}\psi(z)^2 \frac{\partial^2}{\partial z^2}V(t, z, S) \\ &+ \frac{z}{2}\varphi(t, S)^2 \frac{\partial^2}{\partial S^2}V(t, z, S) + \rho\eta(t)\psi(z)\sqrt{z}\varphi(t, S)\frac{\partial^2}{\partial z \partial S}V(t, z, S).\end{aligned} \tag{9.56}$$

This PDE holds for $t \in [0, T]$ and $(S, z) \in \mathbb{R} \times \mathbb{R}^+$.

Fundamentally, (9.56) can be solved numerically by an application of the two-dimensional ADI scheme with a predictor-corrector step, as developed in Section 2.11.2. In an actual implementation of the ADI method, however, several issues in grid design and choice of boundary conditions must be addressed, a task to which we now turn.

9.4.2 Range for Stochastic Variance

Fixing a small probability $q_z > 0$, the range $[z_{\min}, z_{\max}]$ for z in the ADI grid can be set to cover the fraction $(1 - q_z)$ of the range of $z(T)$ in probability, i.e. from the conditions

$$P(z(T) < z_{\min}) = P(z(T) > z_{\max}) = q_z/2.$$

These probabilities are not known in closed form for $z(T)$ satisfying (9.55), so we will often have to resort to approximations. For instance, if ψ is not too different from a square root, we can replace

$$\psi(z) \to \frac{\psi(z_0)}{\sqrt{z_0}}\sqrt{z}, \tag{9.57}$$

to obtain a process of the square root type with time-dependent $\eta(t)$. From this representation, we can find an effective $\bar\eta$ to time horizon T by Theorem 9.3.6 and then apply the exact distribution of $z(T)$ with time-constant parameters from Proposition 8.3.2. Of course an even simpler, Gaussian, approximation is available if ψ is not too different from a constant.

A bit more crudely, but with less effort, we can also attempt to find the range for z from the stationary distribution of $z(t)$. When available, stationary distributions are a good source of approximations for tail probabilities — which is what we are interested in here — as we can often substitute large-z behavior with long-time behavior. The moments $\mathrm{E}(z(T))$, $\mathrm{Var}(z(T))$ of $z(T)$ that follows (9.55) are given by

$$\mathrm{E}\left(z(T)\right) = z_0, \quad \mathrm{Var}\left(z(T)\right) \approx \psi\left(z_0\right)^2 \int_0^T \eta(t)^2 e^{-2\theta(T-t)}\,dt,$$

where we have applied the approximation (9.57). Assuming that (9.57) is reasonable, the stationary distribution of $z(t)$ can be approximated with the Gamma distribution of Proposition 8.3.4; we choose the parameters of the Gamma distribution to match the mean and variance of $z(T)$,

$$\beta = \frac{\mathrm{E}\left(z(T)\right)}{\mathrm{Var}\left(z(T)\right)}, \quad \alpha = \beta \mathrm{E}\left(z(T)\right).$$

The range of z in the ADI scheme can then be established by

$$z_{\min} = F^{-1}\left(q_z/2; \alpha, \beta\right), \quad z_{\max} = F^{-1}\left(1 - q_z/2; \alpha, \beta\right),$$

where $F(q; \alpha, \beta)$ is the Gamma CDF. Finally, we note that we can just use

$$z_{\min} = 0,$$

as long as we use one-sided discretization for boundary conditions at that point, as explained in Section 9.4.4 below.

9.4.3 Discretizing Stochastic Variance

Uniform discretization of z in the PDE (9.56) is rarely the best choice. If we look at the important case of $\psi(z) = \sqrt{z}$, assuming $z(0) = z_0 = 1$, the interval $[z_{\min}, z_{\max}]$ would be something like $[0, 10]$, with the mean of $z(t)$ being 1. Uniformly discretizing the range $[0, 10]$ would tend to put too few points in the interval $[0, 1]$, resulting in poor resolution in an important part of the range (see also Figure 9.2 in Section 9.5.3.1). To provide a remedy, we may recall the discussion in Section 7.4, which considered the transform

$$u(t) = \Psi\left(z(t)\right), \quad \Psi(z) = \int_{z_0}^z \frac{dy}{\psi(y)}. \tag{9.58}$$

Applying Ito's lemma, we get

$$du(t) = \theta \frac{z_0 - \Psi^{-1}(u(t))}{\psi(\Psi^{-1}(u(t)))} dt$$
$$- \frac{\eta(t)^2}{2} \psi'(\Psi^{-1}(u(t))) \, dt + \eta(t) \, dZ(t). \quad (9.59)$$

Noticing that the diffusion coefficient of $u(t)$ is not state-dependent, it appears reasonable to construct the grid in z-space from a uniform discretization in u. For this, suppose $N_z + 1$ points are used for the z-domain. We then define the grid $\{\zeta_n\}_{n=0}^{N_z}$ for z by the condition that $u_n \triangleq \Psi(\zeta_n)$ are spaced uniformly over $[\Psi(z_{\min}), \Psi(z_{\max})]$, so that

$$u_n = \Psi(z_{\min}) + \frac{n}{N_z}(\Psi(z_{\max}) - \Psi(z_{\min})),$$
$$\zeta_n = \Psi^{-1}(u_n)$$
$$= \Psi^{-1}\left(\Psi(z_{\min}) + \frac{n}{N_z}(\Psi(z_{\max}) - \Psi(z_{\min}))\right), \quad n = 0, \ldots, N_z.$$

To give an example, consider the square root case $\psi(z) = \sqrt{z}$ where we have

$$\Psi(z) = \int_{z_0}^{z} \frac{dy}{\sqrt{y}} = 2\left(\sqrt{z} - \sqrt{z_0}\right), \quad \Psi^{-1}(u) = \left(\frac{u}{2} + \sqrt{z_0}\right)^2,$$

such that

$$\zeta_n = \left(\sqrt{z_{\min}} + \frac{n}{N_z}(\sqrt{z_{\max}} - \sqrt{z_{\min}})\right)^2, \quad n = 0, \ldots, N_z. \quad (9.60)$$

Empirically, it appears that concentrating points around the mean $z = z_0$ further improves numerical properties. We can achieve this effect by applying the sinh transform, see p. 167 of Tavella and Randall [2000], and then using (9.60):

$$\zeta_n = \left(z_0 + \sinh\left(\alpha_{\min} + \frac{n}{N_z}(\alpha_{\max} - \alpha_{\min})\right)\right)^2, \quad (9.61)$$
$$\alpha_{\min,\max} = \sinh^{-1}\left(\sqrt{z_{\min,\max}} - z_0\right).$$

To illustrate the discretization strategies above, Figure 9.1 shows the density of grid points over $[z_{\min}, z_{\max}]$ using uniform discretization, quadratic discretization (9.60), and the sinh-quadratic discretization (9.61). As discussed, the quadratic and sinh-quadratic discretizations both increase the density of points in $(0, z_0]$, relative to a uniform discretization. In addition, the sinh-quadratic scheme places more points around z_0 than does the quadratic scheme.

Fig. 9.1. Grid Density

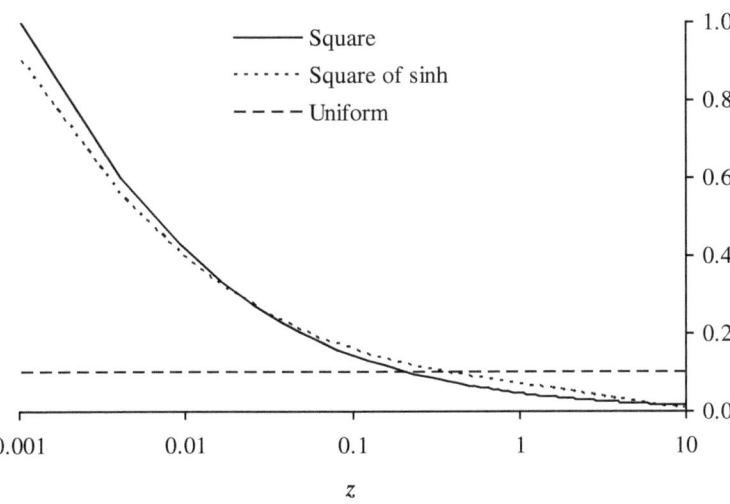

Notes: Density of grid points (number of grid points per unit length) as a function of z for three different discretization schemes for z-domain: uniform, quadratic (9.60), and sinh-quadratic (9.61). We assume $z_{\min} = 0$, $z_0 = 1$, $z_{\max} = 10$. The abscissa axis is in logarithmic scale.

Let us finally note that instead of drawing on (9.58) as an inspiration for grid discretization in z, we could in principle use the variable u directly in the ADI scheme. Indeed, all that would be required is to rewrite (9.56) in terms of u, S and apply a uniform discretization to u. However, the drift of $u(t)$ is rather complicated and, importantly, grows to infinity as $u \to 0$ in the special case of $\psi(z) = \sqrt{z}$, see (9.59). A scheme that can handle large values of the drift robustly, such as the upwinding scheme from Section 2.6.1, would therefore be a necessity.

9.4.4 Boundary Conditions for Stochastic Variance

Practical experience shows that numerical schemes for solving the PDE (9.56) are quite robust with respect to the specifications of boundary conditions for z. Any reasonable choice from Chapter 2 appears to work well, including the standard $\partial^2 V/\partial z^2 = 0$ for $z = z_{\min}$, $z = z_{\max}$. In the case of $\psi(z) = \sqrt{z}$, if $z_{\min} = 0$, i.e. if we use $z = 0$ as the lower bound on the grid, for best results we should derive the boundary conditions for $z_{\min}$ from the PDE itself, see Section 2.2.2. Setting $z = 0$ in (9.56) we obtain

$$0 = \frac{\partial}{\partial t} V(t, 0, S) + \theta z_0 \frac{\partial}{\partial z} V(t, 0, S), \tag{9.62}$$

386   9 Vanilla Models with Stochastic Volatility II

a boundary condition of Neumann type. The validity of this boundary condition is intuitively justified by the fact that the solution to the SDE for $z(t)$ is unique, i.e. the behavior of $z(t)$ at the boundary $z = 0$ is determined by the SDE itself — and hence the boundary condition is determined by setting $z = 0$ in the PDE[5]. Incorporation of (9.62) into the finite difference solver would generally require one to discretize the z-derivative by one-sided differences; see Section 10.1.5.2 for details in a slightly more general setting.

Another, also reasonable, specification for the boundary $z = 0$ is obtained from the fact that the square-root process for $z(t)$ is strongly reflecting at $z = 0$, see Proposition 8.3.1. A reflection at the boundary translates into the boundary condition

$$\frac{\partial}{\partial z} V(t, 0, S) = 0$$

(see Karatzas and Shreve [1991]), which is quite similar to (9.62) and is another reasonable choice.

Interestingly, using the correct boundary conditions for the *forward* PDE, i.e. the forward Kolmogorov equation that the density of the process satisfies, is crucial, especially when the Feller condition (Proposition 8.3.1) is violated. As we have no use for forward PDEs for stochastic volatility processes in this book, we refer the reader to Lucic [2008] for the details.

9.4.5 Range for Underlying

To obtain the range

$$[S_{\min}, S_{\max}]$$

for the underlying S, we need to compute the approximate distribution of $S(T)$. Replacing the stochastic variance process with its expected value $E(z(t)) = z_0$, we obtain

$$dS(t) \approx \varphi(t, S(t)) \sqrt{z_0}\, dW(t).$$

To proceed, we can for example use the connection between option prices and the probability density, and apply various asymptotic results for local volatility models from Section 7.5. In the important special case of a time-dependent linear local volatility

$$\varphi(t, S) = \lambda(t)\left(b(t) S + (1 - b(t)) L\right), \qquad (9.63)$$

a reasonable approach is to replace time-dependent $b(t)$ with the effective time-independent skew $\bar{b}$ via Proposition 9.3.4, and then apply a Gaussian approximation:

---

[5] A formal proof that (9.62) is theoretically correct, at least for payoffs that depend on z only (and not on S), is given in Ekström and Tysk [2008].

$$S(T) \approx \left[\left(\bar{b}S(0) + \left(1 - \bar{b}\right) L\right) e^\xi - \left(1 - \bar{b}\right) L\right] / \bar{b}, \tag{9.64}$$

$$\xi \sim \mathcal{N}\left(-\frac{z_0 \bar{b}^2}{2} \int_0^T \lambda(t)^2 dt, \, z_0 \bar{b}^2 \int_0^T \lambda(t)^2 dt\right).$$

As ξ is Gaussian, it is easy to find $[\xi_{\min}, \xi_{\max}]$ so that

$$P\left(\xi < \xi_{\min}\right) = P\left(\xi > \xi_{\max}\right) = q_S/2$$

for a given small probability $q_S > 0$. This trivially translates into the range for $S(T)$.

9.4.6 Discretizing the Underlying

The representation (9.64) proves useful for discretizing S as well. One approach is to discretize S so that the grid is uniform in ξ,

$$S_n = \left[\left(\bar{b}S(0) + \left(1 - \bar{b}\right) L\right) e^{\xi_n} - \left(1 - \bar{b}\right) L\right] / \bar{b},$$
$$\xi_n = \xi_{\min} + \frac{n}{N_S}\left(\xi_{\max} - \xi_{\min}\right),$$
$$n = 0, \ldots, N_S,$$

where N_S is the grid size. Alternatively, we can apply a transformation

$$y(S) = \ln\left(\frac{\bar{b}S + \left(1 - \bar{b}\right) L}{\bar{b}S(0) + \left(1 - \bar{b}\right) L}\right),$$

rewrite the PDE (9.56) in y instead of S, and discretize in y uniformly.

To conclude we note that even if $\varphi(t, S)$ is not of the form (9.63), we can always approximate it as such in order to compute the effective $\bar{b}$ that is then used in discretization for S or in the mapping $S \to y$. Alternatively, we can always employ the same strategy (integral variable transform) that was advocated in Section 9.4.3 for z — which is what we used in Section 7.4 for discretizing local volatility models as well.

9.5 Monte Carlo Method

For generic stochastic volatility models such as (9.54)–(9.55), little can be said about Monte Carlo simulation that has not already been covered in Chapter 3. For any particular model parameterization, however, special-purpose discretization schemes can be constructed that have significant computational advantages over, say, the general-purpose Ito-Taylor schemes in Section 3.2.6. To demonstrate, we shall here specialize to the standard SV model, i.e. we consider the system

9 Vanilla Models with Stochastic Volatility II

$$dS(t) = \lambda \left(bS(t) + (1-b)L\right)\sqrt{z(t)}\,dW(t), \tag{9.65}$$
$$dz(t) = \theta\left(z_0 - z(t)\right)dt + \eta\sqrt{z(t)}\,dZ(t), \tag{9.66}$$

with $\langle dZ(t), dW(t)\rangle = \rho\,dt$ and $z(0) = z_0$. Our primary objective is to establish a scheme that allows us to time-discretize the SV model dynamics in an efficient manner; as it turns out, this is surprisingly challenging, particularly for the z-process. We shall consequently deal with the Monte Carlo simulation of the SV model in a fairly careful manner, listing a number of schemes with different efficiency/bias trade-offs.

Remark 9.5.1. While we have assumed that parameters in the SV process are constants, all that is ultimately required is that parameters are piecewise constant on the simulation time line. As such, the schemes we suggest will also apply to time-dependent dynamics.

9.5.1 Exact Simulation of Variance Process

According to Proposition 8.3.2, the distribution of $z(t+\Delta)$ given $z(t)$ is known in closed form, and generation of a random sample of $z(t+\Delta)$ given $z(t)$ can be done entirely bias-free by sampling from a non-central chi-square distribution. Using the fact that a non-central chi-square distribution can be seen as a regular chi-square distribution with Poisson-distributed degrees of freedom (see Section 3.1.1.3), the following algorithm can be used.

1. Draw a Poisson random variable N, with mean $\frac{1}{2}z(t)n(t, t+\Delta)$ (here $n(t,T)$ is defined in (8.6)).
2. Given N, draw a regular chi-square random variable χ_v^2, with $v = d+2N$ degrees of freedom (d is defined in (8.6)).
3. Set $z(t+\Delta) = \chi_v^2 \cdot \exp(-\theta\Delta)/n(t, t+\Delta)$.

Steps 1 and 3 of this algorithm are straightforward, and Step 2 can be accomplished using the acceptance-rejection technique discussed in Section 3.1.1.2.

As mentioned in Section 3.1.1.3, if $d > 1$ it may be numerically advantageous to use a different algorithm, based on the relation

$$\chi_d'^2(\gamma) \stackrel{d}{=} (Z + \sqrt{\gamma})^2 + \chi_{d-1}^2, \quad d > 1, \tag{9.67}$$

where $\stackrel{d}{=}$ denotes equality in distribution, $\chi_d'^2(\gamma)$ is a non-central chi-square variable with d degrees of freedom and non-centrality parameter γ, and Z is an ordinary $\mathcal{N}(0,1)$ Gaussian variable. We trust that the reader can complete the details on application of (9.67) in a simulation algorithm for $z(t+\Delta)$.

One might think that the existence of an exact simulation scheme for $z(t+\Delta)$ would settle once and for all the question of how to generate paths of the square-root process. In practice, however, several complications may arise with the application of the algorithm above. Indeed, the scheme is quite complex compared with many standard SDE discretization schemes and may not fit smoothly into existing software architecture for SDE simulation routines. Also, computational speed may be an issue, and the application of acceptance-rejection sampling will potentially cause a "scrambling effect" when process parameters are perturbed[6], resulting in poor convergence of numerically computed sensitivities, see Section 3.3. While caching techniques can be designed to overcome some of these issues, storage, look-up, and interpolation of such a cache pose their own challenges. Further, the basic scheme above provides no explicit link between the paths of the Brownian motion $Z(t)$ and that of $z(t)$, complicating applications in which, say, multiple correlated Brownian motions need to be advanced through time.

In light of the discussion above, it seems reasonable to also investigate the application of simpler simulation algorithms. These will typically exhibit a bias — in the sense discussed in Section 3.2.8 — for finite values of Δ, but convenience and speed may more than compensate for this, especially if the bias is small and easy to control by reduction of step size. We proceed to discuss several classes of such schemes.

9.5.2 Biased Taylor-Type Schemes for Variance Process

9.5.2.1 Euler Schemes

Going forward, let us use $\widehat{z}$ to denote a discrete-time (biased) approximation to z. A classical approach to simulating a path $\widehat{z}$ involves the application of Ito-Taylor expansions, suitably truncated, see Sections 3.2.3 and 3.2.6 for details. The simplest such scheme is the Euler scheme, a direct application of which would here give

$$\widehat{z}(t+\Delta) = \widehat{z}(t) + \theta(z_0 - \widehat{z}(t))\Delta + \eta\sqrt{\widehat{z}(t)}Z\sqrt{\Delta}, \qquad (9.68)$$

where Z is a $\mathcal{N}(0,1)$ Gaussian variable. One immediate (and fatal) problem with (9.68) is that the discrete process $\widehat{z}$ can become negative with non-zero probability. The first time this happens on a path, computation of $\sqrt{\widehat{z}(t)}$ will be impossible and the time-stepping scheme will fail. To get around this problem, several remedies have been proposed in the literature, starting with the suggestion in Kloeden and Platen [2000] that one simply replace $\sqrt{\widehat{z}(t)}$ in (9.68) with $\sqrt{|\widehat{z}(t)|}$. Lord et al. [2006] review a number of similar "fixes" and conclude that the following works best:

$$\widehat{z}(t+\Delta) = \widehat{z}(t) + \theta(z_0 - \widehat{z}(t)^+)\Delta + \eta\sqrt{\widehat{z}(t)^+}Z\sqrt{\Delta}. \qquad (9.69)$$

---

[6] After a perturbation of parameters, the number of rejected samples in the Monte Carlo trial will likely change.

In Lord et al. [2006] this scheme is denoted "full truncation"; its main characteristic is that the process for $\widehat{z}$ is allowed to go below zero, at which point $\widehat{z}$ becomes deterministic with an upward drift of θz_0.

9.5.2.2 Higher-Order Schemes

The scheme (9.69) has first-order weak convergence, i.e. expectations of functions of $\widehat{z}$ will approach their true values as $O(\Delta)$. To improve convergence, it is tempting to apply a Milstein scheme (see Section 3.2.6.3), the most basic of which is

$$\widehat{z}(t+\Delta) = \widehat{z}(t) + \theta(z_0 - \widehat{z}(t))\Delta + \eta\sqrt{\widehat{z}(t)}\, Z\sqrt{\Delta} + \frac{1}{4}\eta^2 \Delta \left(Z^2 - 1\right).$$

As was the case for (9.68), this scheme has a positive probability of generating negative values of $\widehat{z}$ and therefore cannot be used without suitable modifications. Kahl and Jäckel [2006] list several other Milstein-type schemes, some of which allow for a certain degree of control over the likelihood of generating negative values. One interesting variation is the *implicit Milstein scheme*, defined as

$$\widehat{z}(t+\Delta) = \frac{\widehat{z}(t) + \theta z_0 \Delta + \eta\sqrt{\widehat{z}(t)}Z\sqrt{\Delta} + \frac{1}{4}\eta^2\Delta\left(Z^2 - 1\right)}{1 + \theta\Delta}. \qquad (9.70)$$

It is easy to verify that this discretization scheme will result in strictly positive paths for the z process if $4\theta z_0 > \eta^2$. For cases where this bound does not hold, it will be necessary to modify (9.70) to prevent problems with the computation of $\sqrt{\widehat{z}(t)}$. For instance, whenever $\widehat{z}(t)$ drops below zero, we could use (9.69) rather than (9.70).

Under certain sufficient regularity conditions, we have seen in Chapter 3 that Milstein schemes have second-order weak convergence. Due to the presence of a square root in (9.66), these sufficient conditions are violated here, and one should not expect (9.70) to have second-order convergence for all parameter values, even the ones that satisfy $4\theta z_0 > \eta^2$. Numerical tests of Milstein schemes for square-root processes can be found in Kahl and Jäckel [2006] and Glasserman [2004]; overall these schemes perform fairly well in benign parameter regimes, but are typically less robust than the Euler scheme.

9.5.3 Moment Matching Schemes for Variance Process

9.5.3.1 Log-normal Approximation

The simulation schemes introduced in Section 9.5.2 all suffer to various degrees from an inability to keep the path of z non-negative. One, rather obvious, way around this is to draw $\widehat{z}(t+\Delta)$ from a user-selected probability distribution that i) is reasonably close to the true distribution of $z(t+\Delta)$; and ii) is certain not to produce negative values[7]. To ensure that i) is satisfied, it is natural to select the parameters of the chosen distribution to match one or more of the true moments for $z(t+\Delta)$, conditional upon $z(t) = \widehat{z}(t)$. For instance, if we assume that the true distribution of $z(t+\Delta)$ is well approximated by a log-normal distribution with parameters μ and σ^2, we write (see Andersen and Brotherton-Ratcliffe [2005])

$$\widehat{z}(t+\Delta) = e^{\mu+\sigma Z}, \qquad (9.71)$$

where Z is a standard Gaussian random variable, and μ, σ are chosen to satisfy

$$e^{\mu+\frac{1}{2}\sigma^2} = \mathrm{E}\left(z(t+\Delta)|z(t)=\widehat{z}(t)\right), \qquad (9.72)$$

$$e^{2\left(\mu+\frac{1}{2}\sigma^2\right)}\left(e^{\sigma^2}-1\right) = \mathrm{Var}\left(z(t+\Delta)|z(t)=\widehat{z}(t)\right). \qquad (9.73)$$

The results in Corollary 8.3.3 can be used to compute the right-hand sides of this system of equations, which can then easily be solved analytically for μ and σ.

As is the case for many other schemes, (9.71) works best if the Feller condition, as defined in Proposition 8.3.1, is satisfied. If not, the lower tail of the log-normal distribution is often too thin to capture the true distribution shape of $\widehat{z}(t+\Delta)$ — see Figure 9.2 for an example.

9.5.3.2 Truncated Gaussian

Figure 9.2 demonstrates that the density of $z(t+\Delta)|z(t)$ may sometimes be nearly singular at the origin. To accommodate this, one could contemplate inserting an actual singularity through outright truncation at the origin of a distribution that may otherwise go negative. Using a Gaussian distribution for this, say, one could write

$$\widehat{z}(t+\Delta) = (\mu+\sigma Z)^+, \qquad (9.74)$$

where μ and σ are determined by moment-matching, along the same lines as in Section 9.5.3.1 above. While this moment-matching exercise cannot be done in entirely analytical fashion, a number of caching tricks outlined in Andersen [2008] can be used to make the determination of μ and σ essentially instantaneous. As documented in Andersen [2008], the scheme

---

[7] As pointed out in Section 3.2.2, weak consistency — convergence of the first and second moments in the discretization scheme to those of the original SDE — is sufficient (together with some regularity conditions) for weak convergence. Hence, the actual distribution used for time-stepping can be chosen almost arbitrarily. Of course, matching other characteristics of the actual distribution may substantially improve the performance of the scheme.

Fig. 9.2. Cumulative Distribution of z

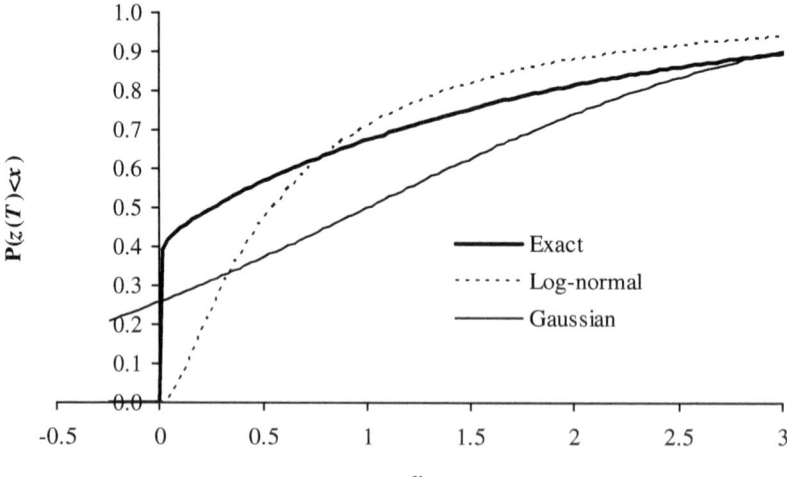

Notes: The figure shows the cumulative distribution function for $z(T)$ given $z(0)$, with $T = 0.1$. Model parameters were $z(0) = z_0 = 1$, $\theta = 50\%$, and $\eta = 100\%$. The log-normal and Gaussian distributions in the graph were parameterized by matching mean and variances to the exact distribution of $z(T)$.

(9.74) is robust and generally has attractive convergence properties when applied to standard option pricing problems. Being fundamentally Gaussian when $\widehat{z}(t)$ is far from the origin, (9.74) is qualitatively similar to the Euler scheme (9.69), although performance of (9.74) is typically somewhat better than (9.69). Unlike (9.69), the truncated Gaussian scheme (9.74) also ensures, by construction, that negative values of $\widehat{z}(t + \Delta)$ cannot be attained.

9.5.3.3 Quadratic-Exponential

We finish our discussion of biased schemes for (9.66) with a more elaborate moment-matched scheme, based on a combination of a squared Gaussian and an exponential distribution. In this scheme, for large values of $\widehat{z}(t)$, we write
$$\widehat{z}(t + \Delta) = a\,(b + Z)^2, \tag{9.75}$$
where Z is a standard Gaussian random variable, and a and b are certain constants, to be determined by moment-matching. The constants a and b will depend on the time step Δ and $\widehat{z}(t)$, as well as the parameters of the SDE for $z(t)$. While based on well-established asymptotics for the non-central chi-square distribution (see Andersen [2008]), formula (9.75) does not work well for low values of $\widehat{z}(t)$ — in fact, the moment-matching exercise fails to

work — so we supplement it with a scheme to be used when $\widehat{z}(t)$ is small. Examination of the true conditional density for $z(t+\Delta)|z(t)$ shows that the upper density tail decays exponentially, so a good choice is to approximate the distribution of $\widehat{z}(t+\Delta)$ with

$$P\left(\widehat{z}(t+\Delta) \in [x, x+dx]\right) = \left(p\delta(x) + \beta(1-p)e^{-\beta x}\right) dx, \quad x \geq 0, \quad (9.76)$$

where δ is the Dirac delta function, and p and β are non-negative constants to be determined. As in the scheme in Section 9.5.3.2, we have a probability mass at the origin, but now the strength of this mass (p) is explicitly specified, rather than implied from other parameters. It can be verified that if $p \in [0, 1]$ and $\beta \geq 0$, then (9.76) constitutes a valid density function.

Assuming that we have determined a and b, Monte Carlo sampling from (9.75) is trivial. To draw samples in accordance with (9.76), we can generate a cumulative distribution function

$$\Psi(x) = P\left(\widehat{z}(t+\Delta) \leq x\right) = p + (1-p)\left(1 - e^{-\beta x}\right), \quad x \geq 0. \quad (9.77)$$

Here, the inverse of Ψ is readily computable:

$$\Psi^{-1}(u) = \Psi^{-1}(u; p, \beta) = \begin{cases} 0, & 0 \leq u \leq p, \\ \beta^{-1} \ln\left(\frac{1-p}{1-u}\right), & p < u < 1. \end{cases} \quad (9.78)$$

By the standard inverse distribution function method from Section 3.1.1.1, we thus get the simple sampling scheme

$$\widehat{z}(t+\Delta) = \Psi^{-1}\left(U_z; p, \beta\right) \quad (9.79)$$

where U_z is a draw from a uniform distribution. Note that this scheme is extremely fast to execute.

Equations (9.75) and (9.79) together define the QE (for *Quadratic-Exponential*) discretization scheme. What remains is the determination of the constants a, b, p, and β, as well as a rule for when to switch from (9.75) to (9.79). The first problem is easily settled by moment-matching techniques, as shown in the following two propositions. We omit their straightforward proofs, which can be found in Andersen [2008].

Proposition 9.5.2. *Let*

$$m \triangleq E\left(z(t+\Delta)|z(t) = \widehat{z}(t)\right), \quad s^2 \triangleq \text{Var}\left(z(t+\Delta)|z(t) = \widehat{z}(t)\right),$$

and set $\psi = s^2/m^2$. *Provided that* $\psi \leq 2$, *set*

$$b^2 = 2\psi^{-1} - 1 + \sqrt{2\psi^{-1}}\sqrt{2\psi^{-1} - 1} \geq 0 \quad (9.80)$$

and

$$a = \frac{m}{1+b^2}. \quad (9.81)$$

Let $\widehat{z}(t+\Delta)$ *be as defined in (9.75); then* $E(\widehat{z}(t+\Delta)) = m$ *and* $\text{Var}(\widehat{z}(t+\Delta)) = s^2$.

Proposition 9.5.3. *Let m, s, and ψ be as defined in Proposition 9.5.2. Assume that $\psi \geq 1$ and set*

$$p = \frac{\psi - 1}{\psi + 1} \in [0, 1), \tag{9.82}$$

and

$$\beta = \frac{1-p}{m} = \frac{2}{m(\psi + 1)} > 0. \tag{9.83}$$

Let $\widehat{z}(t + \Delta)$ be sampled from (9.79); then $\mathrm{E}(\widehat{z}(t+\Delta)) = m$ and $\mathrm{Var}(\widehat{z}(t+\Delta)) = s^2$.

The terms m, s, ψ defined in the two propositions above are explicitly computable from the result in Corollary 8.3.3. For any ψ_c in $[1, 2]$, a valid *switching rule* is to use (9.75) if $\psi \leq \psi_c$ and to sample (9.77) otherwise. The exact value selected for ψ_c is non-critical; $\psi_c = 1.5$ is a natural choice.

9.5.3.4 Summary of QE Algorithm

As the QE algorithm is fairly complex, let us for convenience summarize the entire sampling algorithm step-by-step.

Assume that some arbitrary level $\psi_c \in [1, 2]$ has been selected. The detailed algorithm for the QE simulation step from $\widehat{z}(t)$ to $\widehat{z}(t + \Delta)$ is then:

1. Given $z(t) = \widehat{z}(t)$, compute $m = \mathrm{E}(z(t+\Delta)|z(t) = \widehat{z}(t))$ and $s^2 = \mathrm{Var}(z(t+\Delta)|z(t)\widehat{z}(t))$ from Corollary 8.3.3.
2. Compute $\psi = s^2/m^2$.
3. Draw a uniform random number U_z.
4. **If $\psi \leq \psi_c$:**
    a) Compute a and b from equations (9.81) and (9.80).
    b) Compute $Z = \Phi^{-1}(U_z)$.
    c) Use (9.75), i.e. set $\widehat{z}(t+\Delta) = a(b+Z)^2$.
5. **Otherwise**, if $\psi > \psi_c$:
    a) Compute p and β according to equations (9.82) and (9.83).
    b) Use (9.79), i.e. set $\widehat{z}(t+\Delta) = \Psi^{-1}(U_z; p, \beta)$, where Ψ^{-1} is given in (9.78).

For efficiency, exponentials used in computation of m and s^2 should be pre-cached. The inversion of the Gaussian CDF in Step 4 can be done using the techniques described in Section 3.1.1.1.

The quadratic-exponential (QE) scheme outlined above is typically the most accurate of the biased schemes discussed here. Indeed, in most practical application the bias introduced by the scheme is statistically undetectable at the levels of Monte Carlo noise typically encountered in practical applications; see Andersen [2008] for numerical tests under a range of challenging conditions. Variations on the QE scheme without an explicit singularity in zero can also be found in Andersen [2008].

9.5.4 Broadie-Kaya Scheme for the Underlying

At this point, we are done discussing simulation schemes for the z-process, and now turn to the underlying process (9.65) itself.

For numerical work, it is useful to work with a logarithmic transformation of $S(t)$, rather than $S(t)$ itself. Specifically, we set

$$X(t) = \frac{bS(t) + (1-b)L}{bS(0) + (1-b)L},$$

the logarithm of which, from Proposition 8.3.6, satisfies the SDE

$$d\ln X(t) = -\frac{1}{2}\lambda^2 b^2 z(t)\, dt + \lambda b \sqrt{z(t)}\, dW(t). \tag{9.84}$$

As demonstrated in Broadie and Kaya [2006], it is possible to simulate (9.84) bias-free. To show this, first integrate the SDE for $z(t)$ in (9.66) and rearrange:

$$\int_t^{t+\Delta} \sqrt{z(u)}\, dZ(u) = \frac{1}{\eta}\left(z(t+\Delta) - z(t) - \theta z_0 \Delta + \theta \int_t^{t+\Delta} z(u)\, du \right). \tag{9.85}$$

Performing a Cholesky decomposition we can also write

$$d\ln X(t) = -\frac{1}{2}\lambda^2 b^2 z(t)\, dt + \lambda b \left(\rho \sqrt{z(t)}\, dZ(t) + \sqrt{1-\rho^2}\sqrt{z(t)}\, dB(t) \right),$$

where B is a Brownian motion independent of Z. An integration then yields

$$\ln X(t+\Delta) = \ln X(t) + \frac{\rho \lambda b}{\eta}\left(z(t+\Delta) - z(t) - \theta z_0 \Delta \right)$$
$$+ \left(\frac{\theta \rho \lambda b}{\eta} - \frac{\lambda^2 b^2}{2} \right) \int_t^{t+\Delta} z(u)\, du + \lambda b \sqrt{1-\rho^2} \int_t^{t+\Delta} \sqrt{z(u)}\, dB(u), \tag{9.86}$$

where we have used (9.85). Conditional on $z(t+\Delta)$ and $\int_t^{t+\Delta} z(u)\, du$, it is clear that the distribution of $\ln X(t+\Delta)$ is Gaussian with easily computable moments. After first sampling $z(t+\Delta)$ bias-free from the non-central chi-square distribution (as described in Section 9.5.1), one then performs the following steps:

1. Conditional on $z(t+\Delta)$ (and $z(t)$) draw a bias-free sample of $I = \int_t^{t+\Delta} z(u)\, du$.
2. Conditional on $z(t+\Delta)$ and I, use (9.86) to draw a sample of $\ln X(t+\Delta)$ from a Gaussian distribution.

While execution of the second step is straightforward, the first one is decidedly not, as the conditional distribution of the integral I is not known in closed form. In Broadie and Kaya [2006], the authors instead derive a characteristic function, which they numerically Fourier-invert to generate the cumulative distribution function for I, given $z(t+\Delta)$ and $z(t)$. Numerical inversion of this distribution function over a uniform random variable finally allows for generation of a sample of I. The total algorithm requires great care in numerical discretization to prevent introduction of noticeable biases and is further complicated by the fact that the characteristic function for I contains two modified Bessel functions.

The Broadie-Kaya algorithm is bias-free by construction, but its complexity and lack of speed is problematic in many applications. Smith [2007] and Glasserman and Kim [2008] discuss various techniques to improve computational efficiency of the basic algorithm, but even with such improvements it is safe to say that the method is competitive only for applications that involve long time steps and require very high accuracy (and neither are the norm for fixed income applications).

9.5.5 Other Schemes for the Underlying

9.5.5.1 Taylor-Type Schemes

In their examination of "fixed" Euler-schemes, Lord et al. [2006] suggest simulation of the Heston model by combining (9.69) with the following scheme for $\ln X$:

$$\ln \widehat{X}(t+\Delta) = \ln \widehat{X}(t) - \frac{1}{2}\lambda^2 b^2 \widehat{z}(t)^+ \Delta + \lambda b \sqrt{\widehat{z}(t)^+}\, W \sqrt{\Delta}, \qquad (9.87)$$

where W is a Gaussian $\mathcal{N}(0,1)$ draw, correlated to Z in (9.69) with correlation coefficient ρ. For the periods where $\widehat{z}$ drops below zero in (9.69), the process for $\widehat{X}$ comes to a standstill.

Kahl and Jäckel [2006] examine the usage of Ito-Taylor expansions for joint simulation of $X(t)$ and $z(t)$, proposing several concrete schemes. As these schemes are rather complex, we simply refer the reader to Kahl and Jäckel [2006] for the details. Andersen [2008] tests the most prominent of the schemes in Kahl and Jäckel [2006] (the "IJK" scheme) and concludes that the scheme works well in benign parameter ranges, but has a tendency to deteriorate when parameters are made more extreme.

9.5.5.2 Simplified Broadie-Kaya

We recall from the discussion earlier that the complicated part of the Broadie-Kaya algorithm was the computation of $\int_t^{t+\Delta} z(u)\, du$, conditional on $z(t)$ and $z(t+\Delta)$. Andersen [2008] suggests a naive, but effective, approximation, based on the idea that

$$\int_{t}^{t+\Delta} z(u)\,du \approx \Delta\left[\gamma_1 z(t) + \gamma_2 z(t+\Delta)\right], \tag{9.88}$$

for certain constants γ_1 and γ_2. The constants γ_1 and γ_2 can be found by moment-matching techniques (using calculations similar to those from the proof of Theorem 9.3.6, or results from Dufresne [2001], p. 16), but Andersen [2008] presents evidence that it will often be sufficient to use either an Euler-like setting ($\gamma_1 = 1$, $\gamma_2 = 0$) or a central discretization ($\gamma_1 = \gamma_2 = \frac{1}{2}$). In any case, (9.88) combined with (9.86) gives rise to a scheme for Y-simulation that can be combined with any basic algorithm that can produce $\widehat{z}(t)$ and $\widehat{z}(t + \Delta)$. Andersen [2008] contains numerical results for the case where $\widehat{z}(t)$ and $\widehat{z}(t + \Delta)$ are simulated by the algorithms in Sections 9.5.3.2 and 9.5.3.3; results are excellent, particularly when the QE algorithm in Section 9.5.3.3 is used to sample $\widehat{z}$. Figure 9.3 reproduces some sample convergence results from Andersen [2008].

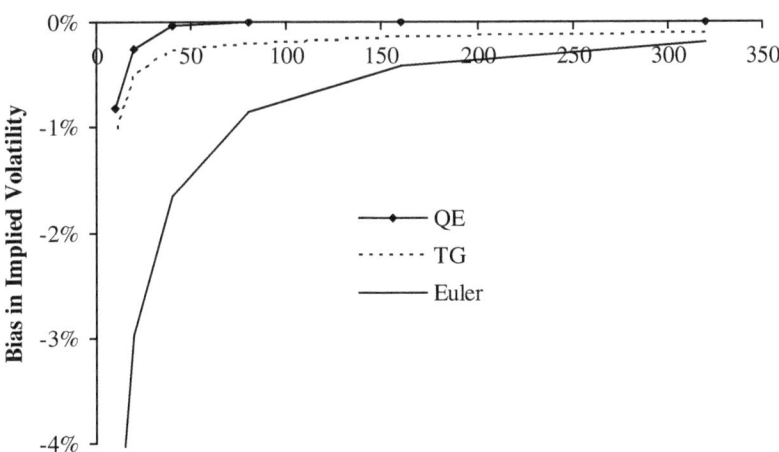

Fig. 9.3. Convergence of Bias

Number of Time Steps

Notes: The figure shows the convergence of the call option price bias in implied volatility terms, as a function of the number of time steps per path ($=T/\Delta$). The Euler scheme graph was computed using the full truncation scheme in (9.69), and the QE scheme used $\gamma_1 = \gamma_2 = 0.5$ and $\psi_c = 1.5$. Model parameters: $S(0) = L = 100$, $b = 1$, $z(0) = z_0 = 1$, $\theta = 0.5$, $\rho = -0.9$, $\eta = 1$, $\lambda = 20\%$. The option maturity is $T = 10$ and the strike is $K = 100$. The bias was estimated from 1,000,000 simulation paths, using the Fourier technique to establish exact prices.

9.5.5.3 Martingale Correction

Finally, let us note that some of the schemes outlined above, including the one in Section 9.5.5.2, will generally not lead to martingale behavior of $\widehat{X}$; that is, $\mathrm{E}(\widehat{X}(t+\Delta)|\widehat{X}(t)) \neq \widehat{X}(t)$. For the cases where the error $e = \mathrm{E}(\widehat{X}(t+\Delta)|\widehat{X}(t)) - \widehat{X}(t)$ is analytically computable, it is, however, straightforward to remove the bias by simply adding $-e$ to the sample value for $\widehat{X}(t+\Delta)$. Andersen [2008] gives several examples of this idea and shows that, for the QE scheme at least, the improvements from martingale correction are minor.

9.A Appendix: Proof of Proposition 9.3.4

Let us fix a time horizon $T > 0$. Let $f(t, x)$ be a local volatility function,
$$f(t, x) \in C^2([0, T] \times \mathbb{R}),$$
satisfying the usual growth requirements. Let $\lambda(t)$, $t \in [0, T]$, be a function of time only. Fix $x_0 \in \mathbb{R}$. For any $\epsilon \geq 0$, define a rescaled local volatility function
$$f_\epsilon(t, x) = f\left(t\epsilon^2, x_0 + (x - x_0)\epsilon\right). \tag{9.89}$$

Without loss of generality we can assume that
$$f(t, x_0) \equiv 1, \quad t \in [0, T],$$
which implies
$$f_\epsilon(t, x_0) \equiv 1, \quad t \in [0, T]. \tag{9.90}$$

Let $w(t)$, $t \in [0, T]$, be a weight function such that
$$\int_0^T w(t)\, dt = 1, \tag{9.91}$$
and let us define an averaged local volatility function
$$\overline{f}_\epsilon(x)^2 = \int_0^T f_\epsilon(t, x)^2 w(t)\, dt. \tag{9.92}$$

Define two families of diffusions indexed by ϵ,
$$dX_\epsilon(t) = f_\epsilon(t, X_\epsilon(t))\sqrt{z(t)}\lambda(t)\, dW(t), \quad X_\epsilon(0) = x_0,$$
$$dY_\epsilon(t) = \overline{f}_\epsilon(Y_\epsilon(t))\sqrt{z(t)}\lambda(t)\, dW(t), \quad Y_\epsilon(0) = x_0,$$

for $t \in [0, T]$, where $z(t)$ is defined by (9.31). The following theorem can be found in Piterbarg [2005b].

9.A Appendix: Proof of Proposition 9.3.4

Theorem 9.A.1. *If the weight function $w(t)$ is set to equal $w_T(t)$, where*

$$w_T(t) \triangleq \frac{v(t)^2 \lambda(t)^2}{\int_0^T v(t)^2 \lambda(t)^2 \, dt}, \tag{9.93}$$

$$v(t)^2 = \mathrm{E}\left(z(t)\left(X_0(t) - x_0\right)^2\right),$$

then, as $\epsilon \to 0$,

$$\mathrm{E}\left((X_\epsilon(T) - x_0)^2\right) - \mathrm{E}\left((Y_\epsilon(T) - x_0)^2\right) = o\left(\epsilon^2\right), \tag{9.94}$$

$$\mathrm{E}\left((X_\epsilon(T) - x_0)^3\right) - \mathrm{E}\left((Y_\epsilon(T) - x_0)^3\right) = o\left(\epsilon^2\right). \tag{9.95}$$

Proof. The stochastic variance process $z(t)$ is Markovian. We denote its infinitesimal generator by L^z,

$$L^z : \phi \longmapsto \theta\left(z_0 - z\right) \frac{\partial \phi}{\partial z} + \frac{1}{2} \eta^2 z \frac{\partial^2 \phi}{\partial z^2}.$$

We note that the process $X_0(t)$ ($\equiv Y_0(t)$) satisfies the following SDE,

$$dX_0(t) = \sqrt{z(t)} \lambda(t) \, dW(t), \quad X_0(0) = x_0.$$

Let us denote the Markov semi-group of operators that corresponds to the process $(X_0(t), z(t))$ by $P_0(s,t)$, and the time-dependent infinitesimal generator by $L_0(t)$,

$$[P_0(s,t)\phi](x,z) = \mathrm{E}_s\left(\phi\left(X_0(t), z(t)\right) | X_0(s) = x, z(s) = z\right),$$

$$L_0(t) : \phi \longmapsto \frac{1}{2} \lambda(t)^2 z \frac{\partial^2 \phi}{\partial x^2} + L^z.$$

Let us denote the same for $(X_\epsilon(t), z(t))$ and for $(Y_\epsilon(t), z(t))$ by $P_\epsilon^X(s,t)$, $L_\epsilon^X(t)$ and $P_\epsilon^Y(s,t)$, $L_\epsilon^Y(t)$, respectively.

From the general operator semigroup theory (see Ethier and Kurtz [1986]) it follows that

$$P_\epsilon^Y(0,T) = P_\epsilon^X(0,T) + \int_0^T P_\epsilon^Y(0,t)\left(L_\epsilon^Y(t) - L_\epsilon^X(t)\right) P_\epsilon^X(t,T) \, dt. \tag{9.96}$$

By Proposition 8.4.13 applied to $f(x) = (x - x_0)^2 / 2$ and $f(x) = (x - x_0)^3 / 6$,

$$\frac{1}{2} \mathrm{E}\left(X_\epsilon(T) - x_0\right)^2 = \int_{-\infty}^{x_0} \mathrm{E}\left(K - X_\epsilon(T)\right)^+ dK + \int_{x_0}^{\infty} \mathrm{E}\left(X_\epsilon(T) - K\right)^+ dK,$$

$$\frac{1}{6} \mathrm{E}\left(X_\epsilon(T) - x_0\right)^3 = \int_{-\infty}^{x_0} (K - x_0) \mathrm{E}\left(K - X_\epsilon(T)\right)^+ dK$$

$$+ \int_{x_0}^{\infty} (K - x_0) \mathrm{E}\left(X_\epsilon(T) - K\right)^+ dK,$$

and the same for Y_ϵ. Expressed in terms of the Markovian semigroup,

$$\frac{1}{(i+2)!} \mathrm{E}\left(X_\epsilon(T) - x_0\right)^{i+2} = \int_{-\infty}^{\infty} \langle \delta_{x_0, z_0}, (K - x_0)^i P_\epsilon^X(0, T) \pi_K \rangle \, dK,$$

(and the same for Y_ϵ) for $i = 0, 1$, where we have defined the payoff π_K by

$$\pi_K(x, z) = \begin{cases} (x - K)^+, & K \geq x_0, \\ (K - x)^+, & K < x_0. \end{cases}$$

Let us denote

$$\Delta(i) = \frac{1}{(i+2)!} \left(\mathrm{E}\left(Y_\epsilon(T) - x_0\right)^{i+2} - \mathrm{E}\left(X_\epsilon(T) - x_0\right)^{i+2} \right), \quad i = 0, 1.$$

To prove the theorem, we need to show that with the appropriate choice of weights $w_T(t)$,

$$\Delta(i) = o\left(\epsilon^2\right), \quad \epsilon \to 0, \quad i = 0, 1. \tag{9.97}$$

Clearly,

$$\Delta(i) = \int_{-\infty}^{\infty} (K - x_0)^i \langle \delta_{x_0, z_0}, (P_\epsilon^Y(0, T) - P_\epsilon^X(0, T)) \pi_K \rangle \, dK.$$

By (9.96) we have,

$$\Delta(i) = \int_{-\infty}^{\infty} (K - x_0)^i \\ \times \left(\int_0^T \langle \delta_{x_0, z_0}, P_\epsilon^Y(0, t) \left(L_\epsilon^Y(t) - L_\epsilon^X(t) \right) P_\epsilon^X(t, T) \pi_K \rangle \, dt \right) dK.$$

After a series of manipulations (see Piterbarg [2005b] for details) we obtain, to order $o(\epsilon^2)$,

$$\Delta(i) = \frac{1}{2} \int_0^T \int \widehat{p}(x, z)(x - x_0)^i \left(\overline{f}_\epsilon(x)^2 - f_\epsilon(t, x)^2 \right) \lambda(t)^2 \, dx \, dt, \tag{9.98}$$

$$\widehat{p}(t, x) \triangleq \mathrm{E}\left(z(t)\delta(X_0(t) - x_0)\right).$$

Expanding $f, \overline{f}$ to the first order around (t, x_0), we obtain

9.A Appendix: Proof of Proposition 9.3.4

$$\delta(t;i) \triangleq \int \widehat{p}(t,x)(x-x_0)^i \left(\overline{f}_\epsilon(x)^2 - f_\epsilon(t,x)^2\right) dx$$

$$= 2\epsilon \left(\int_0^T \frac{\partial f(s\epsilon^2, x_0)}{\partial x} w(s) \, ds - \frac{\partial f(t\epsilon^2, x_0)}{\partial x} \right)$$

$$\times \int \widehat{p}(t,x)(x-x_0)^{i+1} \, dx$$

$$+ \epsilon^2 \left(\int_0^T \left[\frac{\partial f(s\epsilon^2, x_0)}{\partial x} \right]^2 w(s) \, ds - \left[\frac{\partial f(t\epsilon^2, x_0)}{\partial x} \right]^2 \right)$$

$$\times \int \widehat{p}(t,x)(x-x_0)^{i+2} \, dx$$

$$+ \epsilon^2 \left(\int_0^T \frac{\partial^2 f(s\epsilon^2, x_0)}{\partial x^2} w(s) \, ds - \frac{\partial^2 f(t\epsilon^2, x_0)}{\partial x^2} \right)$$

$$\times \int \widehat{p}(t,x)(x-x_0)^{i+2} \, dx$$

$$+ o(\epsilon^2).$$

Calculating the integrals, we obtain to order $o(\epsilon^2)$,

$$\delta(t;0) = \epsilon^2 v(t)^2 \left(\int_0^T \left[\frac{\partial f(s\epsilon^2, x_0)}{\partial x} \right]^2 w(s) \, ds - \left[\frac{\partial f(t\epsilon^2, x_0)}{\partial x} \right]^2 \right.$$

$$\left. + \int_0^T \frac{\partial^2 f(s\epsilon^2, x_0)}{\partial x^2} w(s) \, ds - \frac{\partial^2 f(t\epsilon^2, x_0)}{\partial x^2} \right),$$

$$\delta(t;1) = 2\epsilon v(t)^2 \left(\int_0^T \frac{\partial f(s\epsilon^2, x_0)}{\partial x} w(s) \, ds - \frac{\partial f(t\epsilon^2, x_0)}{\partial x} \right),$$

$$\Delta(i) = \frac{1}{2} \int_0^T \delta(t;i) \lambda(t)^2 \, dt.$$

For $w(t) = w_T(t)$, we obtain $\Delta(i) = 0$, $i = 0, 1$, and the theorem follows. □

Proposition 9.3.4 is proved by applying Theorem 9.A.1 to the equation (9.30). To compute $v(t)^2$, conditioning on $z(t)$ and using conditional independence of $X_0(t)$ and $z(t)$ we obtain,

$$\mathrm{E}\left((X_0(t) - x_0)^2 z(t)\right) = \mathrm{E}\left(z(t) \mathrm{E}\left((X_0(t) - x_0)^2 \big| z(\cdot)\right)\right) \quad (9.99)$$

$$= \mathrm{E}\left(z(t) \int_0^t z(s) \lambda(s)^2 \, ds\right)$$

$$= \int_0^t \lambda(s)^2 \mathrm{E}\left(z(t) z(s)\right) \, ds.$$

Clearly

$$z(t) - z_0 = e^{-\theta(t-s)}(z(s) - z_0) + O(dW),$$

so that

$$E(z(t)z(s)) = z_0^2 + E\left(e^{-\theta(t-s)}(z(s) - z_0)z(s)\right)$$
$$= e^{-\theta(t-s)}E(z(s)^2) + \left(1 - e^{-\theta(t-s)}\right)z_0^2.$$

We also have that

$$E(z(s)^2) = z_0^2 + z_0\eta^2 \frac{1 - e^{-2\theta s}}{2\theta}. \tag{9.100}$$

Substituting into (9.99) yields

$$v(t)^2 = E\left((S(t) - x_0)^2 z(t)\right) \tag{9.101}$$
$$= \int_0^t \lambda(s)^2 \left(e^{-\theta(t-s)} E(z(s)^2) + \left(1 - e^{-\theta(t-s)}\right)z_0^2\right) ds$$
$$= \int_0^t \lambda(s)^2 \left(e^{-\theta(t-s)} z_0^2 + e^{-\theta(t-s)} z_0\eta^2 \frac{1 - e^{-2\theta s}}{2\theta}\right. \tag{9.102}$$
$$\left. + z_0^2 \left(1 - e^{-\theta(t-s)}\right)\right) ds$$
$$= \int_0^t \lambda(s)^2 \left(z_0^2 + z_0\eta^2 e^{-\theta(t-s)} \frac{1 - e^{-2\theta s}}{2\theta}\right) ds$$
$$= z_0^2 \int_0^t \lambda(s)^2 ds + z_0\eta^2 e^{-\theta t} \int_0^t \lambda(s)^2 \frac{e^{\theta s} - e^{-\theta s}}{2\theta} ds.$$

9.B Appendix: Coefficients for Asymptotic Expansion

Set $\Omega = \Omega_0 \bar{v}^{1/2}\tau^{1/2} + \Omega_1 \bar{v}^{3/2}\tau^{3/2}$ where $\bar{v} = \mu_{z\lambda^2}(0,z_0)/T$. Also define the easily computed quantities

$$\Omega_{mn} = \frac{\partial^m \Omega/\partial \bar{v}^m}{\partial^n \Omega/\partial \bar{v}^n}.$$

Then the expansion coefficients in Proposition 9.2.4 are given by

$$\alpha_0 = \tau^{-2} l_{1,2}\left(\Omega_{21} - \frac{1}{4}\Omega^2 \Omega_{10}\right), \quad \alpha_1 = \tau^{-2} l_{1,2} \Omega^{-2} \Omega_{10},$$

and

9.B Appendix: Coefficients for Asymptotic Expansion

$$\beta_0 = \tau^{-2} l_{2,2} \left(\Omega_{21} - \frac{1}{4} \Omega^2 \Omega_{10} \right)$$

$$- \tau^{-3} l_{2,3} \left(\Omega_{31} - \Omega_{21}^2 - \frac{1}{4} \Omega^2 \left(\Omega_{20} + \Omega_{10}^2 \right) + \left(\Omega_{21} - \frac{1}{4} \Omega^2 \Omega_{10} \right)^2 \right)$$

$$+ \frac{1}{2} \tau^{-4} l_{1,2}^2 \left(\Omega_{41} - 3 \Omega_{31} \Omega_{21} + 2 \Omega_{21}^3 - \frac{1}{4} \Omega^2 \Omega_{30} - \frac{3}{4} \Omega^2 \Omega_{10} \Omega_{20} \right)$$

$$+ \frac{3}{2} \tau^{-4} l_{1,2}^2 \left(\Omega_{21} - \frac{1}{4} \Omega^2 \Omega_{10} \right) \left(\Omega_{31} - \Omega_{21}^2 - \frac{1}{4} \Omega^2 \left(\Omega_{20} + \Omega_{10}^2 \right) \right),$$

$$\beta_1 = \Omega^{-2} \tau^{-2} l_{2,2} \Omega_{10}$$

$$- \Omega^{-2} \tau^{-3} l_{2,3} \left(\Omega_{20} - 3 \Omega_{10}^2 + 2 \Omega_{10} \left(\Omega_{21} - \frac{1}{4} \Omega^2 \Omega_{10} \right)^2 \right)$$

$$+ \Omega^{-2} \frac{1}{2} \tau^{-4} l_{1,2}^2 \left(\Omega_{30} - 9 \Omega_{10} \Omega_{20} + 12 \Omega_{10}^3 \right.$$

$$\left. + 3 \Omega_{10} \left(\Omega_{31} - \Omega_{21}^2 - \frac{1}{4} \Omega^2 \left(\Omega_{20} + \Omega_{10}^2 \right) \right) \right)$$

$$+ \Omega^{-2} \frac{3}{2} \tau^{-4} l_{1,2}^2 \left(\Omega_{21} - \frac{1}{4} \Omega^2 \Omega_{10} \right) \left(\Omega_{20} - 3 \Omega_{10}^2 \right),$$

$$\beta_2 = - \Omega^{-4} \tau^{-3} l_{2,3} \Omega_{10}^2 + \Omega^{-4} \frac{3}{2} \tau^{-4} l_{1,2}^2 \Omega_{10} \left(\Omega_{20} - 3 \Omega_{10}^2 \right).$$

References

M. Abramowitz and I. A. Stegun, editors. *Handbook of Mathematical Functions*. Dover, 1965.

P. J. Acklam. An algorithm for computing the inverse Normal cumulative distribution function. Technical report, Unaffiliated, 2003. URL `http://home.online.no/ p̃jacklam/ notes/ invnorm`.

K. J. Adams and D. R. van Deventer. Fitting yield curves and forward rate curves with maximum smoothness. *Journal of Fixed Income*, 4:52–62, 1994.

D.-H. Ahn, R. F. Dittmar, and A. R. Gallant. Quadratic term structure models: Theory and evidence. *Review of Financial Studies*, 15(1):243–288, 2002.

J. Ahrens and U. Dieter. Computer methods for sampling from the gamma, beta, Poisson, and binomial distributions. *Computing*, 12(3):223–246, 1974.

Y. Aït-Sahalia. Testing continuous-time models of the spot interest rate. *Review of Financial Studies*, 9(2):385–426, 1996.

F. Åkesson and J. P. Lehoczky. Discrete eigenfunction-expansion of multi-dimensional Brownian motion and the Ornstein-Uhlenbeck process. Carnegie Mellon University working paper, 1998.

H. Albrecher, P. Mayer, W. Schoutens, and J. Tistaert. The little Heston trap. *Wilmott*, 1(1):83–92, 2007.

L. B. Andersen. Simulation and calibration of the HJM model. General Re Financial Products working paper, 1995.

L. B. Andersen. *Five Essays on Contingent Claims Pricing*. PhD thesis, Aarhus Business School, 1996.

L. B. Andersen. Monte Carlo simulation of options on joint minima and maxima. In B. Dupire, editor, *Monte Carlo: Methodologies and Applications for Pricing and Risk Management*. Risk Books, 1998.

L. B. Andersen. A simple approach to the pricing of Bermudan swaptions in the multi-factor Libor market model. *Journal of Computational Finance*, 3(2):5–32, 2000a.

L. B. Andersen. Separable Libor market models. General Re Financial Products working paper, 2000b.

L. B. Andersen. Yield curve construction with tension splines. *Review of Derivatives Research*, 10(3):227–267, 2005.

L. B. Andersen. Simple and efficient simulation of the Heston stochastic volatility model. *Journal of Computational Finance*, 11:1–42, 2008.

L. B. Andersen. Option pricing with quadratic volatility: A revisit. *Finance and Stochastics*, 2010. Forthcoming.

L. B. Andersen and J. Andreasen. Jump-diffusion processes: Volatility smile fitting and numerical methods for option pricing. *Review of Derivatives Research*, 4(3):231–262, 2000a.

L. B. Andersen and J. Andreasen. Volatility skews and extensions of the Libor market model. *Applied Mathematical Finance*, 7(1):1–32, 2000b.

L. B. Andersen and J. Andreasen. Factor dependence of Bermudan swaption prices: Fact or fiction? *Journal of Financial Economics*, 62(1):3–37, 2001.

L. B. Andersen and J. Andreasen. Volatile volatilities. *Risk*, 15(12):163–168, 2002.

L. B. Andersen and P. Boyle. Monte Carlo methods for interest rate derivatives. In N. Jegadeesh and B. Tuckman, editors, *Advanced Fixed-Income Valuation Tools*. Wiley, 2000.

L. B. Andersen and M. Broadie. A primal-dual simulation algorithm for pricing of multi-dimensional American options. *Management Science*, 50(9):1222–1234, 2004.

L. B. Andersen and R. Brotherton-Ratcliffe. Exact exotics. *Risk*, 9(10):85–89, 1996.

L. B. Andersen and R. Brotherton-Ratcliffe. The equity option volatility smile: An implicit finite difference approach. *Journal of Computational Finance*, 1(2):5–38, 1998.

L. B. Andersen and R. Brotherton-Ratcliffe. Extended Libor market models with stochastic volatility. *Journal of Computational Finance*, 9(1):1–40, 2005.

L. B. Andersen and D. Buffum. Implementation and calibration of convertible bond models. *Journal of Computational Finance*, 7(2):1–34, 2003.

L. B. Andersen and N. Hutchings. Parameter averaging of quadratic SDEs with stochastic volatility. *Journal of Computational Finance*, 2010. Forthcoming.

L. B. Andersen and V. V. Piterbarg. Moment explosions in stochastic volatility models. *Finance and Stochastics*, 11(1):29–50, 2007.

L. B. Andersen, J. Andreasen, and D. Eliezer. Static replication of barrier options: Some general results. *Journal of Computational Finance*, 5(4):1–25, 2002.

L. B. Andersen, J. Sidenius, and S. Basu. All your hedges in one basket. *Risk*, 11:67–72, 2003.

T. G. Andersen and J. Lund. Estimating continuous-time stochastic volatility models of the short-term interest rate. *Journal of Econometrics*, 77(2):343–377, 1997.

S. Andradóttir. A stochastic approximation algorithm with varying bounds. *Operations Research*, 43(6):1037–1048, 1995.

S. Andradóttir. A scaled stochastic approximation algorithm. *Management Science*, 42(4):475–498, 1996.

J. Andreasen. The pricing of discretely sampled Asian and lookback options: A change of numeraire approach. *Journal of Computational Finance*, 1(1):5–30, 1998.

J. Andreasen. Turbo-charging the Cheyette model. Bank of America working paper, 2001.

J. Andreasen. Pricing simple exotics under stochastic volatility. Bank of America working paper, 2002.

References

J. Andreasen. Back to the future. *Risk*, 18(9):104–109, 2005.

J. Andreasen, B. Jensen, and R. Poulsen. Eight valuation methods in financial mathematics: The Black-Scholes formula as an example. *Mathematical Scientist*, 23:18–40, 1998.

A. Antonov and M. Arneguy. Analytical formulas for pricing CMS products in the LIBOR market model with the stochastic volatility. *SSRN eLibrary*, 2009.

A. Antonov and T. Misirpashaev. Markovian projection onto a displaced diffusion: Generic formulas with applications. *International Journal of Theoretical and Applied Finance*, 12(4):507–522, 2009a.

A. Antonov and T. Misirpashaev. Projection on a quadratic model by asymptotic expansion with an application to LMM swaption. *SSRN eLibrary*, 2009b.

A. Antonov, T. Misirpashaev, and V. Piterbarg. Markovian projection onto a Heston model. *Journal of Computational Finance*, 13(1):23–47, 2009.

L. Arnold. *Stochastic Differential Equations: Theory and Practice*. Wiley, 1974.

P. Artzner, F. Delbaen, J.-M. Eber, and D. Heath. Coherent measures of risk. *Mathematical Finance*, 9(3):203–228, 1999.

S. Assefa. Calibration and pricing in a multi-factor quadratic Gaussian model. Research Paper Series 197, Quantitative Finance Research Centre, University of Technology, Sydney, 2007. URL http://ideas.repec.org/p/uts/rpaper/197.html.

M. Attari. Option pricing using Fourier transforms: A numerically efficient simplification. *SSRN eLibrary*, 2004.

P. Austing. Valuing multi-asset options on foreign exchange: A joint density repricing the cross smile. Submitted to Risk, 2010.

M. Avellaneda, R. Buff, C. Friedman, N. Grandechamp, L. Kruk, and J. Newman. Weighted Monte Carlo: A new technique for calibrating asset-pricing models. *International Journal of Theoretical and Applied Finance*, 4(1):91 – 119, 2001.

M. Avellaneda, D. Boyer-Olson, J. Busca, and P. Friz. Reconstructing volatility. *Risk*, 15(10), 2002.

O. Axelsson and V. Barker. *Finite Element Solution of Boundary Value Problems: Theory and Computation*, volume 35 of *Classics in applied mathematics*. SIAM, 1991.

S. Babbs. *The Term Structure of Interest Rates: Stochastic Processes and Contingent Claims*. PhD thesis, University of London, 1990.

Y. Balasanov. A gentle introduction to the BEEMIR model. NationsBanc working paper, 1996.

D. Bang. Numerical computation of Fourier transforms in Heston model. Bank of America working paper, 2009.

G. Barles and H. M. Soner. Option pricing with transaction costs and a nonlinear Black-Scholes equation. *Finance and Stochastics*, 2(4):369–397, 1998.

J. D. Beasley and S. G. Springer. Algorithm AS 111: The percentage points of the Normal distribution. *Applied Statistics*, 26(1):118–121, 1977.

D. Belomestny, C. Bender, and J. Schoenmakers. True upper bounds for Bermudan products via non-nested Monte Carlo. *Mathematical Finance*, 19(1):53–71, 2007.

C. Bender, A. Kolodko, and J. Schoenmakers. Iterating cancellable snowballs and related exotics. *Risk*, 9:126–130, 2006.

References

H. Berestycki, J. Busca, and I. Florent. Computing the implied volatility in stochastic volatility models. *Communications on Pure and Applied Mathematics*, 57(10):1352–1373, 2004.

L. Bergomi. Smile dynamics III. *Risk*, 5:90–96, 2009.

M. Berrahoui. Pricing CMS spread options and digital CMS spread options with smile. In *The Best of Wilmott 2*. Wiley, 2005.

C. Beveridge and M. S. Joshi. Juggling snowballs. *Risk*, 12:100–104, 2008.

C. Beveridge and M. S. Joshi. Practical policy iteration: Generic methods for obtaining rapid and tight bounds for Bermudan exotic derivatives using Monte Carlo simulation. *SSRN eLibrary*, 2009.

C. Beveridge, N. Denson, and M. S. Joshi. Comparing discretization of the Libor market model in the spot measure. *SSRN eLibrary*, 2008.

P. Billingsley. *Probability and Measure*. Wiley Series in Probability and Mathematical Statistics. Wiley, 3d edition, 1995.

T. Björk. A geometric view of interest rate theory. In E. Jouini, J. Cvitanic, and M. Musiela, editors, *Option Pricing, Interest Rates and Risk Management*, pages 241–277. Cambridge University Press, 2001.

F. Black. The pricing of commodity contracts. *Journal of Financial Economics*, 3:167–179, 1976.

F. Black. Interest rates as options. *Journal of Finance*, 50(5):1371–1376, 1995.

F. Black and P. Karasinski. Bond and option pricing when short rates are lognormal. *Financial Analysis Journal*, 47(4):52–59, 1991.

F. Black and M. Scholes. The pricing of options and corporate liabilities. *Journal of Political Economy*, 81(3):637–654, 1973.

F. Black, E. Derman, and W. Toy. A one-factor model of interest rates and its application to Treasury bond options. *Financial Analysts Journal*, 46(1):33–39, 1990.

R. R. Bliss and D. C. Smith. The stability of interest rate processes. Federal Reserve Bank of Atlanta working paper, 1997.

A. N. Borodin and P. Salminen. *Handbook of Brownian Motion — Facts and Formulae*. Probability and Its Applications. Birkhäuser, Basel, 1996.

N. Boyarchenko and S. Z. Levendorski. On errors and bias of Fourier transform methods in quadratic term structure models. *International Journal of Theoretical and Applied Finance*, 10(2):273–306, 2007.

P. Boyle, M. Broadie, and P. Glasserman. Monte Carlo methods for security pricing. *Journal of Economic Dynamics and Control*, 21(8-9):1267–1321, 1997.

A. Brace, D. Gatarek, and M. Musiela. The market model of interest rate dynamics. *Mathematical Finance*, 7:127–154, 1997.

D. T. Breeden and R. H. Litzenberger. Price of state-contingent claims implicit in option prices. *Journal of Business*, 51(4):621–651, 1978.

K. Brekke and B. Øksendal. The high contact principle as a sufficiency condition for optimal stopping. In D. Lund and B. Øksendal, editors, *Stochastic Models and Option Values: Applications to Resources, Environment and Investment Problems*. North-Holland, 1991.

M. J. Brennan and E. S. Schwartz. Analyzing convertible bonds. *Journal of Financial and Quantitative Analysis*, 15(4):907–929, 1980.

D. Brigo and F. Mercurio. *Interest-Rate Models – Theory and Practice*. Springer-Verlag, 2001.

D. Brigo and M. Morini. Efficient analytical cascade calibration of the Libor market model with endogenous interpolation. *Journal of Derivatives*, 14(1): 40–60, 2006.

M. Broadie and M. Cao. Improved lower and upper bound algorithms for pricing American options by simulation. *Quantitative Finance*, 8(8):845–861, 2008.

M. Broadie and J. Detemple. American capped call options on dividend-paying assets. *Review of Financial Studies*, 8(1):161–91, 1995.

M. Broadie and J. Detemple. Valuation of American options on multiple assets. *Review of Financial Studies*, 7(3):241–286, 1997.

M. Broadie and P. Glasserman. Estimating security price derivatives using simulation. *Management Science*, 42(2):269–285, 1996.

M. Broadie and P. Glasserman. Pricing American style securities using simulation. *Journal of Economic Dynamics and Control*, 21(8-9):1323–1352, 1997.

M. Broadie and P. Glasserman. A stochastic mesh method for pricing high-dimensional American options. *Journal of Computational Finance*, 7(4):35–72, 2004.

M. Broadie and O. Kaya. Exact simulation of stochastic volatility and other affine jump diffusion processes. *Operations Research*, 54(2):217–231, 2006.

M. Broadie, P. Glasserman, and S. Kou. A continuity correction for discrete barrier options. *Mathematical Finance*, 7(4):325–349, 1997.

M. Broadie, P. Glasserman, and S. Kou. Connecting discrete and continuous path-dependent options. *Finance and Stochastics*, 3:55–82, 1999.

M. Broadie, D. M. Cicek, and A. Zeevi. General bounds and finite-time improvement for stochastic approximation algorithms. Columbia University working paper, 2009.

R. Brotherton-Ratcliffe. Monte Carlo motoring. *Risk*, 7(12):53–57, 1994.

R. Brotherton-Ratcliffe. The BGM model for path-dependent swaps. General Re Financial Products working paper, 1997.

G. Brunick. *A Weak Existence Result with Application to the Financial Engineer's Calibration Problem*. PhD thesis, Carnegie Mellon University, 2008.

G. Burghardt. *The Treasury Bond Basis*. McGraw-Hill, New York, 2005.

J. Campos, N. R. Ericsson, and D. F. Hendry, editors. *General-to-Specific Modelling*. International Library of Critical Writings in Econometrics. Edward Elgar Publishing Ltd, 2005.

L. Capriotti. Least squares importance sampling for Libor market models. *Wilmott*, 7:100–107, 2007.

P. Carr. Deriving derivatives of derivative securities. *Journal of Computational Finance*, 4(2):5–29, 2000.

P. Carr and R. Jarrow. The stop-loss start-gain paradox: A new decomposition into intrinsic and time value. *The Review of Financial Studies*, 3(3):469–492, 1990.

P. Carr and R. Lee. Put-call symmetry: Extensions and applications. *Mathematical Finance*, 19(4):523–560, 2009a.

P. Carr and R. Lee. Volatility derivatives. *Annual Review Financial Economics*, 1: 1–21, 2009b.

P. Carr and D. Madan. Option valuation using the fast Fourier transform. *Journal of Computational Finance*, 2:61–73, 1999.

P. Carr and L. Wu. What type of process underlies options? A simple robust test. *Journal of Finance*, 58:2581–2610, 2003.

P. Carr, R. Jarrow, and R. Myneni. Alternative characterizations of American puts. *Mathematical Finance*, 2:87–106, 1992.

J. Carrière. Valuation of early-exercise price of options using simulations and nonparametric regression. *Insurance: Mathematics and Economics*, 19(12): 19–30, 1996.

A. P. Carverhill. When is the short rate Markovian? *Mathematical Finance*, 4(4): 305–312, 1994.

A. P. Carverhill and L. J. Clewlow. American options: Theory and numerical analysis. In S. Hodges, editor, *Options: Recent Advances in Theory and Practice*. Manchester University Press, 1990.

J. Casassus, P. Collin-Dufresne, and B. Goldstein. Unspanned stochastic volatility and fixed income derivatives pricing. *Journal of Banking & Finance*, 29(11): 2723–2749, 2005.

E. Catmull and R. Rom. A class of local interpolating splines. In R. Barnhill and R. Reisenfeld, editors, *Computer Aided Geometric Design*, pages 317–326. Academic Press, 1974.

V. Cevher, R. Chellappa, and J. H. McClellan. Gaussian approximations for energy-based detection and localization in sensor networks. In *IEEE Statistical Signal Processing Workshop*, Madison, 2007. URL http://www.umiacs.umd.edu/users/volkan/SSP07.pdf.

K. C. Chan, G. A. Karolyi, F. A. Longstaff, and A. Sanders. An empirical comparison of alternative models of the short-term interest rate. *Journal of Finance*, 47:1209–1227, 1992.

D. Chapman and N. Pearson. Is the short rate drift actually nonlinear? *Journal of Finance*, 55:355–388, 2000.

L. Chen, D. Filipović, and H. V. Poor. Quadratic term structure models for risk-free and defaultible rates. *Mathematical Finance*, 14(4):515–536, 2004.

N. Chen and P. Glasserman. Additive and multiplicative duals for American option pricing. *Finance and Stochastics*, 11:153–179, 2007a.

N. Chen and P. Glasserman. Malliavin greeks without Malliavin calculus. *Stochastic Processes and their Applications*, 117(11):1689–1723, 2007b.

R.-R. Chen and L. Scott. Pricing interest rate options in a two-factor Cox-Ingersoll-Ross model of the term structure. *Review of Financial Studies*, 5:613–636, 1992.

R.-R. Chen and L. Scott. Stochastic volatility and jumps in interest rates: An empirical analysis. Rutgers University working paper, 2001.

Z. Chen and P. A. Forsyth. A semi-Lagrangian approach for natural gas storage valuation and optimal operation. *SIAM Journal of Scientific Computing*, 30 (1):339–368, 2007.

R. C. H. Cheng and G. M. Feast. Gamma variate generators with increased shape parameter range. *Communications of the ACM*, 23(7):389–395, 1980.

T. H. F. Cheuk and T. C. F. Vorst. Complex barrier options. *Journal of Derivatives*, 4:8–22, 1996.

O. Cheyette. Markov representation of the Heath-Jarrow-Morton model. BARRA working paper, 1991.

C. Chiarella, A. Kucera, and A. Ziogas. A survey of the integral representation of American option prices. University of Technology Sydney working paper, 2004.

B. J. Christensen, R. Poulsen, and M. Sørensen. Optimal inference in diffusion models of the short rate of interest. Centre for Analytical Finance Aarhus working paper, 2001.

L. Clewlow and A. P. Carverhill. On the simulation of contingent claims. *Journal of Derivatives*, 999:66–74, 1994.

L. Clewlow and C. Strickland. A note on parameter estimation in the two-factor Longstaff and Schwartz interest rate model. *Journal of Fixed Income*, 3:95–100, 1994.

P. Collin-Dufresne and B. Goldstein. Do credit spreads reflect stationary leverage ratios. *Journal of Finance*, 56:2177–2208, 2001.

P. Collin-Dufresne and R. Goldstein. Pricing swaptions within an affine framework. *Journal of Derivatives*, 10:9–26, 2002a.

P. Collin-Dufresne and R. S. Goldstein. Do bonds span the fixed income markets? Theory and evidence for unspanned stochastic volatility. *The Journal of Finance*, 57(4):1685–1730, 2002b.

G. Courtadon. The pricing of options on default-free bonds. *Journal of Financial and Quantitative Analysis*, 17:75–100, 1982.

T. M. Cover and J. A. Thomas. *Elements of Information Theory*. Wiley, 2006.

J. C. Cox. The constant elasticity of variance option pricing model. *Journal of Portfolio Management*, 22:15–17, 1996.

J. C. Cox, J. E. Ingersoll, and S. A. Ross. The relationship between forward prices and futures prices. *Journal of Financial Economics*, 9:321–346, 1981.

J. C. Cox, J. E. Ingersoll, and S. A. Ross. A theory of the term structure of interest rates. *Econometrica*, 53(2):385–407, 1985.

J. J. D. Craig and A. D. Sneyd. An alternating-direction implicit scheme for parabolic equations with mixed derivatives. *Computers and Mathematics with Applications*, 16(4):341–350, 1988.

P. Craven and G. Wahba. Smoothing noisy data with spline functions: Estimating the correct degree of smoothing by the method of generalized crossvalidation. *Numerische Matematik*, 31:377–403, 1979.

D. Davydov and V. Linetsky. Pricing and hedging path-dependent options under the CEV process. *Management Science*, 47(7):949–965, 2001.

F. Delbaen and W. Schachermayer. A general version of the fundamental theorem of asset pricing. *Mathematische Annalen*, 300:463–520, 1994.

N. Denson and M. S. Joshi. Vega control. *SSRN eLibrary*, 2009.

E. Derman and M. Kamal. When you cannot hedge continuously: The corrections of Black-Scholes. *Risk*, 12:82–85, 1999.

E. Derman and I. Kani. Riding on a smile. *Risk*, 2:32–39, 1994.

J. Dhaene and M. J. Goovaerts. Dependency of risks and stop-loss order. *Astin Bulletin*, 26(2):201–212, 1996.

Y. d'Halluin, P. A. Forsyth, and G. Labahn. A numerical PDE approach for pricing callable bonds. *Applied Mathematical Finance*, 8:49–77, 2001.

Y. d'Halluin, P. A. Forsyth, and K. R. Vetzal. Robust numerical methods for contingent claims under jump diffusion processes. *IMA Journal on Numerical Analysis*, 25:87–112, 2005.

C. G. Ding. Algorithm AS 275: Computing the non-central chi-squared distribution function. *Applied Statistics*, 41:478–482, 1992.

J. Dollard and C. Friedman. *Product Integration with Applications to Differential Equations*. Addison-Wesley, 1979.

L. U. Dothan. On the term structure of interest rates. *Journal of Financial Economics*, 6(1):59–69, 1978.

D. Duffie. *Dynamic Asset Pricing Theory*. Princeton University Press, 2001.

D. Duffie and P. Glynn. Efficient Monte Carlo simulation of security prices. *The Annals of Applied Probability*, 5(4):897–905, 1995.

D. Duffie and M. Huang. Swap rates and credit quality. *Journal of Finance*, 51: 921–949, 1996.

D. Duffie and R. Kan. A yield-factor model of interest rates. *Mathematical Finance*, 6(4):379–406, 1996.

D. Duffie, J. Pan, and K. Singleton. Transform analysis and asset pricing for affine jump-diffusions. *Econometrica*, 68:1343–1376, 2000.

D. Duffie, D. Filipovic, and W. Schachermayer. Affne processes and applications in Finance. *The Annals of Applied Probability*, 13:984–1053, 2003.

D. J. Duffy. Robust and accurate finite difference methods in option pricing: One factor models. DataSim Financial working paper, 2000.

D. Dufresne. The integrated square-root process. University of Montreal working paper, 2001.

B. Dupire. Pricing with a smile. *Risk*, 7(1), 1994.

B. Dupire. A unified theory of volatility. Banque Paribas working paper, 1997.

B. Dupire. Modelling volatility skews. ICBI Conference, Paris, 2006.

P. H. Dybvig. Bond and bond option pricing based on the current term structure. In M. A. H. Dempster and S. Pliska, editors, *Mathematics of Derivative Securities*. Cambridge University Press, 1997.

D. Egloff, M. Kohler, and N. Todorovic. A dynamic look-ahead Monte Carlo algorithm for pricing Bermudan options. *The Annals of Applied Probability*, 17(4):1138–1171, 2007.

E. Ekström and J. Tysk. Existence and uniqueness theory for the term structure equation. Uppsala University working paper, 2008.

S. N. Ethier and T. G. Kurtz. *Markov Processes: Characterization and Convergence*. Wiley, 1986.

I. Evers and F. Jamshidian. Replication of flexi-swaps. *Risk*, 18(3):67–70, 2005.

C.-O. Ewald, R. Poulsen, and K. Schenk-Hopp. Stochastic volatility: Risk minimization and model risk. University of Copenhagen working paper, 2007.

F. J. Fabozzi. *The Handbook of Mortgage-backed Securities*. Wiley, 1985.

F. J. Fabozzi. *Fixed Income Securities*. Wiley, 2nd edition, 2001.

F. J. Fabozzi and T. D. Fabozzi. *Bond Markets, Analysis and Strategies*. Prentice Hall, 1989.

F. J. Fabozzi and F. Modigliani. *Capital Markets: Institutions and Instruments*. Wiley, 1996.

D. Filipovic and J. Teichmann. On the geometry of the term structure of interest rates. In *Proceedings of The Royal Society of London. Series A. Mathematical, Physical and Engineering Sciences*, volume 460, pages 129–167, 2004.

G. S. Fishman. Sampling from the gamma distribution on a computer. *Communications of the ACM*, 19(7):407–409, 1976.

B. Flesaker. Arbitrage free pricing of interest rate futures and forward contracts. *The Journal of Futures Markets*, 13(1):77–91, 1993.

H. Föllmer and M. Schweizer. Hedging of contingent claims under incomplete information. In M. Davis and R. Elliott, editors, *Applied Stochastic Analysis*, pages 389–414. Gordon and Breach, 1990.

H. Fong and O. Vasicek. Fixed income volatility management. *The Journal of Portfolio Management*, 17:41–46, 1991.

P. Forsyth and K. Vetzal. Quadratic convergence of a penalty method for valuing American options. *SIAM Journal of Scientific Computing*, 23:2096–2123, 2002.

E. Fournié, J.-M. Lasry, J. Lebuchoux, P.-L. Lions, and N. Touzi. Application of Malliavin calculus to Monte Carlo methods in finance. *Finance and Stochastics*, 3:391–412, 1999.

S. Frankau, D. Spinellis, N. Nassuphis, and C. Burgard. Going functional on exotic trades. *J. Functional Programming*, 19(1):27–45, 2009.

C. P. Fries. Localized proxy simulation schemes for generic and robust Monte-Carlo Greeks. *SSRN eLibrary*, 2007.

C. P. Fries and M. S. Joshi. Partial proxy simulation schemes for generic and robust Monte-Carlo Greeks. *Journal of Computational Finance*, 11(3):79–106, 2008a.

C. P. Fries and M. S. Joshi. Conditional analytic Monte Carlo pricing scheme for auto-callable products. *SSRN eLibrary*, 2008b.

C. P. Fries and J. Kampen. Proxy simulation schemes for generic robust Monte-Carlo sensitivities, process oriented importance sampling and high accuracy drift approximation. *Journal of Computational Finance*, 10(2):97–128, 2006.

M. Fujii, Y. Shimada, and A. Takahashi. A note on construction of multiple swap curves with and without collateral. *SSRN eLibrary*, 2010.

S. Galluccio and C. Hunter. The co-initial swap market model. *SSRN eLibrary*, 2003.

S. Galluccio, Z. Huang, O. Scaillet, and J.-M. Ly. Theory and calibration of swap market models. *SSRN eLibrary*, 2005.

D. Gatarek. Constant maturity swaps, forward measure and LIBOR market model. *SSRN eLibrary*, 2003.

R. Gâteaux. Sur les fonctionnelles continues et les fonctionnelles analytiques. *Comptes rendus de l'academie des sciences (Paris)*, 157:325–327, 1913.

J. Gatheral. Lecture 2: Fitting the volatility skew. Case Studies in Financial Modelling course notes, Courant Institute, 2001.

J. Gatheral. A parsimonious arbitrage-free implied volatility parameterization with application to the valuation of volatility derivatives. ICBI Conference, Madrid, 2004.

J. Gatheral. *The Volatility Surface: A Practitioner's Guide*. Wiley, 2006.

J. Gatheral and A. Jacquier. Convergence of Heston to SVI. *SSRN eLibrary*, 2010.

J. Gatheral, E. P. Hsu, P. M. Laurence, C. Ouyang, and T.-H. Wang. Asymptotics of implied volatility in local volatility models. *SSRN eLibrary*, 2009.

H. Geman, N. Karou, and J. Rochet. Changes of numeraire, changes of probability measure and option pricing. *Journal of Applied Probability*, 32(2):443–458, 1995.

M. R. Gibbons and K. Ramaswamy. A test of the Cox, Ingersoll, and Ross model of the term structure. *Review of Financial Studies*, 6:619–658, 1993.

M. Giles and R. Carter. Convergence analysis of Rannacher time marching. *Journal of Computational Finance*, 9:89–112, 2006.

M. Giles and P. Glasserman. Smoking adjoints: Fast Monte Carlo Greeks. *Risk*, 19(1):89–112, 2006.

P. Glasserman. *Monte Carlo Methods in Financial Engineering*. Springer-Verlag, New York, 2004.

P. Glasserman and K.-K. Kim. Gamma expansion of the Heston stochastic volatility model. Columbia Business School working paper, 2008.

P. Glasserman and N. Merener. Cap and swaption approximations in LIBOR market models with jumps. *SSRN eLibrary*, 2001.

P. Glasserman and J. Staum. Conditioning on one-step survival in barrier option simulations. *Operations Research*, 49:923–937, 2001.

P. Glasserman and B. Yu. Simulation for American options: Regression now or regression later? In H. Niederreiter, editor, *Monte Carlo and Quasi-Monte Carlo Methods 2002*, pages 213–226, Berlin, 2002. Springer-Verlag.

P. Glasserman and B. Yu. Number of paths versus number of basis functions in American option pricing. *Annals of Applied Probability*, 14(4):2090–2119, 2004.

P. Glasserman and B. Yu. Large sample properties of weighted Monte Carlo estimators. *Operations Research*, 53:298–312, 2005.

P. Glasserman and X. Zhao. Fast greeks in forward Libor models. *Journal of Computational Finance*, 3:5–39, 1999.

P. Glasserman and X. Zhao. Arbitrage-free discretization of lognormal forward Libor and swap rate models. *Finance and Stochastics*, 4:35–68, 2000.

P. Glasserman, P. Heidelberger, and P. Shahabuddin. Asymptotically optimal importance sampling and stratification for pricing path dependent options. *Journal of Mathematical Finance*, 9(2):117–152, 1999.

P. W. Glynn and W. Whitt. Indirect estimation via L = w. *Operations Research*, 37(1):82–103, 1989.

R. Goldstein and W. P. Keirstead. On the term structure of interest rates in the presence of reflecting and absorbing boundaries. Ohio State University working paper, 1997.

G. H. Golub and C. F. van Loan. *Matrix Computations*. The John Hopkins University Press, 1989.

V. Gorovoi and V. Linetsky. Black's model of interest rates as options, eigenfunction expansions and Japanese interest rates. *Mathematical Finance*, 14:49–78, 2004.

J. Gregory. *Counterparty Credit Risk: The New Challenge for Global Financial Markets*. Wiley, 2009.

I. Gyöngy. Mimicking the one-dimensional distributions of processes having an Ito differential. *Probability Theory and Related Fields*, 71:501 – 516, 1986.

P. S. Hagan. Adjusters: Turning good prices into great prices. *Wilmott*, 2:56–59, 2002.

P. S. Hagan and G. West. Interpolation methods for yield curve construction. Working paper, 2004.

P. S. Hagan and D. E. Woodward. Markov interest rate models. *Applied Mathematical Finance*, 6(4):233–260, 1999a.

P. S. Hagan and D. E. Woodward. Equivalent Black volatilities. *Applied Mathematical Finance*, 6:147–157, 1999b.

P. S. Hagan, D. Kumar, A. S. Lesniewski, and D. E. Woodward. Managing smile risk. *Wilmott*, 11:84–108, 2002.

J. M. Hammersley and D. C. Handscomb. *Monte Carlo Methods*. Methuen & Co., 1965.

A. T. Hansen and P. L. Jørgensen. Exact analytical valuation of bonds when spot interest rates are log-normal. *SSRN eLibrary*, 1998.

P. Hansen. Analysis of discrete ill-posed problems by means of the l-curve. *SIAM Review*, 34:561–580, 1992.

M. Harrison and D. Kreps. Martingales and arbitrage in multiperiod securities markets. *Journal of Economic Theory*, 20:381–408, 1979.

M. Haugh and L. Kogan. Pricing American options: A duality approach. *Operations Research*, 52:258–270, 2004.

H. He, W. Keirstead, and J. Rebholz. Double lookbacks. *Mathematical Finance*, 8:201–228, 1998.

D. Heath, R. Jarrow, and A. Morton. Bond pricing and the term structure of interest rates: A new methodology for contingent claims valuation. *Econometrica*, 60(1):77–105, 1992.

V. Henderson and D. Hobson. Local time, coupling, and the passport option. *Finance and Stochastics*, 4:69–80, 2000.

P. Henry-Labordère. A general asymptotic implied volatility for stochastic volatility models. ArXiv working paper, 2005.

P. Henry-Labordère. *Analysis, Geometry and Modeling in Finance*. Chapman and Hall/CRC, 2008.

P. Henry-Labordère. Calibration of local stochastic volatility models to market smiles. *Risk*, 9:112–117, 2009.

S. Heston, M. Loewenstein, and G. Willard. Options and bubbles. *Review of Financial Studies*, 20:359–390, 2007.

S. L. Heston. A closed-form solution for options with stochastic volatility with applications to bond and currency options. *Review of Financial Studies*, 6:327–343, 1993.

N. Higham. Computing the nearest correlation matrix - a problem from finance. *IMA Journal of Numerical Analysis*, 22:329–343, 2002.

T. Ho and S. Lee. Term structure movements and pricing interest rate contingent claims. *Journal of Finance*, 41:1011–1029, 1986.

M. Hogan and J. Weintraub. The lognormal interest rate model and Eurodollar futures. Citibank working paper, 1993.

P. Honore. Maturity induced bias in estimating spot-rate diffusion models. Aarhus School of Business working paper, 1998a.

P. Honore. *Five Essays on Financial Econometrics in Continuous-Time Models*. PhD thesis, Aarhus School of Business, 1998b.

Z. Hu, J. Kerkhof, P. McCloud, and J. Wackertap. Cutting edges using domain integration. *Risk*, 19(11):95–99, 2006.

J. C. Hull. *Options, futures and other derivatives*. Prentice Hall, 2006.

J. C. Hull and A. White. The pricing of options on assets with stochastic volatilities. *Journal of Finance*, 42:281–300, 1987.

J. C. Hull and A. White. Numerical procedures for implementing term structure models I: Single-factor models. *Journal of Derivatives*, 2:7–16, 1994a.

J. C. Hull and A. White. Numerical procedures for implementing term structure models II: Two-factor models. *Journal of Derivatives*, 2:37–48, 1994b.

P. Hunt and J. Kennedy. *Financial Derivatives in Theory and Practice*. Wiley, 2000.

C. Hunter, P. Jäckel, and M. S. Joshi. Drift approximations in a forward-rate based Libor market model. Quarc working paper, 2001.

T. Hyer. *Derivatives Algorithms*. World Scientific, 2010.

J. E. Ingersoll. Valuing foreign exchange rate derivatives with a bounded exchange process. *Review of Derivatives Research*, 1:159–181, 1997.

ISDA. 2005 ISDA collateral guidelines. Technical report, International Swaps and Derivatives Association, Inc., 2005. http://www.isda.org/ publications/ pdf/ 2005isdacollateralguidelines.pdf.

ISDA. ISDA margin survey 2009. Technical report, International Swaps and Derivatives Association, Inc., 2009. http://www.isda.org/ c_and_a/ pdf/ ISDA-Margin-Survey-2009.pdf.

S. D. Jacka. Optimal stopping and the American put. *Mathematical Finance*, 1: 1–14, 1991.

P. Jäckel. *Monte Carlo Methods in Finance*. Wiley, Chichester, U.K., 2002.

P. Jäckel. Weighted sampling for variance reduction. OTC Analytics working paper, 2004.

F. Jamshidian. An exact bond option pricing formula. *Journal of Finance*, 44: 205–209, 1989.

F. Jamshidian. Forward induction and construction of yield curve diffusion models. *Journal of Fixed Income*, 1:62–74, 1991a.

F. Jamshidian. Bond and option evaluation in the Gaussian interest rate model. *Research in Finance*, 9:131–170, 1991b.

F. Jamshidian. An analysis of American options. *Review of Futures Markets*, 11: 72–80, 1992.

F. Jamshidian. Option and futures evaluation with deterministic volatilities. *Mathematical Finance*, 3(2):149–159, 1993.

F. Jamshidian. The duality of optimal exercise and domineering claims: A Doob-Meyer decomposition approach to the Snell envelope. *SSRN eLibrary*, 1995.

F. Jamshidian. Libor and swap market models and measures. *Finance and Stochastics*, 1(4):293–330, 1997.

C. Jeffery. Reverse cliquets: End of the road? *Risk*, 17(2):20–22, 2004.

M. Jensen and M. Svenstrup. Efficient control variates and strategies for Bermudan swaptions in a Libor market model. *SSRN eLibrary*, 2003.

N. L. Johnson, S. Kotz, and N. Balakrishnan. *Continuous Univariate Distributions, Volume 2*. Wiley, 1995.

S. P. Jones, J.-M. Eber, and J. Seward. Composing contracts: An adventure in financial engineering. ICFP, 2000. URL http:// www.lexifi.com/ downloads/ MLFiPaper.pdf.

F. D. Jong, J. Driessen, and A. Pelsser. Libor market models versus swap market models for pricing interest rate derivatives: An empirical analysis. *European Finance Review*, 5(3):201–237, 2001.

M. S. Joshi. *C++ Design Patterns and Derivatives Pricing*. Cambridge University Press, 2004.

M. S. Joshi and A. Stacey. New and robust drift approximations for the Libor market model. *Quantitative Finance*, 8(4):427–434, 2008.

C. Joy, P. P. Boyle, and K. S. Tan. Quasi-Monte Carlo methods in numerical finance. *Management Science*, 42:926–938, 1996.

N. Ju. Pricing an American option by approximating its early exercise boundary as a multi-piece exponential function. *Review of Financial Studies*, 11:627–646, 1998.

S. Juneja and H. Kalra. Variance reduction techniques for pricing American options using function approximations. *Journal of Computational Finance*, 12 (3):79–102, 2009.

Y. M. Kabanov and M. M. Safarian. On Lelands strategy of option pricing with transaction costs. *Finance and Stochastics*, 1:239–250, 1997.

C. Kahl and P. Jäckel. Not-so-complex logarithms in the Heston model. *Wilmott*, 19(9), 2005.

C. Kahl and P. Jäckel. Fast strong approximation Monte-Carlo schemes for stochastic volatility. *Quantitative Finance*, 6:513–536, 2006.

D. Kainth and N. Saravanamuttu. Multifactor stochastic volatility models. ICBI Conference, Paris, 2007.

R. Kangro and R. Nicolaides. Far field boundary conditions for Black-Scholes equations. *SIAM Journal on Numerical Analysis*, 38:1357–1368, 2000.

I. Karatzas and S. E. Shreve. *Brownian Motion and Stochastic Calculus*. Springer-Verlag, 2nd edition, 1991.

S. Karlin and H. Taylor. *A Second Course in Stochastic Processes*. Academic Press, 1981.

J. Kennedy, P. Hunt, and A. Pelsser. Markov-functional interest rate models. *Finance and Stochastics*, 4:391–408, 2000.

F. Kilin. Accelerating the calibration of stochastic volatility models. Frankfurt School of Finance & Management working paper, 2007.

I. J. Kim. The analytical valuation of American options. *The Review of Financial Studies*, 3:547–572, 1990.

P. E. Kloeden and E. Platen. *Numerical Solution of Stochastic Differential Equations (Stochastic Modelling and Applied Probability)*. Springer-Verlag, 2000.

A. Kolodko and J. Schoenmakers. Iterative construction of the optimal Bermudan stopping time. *Finance and Stochastics*, 10:27–49, 2006.

S. Kotz, N. Balakrishnan, and N. L. Johnson. *Continuous Multivariate Distributions. Volume 1: Models and Applications*. Wiley, 2000.

J. F. B. M. Kraaijevanger, H. W. J. Lenferink, and M. N. Spijker. Stepsize restrictions for stability in the numerical solution of ordinary and partial differential equations. *Journal of Computational and Applied Mathematics*, 20: 67–81, 1987.

M. Kramin. A multi-factor Markovian HJM model for pricing exotic interest rate derivatives. ICBI Conference, Paris, 2008.

H. Kreiss, V. Thomee, and O. Widlund. Smoothing of initial data and rates of convergence for parabolic difference equations. *Communications on Pure and Applied Mathematics*, 23:241–259, 1970.

H. Kunita. *Stochastic Flows and Stochastic Differential Equations*. Cambridge University Press, 1990.

B. Kvasov. *Methods of Shape-Preserving Spline Approximation*. World Scientific, 2000.

P. K. Kythe and M. R. Schäferkotter. *Handbook of Computational Methods for Integration*. Routledge, USA, 2004.

M. Leclerc, Q. Liang, and I. Schneider. Fast Monte Carlo Bermudan greeks. *Risk*, 22(7):84–88, 2009.

P. L'Ecuyer. Uniform random number generation. *Annals of Operations Research*, 53:77–120, 1994.

R. W. Lee. Implied and local volatilities under stochastic volatility. *International Journal of Theoretical and Applied Finance*, 4(1):45–89, 2001.

R. W. Lee. Option pricing by transform methods: Extensions, unification, and error control. *Journal of Computational Finance*, 7(3):51–86, 2004.

R. W. Lee. Implied volatility: Statics, dynamics, and probabilistic interpretation. In R. Baeza-Yates, J. Glaz, H. Gzyl, J. Husler, and J. L. Palacios, editors, *Recent Advances in Applied Probability*, Berlin, 2005. Springer-Verlag.

M. Leippold and L. Wu. Asset pricing under the quadratic class. *Journal of Financial and Quantitative Analysis*, 37:271–295, 2002.

H. Leland. Option pricing and replication with transaction costs. *Journal of Finance*, 40(5):1283–1301, 1985.

H. W. Lenferink and M. N. Spijker. On the use of stability regions in the numerical analysis of initial value problems. *Mathematics of Computation*, 57:221–237, 1991.

A. L. Lewis. *Option Valuation under Stochastic Volatility: with Mathematica Code*. Finance Press, 2000.

A. L. Lewis. A simple option formula for general jump-diffusion and other exponential Levy processes. *SSRN eLibrary*, 2001.

Q. Li and H. Qi. A sequential semismooth Newton method for the nearest low-rank correlation matrix problem. Working paper, 2009.

E. Liebscher. Construction of asymmetric multivariate copulas. *Journal of Multivariate Analysis*, 99(10):2234–2250, 2008.

A. Lipton. *Mathematical Methods For Foreign Exchange: A Financial Engineer's Approach*. World Scientific, 2001.

A. Lipton. The vol smile problem. *Risk*, 15(2):61–65, 2002.

F. A. Longstaff and E. S. Schwartz. Interest rate volatility and the term structure: A two-factor general equilibrium model. *The Journal of Finance*, 4:1259–1282, 1992.

F. A. Longstaff and E. S. Schwartz. Implementation of the Longstaff-Schwartz interest rate model. *Journal of Fixed Income*, 3:7–14, 1993.

F. A. Longstaff and E. S. Schwartz. Valuing American options by simulation: A simple least-squares approach. *The Review of Financial Studies*, 14(1):113–147, 2001.

R. Lord and C. Kahl. Optimal Fourier inversion in semi-analytical option pricing. *SSRN eLibrary*, 2007.

R. Lord, R. Koekkoek, and D. van Dijk. A comparison of biased simulation schemes for stochastic volatility models. Tinbergen Institute working paper, 2006.

V. Lucic. On singularities in the Heston model. *SSRN eLibrary*, 2007.

V. Lucic. Boundary conditions for computing densities in hybrid models via PDE methods. *SSRN eLibrary*, 2008.

R. W. Lynch. A method for choosing a tension factor for spline under tension interpolation. University of Texas working paper, 1982.

W. Margrabe. The value of an option to exchange one asset for another. *Journal of Finance*, 33:177–186, 1978.

M. Matsumoto and T. Nishimura. Mersenne twister: A 623-dimensionally equidistributed uniform pseudo-random number generator. *ACM Transactions on Modeling and Computer Simulation*, 8:3–30, 1998.

J. Mayle. *Standard Securities Calculation Methods: Fixed Income Securities Formulas for Price, Yield and Accrued Interest*. SIFMA, 1993.

R. Merton. The theory of rational option pricing. *Bell Journal of Economics and Management Science*, 4:141–183, 1973.

S. Meyers. *Effective C++: 55 Specific Ways to Improve Your Programs and Designs*. Addison-Wesley, 2005.

K. Miltersen, K. Sandmann, and D. Sondermann. Closed form solutions for term structure derivatives with lognormal interest rates. *Journal of Finance*, 52(1): 409–430, 1997.

P. Miron and P. Swannell. *Pricing and hedging swaps*. Euromoney Institutional Investor PLC, 1991.

A. Mitchell and D. Griffiths. *The Finite Difference Method in Partial Differential Equations*. Wiley, New York, 1980.

C. Moni. Fast American Monte Carlo. *SSRN eLibrary*, 2005.

B. Moro. The full monte. *Risk*, 8(2):57–58, 1995.

V. Morozov. On the solution of functional equations by the method of regularization. *Soviet Mathematics Doklady*, 7:414–417, 1966.

A. Morton. Arbitrage and martingales. Working paper, 1988.

B. Moskowitz and R. Caflisch. General framework for pricing derivative securities. *Mathematical and Computer Modeling*, 23:37–54, 1996.

M. Musiela and M. Rutkowski. *Martingale Methods in Financial Modeling*. Applications of Mathematics. Springer-Verlag, 1997.

R. B. Nelsen. *An Introduction to Copulas (Lecture Notes in Statistics)*. Springer-Verlag, 2nd edition, 2006.

B. L. Nelson. Control variate remedies. *Operations Research*, 38(6):974–992, 1990.

C. R. Nelson and A. F. Siegel. Parsimonious modeling of yield curves. *Journal of Business*, 60:473–489, 1987.

I. E. Nesterov, A. Nemirovskii, and Y. Nesterov. *Interior-Point Polynomial Algorithms in Convex Programming*, volume 13 of *Siam Studies in Applied Mathematics*. Society for Industrial & Applied Mathematics, 1994.

J. Obloj. Fine-tune your smile. Imperial College London working paper, 2008.

M. K. Ochi. *Applied Probability and Stochastic Processes*. Wiley, 1990.

B. Øksendal. *Stochastic differential equations: an introduction with applications*. Springer-Verlag, New York, 1992.

Y. Osajima. General asymptotics of Wiener functionals and application to mathematical finance. Mitsubishi UFJ Securities working paper, 2007.

S. Paskov and J. Traub. Faster valuation of financial derivatives. *Journal of Portfolio Management*, 22:113–120, 1995.

M. B. Pedersen. Calibrating Libor market models. *SSRN eLibrary*, 1998.

M. B. Pedersen. Bermuda swaptions in the LIBOR market model. *SSRN eLibrary*, 1999.

J. Perello, J. Masoliver, and J.-P. Bouchaud. Multiple time scales in volatility and leverage correlations: A stochastic volatility model. *Applied Mathematical Finance*, 11:1–24, 2004.

R. Pietersz. Importance sampling in market models for targeted accrual redemption notes (TARNs). ABN AMRO working paper, 2005.

R. Pietersz and P. Groenen. Rank reduction of correlation matrices by majorization. *Quantitative Finance*, 4:649–662, 2004.

R. Pietersz and A. Pelsser. Swap vega in BGM: Pitfalls and alternatives. *Risk*, 17(3):91–93, 2004.

R. Pietersz and M. van Regenmortel. Importance sampling for stable Greeks in the LIBOR market model for targeted accrual redemption notes (TARNs). 3d WBS Fixed Income Conference, Amsterdam, 2006.

R. Pietersz, A. Pelsser, and M. van Regenmortel. Fast drift approximated pricing in the BGM model. *Journal of Computational Finance*, 8(1):93–124, 2004.

V. V. Piterbarg. A note on pricing weakly-path-dependent American-style options by backward induction. *SSRN eLibrary*, 2002.

V. V. Piterbarg. A Practitioner's guide to pricing and hedging callable LIBOR exotics in forward LIBOR models. *SSRN eLibrary*, 2003.

V. V. Piterbarg. Risk sensitivities of Bermuda swaptions. *International Journal of Theoretical and Applied Finance*, 7(4):465–510, 2004a.

V. V. Piterbarg. Computing deltas of callable Libor exotics in forward Libor models. *Journal of Computational Finance*, 7(3):107–144, 2004b.

V. V. Piterbarg. TARNs: Models, valuation, risk sensitivities. *Wilmott*, 14(11): 62–71, 2004c.

V. V. Piterbarg. Pricing and hedging callable Libor exotics in forward Libor models. *Journal of Computational Finance*, 8(2):65–119, 2005a.

V. V. Piterbarg. Stochastic volatility model with time-dependent skew. *Applied Mathematical Finance*, 12(2):147–185, 2005b.

V. V. Piterbarg. Time to smile. *Risk*, 18(5):71–75, 2005c.

V. V. Piterbarg. Smiling hybrids. *Risk*, 19(5):66–71, 2006.

V. V. Piterbarg. Practical multi-factor quadratic Gaussian models of interest rates. 5th WBS Fixed Income Conference, Budapest, 2008.

V. V. Piterbarg. Rates squared. *Risk*, 22(1):100–105, 2009a.

V. V. Piterbarg. Quadratic Gaussian models for CMS spread options. ICBI Conference, Rome, 2009b.

V. V. Piterbarg. Funding beyond discounting: Collateral agreements and derivatives pricing. *Risk*, 2:97–102, 2010.

V. V. Piterbarg and L. B. Andersen. Bermudan swaptions and callable Libor exotics. In R. Cont, editor, *Encyclopedia of Quantitative Finance*, pages 177–181. Wiley, 2010a.

V. V. Piterbarg and L. B. Andersen. Libor market model. In R. Cont, editor, *Encyclopedia of Quantitative Finance*, pages 1031–1036. Wiley, 2010b.

V. V. Piterbarg and M. A. Renedo. Eurodollar futures convexity adjustments in stochastic volatility models. *Journal of Computational Finance*, 9(3):71–94, 2006.

D. M. Pooley, K. R. Vetzal, and P. A. Forsyth. Convergence remedies for non-smooth payoffs in option pricing. *Journal of Computational Finance*, 6(4):25 – 40, 2003.

W. H. Press, B. P. Flannery, S. A. Teukolsky, and W. T. Vetterling. *Numerical Recipes in C: The Art of Scientific Computing*. Cambridge University Press, 1992.

P. E. Protter. *Stochastic Integration and Differential Equations*. Springer-Verlag, 2005.

S. Pruess. Properties of splines in tension. *Journal of Approximation Theory*, 17: 86–96, 1976.

R. Rannacher. Finite element solution of diffusion problems with irregular data. *Numerische Mathematik*, 43(2):309–327, 1984.

N. S. Rasmussen. Control variates for Monte Carlo valuation of American options. *Journal of Computational Finance*, 9(1):83–118, 2005.
R. Rebonato. *Interest-Rate Option Models*. Wiley, 1998.
R. Rebonato. *Modern pricing of interest rate derivatives: The Libor market model and beyond*. Princeton University Press, 2002.
Y. Ren, D. Madan, and M. Q. Qian. Calibrating and pricing with embedded local volatility models. *Risk*, 9:138–143, 2007.
R. Rendleman and B. Bartter. The pricing of options on debt securities. *Journal of Financial and Quantitative Analysis*, 15:11–24, 1980.
R. J. Renka. Interpolatory tension splines with automatic selection of tension factors. *SIAM Journal of Scientific and Statistical Computing*, 8(3):393–415, 1987.
P. Rentrop. An algorithm for the computation of exponential splines. *Numerische Matematik*, 35:81–93, 1980.
D. Revuz and M. Yor. *Continuous Martingales and Brownian Motion*. Springer-Verlag, 3d edition, 1999.
P. Ritchken and L. Sankarasubramanian. Volatility structure of forward rates and the dynamics of the term structure. *Mathematical Finance*, 5:55–72, 1995.
L. C. G. Rogers. Which model for term-structure of interest rates should one use? In *Proceedings of IMA Workshop on Mathematical Finance, IMA Vol 65*, pages 93–116, New York, 1995. Springer-Verlag.
L. C. G. Rogers. Gaussian errors. *Risk*, 9(1):42–45, 1996.
L. C. G. Rogers. Monte Carlo valuation of American options. University of Bath working paper, 2001.
L. C. G. Rogers and Z. Shi. The value of an Asian option. *Journal of Applied Probability*, 32:1077–1088, 1995.
M. Romano and N. Touzi. Contingent claims and market completeness in a stochastic volatility model. *Mathematical Finance*, 7(4):399–410, 1997.
C. Rouvinez. Going Greek with VaR. *Risk*, 10:57–65, 1997.
Y. Saad. *Iterative methods for sparse linear systems*. SIAM, 2003.
K. Sandmann and D. Sondermann. A note on the stability of lognormal interest rate models and the pricing of eurodollar futures. *Mathematical Finance*, 7(2): 119–125, 1997.
E. Schlögl. Arbitrage-free interpolation in models of market observable interest rates. In K. Sandmann and P. Schnbucher, editors, *Advances in Finance and Stochastics: Essays in Honour of Dieter Sondermann*, pages 197–218. Springer-Verlag, 2002.
J. Schoenmakers and B. Coffey. Stable implied calibration of a multi-factor Libor model via a semi-parametric correlation structure. WIAS Preprint No. 611, 2000.
J. Schoenmakers and A. Heemink. Fast valuation of financial derivatives. *Journal of Computational Finance*, 1:47–62, 1997.
M. Schroder. Computing the Constant Elasticity of Variance option pricing formula. *J. Finance*, 44:211–219, 1989.
D. G. Schweikert. An interpolating curve using a spline in tension. *Journal of Mathematics and Physics*, 45:312–317, 1966.
M. Selby and C. Strickland. Computing the Fong and Vasicek pure discount bond price formula. *Journal of Fixed Income*, 5:78–85, 1995.

C. E. Shannon. Communication in the presence of noise. *Proceedings of Institute of Radio Engineers*, 37:10–21, 1949.

W. F. Sharpe. Capital asset prices: A theory of market equilibrium under conditions of risk. *Journal of Finance*, 19:425–442, 1964.

G. S. Shea. Pitfalls in smoothing interest rate terms structure data: Equilibrium models and spline approximation. *Journal of Financial and Quantitative Analysis*, 19:253–269, 1984.

J. Sidenius. Libor market model in practice. *Journal of Computational Finance*, 3(3):75–99, 2000.

C. Sin. Complications with stochastic volatility models. *Advances in Applied Probability*, 30:256–268, 1998.

R. Smith. An almost exact simulation method for the Heston model. *Journal of Computational Finance*, 11:115–125, 2007.

H. Soner, S. Shreve, and J. Cvitanic. There is no nontrivial hedging portfolio for option pricing with transaction costs. *Annals of Applied Probability*, 5:327–355, 1995.

M. N. Spijker and F. A. J. Straetemans. Error growth analysis via stability regions for discretizations of initial value problems. *BIT*, 37:442–464, 1997.

M. L. Stigum and F. L. Robinson. *Fixed Income Calculations: Money Market Paper and Bonds*. Irwin Professional Publishing, 1996.

G. Stoyan. Monotone difference schemes for diffusion-convection problems. *ZAMM*, 59:361–372, 1979.

Q. Su and C. Randall. General market greeks in practice. *Wilmott*, 8, 2008.

Y. Su and M. C. Fu. Optimal importance sampling in securities pricing. *Journal of Computational Finance*, 5:27–50, 2002.

H. Sutter and A. Alexandrescu. *C++ Coding Standards: 101 Rules, Guidelines, and Best Practices*. Addison-Wesley, 2004.

K. Sydsaeter and P. Hammond. *Essential Mathematics for Economic Analysis*. Prentice Hall, 2008.

D. Talay. Efficient numerical schemes for the approximation of expectations of functionals of an SDE, and applications. In *Lecture Notes in Control and Information Sciences, Vol. 61*, pages 294–313. Springer-Verlag, 1984.

D. Talay and L. Tubaro. Romberg extrapolations for numerical schemes solving stochastic differential equations. *Structural Safety*, 8, 1990.

N. Taleb. *Dynamic Hedging*. Wiley, 1997.

C. Tanggaard. Nonparametric smoothing of yield curves. *Review of Quantitative Finance and Accounting*, 9:251–267, 1997.

D. Tavella and C. Randall. *Pricing Financial Instruments – The Finite Difference Method*. Wiley, 2000.

C. Tezier. Short rate models. Linear and quadratic Gaussian models. Barclays Capital working paper, 2005.

H. Theil. *Principles of Econometrics*. Wiley, Amsterdam, 1971.

S. Traven. Pricing linear derivatives with a single discounting curve. Barclays Capital working paper, 2008.

J. N. Tsitsiklis and B. V. Roy. Regression methods for pricing complex American-style options. *IEEE Transactions on Neural Networks*, 12(4):694–703, 2001.

R. J. Van Steenkiste. Term structure model perturbation. Barclays Capital working paper, 2009.

O. Vasicek. An equilibrium characterization of the term structure. *Journal of Financial Economics*, 5:177–188, 1977.

M. Wichura. Algorithm AS 241: The percentage points of the normal distribution. *Applied Statistics*, 37:477–484, 1988.

P. Wilmott, J. Dewynne, and J. Howison. *Option Pricing: Mathematical Models and Computation*. Oxford Financial Press, 1993.

H. Windcliff, P. A. Forsyth, and K. R. Vetzal. Shout options: A framework for pricing contracts which can be modified by the investor. *Journal of Computational and Applied Mathematics*, 134:213–241, 2001.

L. Withington and L. Lucic. Noisy hedges and fuzzy payoffs: Using soft computing to improve risk stability. RBC Capital Markets working paper, 2009.

F. Wu, E. A. Valdez, and M. Sherris. Simulating exchangeable multivariate Archimedean copulas and its applications. UNSW working paper, 2006.

N. Yoshida. Asymptotic expansion for small diffusions via the theory of Malliavin-Watanabe. *Probability Theory and Related Fields*, 92:275–311, 1992.

L. Zadeh. Fuzzy sets. *Information and Control*, 8:338–353, 1965.

Z. Zhang and L. Wu. Optimal low-rank approximation to a correlation matrix. *Linear Algebra and its Applications*, 364:161–187, 2003.

R. Zvan, P. A. Forsyth, and K. R. Vetzal. Robust numerical methods for PDE models of Asian options. *Journal of Computational Finance*, 1(2):39–78, 1998.

R. Zvan, P. A. Forsyth, and K. R. Vetzal. Discrete Asian barrier options. *Journal of Computational Finance*, 3(1):41–67, 1999.

Index

absorbing boundary, *see* diffusion, absorbing barrier
accrual factor, *see* year fraction
ADI, *see* PDE, ADI scheme
adjusters method, *see* out-of-model adjustment, adjusters method
affine short rate model, 431–444, 512–520
  bond reconstitution formula, 433, 515–517
  calibration, 441–443
    multi-pass bootstrap, 442
  calibration to yield curve, 437–439
  characteristic function, 434
  European swaption, 439
    Fourier integration, 439
    Gram-Charlier expansion, 439
  extended transform, 433
    constant parameters, 434, 436
    piecewise constant parameters, 436
  Feller condition, 319, 432
  importance sampling, 1065
  moment-generating function, 434, 439
  Monte Carlo, 444
  multi-factor, 512–520
    bond dynamics, 516
    bond reconstitution formula, 515–517
    existence and uniqueness, 514
    exponential affine, 513
    Feller condition, 515
    forward rate correlation, 516
    forward rate dynamics, 516
    regularity issues, 514–515
    short rate state dynamics, 513
  one-factor, 431–444
  PDE, 444
  regularity issues, 432
  short rate domain, 432
  short rate dynamics, 431
  short rate state dynamics, 437
  swap rate volatility, 440
    affine approximation, 440
    time averaging, 440
  time-dependent, 433
  volatility skew range, 433
  volatility smile, 432
almost surely, 4
American capped straddle, 936
American swaption, 895–899
  accrued current coupon, 895
  approximating with Bermudan swaption, *see* Bermudan swaption, approximating American swaption
  discontinuity of exercise value in time, 895
  PDE, 897–899
    extra state variable, 898–899
    proxy Libor rate method, 897–898
American/Bermudan option, 30–41
  Bellman principle, 32, 33, 69
  Black-Scholes model, 839
  capped, 936
  conditional on no exercise, 31

xxii Index

continuation region, 33
discontinuity at expiry, 39
duality, 35
early exercise boundary, 37
early exercise premium, 36, 39, 41
exercise never optimal, 36
exercise policy, 30
exercise region, 33
exercise value, 30
high contact condition, 37
hold value, 32
integral representation, 39, 40
lower bound, *see* Monte Carlo, lower bound for American option
marginal exercise value decomposition, 41
Monte Carlo, 158–165
  confidence interval for value, 164
  random tree, 164
  stochastic mesh, 165
PDE jump condition, 34
perfect foresight bias, 160
short-maturity asymptotics, 38
smooth pasting condition, 37
supermartingale, 31
upper bound, *see* Monte Carlo, upper bound for American option
annuity mapping function, *see* terminal swap rate model, annuity mapping function
annuity measure, *see* measure, annuity
arbitrage opportunity, 8
arbitrage pricing, 11
arithmetic put-call symmetry, 940
Arrow-Debreu security, 21, 76, 77, 79, 458, 462, 1048
  backward Kolmogorov equation, 458
  forward Kolmogorov equation, 458
art of derivatives trading, 980
Asian option, 70
  Black model, 922
  Monte Carlo, *see* Monte Carlo, Asian option
  PDE, *see* PDE, Asian option
ATM backbone, *see* volatility smile, ATM backbone
autocorrelation, *see* inter-temporal correlation

averaging, *see* calibration, time averaging
averaging cash flow, 201, 722–723
  convexity adjustment, 722
averaging swap, *see* averaging cash flow

Bachelier model, *see* Normal model
backbone, *see* volatility smile, backbone
backward Kolmogorov equation, *see* Kolmogorov backward equation
balance-guarantee swap, 900
band swap, *see* flexi-swap
"bang-bang", 902
barrier option, 44
  Broadie adjustment for sampling frequency, *see* Monte Carlo, sampling extremes, adjusting barrier for sampling frequency
  continuous barrier, 64
  discrete barrier, 66
  importance sampling, 1074–1077
  Markovian projection, *see* Markovian projection, barrier option
  Monte Carlo, *see* Monte Carlo, barrier option
  on capped straddle, 937
  one-touch, 939
  pathwise differentiation method, 1041–1044
    recursion, 1043
  payoff smoothing, *see* payoff smoothing, barrier option
  PDE jump condition, 66
  rebate, 64
  semi-static replication, 939
  step-down, 64
  step-up, 64
  tube Monte Carlo, 1024
  up-and-out, 44, 64, 66, 124, 126, 1134
basis point, 169
basis risk, *see* yield curve, basis risk
basket option, 205, 1146
  Black model, 924
  displaced log-normal approximation, 1147
  local volatility model, 1145
  Monte Carlo, *see* Monte Carlo, Asian option on basket
  slope of volatility smile, 1148

Index   xxiii

stochastic volatility model, 1149
BDT model, *see* Black-Derman-Toy model
Bermudan cancelable swap, *see* Bermudan swaption; cancelable note
Bermudan option, *see* American/Bermudan option
Bermudan swaption, 207, 875–920
  accreting, *see* Bermudan swaption, non-standard
  American, *see* American swaption
  amortizing, *see* Bermudan swaption, non-standard
  approximating American swaption, 896
  bullet, *see* Bermudan swaption, vanilla
  carry, 908, 914
  impact on exercise decision, 914
  exercise fee, 899
  exercise value, XXXIV, 208, 875
  flexi-swap, *see* flexi-swap
  gamma-theta mismatch, 914
  hold value, XXXIV, 208
  lockout, 207, 875
  mid-coupon, 897, 899
  no-call, *see* Bermudan swaption, lockout
  non-standard, 880–899
    calibration by payoff matching, 884–886
    calibration by PVBP matching, 884, 886
    calibration by tenor matching, 883
    calibration to basket, 887–889
    calibration to representative swaption, 884
    calibration to row of European swaptions, 888
    Gaussian short rate model, 888
    global calibration, 881, 883
    Libor market model, 887
    local projection method, 881, 883
    lower bound, 893, 909
    Markov-functional model, 881
    quadratic Gaussian model, 888
    quasi-Gaussian model, 881, 888, 891
    representative swaption for accreting Bermudan, 886
    representative swaption for amortizing Bermudan, 885, 886
    super-replication, 890–894
    upper bound, 891, 892, 909
  non-vanilla, *see* Bermudan swaption, non-standard
  PDE jump condition, *see* American/Bermudan option, PDE jump condition
  strike, 875
  survival measure, 1047
  vanilla, 880
  zero-coupon, 894–895
Bermudan swaption calibration
  adjusters method, 955
  local projection method, 554, 876–880
    Gaussian short rate model, 877
    non-standard Bermudan, *see* Bermudan swaption, non-standard
    quadratic Gaussian model, 877
    quasi-Gaussian model, 877
  smile calibration, 878–880
    at-the-money, 878
    exercise boundary, 879
    strike, 878
Bermudan swaption greeks
  pathwise differentiation method, 1044–1050
    forward induction, 1049–1050
    performance, 1050
    survival density, 1048
    survival measure, 1047
  portfolio replication for hedging, 913
  Principal Components Analysis, 913
  robust hedging, 912–914
  static hedging, 912
Bermudan swaption valuation, 822–873
  control variate, 1086–1090
    non-linear, 1088
    sampled at exercise time, 1087
  fast pricing, 916
  impact of forward volatilities, 876
  impact of inter-temporal correlation, 554, 877
  impact of mean reversion, 554, 876

xxiv   Index

impact of the number of factors, 877
Monte Carlo, 905–912
   exercise strategy, 906
   explanatory variables, 905
   parametric lower bound, 906–912
   regression lower bound, 905
Bermudanality, 879
Bessel function of the first kind, 282
Bessel process, 281, 282
best-of option, see MAX-option
best-of-calls option, 782
BGM model, see Libor market model
Black model, XXXIV, 21, 28, 202, 279, 283
   Asian option, see Asian option, Black model
   basket option, see basket option, Black model
   call option, 24
   CMS spread, 776
   delta, 350, 698
   effects of volatility mis-specification, 987
   Fourier integration, 329
   gamma-vega, 981
   log-likelihood ratio, 1060
   moment-generating function, 329
   PDE, 25
   stochastic interest rates, 28, 30
   strike-specific volatility, 698
   time-dependent parameters, 27, 983–985
   vega, 698
      use in calibration, 704
   with dividends, 27
Black shadow rate model, 452
Black-Derman-Toy model, 445–447
   mean-fleeting, 447
   short rate dynamics, 446
Black-Karasinski model, 447
Black-Scholes model, see Black model
Black-Scholes-Merton model, see Black model
BMA index, 192, 265
BMA rate, 192
Boltzman-Gibbs distribution, see out-of-model adjustment, path re-weighting method, Boltzman-Gibbs distribution

Bond Market Association, see BMA index
box smoothing method, see payoff smoothing, box smoothing
break-even rate, see forward swap rate
Broadie adjustment for sampling frequency of barriers, see Monte Carlo, sampling extremes, adjusting barrier for sampling frequency
Brownian bridge, 125, 647, 648
   conditional moments, 129
   Libor market model, see Libor market model valuation, Monte Carlo, Brownian bridge
   path construction, see Brownian motion, path construction by Brownian bridge
   sampling extremes, see Monte Carlo, sampling extremes, with Brownian bridge
Brownian motion, 4
   geometric, 16
   Haar function decomposition, see Brownian motion, path construction by Brownian bridge
   Ito integral, see Ito integral
   Karhunen-Loeve decomposition, see Brownian motion, path construction by Principal Components
   path construction, 106
   path construction by Brownian bridge, 128, 129
   path construction by Principal Components, 130
   Stratonovich integral, see Stratonovich integral
BSM model, see Black model

C^0, XXXIV
C^1, XXXIV
C^2, XXXIV
C^n, XXXIV
calibration, 299
   calibration norm, 630–633
      fit, 634
      regularity, 634
   cold start, 633

forward induction, 445, 458, 953
Levenberg-Marquardt, 633
local projection method, *see* local projection method
Markovian projection method, *see* Markovian projection
most likely path, 990
stochastic optimization method, 953
time averaging, 301, 307, 363, 371–381, 550, 584, 667
  algorithm, 377–381
  non-zero correlation, 376
  skew, 373–374
  volatility, 371–373
  volatility of variance, 374–377
callable Libor exotic, *see* CLE
callable zero, *see* Bermudan swaption, zero-coupon
cancelable note, 214, 829, 830
  ATM, 859
  carry, 858, 914
cancelable swap, *see* cancelable note
cap, 186, 202
  caplet volatility from cap volatility, 706
    interpolation, 707
    precision norm, 707
    relaxation, 708
    smoothness norm, 708
    splitting scheme, 708
  digital, 203, 209
  valuation formula, 202
Capital Asset Pricing Model, 357
capped floater, 209
Cauchy distribution, 98, 101
  Monte Carlo, 98
certificate of deposit, 194
CEV model, 280–286
  attainability of zero, 280
  displaced, 285
  European call option value, 282, 283
  explosion, 280
  regularization, 284
  relation to Bessel process, 281
  strict supermartingale, 280
  time-dependent, 304
    effective parameter, 305
  volatility skew, 283
characteristic function, 20

Cheyette model, *see* quasi-Gaussian model
chi-square distribution, 100
  Monte Carlo, 100, 102
  non-central, *see* non-central chi-square distribution
  PDF, 100
chooser cap, *see* flexi-cap
chooser swap, *see* flexi-swap
CIR model, *see* Cox-Ingersoll-Ross model
CLE, 213, 216, 628, 817–873, 875
  accreting at coupon rate, 216, 870
  carry, 858, 908, 915
    impact on exercise decision, 849, 858
  definition, 822
  exercise value, XXXIV, 215, 822
  hold value, XXXIV, 215, 822
  lockout, 213
  marginal exercise value decomposition, 824
  multi-tranche, 217
  no-call, *see* CLE, lockout
  optimal exercise, 823
  single-rate, 864
  smooth function of Monte Carlo path, 1029
  snowball, 216, 872
CLE calibration, 817–822
  local projection method, 864–869
    calibration targets, 865
    core swap rate analog, 867
    local models, 866–867
    quadratic Gaussian model, 866
    quasi-Gaussian model, 866
    two-factor Gaussian model, 866
    two-strike calibration, 867
    vega, 869
  low-dimensional models, 864–869
  model choice, 821
  single-rate, 864
  to forward volatility, 821
CLE greeks, 1036–1040
  as sum of coupon greeks, 1037
  discontinuity in Monte Carlo, 1041
  freezing exercise boundary, 835, 1039, 1040
  freezing exercise time, 1038–1040

likelihood ratio method, *see* likelihood ratio method
pathwise differentiation method, 1035–1040, 1058–1060
  computational complexity, 1052
  forward induction, 1049–1050
  survival density, 1048
  survival measure, 1047
  perturbation method, 1040, 1059
  computational complexity, 1053
  portfolio replication for hedging, 913
  recursion, 1036
  source of noise, 1040
  tube Monte Carlo, 1029
CLE regression, 825–864
  automatic selection of regression variables, 857
  boundary optimization, 832
  cancelable note, 829–830
  choice of regression variables, 850–856
  decision only, 830–832
  discrepancy principle, 861
  excluding suboptimal points, 858
  exercise value, 827–829
  explanatory variables, 851–856
    classification, 853
    CMS spread, 853
    core swap rate, 853
    stochastic volatility, 856
    with convexity, 854–856
  general-to-specific approach, 857
  generalized cross-validation, 861
  L-curve method, 861
  Libor market model, 851, 852
    state variables, 851
  lower bound, 833–835
  perfect foresight bias, 834
  pseudo-inverse method, 862
  quadratic Gaussian model, 851
  quasi-Gaussian model, 851
  regression operator, 826
  regression variables, 825
  rescaling, 863
  reuse exercise boundary, *see* CLE greeks, freezing exercise boundary
  ridge regression, *see* CLE regression, Tikhonov regularization
  robust implementation, 860–864
  singular value decomposition, 104
  stabilization, 861
  state variables, 850–851
    Libor market model, 851
  SVD decomposition, 862, 863
    connection to Tikhonov regularization, 863
  Tikhonov regularization, 162, 255, 861–863
    connection to TSVD, 863
  truncated SVD decomposition, 162, 862, 863
  two-step, 859
  upper bound, 839–850
    alternative methods, 849
    computational cost, 843
    improvements to algorithm, 847–849
    nested simulation algorithm, 839–849
    non-analytic exercise values, 845–847
    simulation within a simulation, *see* CLE regression, upper bound, nested simulation algorithm
CLE valuation, 215, 822–873
  as cancelable note, 829
  boundary optimization, 832
  confidence interval for value, 844
  control variate, *see* Bermudan swaption valuation, control variate
  discontinuous function of Monte Carlo path, 1041
  duality, 838, 1093
    multiplicative, 1093
  duality gap, 841, 844, 910
    in stochastic volatility models, 912
  exercise policy consistency conditions, 835
  fast pricing, 918
  Hamilton-Jacobi-Bellman equation, 823
  impact of forward volatility, 820
  impact of inter-temporal correlation, 865
  impact of volatility smile dynamics, 820, 821
  Libor market model, 826

lower bound, 836, 843, 847, 850
  by regression, *see* CLE regression, lower bound
  iterative improvement, 835
  iterative improvement by nested simulation, 837
  quality test, 1060
LS method, *see* CLE regression
Monte Carlo, 825–864, 905
optimal exercise policy, 835, 837, 1039
PDE, 870–873
  accreting at coupon rate, 870
  path-dependent, 870–873
  similarity reduction, 871
  snowball, 872
perfect foresight bias, 834
policy fixing, 847
recursion, 823
regression method, *see* CLE regression
tube Monte Carlo, 1029
upper bound, 838–850
  cancelable note, 846
  nested simulation algorithm, 841, 909
  non-analytic exercise values, 845–847
weighted coupon decomposition, 918
CMS, 206
  annuity to forward measure change, 737–739
  convexity adjustment, 723–745
    annuity mapping function, *see* terminal swap rate model, annuity mapping function
    correcting arbitrage, 733–735
    density integration method, 738
    impact of mean reversion, 735–736
    impact of volatility smile, 735
    impact on implied volatility, 776
    Libor market model, 731–733
    linear TSR model, 728–730
    out-of-model adjustment, 963, 964
    quasi-Gaussian model, 730–731
    replication method, 723–725
    stochastic volatility model, 739
    swap-yield TSR model, 727–728

  vega hedging, *see* terminal swap rate model, linear TSR model, vega hedging
  hedging portfolio, 725
  quanto, *see* quanto CMS
CMS cap, 207, 697
  impact of CMS convexity on volatility smile, 741
  link to European swaptions, 741
CMS digital spread option, 791
  dimensionality reduction, 791
CMS floor, 207
CMS rate, 206
  distribution in forward measure, 737–739
CMS spread option, 210, 211, 621, 690, 765, 776
  by integration, 777
  copula method, 776–784
  dimensionality reduction, 789
  floating digital, 792
  Gaussian copula, 777
    correlation impact, 778
    vega to swaptions, 778
  implied copula, 781
  implied correlation, 778
  Libor market model, 619–621, 636, 692, 808
    closed-form approximation, 810
  Libor market model calibration, 636
  local volatility model, 1145
  Margrabe formula, 812
  Markovian projection, 1145, 1149
  multi-stochastic volatility, *see* multi-stochastic volatility model
  non-standard gearing, 777, 792
    dimensionality reduction, 792
  Normal spread volatility, 776
  one-dimensional integration, 789
  out-of-model adjustment, 964, 966
  power Gaussian copula, 781
  quadratic Gaussian model, 810
    closed-form approximations, 810
  risk management with one-factor model, 971
  stochastic volatility
    correlation impact, 807
  stochastic volatility de-correlation, 962

xxviii   Index

   stochastic volatility model, 1149
      correlation impact, 805
   vega in Libor market model, 1116
CMS swap, 206, 697
   valuation formula, 207
CMS-linked cash flow, 723–745
   direct integration method, 737
   replication method, 725
coherent risk measure, see risk measure,
      coherent
collateral, 192, 266
complementary Gamma function, 281
complete market, 11
compounded rate, 200
conditional expected value, 19
   iterated conditional expectations, see
      iterated conditional expectations
   projection approximation, see
      Markovian projection, conditional
      expected value by projection
constant elasticity of variance model,
      see CEV model
constant maturity swap, see CMS swap
contingent claim, see derivative security
continuity correction, see payoff
      smoothing, continuity correction
control variate, 145–148, 330, 654,
      1077–1094
   adjusters method, 955
   construction from MC upper bound,
      1093
   dynamic, 148, 654, 1090–1093
      regression-based, 1091
   efficiency, 147
   impact on risk stability, 1093
   instrument-based, 1086–1090
   model-based, 676, 1077–1085
   non-linear controls, 147–148
   path re-weighting method, 961
   proxy Markov LM model, 1078
   proxy model, see control variate,
      model-based
convexity adjustment
   averaging swap, see Libor-with-delay,
      convexity adjustment
   CMS, see CMS, convexity adjustment
   futures, see ED future, convexity
      adjustment

   Libor-in-arrears, see Libor-in-arrears,
      convexity adjustment
   Libor-with-delay, see Libor-with-
      delay, convexity adjustment
   moment explosion, 760–763
   second moment, 760
copula, 770
   Archimedean, 772
      Monte Carlo, 800
   Clayton, 773
   conditional CDF, 792
   Frechet bounds, 771
   Gaussian, 768
      CMS spread option, see CMS
         spread option, Gaussian copula
      integration, 789
      joint CDF, 769
      joint PDF, 769, 777
      mixture, 774
      Monte Carlo, 799
   Gumbel, 773
   implied, 781
   independence, 770
   mixture, 774
      Monte Carlo, 800
   perfect anti-dependence, 771
   perfect dependence, 770
   power Gaussian, 775, 780
      parameter impact, 781
   product, 775
      Monte Carlo, 800
   reflection, 773
      Monte Carlo, 800
   Sklar's theorem, 771
copula density, 772
copula method, 768
   CMS spread option, see CMS spread
      option, copula method
   dimensionality reduction, 789–799
      by conditioning, 793–797
      by measure change, 797–799
   forward swaption straddle, 949
   integration, 786–799
      inverse CDF caching, 787
      singularities, 788
   limitations, 801–802
   mapping function, 795
   Monte Carlo, 799–801
   observation lag, 784

quanto options, 748
volatility swap, 934
core correlations, *see* inter-temporal correlation
core volatilities, 865, 876
correlation extractor, *see* Libor market model, correlation extractor
correlation risk sensitivity, 1119
correlation smile, 778
Cox-Ingersoll-Ross model, 432
  multi-factor, 519
  two-factor, 517
Crank-Nicolson scheme, *see* PDE, Crank-Nicolson scheme
credit risk, 260, 975
credit value adjustment, 266, 916
cross-currency basis swap, *see* floating-floating cross-currency basis swap
cross-currency basis swap spread, 261, 265
CRX basis swap, *see* floating-floating cross-currency basis swap
CRX spread, *see* cross-currency basis swap spread
cumulant-generating function, 154
curve cap, 211, 766
  range accrual, *see* range accrual, curve cap
CVA, *see* credit value adjustment

date rolling convention, 224
day count convention, 224–226
  30/360, 226
  Actual/360, 225
  Actual/365.25, 224
day count fraction, *see* year fraction
deflator, 9
delta, 18, 132, 355, 980
  bucketed interest rate deltas, 251, 1045
  forward rate, 253
  Jacobian method, *see* risk sensitivities, Jacobian method
  par-point, 251, 252, 256, 257, 993
  parallel, 257
  with backbone, 1120–1122
delta hedge, 18
density process, 9

derivative security, 11
  attainable, 11
  pricing, 11
diffusion, 4, 14
  absorbing barrier, 281, 289
  displaced, 285
  Feller boundary classification, 280
  Feller condition, 319
  Fubini's theorem, 409
  integration by parts, 120
  Ito integral, *see* Ito integral
  Ito process, 4
  local time, 25, 26, 294
  Ornstein-Uhlenbeck process, 413
  polynomial growth condition, 19
  predictable process, 7
  scale measure, 280
  SDE, 14
    generator, 19
    linear, 16
    locally deterministic, 172, 541
    strong Markov, 15
    strong solution, 15
    weak solution, 15
  speed measure, 280
diffusion invariance principle, 14
discount bond, XXXIV, 23, 167
  valuation formula, 172
discount curve, *see* yield curve
displaced CEV model, *see* CEV model, displaced
displaced log-normal model, 285
  basket option, 1147
  canonical form, 286
  explicit solution to SDE, 312
  Fourier integration, 328
  implied correlation, 811
  moment matching, 922
  moment-generating function, 329
  time-dependent, 304
    effective skew, 305
    explicit solution to SDE, 307
    range for process, 306
Dupire local volatility, 1131
  proof by Tanaka extension, 294, 1131
duration, 246
DVF model, *see* local volatility model
Dybvig parameterization, *see* short rate model, Dybvig parameterization

xxx   Index

early exercise, 30
ED future, 168–170, 196–197, 697,
        750–760
    convexity adjustment, 187, 197,
            750–760
        from market inputs, 752
        Gaussian HJM model, 186
        impact of volatility smile, 752, 757
        Libor market model, 753, 757
        replication method, 753, 757
    delivery arbitrage, 170
    futures rate, 169
        definition, 196
        instantaneous, 170, 172, 173
        martingale in risk-neutral measure,
            173, 750
        martingale in spot Libor measure,
            751
        simple, 170
        to forward rate, 756, 760
    mark to market, 170
    yield curve construction, 231, 992
ED futures contract, see ED future
effective volatility
    local volatility model, see local
        volatility model, effective
        volatility
    stochastic volatility model, see
        stochastic volatility model,
        effective volatility
envelope theorem, 1038
Eonia, 193, 200
equivalent martingale measure, see
    measure, equivalent martingale
Esscher transform, see exponential
    twisting
Eurodollar futures contract, see ED
    future
European call option, 23
    at-the-money, 24
    Fourier integration, 324
    in-the-money, 24
    out-of-the-money, 24
    probability density from, see
        volatility smile, probability
        density from
European digital call option, 59
European option
    Fourier integration, 326

European put option, 24
    at-the-money, 24
    in-the-money, 24
    out-of-the-money, 24
European swaption, 203, 697–705
    cash-settled, 205, 744–745
        payoff, 744
        put-call parity, 745
        replication method, 744, 745
    core swaptions, 424, 819
    coterminal swaptions, see European
        swaption, core swaptions
    diagonal swaptions, see European
        swaption, core swaptions
    forward swaption straddle, see
        forward swaption straddle, 943
    midcurve, 223
    non-standard, see Bermudan
        swaption, non-standard
        Black formula, 889
    physically-settled, 205
    SV model calibration, 703–704
    swap-settled, 205, 744, 745
    swaption grid, 205, 703
    swaption strip, 423
    tenor, 204
    valuation formula, 204
    volatility cube, 697
European-style option, 95
    replication method, 337
    valuation by volatility mixing, 339
exchange market, 193
    Chicago Mercantile Exchange, 196
    London International Financial
        Futures and Options Exchange,
        196
    Marché à Terme International de
        France, 196
exotic swap, 205, 208, 209, 822, 951
    CMS spread, 766
    CMS-based, 210
    digital CMS spread, 766
    global cap, 219
    global floor, 219
    knock-out, 218
    Libor-based, 209
    multi-rate, 210, 766
    path-dependent, 212
    principal amount, 208

range accrual, see range accrual
snowball, 212
spread-based, 210
structured coupon, 208–211
expectations hypothesis, 173
expected hedging P&L, 988
exponential distribution, 98
    Monte Carlo, 98
exponential integral, 334
exponential twisting, 154
extra state variable method, see PDE, path-dependent options

"The Fed Experiment", 452
Federal funds future, 201
Federal funds rate, 192, 200, 201, 266
    effective, 192
    target, 192
Federal funds/Libor basis swap, 201, 266
Feller condition, see diffusion, Feller condition
Feynman-Kac solution, 21
FFT, see stochastic volatility model, Fourier integration
filtration, 3, 4
    usual condition, 3
flexi-cap, 71
flexi-swap, 900–904
    decomposition into Bermudan swaptions, 901
    local projection method, 901
    marginal exercise value decomposition, 903
    narrow band limit, 904
    PDE, 900, 902
    purely local bounds, 901
"flip-flop", 210
floating digital, 792, 794
    dimensionality reduction, 792
floating digital spread option, 792
    dimensionality reduction, 792
floating-floating cross-currency basis swap, 262, 264, 265
floating-floating single-currency basis swap, 201, 268
floor, see cap
Fokker-Plank equation, see Kolmogorov forward equation

Fong-Vasicek model, 454–455, 517
    bond reconstitution formula, 454
forward CMS straddle, 941, 944, 945
    swaption, see forward swaption straddle
    volatility, see forward volatility
forward contract, 195
forward Kolmogorov equation, see Kolmogorov forward equation
forward Libor model, see Libor market model
forward Libor rate, XXXIV, 168, 191, 192, 196
    accrual end date, 224
    accrual period, 224
    accrual start date, 224
    martingale in forward measure, 174
    tenor, 168
    variance by replication method, 757
    year fraction, see year fraction
forward par rate, see forward swap rate
forward price, 23, 168
forward rate, 167
    continuously compounded, XXXIV, 168
    instantaneous, XXXIV, 169
    simple, 168
    tenor, 168
    volatility hump, 418, 494
forward rate agreement, see forward contract
forward starting option, 222
forward swap rate, XXXIV, 171, 199
    distribution in forward measure, see CMS rate, distribution in forward measure
    expiry, 171
    fixing date, 171
    linking forward and annuity measure, 737
    market-implied variance, 557
    martingale in swap measure, 178
    non-standard, 882
        decomposition, 882
    tenor, 171
    weighted average of Libor rates, 171, 256
forward swaption straddle, 223, 945–950

copula method, 949
  relation to CMS spread option, 948
  triangulation, *see* forward volatility, triangulation
  vanilla model, 946
  vega exposure, 948
  volatility, *see* forward volatility
forward volatility, 223
  connection to inter-temporal correlations, *see* inter-temporal correlation, connection to forward volatilities
  hedging, 914
  impact of rate correlation, 919
  impact of volatility smile, 945
  Libor rate, *see* volatility, forward volatility of Libor rate
  triangulation, 948
forward volatility derivative, 220, 222
  forward swaption straddle, *see* forward swaption straddle
  implied Normal volatility contract, 223
  midcurve swaption, *see* European swaption, midcurve
  volatility swap, *see* volatility swap
forward yield, *see* forward rate
Fourier transform, 325
  inverse, 325
FRA, *see* forward contract
Frobenius norm, *see* matrix, Frobenius norm
fundamental matrix, 486
fundamental theorem of arbitrage, 10
fundamental theorem of derivatives trading, 987
futures contract, *see* ED future
futures rate, *see* ED future, futures rate
fuzzy logic, *see* payoff smoothing, fuzzy logic
FX rate, 179, 746, 748
  dynamics in domestic risk-neutral measure, 180
  forward, 178
    martingale in domestic forward measure, 180

Gâteaux derivative, 253

gamma, 980
  pathwise differentiation method, *see* pathwise differentiation method, gamma
  payoff smoothing, 1019
  relationship to vega, 981
gamma distribution, 100
  Monte Carlo, 100, 102
  PDF, 100
Gamma function, XXXIII
  incomplete, *see* incomplete Gamma function
  quick approximation, 1153
Gauss-Hermite quadrature, *see* quadrature, Gauss-Hermite
Gaussian copula, *see* copula, Gaussian
Gaussian distribution, XXXIII
  conditional distribution, 648
  cumulant-generating function, 154
  imaginary mean, 798
  inverse CDF, 99, 165
  linear transform, 103
  measure change, 797
  multi-dimensional PDF, 103
  quadratic form, 523
    moment-generating function, 524, 535
    moments, 535
Gaussian HJM model, 184–187
  caplet, 186
  ED future convexity adjustment, *see* ED future, convexity adjustment, Gaussian HJM model
  time-stationary, 418
  zero-coupon bond option, 185
Gaussian multi-factor short rate model, *see* Gaussian short rate model, multi-factor
Gaussian one-factor short rate model, *see* Gaussian short rate model
Gaussian short rate model, 408, 415–431, 480–512
  as special case of affine model, 432
  Bermudan swaption, *see* Bermudan swaption calibration, local projection method, Gaussian short rate model
  bond dynamics, 417
  bond reconstitution formula, 416

efficient calculation, 417
calibration, 423
  bootstrap, 424
calibration to yield curve, 416
European swaption, 420, 423
  Jamshidian decomposition, 420
fast pricing of Bermudan swaptions, 916
forward rate dynamics, 415
forward rate volatility, 415
  dynamics, 419
humped volatility structure, 418
in spot measure, 430
in terminal measure, 430
mean reversion, *see* mean reversion
mean reversion calibration, *see* mean reversion calibration
Monte Carlo, 427–431
  approximate, 429
  Euler scheme, 429
  exact, 427
  other measures, 430
multi-factor, 480–512
  benchmark rate parameterization, 507–510
  benchmark rates, 508
  benchmark tenors, 508
  bond reconstitution formula, 480, 483, 485
  bond volatility, 481
  calibration, 507
  classic development, 487–490
  correlated Brownian motions, 490
  correlation stationarity, 490
  European swaption, 501–507
  European swaption by Jamshidian decomposition, 505
  factors and loadings, *see* Gaussian short rate model, multi-factor, statistical approach
  forward rate correlation, 490–491
  forward rate volatility, 484
  Gaussian swap rate approximation, 506–507
  loadings, 500
  mean reversion matrix diagonalization, 488–490
  Monte Carlo, 510–511
  PDE, 511–512

rotations, 486
separability, 480–487
short rate dynamics, 481
short rate state distribution, 487, 510
short rate state dynamics, 481–487
short rate state dynamics, integrated, 487, 510
single Brownian motion, 498
statistical approach, 497–501
swap rate volatility, 506
PDE, 425–427
  boundary conditions from PDE, 426
short rate distribution, 428
short rate dynamics, 415
short rate state dynamics, 416, 427
  integrated, 427
swap rate dynamics in annuity measure, 422
swap rate volatility, 422
time-stationary, 418
two-factor, 491–497
  bond reconstitution formula, 492, 502
  CLE, *see* CLE calibration, local projection method, two-factor Gaussian model
  correlated Brownian motions, 492
  correlation stationarity, 493
  doubly mean-reverting form, 494
  European swaption by Jamshidian decomposition, 502–506
  forward rate correlation, 492–494
  forward rate dynamics, 492
  forward rate volatility, 492–496
  short rate state conditional distribution, 504
  short rate state correlation, 491
  short rate state dynamics, 491
  single Brownian motion, 497
  volatility hump, 494
Gaussian two-factor short rate model, *see* Gaussian short rate model, two-factor
generalized trigger product, 1074
  importance sampling, 1074–1077
  pathwise differentiation method, 1041–1044

payoff smoothing, 1074–1077
trigger variable, 1074
tube Monte Carlo, see barrier option, tube Monte Carlo
Girsanov's theorem, 12, 13
Gaussian distribution, 797
Gram-Charlier expansion, 368, 439
greeks, see risk sensitivities
Green's function, 20
grid shifting, see payoff smoothing, grid shifting
GSR model, see Gaussian short rate model
Gyöngy theorem, see Markovian projection, Gyöngy theorem

H^2, 5
Hagan and Woodward parameterization, see short rate model, Hagan and Woodward parameterization
hat smoothing method, see payoff smoothing, hat smoothing
Heath-Jarrow-Morton model, see HJM model
hedge, 251
  best hedging strategy, 355
  beta, 357
  minimum variance, 355–357
  model-independent, 718
  semi-static, see replication method, semi-static
  shadow delta, see volatility smile, shadow delta hedging
  sub-replicate, 719
  super-replicate, 719, 979
  zero-beta, 357
Hermite matrix, 270
Heston model, see stochastic volatility model
HJM model, 181–190
  bond dynamics, 181
  forward bond dynamics, 182
  forward rate dynamics, 182
  Gaussian, see Gaussian HJM model
  Gaussian Markov, 187–189
    short rate dynamics, 188
  log-normal, 189–190
  Markovian, 407
  separable, 415

  short rate dynamics, 183
  stochastic basis, see HJM model, two-curve
  two-curve, 680–683
    forward rate spread dynamics, 681
    Gaussian basis spread, 683
    index bond dynamics, 682
    index forward rate dynamics, 681
    index short rate dynamics, 681
    quanto correction, 682
Ho-Lee model, 408–412
  bond dynamics, 411
  bond reconstitution formula, 410
  calibration to yield curve, 409
  drawbacks, 412
  forward rate dynamics, 411
  short rate dynamics, 410
hybrid differentiation method, 1061

implied volatility, see volatility, implied
importance sampling, 146, 149–158, 1063–1077
  application to payoff smoothing, 1067
  barrier option, see barrier option, importance sampling
  density formulation, 149
  efficiency, 150
  generalized trigger product, see generalized trigger product, importance sampling
  least-squares, 153
  likelihood ratio, 150, 153, 155
  rare events, 154
    approximately optimal mean shift in multi-variate case, 157
    asymptotic optimality, 158
    efficiency, 156
    minimal variance, 155
    multi-variate, 156
  SDE, 151–154
  short rate model, see short rate model, importance sampling
  survival measure, 1067
    simulation under, 1072, 1074, 1076
  TARN, see TARN, importance sampling
incomplete Gamma function, XXXIII, 281

index, 206
index option, *see* basket option
infinitesimal operator of SDE, *see* diffusion, SDE, generator
infinitesimal perturbation analysis, 136
information theory, 957
instantaneous futures rate, *see* ED future, futures rate, instantaneous
integration by parts for diffusion process, *see* diffusion, integration by parts
inter-temporal correlation, 424, 554, 820, 865, 876
   connection to forward volatilities, 820
   hedging, 914
   impact of mean reversion, 554
   impact of volatility smile, 945
   impact on Bermudan swaption, *see* Bermudan swaption valuation, impact of inter-temporal correlation
   impact on CLEs, *see* CLE valuation, impact of inter-temporal correlation
   impact on TARNs, 929
   mean reversion calibration to, *see* mean reversion calibration, to inter-temporal correlations
interbank money market, 192
International Swaps and Derivatives Association, 192, 266
intrinsic value, 26
inverse floater, 209
iterated conditional expectations, 176
Ito integral, 4, 5
Ito isometry, 5
Ito's lemma, 6
Ito-Taylor expansion, 118

Jacobian, *see* risk sensitivities, Jacobian method
Jamshidian decomposition
   American/Bermudan option, *see* American/Bermudan option, Jamshidian decomposition
   European swaption, *see* Gaussian short rate model, European swaption, Jamshidian decomposition

Kolmogorov backward equation, 19, 20
Kolmogorov forward equation, 20, 386, 459, 1048
   correct boundary conditions, 386
   discrete consistency with backward equation, 460
Kullback-Leibler relative entropy, 957
kurtosis, 375

L^1, XXXIV, 4
L^2, XXXIV, 4
ladder, 985
ladder swap, *see* ratchet swap
Lagrange basis functions, *see* PDE, Lagrange basis; payoff smoothing, Lagrange basis
Lagrange multiplier, 249, 958
least squares method, *see* CLE regression
LIA, *see* Libor-in-arrears
Libor curve, *see* yield curve
Libor market model, 451, 591–694, 731, 868, 911
   annuity mapping function, 732, 733
   asset-based adjustment, 963
   back stub, 657–662
      arbitrage-free, 659–661
      from Gaussian model, 661–662
      simple, 658–659
   choosing number of factors, 614
   CLE, 821
   CMS convexity adjustment, 964
   correlation extractor, 865
   deflated bond dynamics, 651
   delta with backbone, 1120–1122
   drift approximation, 646
      Brownian bridge, 1079
   drift freezing, 1052
   exercise boundary, 911
   exercise strategy, 908
   expected value of Libor rate in annuity measure, 671
   front stub, 662–667
      exogenous volatility, 663–665
      from Gaussian model, 666–667
      simple interpolation, 666
      zero volatility, 662–663
   in hybrid measure, 642

index function, *see* tenor structure, index function
Libor rate correlation, 603–614, 758
  correlation PCA, 611
  covariance PCA, 626
  historical estimation, 606
  majorization, 613
  parametric form, 608, 609
  PCA, 604–606
  poor man's correlation PCA, 614
  regularization, 610
Libor rate dynamics, 593–603
  annuity measure, 733
  in forward measures, 594–595
  in hybrid measure, 597
  in spot measure, 596
  in terminal measure, 595, 641
Libor rate inter-temporal correlation, 758
Libor rate volatility
  from volatility norm, 625–627
  functional form, 622
  grid-based, 622–623
  interpolation, 624–625
Libor rate volatility link to HJM forward rate volatility, 598
link to HJM, 597
local volatility, 598–600
  CEV, 599
  displaced log-normal, 599
  existence and uniqueness, 599
  LCEV, 599
  log-normal, 599
Markov, 676, 1078–1085
  as control variate, 1084
  Brownian bridge, 1079
  calibration, 1082
  one-factor, 1079
  one-factor reconstitution formula, 1080
  separable volatility, 1080
  two-factor, 1081
  two-factor reconstitution formula, 1081
Markovian projection, 667, 669, 1139
model risk, 629
multi-stochastic volatility, 690–694, 962
  caplet, 691

CMS spread option, 692
European swaption, 691
moment-generating function, 692
Musiela parameterization, 604
pathwise derivative
  forward Libor rate, 1051
  forward swap rate, 1055
  numeraire, 1054
  structured coupon, 1055
  stub bond, 1054
pathwise differentiation method, 1051–1058
  computational complexity, 1052
PCA, *see* Principal Components Analysis
portfolio replication, 914
stochastic basis, *see* Libor market model, two-curve
stochastic variance dynamics, 689
stochastic volatility, 600–603
  moment-generating function, 689
  non-zero correlation, 687
stub volatility, 664, 667
swap rate correlation, 620–621
swap rate dynamics, 617, 669
  approximate, 618
time-stationary, 623
tool to extract forward volatility, 821
two-curve, 683–687
  deterministic spread, 687
  European swaption, 686
  Libor rate dynamics, 685
  Monte Carlo, 685
  swap rate dynamics, 686
vega, *see* vega, Libor market model
Libor market model calibration, 622–637
algorithm, 633, 636, 675
bootstrap, 635
  for vega, 1111
cascade, *see* Libor market model calibration, bootstrap
choice of instruments, 627
effective skew, 672
effective volatility, 670
global, 628
grid-based, *see* Libor market model calibration, global
local, 628

Index   xxxvii

objective function, 630
PCA, 626
row-by-row, 633, 634
to spread options, 635, 808
volatility skew, 637
volatility smile, 674
Libor market model valuation
   Bermudan swaption, *see* Bermudan swaption valuation, Monte Carlo
   caplet, 615
   CLE, *see* CLE valuation, Libor market model
   CMS convexity adjustment, *see* CMS, convexity adjustment, Libor market model
   CMS spread option, *see* CMS spread option, Libor market model
   curve interpolation, 657–667
   European swaption, 616, 618, 667
   Libor-with-delay, *see* Libor-with-delay, Libor market model
   Monte Carlo, 637
      analysis of computational effort, 639
      antithetic variates, 653
      Brownian bridge, 647
      choice of numeraire, 642
      control variate, 654
      discretization bias, 639
      Euler scheme, 638
      front stub, 664
      high-order schemes, 650
      importance sampling, 655
      lagging predictor-corrector, 644
      large time steps, 641, 646–649
      log-Euler scheme, 638
      martingale discretization, 650–653
      Milstein scheme, 650
      predictor-corrector, 643, 644, 647, 653
      survival measure, 1072, 1075
      two-curve, 685
      variance reduction, 653–655
   multi-rate vanilla derivative, 807
   PDE, *see* Libor market model, Markov
   TARN, *see* TARN, Libor market model

volatility swap, *see* volatility swap, Libor market model
Libor rate, *see* forward Libor rate
Libor-in-arrears, 200, 716–719
   convexity adjustment, 717
   replication method, 718
   sub-replicating portfolio, 719
   super-replicating portfolio, 719
Libor-with-delay, 719–723
   convexity adjustment, 719
   Libor market model, 720, 722
   quasi-Gaussian model, 720, 721
   replication method, 720, 722
   swap-yield TSR model, 720
likelihood ratio method, 139–142, 1060–1061
   discontinuous payoff, 138
   exploding variance, 1061
   for Euler scheme, 140–141
   for Milstein scheme, 141
   log-likelihood ratio, 140
   score function, 140
   vega, 1124
linear regression, 146
Lipschitz function, 137
LM model, *see* Libor market model
local projection method, 560, 864, 953, 1097
   Bermudan swaption, *see* Bermudan swaption calibration, local projection method
   CLE, *see* CLE calibration, local projection method
   non-standard Bermudan swaption, *see* Bermudan swaption, non-standard, local projection method
   TARN, *see* TARN, local projection method
   volatility swap, *see* volatility swap, local projection method
local stochastic volatility model, 316, 1137–1145
   calibration, *see* Markovian projection, LSV calibration
   Markovian projection, *see* Markovian projection, LSV calibration
local time, *see* diffusion, local time
local volatility model, 277–312

approximation with displaced log-normal model, 286
asymptotic expansion, 295–299
basket option, see Markovian projection, basket option in LV model
CEV, see CEV model
displaced log-normal, see displaced log-normal model
effective convexity, 307–312
effective skew, 301–312
effective volatility, 301
expansion around displaced log-normal model, 296
expansion around Gaussian model, 298
forward equation for call options, 293
PDE, 292–295
  simultaneous for multiple parameters, 293
  space discretization, 292
  transform to constant diffusion coefficient, 87, 292
quadratic volatility, see quadratic volatility model
range-bound, 287
small-noise expansion, see volatility, small-noise expansion
smile dynamics, 279, 350, 352
time-dependent, 299–312
  separable, 300
log-normal distribution, XXXIII, 16
  moment matching, see moment matching
  moments, 16
  Monte Carlo, 101
Longstaff-Schwartz method, see CLE regression
Longstaff-Schwartz model, 517–519
  bond reconstitution formula, 518
lookback option, 124
  Monte Carlo, see Monte Carlo, lookback option
LS method, see CLE regression
LSV model, see local stochastic volatility model
LVF model, see local volatility model

Malliavin calculus, 142, 1042, 1060

Margrabe formula for spread option, 812
mark-to-model, 818
Markov process, 15
  Feynman-Kac theorem, see Feynman-Kac solution
  strong, 15
  transition density, 20
Markov-functional model, 472–478
  calibration to yield curve, 475
  criticism, 478
  Libor parameterization, 473
  log-normal, 474
  no-arbitrage condition, 473
  non-standard Bermudan swaption, 881
  numeraire, 472
  numeraire mapping, 472
    Libor parameterization, 473
    non-parametric, 476
    swap parameterization, 475
  PDE, 477
  state process, 472
  swap parameterization, 475
  transition density, 472
Markovian projection, 805, 1129–1156
  average option, 1133
  barrier option, 1134
  basket option in LV model, 1145–1148
  basket option in SV model, 1149–1152
  CMS spread option, 1145
  conditional expected value by Gaussian approximation, 1134–1135
  conditional expected value by projection, 726, 727, 1136–1137
  displaced Heston model, 1149, 1151
    non-perturbative approximation, 1151
  displaced log-normal model, 1136, 1146
  Gyöngy theorem, 1130
  LSV calibration, 1139–1145
    mapping function, 1142
    proxy model, 1143–1145
  quadratic volatility model, 1137, 1148

quasi-Gaussian model, *see* quasi-Gaussian model, Markovian projection
spread option, 1151
stochastic volatility model, 1138
martingale, 5
  Doob-Meyer decomposition, 35
  exponential, 12
    Doleans exponential, XXXIII, 12
  local, 5
    bounded, 288
  martingale representation theorem, 6
  Novikov condition, 12
  optional sampling theorem, 35
  Snell envelope, 31, 823
  square-integrable, 5
  stopping time, *see* stopping time
  submartingale, 5
  supermartingale, 5, 360
    CEV, *see* CEV model, strict supermartingale
    quadratic volatility, *see* quadratic volatility model, strict supermartingale
    SV model, *see* SV model with general variance process, strict supermartingale
matrix
  exponential, 486
  Frobenius norm, 105, 609–611, 626, 627, 851
  infinity norm, 53
  positive semi-definite, 103
    Cholesky decomposition, 103
  rank-deficient, 106
  spectral norm, 53
  stiffness, 1111
  tri-diagonal, 47
MAX-option, 908
mean reversion, 316, 413, 552, 573
  effects, 552–554
    inter-temporal correlation, 554
    swaption volatility ratio, 553
mean reversion calibration, 552–560, 573
  to inter-temporal correlations, 557–559
  to row of European swaptions, 555, 888

  to volatility ratios, 554–557
mean-reverting square-root process, *see* square-root process
measure, XXXIII
  absolutely continuous, 1067
  annuity, 178, 204
  change of numeraire, *see* numeraire, change of numeraire
  domestic, 746
  equivalent, 9, 1067
  equivalent martingale, 8, 9, 14, 171
  foreign, 746
  hybrid, 176
  local martingale, 10
  risk-neutral, XXXIII, 22, 172
    domestic and foreign, 179, 180
  spot, XXXIII, 175
  survival density, 1047
  survival for Bermudan swaption, *see* Bermudan swaption, survival measure
  survival in importance sampling, *see* importance sampling, survival measure
  T-forward, XXXIII, 29, 174
    domestic and foreign, 180
  terminal, 176
min-max volatility swap, 222, 938
  capped, 940
  semi-static replication, 939
moment explosion, 323, 343, 344, 361, 760, 761
  impact on convexity adjustment, *see* convexity adjustment, moment explosion
  SABR model, *see* SABR model, moment explosion
  stochastic volatility model, *see* stochastic volatility model, moment explosion
  SV model with general variance process, *see* SV model with general variance process, moment explosion
moment matching, 889, 920–924
  Asian option, 922
  basket option, 924
moment-generating function, 13
Monte Carlo, 95–165

A-stable scheme, 110
Asian option, 107
Asian option on basket, 107
average rate option, see Monte Carlo, Asian option
barrier option, 124–128
  adjusting barrier for sampling frequency, 128
  double-barrier knock-out, 124
bias, 122
bias/standard error trade-off, 123
Brownian motion, see Brownian motion
calibration by stochastic optimization method, 953
central limit theorem, 96
convergence rate, 97
discretization bias, 428
efficiency, 143
Euler scheme, 110, 111
  linear SDE, 112
  region of stability, 111
  weak convergence order, 111
Euler-Maruyama scheme, see Monte Carlo, Euler scheme
Heun scheme, 116
higher-order schemes, 116
implicit Euler scheme, 113
  region of stability, 114
implicit Milstein scheme, 390
log-Euler scheme, 112, 113
lookback option, 125
low-discrepancy sequence, see Monte Carlo, random number generation, quasi-random
lower bound for American option, 34, 35, 164
  parametric, 159, 160
  regression-based, 161
mean-square error, 123
Milstein scheme, 119, 121
  multi-dimensional, 121
modified trapezoidal scheme, see Monte Carlo, Heun scheme
optimal root-mean-square error, 123
perfect foresight bias, see American/Bermudan option, perfect foresight bias
predictor-corrector, 115, 116

convergence order, 116
random number generation, 97
  acceptance-rejection method, 99–101
  Box-Muller method for Gaussian distribution, 99
  composition method, 101–102
  conditional Gaussian, 1066
  correlated Gaussian, 103
  correlated Gaussian by Cholesky decomposition, 103
  correlated Gaussian by eigenvalue decomposition, 104
  inverse transform method, 98
  linear congruential generator, 97
  Marsaglia polar method for Gaussian distribution, 99
  Mersenne twister, 98
  period, 98
  pseudo-random, 97, 130
  quasi-random, 129
  Sobol, 129
region of stability, 110
Richardson extrapolation, 122, 470
sample mean, 96
sampling extremes, 124–128
  adjusting barrier for sampling frequency, 128, 937, 970
  with Brownian bridge, 125
SDE discretization, 108
second-order scheme, 119, 121
seed, 97
standard error, 97, 122
  for digital option, 133
  for greeks, 132, 135
strong convergence order, 111
strong law of large numbers, 96
strongly consistent, 109
third-order scheme, 470
upper bound for American option, 34–36, 163, 164
variance reduction, see variance reduction
weak convergence, 109
weak convergence order, 110
weakly consistent, 109
most likely path, see volatility, implied, most likely path approximation
multi-rate vanilla derivative, 765–815

copula method, *see* copula method
Libor market model, 809
observation lag, 784
stochastic volatility, *see* multi-stochastic volatility model
term structure models, 806
multi-stochastic volatility model, 803–808, 1149
  correlation impact, 805
  measure change by CMS caplet calibration, 804
  measure change by drift adjustment, 803
  Monte Carlo
    Quadratic-Exponential scheme, 805
  multi-rate vanilla derivative, 803–808
multi-tranche, *see* CLE, multi-tranche

non-central chi-square distribution, 283
  asymptotics, 392
  CDF, 102, 319
  in CEV model, 283
  in delta-gamma VaR/cVaR, 998
  in LS model, 519
  two-dimensional, 519
Normal model, XXXIV, 283
  CMS spread, 776
    vega to swaptions, 777
numeraire, 10, 171
  change of numeraire, 11
    Girsanov's theorem, *see* Girsanov's theorem
  discrete money market account, XXXIV, 175, 176
  money market account, XXXIV, 22, 28, 172

OIS, *see* overnight index swap
one-dimensional integral for spread option, 789
operator calculus, 998–999
OTC market, *see* over-the-counter market
out-of-model adjustment, 951–971
  adjusters method, 954–956
    algorithm, 955
    as control variate, 955
    volatility adjustment, 956
  asset-based adjustment, 963–964

CMS spread option, 964
coupon calibration, 952–954
delta-adjustment method, 956
extended calibration, 953
fee adjustment method, 967–969
  additive, 967
  blended, 968
  impact on derivatives, 968
  multiplicative, 968
issues, 961, 964
mapping function adjustment, 965
market adjustment, 965
path re-weighting method, 956–961
  as control variate, 961
  Boltzman-Gibbs distribution, 959
  Boltzman-Gibbs weights, 959
  dual, 961
  inappropriate use, 958
  partition function, 958
  risk sensitivities, 961
PDE for coupon values, 953
proxy model method, 961
spread adjustment method, 966
strike adjustment method, 969–971
  impact on derivatives, 970
over-the-counter market, 193
overhedge, 1023
overlay curve, *see* yield curve, overlay curve
overnight index swap, 193, 200, 266

P&L, 698, 990–995
P&L analysis, 986
P&L attribution, *see* P&L explain
P&L explain, 993–995
  bump-and-do-not-reset explain, *see* P&L explain, waterfall explain
  bump-and-reset explain, 994–995
  waterfall explain, 993–994
P&L explanation, *see* P&L explain
P&L of hedged book, 987–990
P&L predict, *see* P&L prediction analysis
P&L prediction analysis, 258, 991–993
  first-order, 991
  second-order, 991
  unpredicted P&L, 991
par rate, *see* forward swap rate

parameter averaging, *see* calibration, time averaging
partial differential equation, *see* PDE
partition function, 958
pathwise delta approximation, *see* pathwise differentiation method, pathwise delta approximation
pathwise differentiation method, 135–138, 1035–1060
  adjoint method, 1056
    computational complexity, 1053, 1057
  barrier option, *see* barrier option, pathwise differentiation method
  Bermudan swaption, *see* Bermudan swaption greeks, pathwise differentiation method
  CLE, *see* CLE greeks, pathwise differentiation method
  computational complexity, 1052, 1053
  discontinuous payoff, 1042, 1061
  European option, 1054
  gamma, 1050, 1056
  generalized trigger product, *see* generalized trigger product, pathwise differentiation method
  Libor market model, *see* Libor market model, pathwise differentiation method
  money market account, 1046
  Monte Carlo models, 1051–1060
  pathwise delta approximation, 1059
  PDE models, 1044–1050
  sensitivity path generation, 138
  TARN, *see* TARN, pathwise differentiation method
  vega, 1050, 1056
payoff smoothing, 1001–1033
  adaptive integration, 1006
  adding singularity to grid, 78, 1007
  barrier option, 1074–1077
  benefits, 1012
  Bermudan swaption, *see* CLE greeks, tube Monte Carlo
  box smoothing, 1015–1018
    multiple dimensions, 1020
    on discrete grid, 1015
  by importance sampling, 1065–1077
  CLE, *see* CLE greeks, tube Monte Carlo
  continuity correction, 59, 1012
  fuzzy logic, 1028
  gamma, 1019
  grid shifting, 1007
  hat smoothing, 1019
  integration, 1012
  Lagrange basis, 59, 1018
  locality, 1019
  Monte Carlo, 1022–1030
  moving average, 1012
    choice of window, 1014
  multiple dimensions, 1019–1022
    box smoothing, 1020
    dominant dimension, 1022
  one dimension, 1014
  partial analytical integration, 76–78, 1010
  partial coupons, 1028
  PDE, 1012
  piecewise smooth function on a grid, 1016
  singularity removal, 1009
  TARN, *see* TARN, payoff smoothing; TARN, tube Monte Carlo
  tube Monte Carlo, *see* tube Monte Carlo
PCA, *see* Principal Components Analysis
PDE, 18, 43–93
  A-stable scheme, 55
  ADI scheme, 43, 82–85
    boundary conditions, 85
  Asian option, 70
  backward induction, 51
  Black-Scholes, *see* Black model, PDE
  boundary conditions
    for barrier options, 64
    from PDE itself, 385, 426
    linear at boundary, 48
    log-linear at boundary, 48
  Cauchy problem, 18, 44
  centering, 563
  conditional stability, 55
  consistent scheme, 56
  convection-dominated, 61–63
  convergent scheme, 56
  coupon-paying, 67

Craig-Sneyd scheme, *see* PDE, predictor-corrector scheme
Crank-Nicolson scheme, 50
  American options, 69
  not strongly A-stable, 55
  oscillations, 55, 58
Dirichlet problem, 44, 64
  space discretization, 46
dividends, 67, 68
domain truncation, 44
  stability of greeks, 1002
Douglas-Rachford scheme, 84, 91
  boundary conditions, 85
early exercise, 68
exponentially fitted schemes, 63
extra state variable method, *see* PDE, path-dependent options
for implied volatility, *see* volatility, implied, PDE for
forward equation, *see* Kolmogorov forward equation
fully implicit scheme, 50
greeks off grid, 1005
L-stable scheme, 55
Lagrange basis, 58, 59
Lax equivalence theorem, 56
local volatility model, *see* local volatility model, PDE
mesh refinement, 72, 79
  equidistant blocks, 73, 74
  non-equidistant, 74
multi-dimensional, 91
multi-exercise, 71
multi-level time-stepping, 58
non-equidistant discretization, 56
Nyquist frequency, 59
odd-even effect, 59
operator splitting, 82
orthogonalization, 86
  drawbacks, 88
partial analytical integration, *see* payoff smoothing, partial analytical integration
path-dependent options, 69, 71, 870, 872, 898, 900, 932, 934
Peaceman-Rachford scheme, 84
  boundary conditions, 85
predictor-corrector scheme, 89–92
quantization error, 59

Rannacher stepping, 58–61, 67, 459
semi-Lagrangian methods, 63
Shannon Sampling Theorem, 59
similarity reduction, 71
sinh transform, 384
smoothing, 58–60
  continuity correction, 59
  grid dimensioning, 1002
  grid shifting, 59, 1002
  space discretization, 45
  stable scheme, 53
  strongly A-stable scheme, 55
  time discretization, 49
    theta scheme, 50
  two-dimensional, 80
  two-dimensional with mixed derivatives, 85, 86, 89
  upwinding, 62
  variable transform, 44
  von Neumann method, 53–56
    amplification factor, 54
    stability criterion, 54
    well-posed, 56
Poisson distribution, 102
portfolio replication, *see* Bermudan swaption greeks, portfolio replication for hedging
power Gaussian copula, *see* copula, power Gaussian
predictor-corrector, 89, 115, 382, 643
  Monte Carlo, *see* Monte Carlo, predictor-corrector
  PDE, *see* PDE, predictor-corrector scheme
present value of a basis point, *see* swap, annuity
principal component, 105
Principal Components Analysis, 105, 106, 500, 604–606
principal factor, 105
product integral, 486
Profit-And-Loss, *see* P&L
pseudo-Gaussian model, *see* quasi-Gaussian model
pseudo-random number generator, *see* Monte Carlo, random number generation, pseudo-random
put-call parity, 24
PVBP, *see* swap, annuity

QG model, *see* quadratic Gaussian model
qG model, *see* quasi-Gaussian model
quadratic covariation, XXXIII, 7
quadratic Gaussian model, 443, 520–534
  as affine model, 521
  benchmark rate parameterization, 527
  Bermudan swaption, *see* Bermudan swaption calibration, local projection method, quadratic Gaussian model
  bond dynamics, 522
  bond reconstitution formula, 522
  calibration, 533–534
    multi-pass bootstrap, 533
  CLE, *see* CLE calibration, local projection method, quadratic Gaussian model
  CMS spread option, *see* CMS spread option, quadratic Gaussian model
  curve factor, 525
  European swaption, 528–533
    approximations, 530
    exact, 529
    Fourier integration, 531
    rank-2 approximation, 532
  Fourier integration, 531
  mean-reverting state variables, 521
  moment-generating function, 531
  Monte Carlo, 534
  one-factor, 443
  parameterization, 524–527
  PDE, 534
  quadratic approximation to swap rate, 530
  short rate, 520
  short rate in SV form, 527
  short rate state distribution
    in annuity measure, 528
    in forward measure, 523
  short rate state dynamics, 443, 520
    in forward measure, 522
    in annuity measure, 528
  smile generation, 524–525
  spanned stochastic volatility, 525, 534

TARN, *see* TARN, local projection method, quadratic Gaussian model
volatility factor, 525
volatility smile, 533
volatility swap, *see* volatility swap, quadratic Gaussian model
quadratic variation, XXXIII, 7
quadratic volatility model, 287–291
  European call option value, 290
  European put option value, 290, 291
  Markovian projection, 1137
  measure change, 289
  small-noise expansion, 308
  smile dynamics, 350
  strict supermartingale, 288
  time-dependent, 308
Quadratic-Exponential scheme, *see* square-root process, Monte Carlo, Quadratic-Exponential scheme
  multi-dimensional, *see* multi-stochastic volatility model, Monte Carlo, Quadratic-Exponential scheme
quadrature, 533, 788
  Gauss-Hermite, 533, 789
  Gauss-Legendre, 788
  Gauss-Lobatto, 788
quanto CMS, 745–750
  annuity mapping function, 749
  convexity adjustment, 749–750
  copula method, 748
  quanto adjustment, 747
  replication method, 747
quasi-Gaussian model, 539–590
  Bermudan swaption, *see* Bermudan swaption calibration, local projection method, quasi-Gaussian model
  bond reconstitution formula, 540
  calibration, 584
  CEV local volatility, 547
  CLE, *see* CLE calibration, local projection method, quasi-Gaussian model
  CMS convexity adjustment, *see* CMS, convexity adjustment, quasi-Gaussian model
  density approximation, 585

Index   xlv

direct integration, 560, 585
Libor-with-delay, *see* Libor-with-delay, quasi-Gaussian model
linear local volatility, 547–550
  calibration, 550
  European swaption, 549
  for swaption strip, 549
  swap rate dynamics, 548
  swap rate inter-temporal correlation, 557
  swap rate variance ratio, 555
Markovian projection, 543, 566, 579, 1139
mean reversion, *see* mean reversion
mean reversion calibration, *see* mean reversion calibration
Monte Carlo, 565
  Euler scheme, 565
multi-factor, 575–585
  benchmark rate correlations, 584
  benchmark rate parameterization, 577
  bond reconstitution formula, 576
  calibration to spread options, 584
  correlation smile, 585
  loadings, 584
  local volatility, 576
  Monte Carlo, 585
  PDE, 585
  short rate state distribution in annuity measure, 580
  short rate state dynamics, 576
  stochastic volatility, 576–585
  swap rate dynamics, 579–583
  swap rate dynamics by Markovian projection, 579
one-factor local volatility, 541
  short rate state dynamics, 541
PDE, 562–565
  convection-dominated, 563
  domain truncation, 564
  space discretization, 563
short rate state distribution, 561
short rate state dynamics, 540
  in annuity measure, 544, 545
  in forward measure, 586
single-state approximation, 565–569
small-time asymptotics, 561

stochastic volatility, 569–574
  bond reconstitution formula, 570
  calibration, 572–573
  Monte Carlo, 574
  non-zero correlation, 574
  PDE, 574
  swap rate dynamics, 571–572
  unspanned, 570
swap rate dynamics, 542–547, 551
  approximate, 543–547
  approximate linear, 544
  approximate quadratic, 547
swap rate variance, 546
swap rate volatility, 542
TARN, *see* TARN, local projection method, quasi-Gaussian model
volatility swap, *see* volatility swap, quasi-Gaussian model

Radon-Nikodym derivative, 9, 1067
range accrual, 211
  CMS, 211
  CMS spread, 211, 766
  curve cap, 212, 766
  dual, 212, 766
  floating, 766
  product-of-ranges, 212
ratchet swap, 212
relative entropy, 957
replication method, 337, 724
  CMS, *see* CMS, convexity adjustment, replication method
  European option, *see* European-style option, replication method
  Libor-in-arrears, *see* Libor-in-arrears, replication method
  Libor-with-delay, *see* Libor-with-delay, replication method
  semi-static, 939
reserve, 986
rho, 980
Riccati, 364
Riemann zeta function, 128
risk limit, 985
risk measure, 996
  coherent, 996
risk sensitivities, 1093
  common definitions, 980
  delta, *see* delta

grid dimensioning for stability, 1002
grid shifting for stability, 1002
Jacobian method, 254–258, 985, 986, 1105, 1106, 1111, 1118, 1119, 1121
off PDE grid, 1005
perturbation approach, 1050
vega, *see* vega
root search, 99
   Newton-Raphson method, 99, 116, 235
   secant method, 235
Runge-Kutta method, 116, 365, 434, 436, 516
running maximum, 124
running minimum, 124

SABR model, 343–345, 357, 951, 1121
   ad-hoc improvements, 705
   density tail, 761
   moment explosion, 344
   volatility smile expansion, 345
SALI tree, *see* tree, SALI
sausage Monte Carlo, *see* tube Monte Carlo
scripting language for trades, 220
SDE, *see* diffusion, SDE
SDE discretization, *see* Monte Carlo, SDE discretization
Sharpe ratio, 22
shifted log-normal model, *see* displaced log-normal model
short rate, 169
short rate model, 172
   affine, *see* affine short rate model
   affine one-factor, *see* affine short rate model, one-factor
   Black-Derman-Toy, *see* Black-Derman-Toy model
   calibration to yield curve, 457
      forward induction, 458
      forward-from-backward induction, 460
   Cox-Ingersoll-Ross, *see* Cox-Ingersoll-Ross model
   Dybvig parameterization, 463–465, 468
      HJM representation, 464
   econometric, 451
   empirical estimation, 451

   forward volatility impact on Bermudan swaption, 878
   Gaussian approximation, 1064
   Gaussian model for basis spread, 683
   Gaussian short rate, *see* Gaussian short rate model
   Hagan and Woodward parameterization, 465–468
   Ho-Lee, *see* Ho-Lee model
   importance sampling, 1063–1065
   log-normal, 445–451
      issues, 447
      Sandmann-Sondermann transform, 448
   Monte Carlo, 469–471
      Euler scheme, 469
      Milstein scheme, 469
      payoff construction issues, 470
      SDE discretization, 469
      variance reduction, 470
   multi-factor, 479
   path independence, 446
   PDE, 456–457
      domain truncation, 456
   power-type, 451
   quadratic Gaussian, *see* quadratic Gaussian model
   quasi-Gaussian, *see* quasi-Gaussian model
   time-stationary, 418
   volatility calibration, 461–463
      multi-pass bootstrap, 463
shout option, 935
   on capped coupon, 935
   optimal stopping time, 936
similarity reduction, 71, 871
   CLE, *see* CLE valuation, PDE, similarity reduction
   PDE, *see* PDE, similarity reduction
single-rate vanilla derivative, 697–763
   approximately single-rate, 709
   cap, *see* cap
   CMS cap, *see* CMS cap
   CMS floor, *see* CMS floor
   CMS swap, *see* CMS swap
   ED future, *see* ED future
   European swaption, *see* European swaption
   futures contract, *see* ED future

Libor-in-arrears, *see* Libor-in-arrears
Libor-with-delay, *see* Libor-with-delay
range accrual, *see* range accrual
singular value, 862
singular value decomposition, *see* CLE regression, SVD decomposition
    truncated, *see* CLE regression, truncated SVD decomposition
singularity removal, *see* payoff smoothing, singularity removal
skew vega, *see* vega, skew vega
smile vega, *see* vega, smile vega
snowball, *see* CLE, snowball
snowbear, 213
snowrange, 213
snowstorm, 213
Sonia, 193, 200
spline, 229, 270–275
    Catmull-Rom, 238, 240, 271, 272
    cubic C^2, 273
    cubic smoothing, 248
    exponential tension spline, 243
    Hermite cubic, 238, 270–273
    interpolating, 248
    Kochanek-Bartels, 272
    least-squares regression, 248
    natural, 241
    natural cubic, 273
    shape preserving, 275
    smoothing, 234
    TCB, *see* spline, Kochanek-Bartels
    tension, 240, 243, 244, 246, 247, 250, 272, 274–275
        convergence to piecewise linear, 275
    tension factor, 243
spot Libor measure, *see* measure, spot
spot rate, *see* short rate
square-root process, 315
    $E(\sqrt{z})$, 1153, 1155
    basic properties, 319–320
    boundary behavior, 319
    conditional CDF, 319
    conditional moments, 319
    Feller condition, 319
    moment-generating function, 322, 342, 364, 372
        time-dependent parameters, 364
    moments, 375

Monte Carlo, 388–394
    Euler scheme, 389
    exact simulation, 388
    full truncation scheme, 390
    higher-order schemes, 390
    log-normal approximation, 390
    moment-matching schemes, 390
    Quadratic-Exponential scheme, 392, 394
    truncated Gaussian scheme, 391
    multi-dimensional, 1152
    PDF, 1153, 1155
    stationary distribution, 320, 383
static replication, 210, 718
    CMS, *see* CMS, convexity adjustment, replication method
    European option, *see* European-style option, replication method
    Libor-in-arrears, *see* Libor-in-arrears, replication method
    Libor-with-delay, *see* Libor-with-delay, replication method
stochastic optimization method, 953
stochastic volatility model, 315–403, 571, 572, 1140
    as interpolation rule, 703
    ATM volatility, 348
    basket option, *see* Markovian projection, basket option in SV model
    calibration, 703–704
        calibration norm, 704
        normalization, 704
    caplet calibration, 707
    CEV type, *see* SABR model
    CMS convexity adjustment, 739
    correlation, 347
    dampening constant, 325
    delta, 699
    effective skew, 373
    effective volatility, 371, 372
    effective volatility of variance, 375
    European option, 327
        control variate, 328
        volatility mixing, 339
    explicit solution, 320
    for CMS rate, 739–744
        dynamics in forward measure, 741
    Fourier integration, 324–339

arbitrary European payoffs, 336, 338
convolution, 325
direct integration, 330
discrete, 330
FFT, 330
for variance, 339–343
integration bounds, 330
strip of convergence, 329
with control variate, 328, 330
hedging, 353–358
level parameter, 318
link between forward and annuity measures, 740
LSV, see local stochastic volatility model
martingale property, 320
mean reversion speed, 316, 317, 348
half-life, 318
measure change, 322
moment explosion, 323
moment-generating function, 321, 324, 327
branch cut, 330
singularities, 329
time-dependent parameters, 364
Monte Carlo, 387–398
Broadie-Kaya scheme, 395
Broadie-Kaya simplified scheme, 396
exact scheme, 395
martingale correction, 398
Taylor-type schemes, 396
variance process, see square-root process, Monte Carlo
multi-dimensional, see multi-stochastic volatility model
PDE, 381–387
boundary conditions for stochastic variance, 385
boundary conditions from PDE itself, 385
discretizing spot, 387
discretizing stochastic variance, 383
for forward Kolmogorov equation, 386
predictor-corrector, 382
quadratic discretization, 384
range for spot, 386
range for stochastic variance, 382
sinh transform, see PDE, sinh transform
sinh-quadratic discretization, 384
variable transform, 383, 385
process for variance, see square-root process
skew, 317, 346
smile dynamics, 347–349, 351, 353, 354
SV volatility, 317
time-dependent, 363–403
asymptotic expansion, 366–370
averaging, see calibration, time averaging
Fourier integration, 363, 366
volatility of variance, 316, 317, 346
volatility of volatility, 318
stopping time, 6
straddle, 223
strategy, 7
doubling, 9
gains process, 8
permissible, 9
replicating, 11
self-financing, 8, 17
Stratonovich integral, 5
strike price, 23
structured note, see exotic swap
structured swap, see exotic swap
Student's t-distribution, 101
Monte Carlo, 101
survival measure
Bermudan swaption, see Bermudan swaption, survival measure
importance sampling, see importance sampling, survival measure
SV model, see stochastic volatility model
SV model with general variance process, 359–361
martingale properties, 360
moment explosion, 361
properties, 359
stationary distribution, 360
strict supermartingale, 360
SVD, see CLE regression, SVD decomposition
SVI model, see volatility smile, SVI

swap, 197
  accreting, 199
  amortizing, 199
  annuity, XXXIV, 199
  annuity factor, 170
  averaging, see averaging cash flow
  cash-settled, 745
  CMS, see CMS swap
  effective date, 225
  fixed-floating, 198, 199, 230
    valuation formula, 199
  fixing dates, 198
  legs, 197
  Libor-in-arrears, see Libor-in-arrears
  Libor-with-delay, see Libor-with-delay
  par rate, see forward swap rate
  payer, 203
  payment dates, 198
  receiver, 203
  swap rate, see forward swap rate
swap market model, 619, 676–678
swap measure, see measure, annuity
swap rate, see forward swap rate
swaption grid, see European swaption, swaption grid

Tanaka extension of Ito's lemma, 7, 26, 294, 1131
targeted redemption note, see TARN
TARN, 217, 218, 925–933
  cap at trigger, 219
  global model, 927
  impact of inter-temporal correlation, see inter-temporal correlation, impact on TARNs
  importance sampling, 1068–1077
    one-step survival conditioning, 1069
    removing first digital, 1068
  leverage, 927
  Libor market model, 927
  lifetime cap, see TARN, cap at trigger
  lifetime floor, see TARN, make whole
  local projection method, 928–931
    Gaussian short rate model, 929
    Markov-functional model, 931
    quadratic Gaussian model, 931
    quasi-Gaussian model, 931

  make whole, 219
  Markov-functional model, 475
  multi-factor quasi-Gaussian model, 927
  partial analytical integration, 1011
  pathwise differentiation method, 1044
  payoff smoothing, 1011, 1029, 1068–1077
  PDE, 931–933
    cap at trigger, 933
    make whole, 933
    Monte Carlo pre-simulation, 933
    upper bound for extra state variable, 932
  tube Monte Carlo, 1029
  valuation formula, 218
  volatility smile, 927, 929–931
tenor structure, XXXIV, 170
  index function, 593
tension spline, see spline, tension
term parameters, 378
term structure model, 202, 277
terminal swap rate model, 709–716
  annuity mapping function, 710, 715, 724, 726–727, 730, 731, 733
    as conditional expected value, 726–727
  calibration to market, 730
  forward swap rate condition, 734
  forward value condition, 734
  in measure change, 737
  linear approximation, 729
  LM model, see Libor market model, annuity mapping function
  mean reversion, see CMS, convexity adjustment, impact of mean reversion
  multi-rate, 767
  swap rate squared condition, 735
  CMS convexity adjustment, see CMS, convexity adjustment, linear TSR model
  consistency condition, 710
  exponential TSR model, 714–715
  Libor-with-delay, see Libor-with-delay, swap-yield TSR model
  linear TSR model, 711

Index

CMS convexity adjustment, *see* CMS, convexity adjustment, linear TSR model
  forward CMS straddle, 941
  mean reversion parameterization, 712
  swap rate distribution in forward measure, 738, 739
  vega hedging, 714
 loading from Gaussian model, 714
 no-arbitrage condition, 710
 PDF of swap rate in forward measure, 739
  from CMS caplets, 739
 reasonableness, 710
 swap rate distribution in forward measure, 738
 swap-yield TSR model, 715–716
  CMS convexity adjustment, *see* CMS, convexity adjustment, swap-yield TSR model
theta, 980, 991
 rolling yield curve, 992
Tikhonov regularization, *see* CLE regression, Tikhonov regularization
time decay, 52
time value, 26
"tip-top", *see* "flip-flop"
tower rule, *see* iterated conditional expectations, 176
tree, 425
 binomial, 446, 458
 SALI, 78
 trinomial, 51, 458
truncated Gaussian scheme, *see* square-root process, Monte Carlo, truncated Gaussian scheme
TSR model, *see* terminal swap rate model
tube Monte Carlo, 1022–1030
 barrier option, *see* barrier option, tube Monte Carlo
 Bermudan swaption, *see* CLE greeks, tube Monte Carlo
 CLE, *see* CLE greeks, tube Monte Carlo
 digital option, 1024
 discrete knock-in barrier, 1028
 generalized trigger product, *see* barrier option, tube Monte Carlo
 partial coupons, 1028
 TARN, *see* TARN, tube Monte Carlo

underhedge, 1023
uniform distribution, XXXIII, 770
universal law of volatility, 1137
upwinding, *see* PDE, upwinding

value-at-risk, 501, 975, 995–998
 conditional, 996
 delta VaR, 998
 delta-gamma VaR/cVaR, 998
 Gaussian, 997
 historical, 996
vanilla derivative, 697–815
 multi-rate, *see* multi-rate vanilla derivative
 single-rate, *see* single-rate vanilla derivative
vanilla model, 202, 277, 315, 1121, 1129
 for multi-rate derivative, *see* multi-rate vanilla derivative
 for single-rate derivative, *see* single-rate vanilla derivative
 local volatility model, *see* local volatility model
 stochastic volatility model, *see* stochastic volatility model
vanna, 980
VaR, *see* value-at-risk
variance reduction, 142–158
 antithetic variates, 144
  efficiency, 144
  non-Gaussian, 145
 common random number scheme, 132, 134
 conditional Monte Carlo, 127
 control variate, *see* control variate
  from hedging strategy, *see* control variate, dynamic
 importance sampling, *see* importance sampling
 moment matching, 146
 systematic sampling, 145
Vasicek model, 413–415
 bond reconstitution formula, 414
 bond volatility, 415

forward rate volatility, 415
short rate distribution, 413
short rate dynamics, 413
yield curve shapes, 414
vega, 355, 980, 1095–1125
  additivity, 1103
  Bermudan swaption, 1114
  bucketed shocks, 1099
  CMS spread option, 1116, 1120
    constant Libor correlations, 1120
  constant Libor correlations, 1115, 1120
  constant term swap correlations, 1116, 1118–1120
  cumulative shocks, 1099
  direct method, 1098–1102, 1110
    Bermudan swaption, 1103
    European swaption, 1102
    second-order effects, 1110
  European swaption, 1113
  flat shock, 1099
  forward swaption straddle, 948
  "good", 1102–1105
  hybrid method, 1111–1114
    algorithm, 1112
    Bermudan swaption, 1114
    CMS spread option, 1116
    European swaption, 1113
  in LM model
    coverage, 886
  indirect method, 1105–1110, 1121
    Bermudan swaption, 1109
    European swaption, 1108
    least-squares problem, 1106
    locality, 1107
    smoothing, 1107
  Jacobian method, *see* vega, indirect method; risk sensitivities, Jacobian method
  Libor market model, 1095–1125
    bootstrap calibration, 1111
    multi-factor, 1115
    projection, 1123
  local projection method, 869
  local vs. global, 1096
  locality, 1104
    benchmark set locality, 1104
    exotic locality, 1104
    full set locality, 1104
  market vega, 984, 1096, 1110
  model vega, 984, 1096, 1124–1125
  pathwise differentiation method, *see* pathwise differentiation method, vega
  projection, 1122–1124
  relationship to gamma, 981
  row shocks, 1099
  running cumulative shocks, 1099
  scaling, 1103
  skew vega, 1114–1115
  smile vega, 1114–1115
volatility, 27
  average convexity, 307
  Bachelier, *see* volatility, Normal
  basis point, *see* volatility, Normal
  Black, XXXIV, 204
  bp, *see* volatility, Normal
  CEV, 280, 625
  Dupire's, *see* Dupire local volatility
  factor volatility, 501
  forward volatility of Libor rate, 819
  Gaussian, *see* volatility, Normal
  implied, 278
    as average of realized, 989
    effects of mis-specification, 987
    most likely path approximation, 990
    PDE for, 296
  local, *see* Dupire local volatility
  Normal, 204, 283, 625
  Normal for CMS spread option, 776
  separable, 300
  small-noise expansion, 307
  spanned stochastic volatility, 454
  spot volatility, 819
  spread, 776
    strike-dependent, 777
  stochastic, *see* stochastic volatility model
  unspanned stochastic volatility, 445
  "volatility squeeze", 424
volatility cube, *see* European swaption, volatility cube
volatility derivative, *see* forward volatility derivative
volatility skew, 279
volatility smile, 279, 315
  ATM backbone, 701, 702

backbone, 698
  adjustable, 699–702
  curvature, 1138
  dynamics, 279, 348, 698–702, 820
    sticky delta, 350, 352, 699
    sticky strike, 699
  forward skew, 944
  Gaussian backbone, 700
  impact on forward volatilities, *see* forward volatility, impact of volatility smile
  impact on inter-temporal correlations, *see* inter-temporal correlation, impact of volatility smile
  probability density from, 278
  SABR, *see* SABR model
  shadow delta hedging, 699
  skew vega, 1114
  skew-dominated, 352
  slope, 279
  smile vega, 1114
  SVI, 705, 951, 1121
  upward sloping, 281
  vega, 1114
volatility structure, 817
volatility swap, 220, 221, 933–945
  capped, 937
  CMS spread, 221
  copula method, *see* copula method, volatility swap
  fixed-expiry, 221, 940
  fixed-tenor, 221, 940
  impact of forward volatility, 944
  impact of volatility smile dynamics, 941
  Libor market model, 933, 934
  local projection method, 934
  min-max, *see* min-max volatility swap
  PDE, 934
  quadratic Gaussian model, 941
  quasi-Gaussian model, 941
  with barrier, 222
  with shout, 221, 935
volga, 980
Volterra integral equation, 438
vomma, 980

Wiener process, *see* Brownian motion

year fraction, 224
yield curve, 191, 230, 231, 233
  base index curve, 268
  basis risk, 270
  benchmark set, 230
  forecasting curve, *see* yield curve, index curve
  index curve, 261, 267, 679
  index-discounting basis, 197, 261
  instantaneous forward curve, 233
  joint evolution of discount and forward curves, 678
  multi-index curve group, 267–270
  overlay curve, 258
  perturbation locality, 230, 251–253, 258
  Principal Components Analysis, *see* Principal Components Analysis
  ringing, 235, 242, 243, 252
  smooth, 258
  spread curve, 269, 886
  tenor basis, 230, 267
  TOY effect, 258
yield curve construction, 229–275
  benchmark set, 231
  bootstrapping, 234
    flat forward, 236
    linear yield, 235
  constrained optimization, 248
  cross-currency, 259
  cross-currency arbitrage, 260
  cubic spline C^2, 240–243
    problems, 242
  curve overlays, 258
  FX forwards, 259
  Hermite spline, 238–240
    iterative solution, 239
  Jacobian rebuild, 256
  multi-index curve group, 230, 265
  non-parametric fitting, 245–250
    norm specification, 245
    optimization algorithm, 245
  separate discount and forward curves, 260
  spline, *see* spline
  spline fitting, 234–244
  tension spline, 243–244

yield curve risk, 250–258
   cumulative shifts, 256, 257
   forward rate approach, 252
   Jacobian method, *see* risk sensitivities, Jacobian method
   par-point approach, 251
   rolling for theta, 992
   waterfall approach, *see* yield curve risk, cumulative shifts
yield curve spread option, *see* CMS spread option

zero-coupon bond, *see* discount bond
zero-coupon bond option, 185

Lightning Source UK Ltd.
Milton Keynes UK
UKHW022309060223
416578UK00006B/812/J